Emerging Nanomaterials for Catalysis and Sensor Applications

This book reviews emerging nanomaterials in catalysis and sensors. The catalysis section covers the role of nano-photocatalysts in organic synthesis and health care application, oxidation and sulphoxidation reactions, liquid phase oxidation, hydrogen evolution and environmental remediation. It highlights the correlation of surface properties and catalytic activity of the mesoporous materials. The sensor section discusses the fabrication and development of various electrochemical, chemical, and biosensors.

Features:

- Combines catalysis and sensor applications of nanomaterials, including detailed synthesis techniques of these materials.
- Explores methods of designing, engineering, and fabricating nanomaterials.
- Covers material efficiency, their detection limit for sensing different analytes and other properties of the materials.
- Discusses sustainability of nano materials in the industrial sector.
- Includes case studies to address the challenges faced by research and development sectors.

This book is aimed at researchers and graduate students in Chemical Engineering, Nanochemistry, Water Treatment Engineering and Labs, Industries, Research Labs in Catalysis and Sensors, Environmental Engineering, and Process Engineering.

Emerging Materials and Technologies

Series Editor: Boris I. Kharissov

The *Emerging Materials and Technologies* series is devoted to highlighting publications centered on emerging advanced materials and novel technologies. Attention is paid to those newly discovered or applied materials with potential to solve pressing societal problems and improve quality of life, corresponding to environmental protection, medicine, communications, energy, transportation, advanced manufacturing, and related areas.

The series takes into account that, under present strong demands for energy, material, and cost savings, as well as heavy contamination problems and worldwide pandemic conditions, the area of emerging materials and related scalable technologies is a highly interdisciplinary field, with the need for researchers, professionals, and academics across the spectrum of engineering and technological disciplines. The main objective of this book series is to attract more attention to these materials and technologies and invite conversation among the international R&D community.

Sustainable Nanomaterials for the Construction Industry
Ghasan Fahim Huseien and Kwok Wei Shah

4D Imaging to 4D Printing
Biomedical Applications
Edited by Rupinder Singh

Emerging Nanomaterials for Catalysis and Sensor Applications
Edited by Anitha Varghese and Gurumurthy Hegde

Advanced Materials for a Sustainable Environment
Development Strategies and Applications
Edited by Naveen Kumar and Peter Ramashadi Makgwane

Nanomaterials from Renewable Resources for Emerging Applications
Edited by Sandeep S. Ahankari, Amar K. Mohanty, and Manjusri Misra

Multifunctional Polymeric Foams
Advancements and Innovative Approaches
Edited by Soney C. George and Resmi B. P.

Nanotechnology Platforms for Antiviral Challenges
Fundamentals, Applications and Advances
Edited by Soney C George and Ann Rose Abraham

Carbon-Based Conductive Polymer Composites
Processing, Properties, and Applications in Flexible Strain Sensors
Dong Xiang

Nanocarbons
Preparation, Assessments, and Applications
Ashwini P. Alegaonkar and Prashant S. Alegaonka

For more information about this series, please visit:
www.routledge.com/Emerging-Materials-and-Technologies/book-series/CRCEMT

Emerging Nanomaterials for Catalysis and Sensor Applications

Edited by
Anitha Varghese and Gurumurthy Hegde

CRC Press
Taylor & Francis Group
Boca Raton London New York

CRC Press is an imprint of the
Taylor & Francis Group, an **informa** business

Designed cover image: © Shutterstock

First edition published 2023
by CRC Press
6000 Broken Sound Parkway NW, Suite 300, Boca Raton, FL 33487-2742

and by CRC Press
4 Park Square, Milton Park, Abingdon, Oxon, OX14 4RN

CRC Press is an imprint of Taylor & Francis Group, LLC

ISBN: 9781032111711 (hbk)
ISBN: 9781032111728 (pbk)
ISBN: 9781003218708 (ebk)

DOI: 10.1201/9781003218708

Typeset in Times
by Newgen Publishing UK

Contents

Editor Biographies vii
List of Contributors ix

SECTION I Emerging Materials in Nanocatalysis

Chapter 1 The Role of Nanomaterials in Sustainable Organic Synthesis 3

Dileep Ramakrishna and Thitima Rujiralai

Chapter 2 Nanocatalysts in Oxidation and Sulfoxidation Reactions 25

Hanna S. Abbo, Tariq Shah, Divine M. Yufanyi, Nader G. Khaligh and Salam J.J. Titinchi

Chapter 3 Correlation of Surface Properties and Catalytic Activity of Metal Aluminophosphates 49

C.J. Binish and A.V. Vijayasankar

Chapter 4 Carbon Supported Noble Metal Nanocatalysts for Liquid Phase Oxidation Reactions 65

Mabuatsela V. Maphoru

Chapter 5 Metal Oxide Nanomaterials for Visible Light Photocatalysis 103

G. Reenamole

Chapter 6 Progress in Photocatalysis for Hydrogen Evolution and Environmental Remediation 131

R. Shwetharani, M.S. Jyothi, M. Dinamani and S. Radoor

SECTION II Emerging Materials in Nanosensors

Chapter 7 Nanostructured Materials for Sensors Applications 159

Saravanan Chandrasekaran and Anu Sukhdev

Chapter 8 Synthesis of Fluorescent Nanosensor for Biomedical Engineering 181

Saba Farooq and Zainab Ngaini

Chapter 9 Applications of Peptide Luminescent Nanosensors 203

Tatiana Duque Martins Ertner de Almeida, Diéricon Sousa Cordeiro, Ramon Miranda Silva, and Igor Lafaete de Freitas Pereira Morais

Chapter 10 Graphene-Based Hybrid Nano Composites for Bio/Chemical Sensors 217

C. Kavitha and Rapela R. Maphanga

Chapter 11 Nanomaterial-Based Electrochemical Sensors for Vitamins and Hormones 249

Anila Rose Cherian, Rijo Rajeev, M. Nidhin, Anitha Varghese, and Gurumurthy Hegde

Chapter 12 Economic Analysis, Environmental Impact, Future Prospects and Mechanistic Understandings of Nanosensors and Nanocatalysis 271

Libina Benny, Vinay Bhat, Ashlay George, Anitha Varghese, and Gurumurthy Hegde

Index 289

Editor Biographies

Anitha Varghese obtained her PhD from Mangalore University, Karnataka, India in 2007, under the supervision of Dr A.M.A. Khader. Her PhD work concerned the sensing of heavy metal ions in various alloy steels, minerals, biological, soil, and water samples. In 2008, she started her career in CHRIST (Deemed to be University), Bengaluru, where she is a professor. Her current research interests include electrocatalysis for organic transformations and developing electrochemical sensors for biomolecules, flavonoids, pollutants, and food adulterants.

Gurumurthy Hegde obtained his PhD in liquid crystals in 2007 from the Centre for Nanomaterials and Soft Matter Science, Bengaluru. Expert in photoalignment and waste to wealth technologies, Dr Hegde has more than 18 years of R and D experience. His major research is in the field of liquid crystals, LCDs, nanoparticles, energy, water, and health. He has invented many concepts for aligning materials, waste to wealth technologies, and so forth. Presently he is working as the Director, Centre for Advanced Research and Development (CARD), CHRIST (Deemed to be University) and also as the Professor of Materials Science, Dept of Chemistry, CHRIST (Deemed to be University), Bengaluru. He has published more than 220 scientific publications and has more than 25 patents. He also has obtained more than 60 international awards and guided many PhD and MSc, MTech students.

Contributors

Hanna Abbo
Department of Chemistry, University of the Western Cape, Cape Town, South Africa

Libina Benny
Department of Chemistry, CHRIST (Deemed to be University), Hosur Road, Bengaluru, India

Vinay Bhat
Centre for Nano-materials and Displays, Bhusanayana Mukundadas Sreenivasaiah College of Engineering, Bull Temple Road, Bengaluru, India

C. J. Binish
Department of Sciences & Humanities, School of Engineering and Technology, CHRIST (Deemed to be University), Kengeri Campus, Bengaluru, India

Saravanan Chandrasekaran
Department of Chemistry, School of Engineering, Presidency University, Rajanukunte, Itgalpura, Bangalore, India

Anila Rose Cherian
Department of Chemistry, CHRIST (Deemed to be University), Hosur Road, Bengaluru, India

Diéricon Sousa Cordeiro
Chemistry Institute, Federal University of Goias, Av. Esperança s/n, Vilaltatiaia, Goiânia, Brazil

M. Dinamani
Department of Chemistry, Dayananda Sagar College of Engineering, Bengaluru, India

Tatiana Duque Martins Ertner de Almeida
Chemistry Institute, Federal University of Goias, Av. Esperança s/n, Vila Itatiaia, Goiânia, Brazil

Saba Farooq
Faculty of Resource Science and Technology, Universiti Malaysia Sarawak, 94300 Kota Samarahan, Sarawak, Malaysia

Ashlay George
Department of Chemistry, CHRIST (Deemed to be University), Hosur Road, Bengaluru, India

Gurumurthy Hegde
Department of Chemistry, CHRIST (Deemed to be University), Hosur Road, Bengaluru, India

M. S. Jyothi
Department of Chemistry, AMC Engineering College, Bannerghatta Main Road, Bengaluru, India

C. Kavitha
Department of Physics, Bhusanayana Mukundadas Sreenivasaiah Institute of Technology and Management, Avalahalli, Yelahanka, Bangalore, India

Nader G. Khaligh
Department of Chemistry, University of the Western Cape, Cape Town, South Africa

Igor Lafaete de Freitas Pereira Morais
Chemistry Institute, Federal University of Goias, Av. Esperança s/n, Vila Itatiaia, Goiânia, Brazil

Rapela R. Maphanga
Council for Scientific and Industrial Research, Next generation Enterprises and Institutions, Pretoria, South Africa

Mabuatsela V. Maphoru
Department of Chemistry, Tshwane University of Technology, Private Bag X680, Pretoria, South Africa

Ramon Miranda Silva
Chemistry Institute, Federal University of Goias, Av. Esperança s/n, Vila Itatiaia, Goiânia, Brazil

Zainab Ngaini
Faculty of Resource Science and Technology, Universiti Malaysia Sarawak, 94300 Kota Samarahan, Sarawak, Malaysia

M. Nidhin
Department of Chemistry, CHRIST (Deemed to be University), Hosur Road, Bengaluru, India

S. Radoor
Department of Polymer-Nano Science and Technology, Jeonbuk National University, 567 Baekje-daero, Deokjin-gu, Jeonju-si, Korea

Rijo Rajeev
Department of Chemistry, CHRIST (Deemed to be University), Hosur Road, Bengaluru, India

Dileep Ramakrishna
Department of Chemistry, Presidency University, Yelahanka, Bangalore, India

G. Reenamole
Department of Chemistry, Mar Thoma College, Thiruvalla, Kerala, India

Thitima Rujiralai
Department of Chemistry and Centre of Excellence for Innovation in Chemistry, Faculty of Science, Prince of Songkla University, Hat Yai, Songkla, Thailand

Tariq Shah
Department of Chemistry, University of the Western Cape, Robert Sobukwe Rd Bellville 7535, Cape Town, South Africa

Rani R. Shwetha
Center for Nano and Material Sciences, Jain University, Kanakapura Road, Bangalore, India

Anu Sukhdev
Department of Chemistry, School of Engineering, Presidency University, Rajanukunte, Itgalpura, Bangalore, India

Salam Titinchi
Department of Chemistry, University of the Western Cape, Robert Sobukwe Rd Bellville 7535, Cape Town, South Africa

Anitha Varghese
Department of Chemistry, CHRIST (Deemed to be University), Hosur Road, Bengaluru 560029, India

A. V. Vijayasankar
Department of Sciences & Humanities, School of Engineering and Technology, CHRIST (Deemed to be University), Kengeri Campus, Bengaluru, India

Divine M. Yufanyi
Department of Chemistry, University of the Western Cape, Robert Sobukwe Rd Bellville 7535, Cape Town, South Africa

Section I

Emerging Materials in Nanocatalysis

This section is dedicated to explaining the important role of nanomaterials in the area of catalysis. Here, various kinds of catalysts were used, ranging from photocatalysts, nanocatalysts, metal nanocatalysts, and so forth. Using such nanocatalysts, different kinds of innovative applications were highlighted and explained, such as healthcare, oxidation, and sulphoxidation reactions, the catalytic activity of mesoporous metal aluminophosphates, liquid-phase oxidation, visible light photocatalysis, hydrogen evolution, and environmental remediation, and so forth. This section also provides information about recent trends in this area of research.

DOI: 10.1201/9781003218708-1

1 The Role of Nanomaterials in Sustainable Organic Synthesis

Dileep Ramakrishna and Thitima Rujiralai

CONTENTS

1.1 Nanomaterial as Catalysts ... 3
1.2 Characteristics of Nanomaterials ... 4
1.3 Properties of Nanomaterials as Catalysts ... 5
1.4 Nanocatalyst in Organic Transformations ... 6
1.5 Recent Developments in Advanced Nanocatalysts ... 13
1.6 Nanostructured Catalysts for Greener and Sustainable Organic Processes ... 15
1.7 Outlook ... 16
References ... 16

1.1 NANOMATERIAL AS CATALYSTS

The catalytic applications of nanoparticles are being studied due to their small particle size. On a general note, the nanoparticle-based technology is aimed at enhancing existing processes in terms of their efficiency, durability and speed. This is conceivable because of the materials employed in industrial operations in the past (e.g., industrial catalysis). The duty of a catalyst is to speed up a chemical process without being spent in the process. Nanocatalysts are nanoscale materials with at least one nanoscale dimension or nanoscale structural alteration to promote catalytic activity. The following are some of the important catalysts with at least one nano level property/structure modification.

- Nanoporous Catalysts Microporous (4 – 14 Angstrom) – Mesoporous (15 – 250 Angstrom)
- Nanoparticle Catalysts (ex. Au/TiO2, Pt-Pd-Rh Three-way catalyst)
- Nanocomposite Catalysts
- Nanocrystalline Catalysts (ex. Nanocrystalline CeO_2-x, TiO_2)
- Supramolecular Catalysts

It has become apparent that the nanoparticles can replace traditional materials, such as heterogeneous catalysts, which normally have large surface areas and also sometimes supported catalysts as well. The nano-sized particles increase the active component's accessible surface area, greatly enhancing reactant–catalyst interaction and emulating homogeneous catalysts. Their insolubility in reaction solvents, on the other hand, makes them easily separate from the reaction mixture, similar to heterogeneous catalysts, making the product separation stage simple. In addition, chemical and physical parameters such as size, shape, and composition can be tailored to control the activity and selectivity of nanocatalysts.

The scientific challenge is to synthesize nanocatalysts of appropriate size and shape to support the easy mobility of the materials in the reaction medium and to have control over nanostructure

DOI: 10.1201/9781003218708-2

morphology to adjust physical and chemical capabilities. The broad landscape of heterogeneous catalysis is dominated by metals and their oxides. In general, metal oxides have both electron transport and surface polarizsing properties, which are important in redox and acid-base catalytic reactions. Selective organic transformations have been carried out using the surface acid-base characteristics of oxides. A variety of metal oxides can be used as precious metal supports. The conventional duty of a support in high-temperature catalytic applications is to finely scatter and stabilize the active metal particles by preventing agglomeration. The use of metal oxides as catalysts and supports is a well-studied area of catalysis research. The most common metal oxides used as catalysts and support are silica, alumina, and zirconium dioxide.

1.2 CHARACTERISTICS OF NANOMATERIALS

Nanomaterials are made up of organized components with a minimum dimension of 100 nanometers. Materials like thin films or surface coatings are those that have one dimension in the nanoscale. Sometimes they may be extended in two other dimensions as well. This category includes some of the characteristics of computer chips. Nanowires and nanotubes are two-dimensional nanoscale materials that can be expanded in one dimension. Particles in three-dimensional nanoscale materials include precipitates, colloids, and quantum dots, the tiny particles found in semiconductor materials. This category also includes nanocrystalline materials, which are granular in nature and are nanometer-sized. Some of these materials have been around for a while, while others are completely new [1–3].

Nanomaterials' characteristics differ greatly from those of other materials due to two main factors: One is due to the greater relative surface area and the other, most importantly, to the quantum effects. These factors help in altering and even improving the properties like reactivity, strength, and electrical characteristics of nanomaterials. A greater number of atoms are located at the surface of a particle as it shrinks in size, compared to those inside. For example, a particle with a diameter of 30 nm has 5 percent of its atoms on its surface, a particle with a diameter of 10 nm has 20 percent of its atoms on its surface, and a particle with a diameter of 3 nm has 50 percent of its atoms on its surface. The nanoparticles have substantially a larger surface area per unit mass compared to bigger particles. As the growth and catalytic chemical reactions occur at surfaces, a given quantity of material in nano particulate form will be far more reactive than that of material of same quantity, composed of larger particles [1–3].

For the synthesis of the nanomaterials, two strategic methodologies are employed: The "bottom-up" method and the "top-down" method. In the "bottom-up" method the materials and devices are produced from molecular components that assemble themselves chemically using molecular recognition principles. In the "top-down" approach nano-objects are built from larger things without atomic-level control.

Big to small: A materials perception: As the size of the system shrinks, a number of physical phenomena become more prominent. These include statistical, mechanical, and quantum mechanical phenomena, such as the "quantum size effect," in which the electrical characteristics of solids are altered when particle size is reduced dramatically. Going from macro to micro dimensions has no influence on this effect. When the nanoscale size range is reached, however, it becomes dominating. When contrasted to macroscopic systems, a multitude of physical features alter (mechanical, electrical, optical, etc.). A change in the thermal, mechanical and in turn the catalytic properties of materials can be observed when there is an increase in the surface area-to-volume ratio. Nanoionics refers to diffusion and the physical and/or chemical changes happening at the nanolevel. This includes nanostructured materials and nanodevices as well with rapid ion mobility. Nanomechanics research is interested in novel mechanical properties of nanosystems. Nanomaterials' catalytic activity creates possible dangers when they interact with biomaterials [2].

Materials reduced to the nanoscale can display different properties than those observed at the macroscale, allowing for novel uses. For example, from opaque to transparency (copper); stability to combustibility (aluminum); change in the state at room temperature (gold); insulation to electrical conduction (silicon). Substances such as gold, which are chemically inert at a larger dimension, can act as powerful chemical catalysts on the nanoscale. Ultimately the two factors, that is, quantum and surface phenomena, are those that matter and prove that materials at the nanoscale are at the root of development with nanotechnology [2].

Small to big: A molecular approach: Synthetic chemistry nowadays has advanced to a point where tiny molecules may be prepared to nearly any structure. Today, these techniques are employed to make a wide range of valuable compounds, including medicines and commercial polymers. Through a bottom-up approach, these systems leverage molecular self-assembly and/or supramolecular chemistry ideas to spontaneously arrange themselves into some useful configuration. The concept of molecular recognition is particularly important: Non-covalent intermolecular forces can be used to design molecules such that a particular structure or arrangement could be preferred. The base pairing rules by Watson and Crick, the specificity of an enzyme targeting a substrate, the particular folding of a protein, and so forth are all the direct results of this. Hence two or more components can be built to be mutually attracted and complemented, and that result in more complex and valuable products [1, 3].

Bottom-up methods should be able to develop gadgets more quickly and for less cost than top-down ones. However, when the size and complexity of the planned assembly develops, they risk becoming overloaded. The majority of useful structures necessitate sophisticated and thermodynamically improbable atom configurations. However, in biology, numerous examples of self-assembly can be given that are based on molecular recognition, which include base pairing (Watson–Crick model) and interactions between enzyme and the substrate. The question for nanotechnology is whether these principles can be employed to create innovative structures in addition to those found in nature [4].

1.3 PROPERTIES OF NANOMATERIALS AS CATALYSTS

In parallel with surface-area effects, quantum phenomena can begin to dominate the properties of matter as size is reduced to the nanoscale. These can have an impact on a material's optical, electrical, and magnetic properties, especially as the structure or particle size approaches the nanoscale. Optoelectronic materials like quantum dots, quantum well lasers, utilize the advantages of these characteristic features. Other materials, such as crystalline solids, have a substantially larger interface area within the material as the size of its structural components reduces; this can have a significant impact on both mechanical and electrical properties. Most metals are, for example, made up of minute crystalline grains; when the material is strained, the boundaries between the grains slow or stop the spread of faults, providing it strength. The contact area within the material considerably rises if these grains are made very small, or even nanoscale in size, which boosts its strength. Nanocrystalline nickel, for example, is as strong as hardened steel [3].

Catalytic technologies are crucial in the energy, chemical process, and environmental industries, both now and in the future. Catalytic technologies are used to transform fossil fuels like coal and petroleum as well as natural gas into fuels and chemical products. Especially a variety of chemicals and petrochemicals can be manufactured by employing the catalytic approach. The catalytic activity in these processes also improves the efficiency by reducing hydrocarbons, CO and NO emissions, into useful products. Electrodes used in fuel cells that contain either an ionic solid oxide or a polymeric proton electrolyte that act as catalysts. The development of improved catalysts is driven by a number of factors, including (i) using low-cost basic resources to create high-value products, (ii) chemical conversion technologies that are both energy efficient and ecologically friendly, (iii) environmental laws that are becoming increasingly strict, and (iv) inexpensive catalysts such as with reduction or replacement of precious metals [5].

Catalysis is the process of accelerating chemical processes in solids, gases, or liquids by adding a solid phase containing large concentrations of the suitable type of site for chemical reactants to either adsorb or react and/or desorb. It is necessary to lower the size of the catalytic particle since increasing the number of sites in order to increase surface area is necessary for optimizing the catalyst. Active catalysts in modern laboratories are typically composed of particles with nano-level structural characteristics. Modern catalysts generally have many active phases, each of which may include a support designed to distribute, isolate, or otherwise improve the structure or characteristics of individual catalytic particles. One goal of catalysis research is to figure out how shrinking catalytic particles affects intrinsic catalytic performance in ways other than just increasing surface area. Also on the agenda is the better understanding of catalyst design and synthesizing with the most effective size and structure.

Nanoscience's potential application in practically every domain is an exciting promise. Nanotechnology breakthroughs will assist a wide range of industries, including medical, electronics, manufacturing, and fashion. While nanotechnology has a wide range of applications – the use of catalysts in the form of nanocrystals, is most exciting. The ratio of surface area to volume is an important notion in understanding nanocrystal catalysis. The surface area of an object grows less in proportion to its volume as it grows larger. As a result, smaller things have more surface area per unit of volume. This has far-reaching consequences for chemical reactions. Chemical processes benefit from high surface area-to-volume ratios. To return to the campfire, as an example, the fire is started with kindling. The surface area of little bits of wood is bigger than the surface area of larger logs. As a result, lighting the kindling speeds up the combustion process. A big flare is also produced when some amount of sawdust is thrown onto a burning flame. This process has the same chemical make-up as regular wood burning, but it happens significantly faster. Catalysts have the common function of increasing the rate of a reaction. This is accomplished by kinetic mechanisms and has no direct impact on a chemical system's thermodynamic properties. A catalyst can speed up a reaction in one of three ways: By lowering the activation energy, acting as a facilitator and bringing the reactive species together more effectively, or by increasing the yield of one species when two or more products are generated. Nanocatalysts could be employed in any of the ways indicated above, depending on the application. For two reasons, nanomaterials are more effective than traditional catalysts. To begin with, their incredibly minute size (usually 10–80 nm) results in a huge surface area/volume ratio. Furthermore, materials created at the nanoscale have characteristics not present in their macroscopic counterparts. Nanocatalysts' adaptability and efficacy are due to both of these factors.

1.4 NANOCATALYST IN ORGANIC TRANSFORMATIONS

Nanotechnology has opened up exciting new possibilities in the science of catalysis. As about 80 percent of chemical industry processes rely on catalysts to function efficiently, new understanding of catalysts, as well as their acquisition, could have far-reaching societal ramifications. Although new heterogeneous catalysts have been created to tackle surface science, until recently empirical testing in trial-and-error tests remained the only approach to producing new or superior catalysts. Now, this time-consuming and expensive process is being increasingly superseded by rational design methods that take advantage of basic nanoscale catalyst understanding. The advent of nanoscience and nanotechnology has enabled the investigation and optimization of a wide range of catalytic processes by generating regulated and geometric structures. As a result, researchers have gained a fundamental understanding of critical functions that determine nanocatalyst activity, selectivity, and lifespan.

Recently, mixed metal oxides in the form of transition metals, metal oxides, and metal salt nanoparticles have emerged as viable alternatives to conventional materials in various fields of chemistry [6]. Nanocatalysts have useful physical and chemical properties that serve the dual role of heterogeneous catalysts and homogeneous catalysts. In addition to their traditional importance in the development of the petroleum field, nanocatalysts are a phenomenon of important basic research and important practical application in various fields, such as chemistry, physics, materials science, environmental science, and atmospheric science. It includes supported or unsupported nanometer scale metal catalytic structures (spherical nanoparticles, nanorods, nanoplates, nanocubes, nanotubes, and so on) where the catalytic phenomena are particular to that length scale and are in general associated to high surface area [7]. Catalysis of nanophase metal and metal oxide with controlled particle size, a high surface area and more densely populated, coordinated sites of the surface could potentially greatly increase catalytic performance in comparison to conventional catalysts.

The green approach to the process increases with greater selection, return and catalytic recovery being suggested, and targets, in combination with advantages of such heterogeneous assistance. One of the world's fastest expanding fields of research is metal nanoparticles supported with its potential in many applications now and future [8]. Due to the potential for improving selectivity and the potential for size-/form selective catalysis, the control of nanoparticles size, shape and activity as well as the benefit of nanoparticles aggregated inhibition by immobilization/separation on heterogeneous support, such supported nanoparticle materials have attracted considerable attention in recent decades.

In a variety of organic transformations, such as oxidation of CO, alcohols, alkenes and olefins, hydrogenating of alkenes and diens, alkyl and allylic compounds, C-C combined reactions, like Heck, Suzuki, Sonogashira, Neguishi, and Kumanda reactions and hydro chlorination, metal nanoparticles are reported as promising materials for the growth of heterogeneous catalysts in the area [9–11]. There are several more chemical reactions that have been carried out in heterogeneous catalysis employing supported metal transition nanoparticles [12, 13].

Important catalysts, along with operative simplicity, high reactivity, and eco-friendly characteristics, shorten the reaction time and re-use nanoparticles to make nanocatalysis more effective. In our thesis, we focused on the organic processes catalysed by heterogeneous nanocatalysts in TiO_2, Ni, and Ti-Ni metals.

The new age of chemistry is moving toward creative technology, focusing on environmental factors in particular [14, 15]. Each element of the reaction is examined in a context of environmentally friendly notions, such as the use of safe solvents like water or solvent free synthesis and low-cost catalysts, without compromising its output and quality. The heterocyclic core synthesis is the major part of organic synthesis, as it has several pharmacological activities [16–19]. For the synthesis, many approaches have been used, including the use of catalysts [20, 21], ultrasound irradiation [22–24], and microwave irradiation [25, 26]. Although such techniques have their own advantages, they have certain downsides such as costly instruments, non-recyclable materials, non-selectivity, and so forth The role of nanocatalyst holds its applicability in overcoming them [27]. Nanoscience is the study of phenomena that occur at the nanoscale scale. Atmospheric and molecular diameters are commonly measured in nanometers. Nanometer-sized structures were produced by humans and will continue to be created by humans in the future. Due to the fact that a nanometer-sized structure is formed as soon as a few atoms are positioned adjacent to each other. Practically all of nanoscience can be simplified to chemistry because the particles are characteristically a few nanometers in size. The carbon based nanomaterials like nanospheres, nanotubes, nanofibers, C60 molecules and DNA-based structures are all topics of current research in nanotechnology. A nanostructure's chemical description may not always be sufficient to define its function. The ancient colloid science has been given a fresh lease on life as a result of the rapid advancement of nanoscience and nanotechnology. Nanoscale materials, including colloids, have gotten a lot of interest in the last decade because of their significant distinctions between single molecules and bulk materials, particularly in the field of catalysis. So, too, have catalysts and catalytic reactions received a lot of attention in recent decades, with the goal of discovering useful uses in the pharmaceutical and fine chemical industries. Nanocatalysts outperform regular catalysts because they are more selective, reactive, and stable In many circumstances, the activity increases as the particle size decreases because the electronic characteristics of surface atoms are favorable, situated largely on small particles' borders and corners. The reactivity and selectivity of metal nanocatalysts depends significantly on various crystallographic planes present on the nanoparticles that can be regulated by morphology. The size and surface of nanocatalysts are important since they determine its selectivity and reactivity. Additionally, doping and surface chemical changes may be used to improve performance in specific circumstances [28]. Nanocatalysts are employed in a variety of applications in addition to organic transformation [29, 30]. Thermal decomposition, microarc oxidation irradiation, chemical vapour synthesis, nonsono and sonoelectro oxidation, sol-gel technique, chemical precipitation, photochemical method, hydrothermal method, antisolvent precipitation, glow discharge plasma electrolysis, wet-chemical method, microwave irradiation, and sonochemical methods are some of the methods used to make these nanocatalysts [29–34]. The size and nature of nanocatalysts differ depending on the technique of synthesis [35–40]. The method of preparation can be chosen based on the requirements. In this chapter, a review of recent examples of nanoparticles used in organic transformations are been discussed.

In various fields of research and industry, metal nanoparticles (M-NPs), particularly due to their large activity, are of major interest in technological applications. For a variety of applications, the regulated and repeatable synthesis of defined and stable M-NPs with a limited size dispersion is critical [41–46]. Note that over the years, metal-based nanoparticles, metal nano crystals and metal colloids were also mentioned as nanophase metal clusters. For simplicity, nanoparticles are primarily used in the following phrase (metal). MNPs are only kinetically stable, combining to favor larger metal particles through agglomeration thermodynamically. The high surface energy and huge surface areas are responsible for this M-NP inclination to aggregate. M-NPs need to be stabilized by strongly coordinating ligand protective layers which provide electrostatic and/or steric protection such as polymers and surfactants to prevent this agglomeration [47–49]. M-NP immobilization on a surface is a stabilizing process that does not need ligand layers [50].

A summary of the literature review on the catalytic applications of nanomaterials/nanoparticles is given below.

The total synthesis of substituted indoles was carried out in presence of nano-TiO_2 (Scheme 1.1). A lot of researchers are working on this nanomaterial as the same supports photocatalytic activity as well [51].

Copper-based nanocatalysts have been explored for different organic synthesis, including the formation of biaryls, synthesis of benzazoles, Spirooxindoles, and so forth (Schemes 1.2–1.5) [52–55].

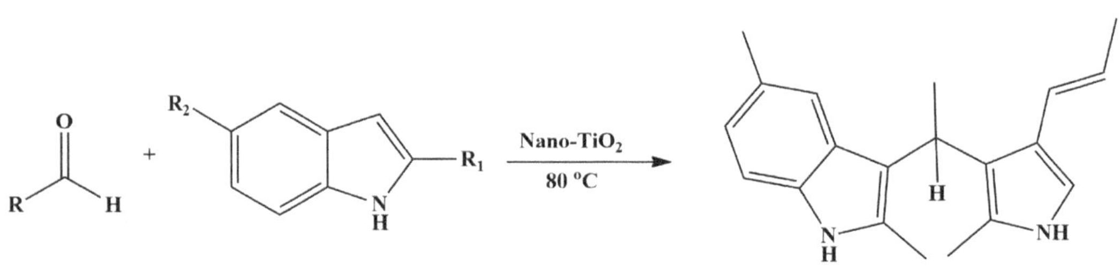

SCHEME 1.1 Synthesis of substituted indoles.

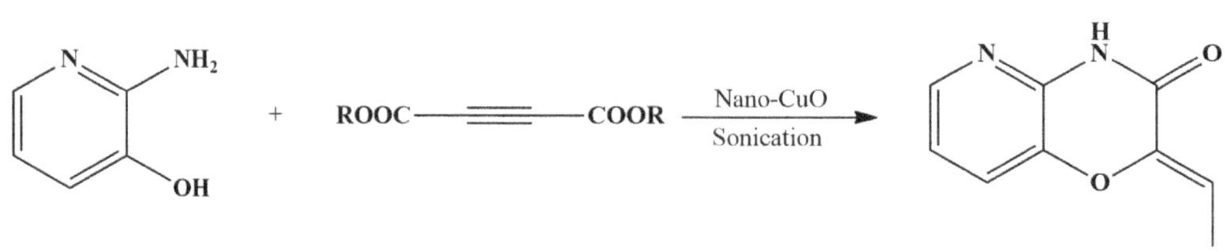

SCHEME 1.2 Synthesis of Spirooxindoles.

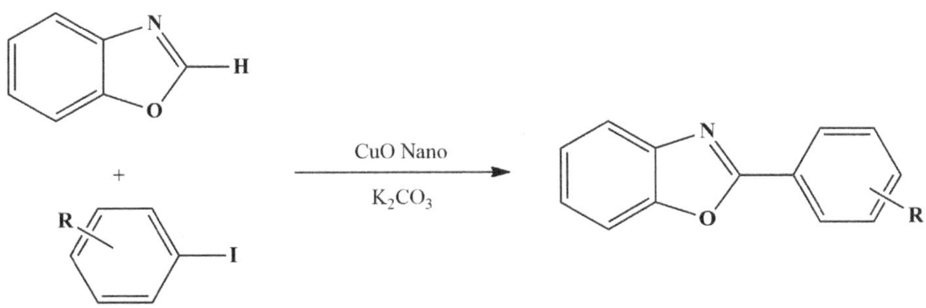

SCHEME 1.3 Synthesis of benzazoles derivatives.

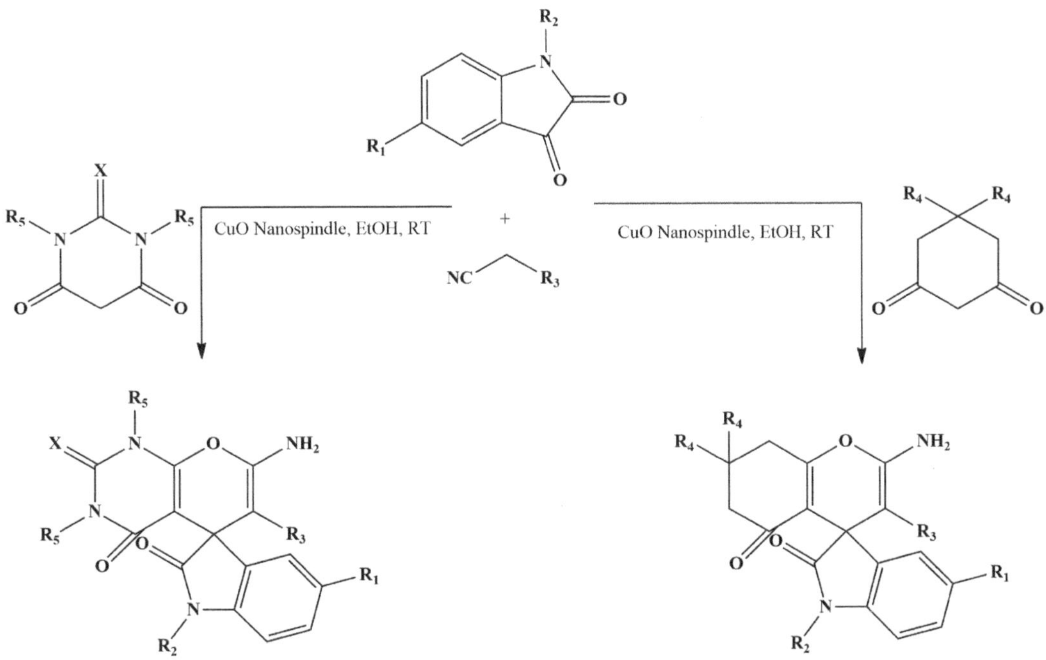

SCHEME 1.4 Synthesis of Spirooxindoles.

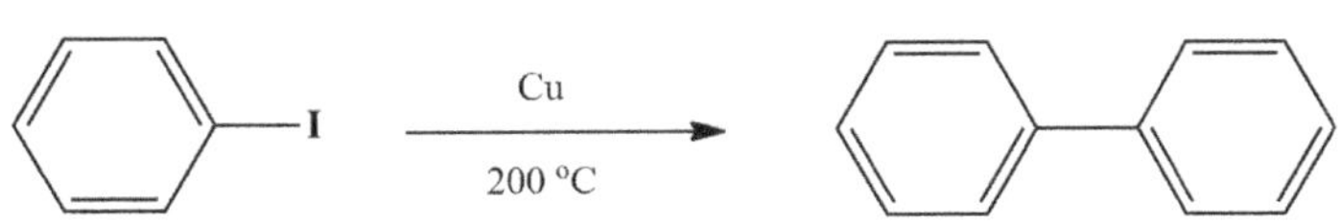

SCHEME 1.5 Synthesis of biaryls.

There are many uses for commercialization of silver nanoparticles, such as sterilizing nanomaterials in food and medical items, fabrics, personal care products and food storage bags [58–63]. They also exhibit antimicrobial properties, as well as optical, thermal, and catalytic capabilities [64–67]. Silver nanoparticles have been prepared using both seed-mediated growth in ionic liquid media (ionotherml method), as well as conventional reduction methods using hydrazine, sodium borohydride and so forth. A green photocatalytic approach was used to make Ag nanoparticles, with

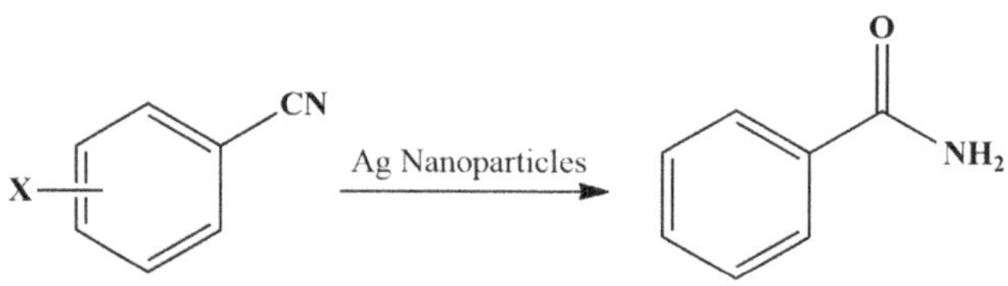

SCHEME 1.6 Synthesis of benzamide.

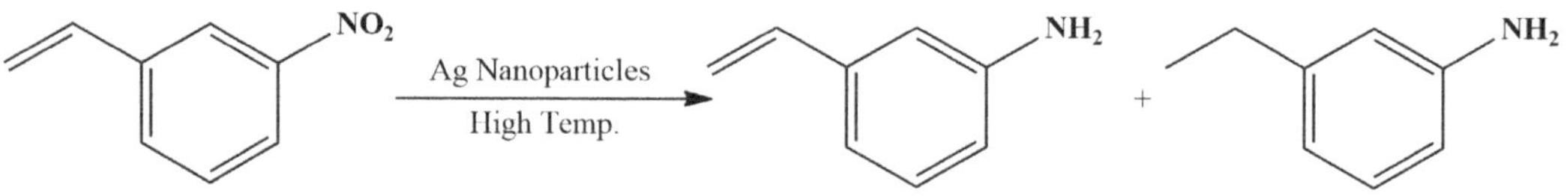

SCHEME 1.7 Synthesis of benzamide.

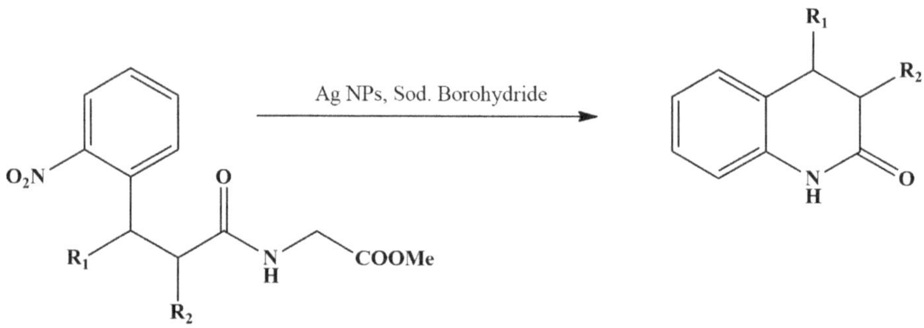

SCHEME 1.8 Synthesis of MCR Scaffolds Substituted by Amines and the derivatives of Dihydroquinoxalinone.

the process taking place in aqueous medium [56]. A photochemical green synthesis approach is used to make calciumalginate stabilized silver nanoparticles [57]. A process involving 4-nitrophenol is catalyzed by these nanoparticles. A few reaction schemes reported with silver nanoparticles are given below (Schemes 1.6–1.8) [68–70]. Reduction reaction of methylene blue (oxidized form) by Ag nanoparticles is also reported [71].

AuNPs have piqued the interest of many scientists in recent years due to their distinct properties, such as controlling physicochemical properties, strong affinity to bind compounds like thiols, disulfides, and amines high X-ray absorption coefficient, ease of synthesis, unique tunable optical and distinct electronic properties, and unique tunable optical and distinct electronic properties [72–78]. Optoelectronic properties of gold nanoparticles for hi-tech applications, such as sensory probes, targeted drug delivery, that is drug delivery in biological and medical applications electronic conductors, medicines, organic photovoltaics, fuel cells and of course catalysis are being studied extensively [79–83]. Some of the most recent works employing gold nanoparticles as catalysts are depicted in Schemes 1.9–1.14 [84–93].

Calcium oxide nanoparticles have gotten a lot of attention recently because of their remarkable features and possible applications in a variety of sectors [94]. Calcium oxide (CaO) is a low-cost, very basic, totally non-corrosive, environmentally benign, and cost effective material that may be recycled and reused. As a bonus, they create substantial quantities of products in a short amount of time compared to normal catalysts [95–97]. For example, the adsorption of Cr (VI) from water has been reported by several studies to use calcium oxide nanoparticles as an active catalyst in

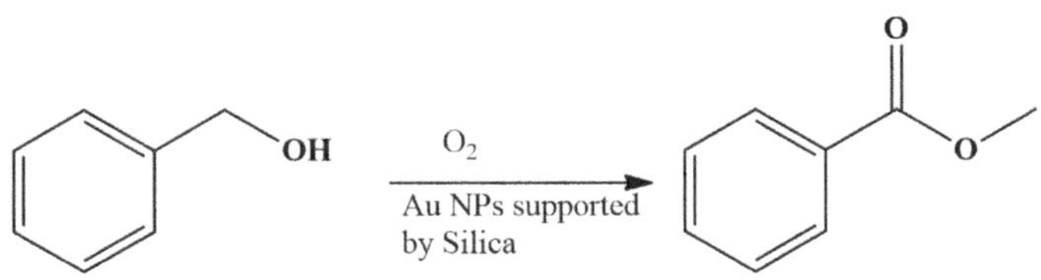

SCHEME 1.9 Synthesis of methyl esters.

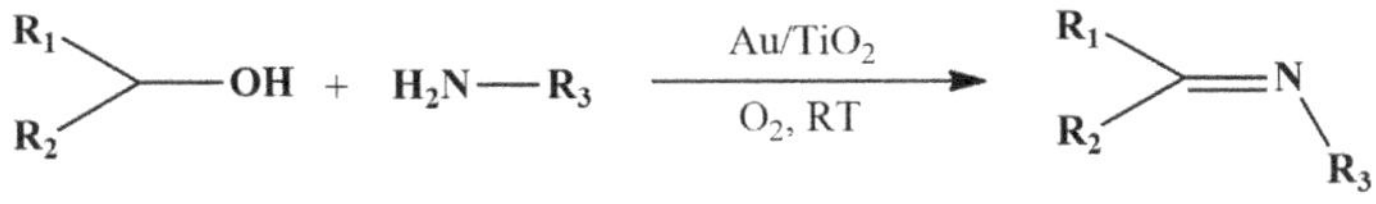

SCHEME 1.10 Synthesis of enamines.

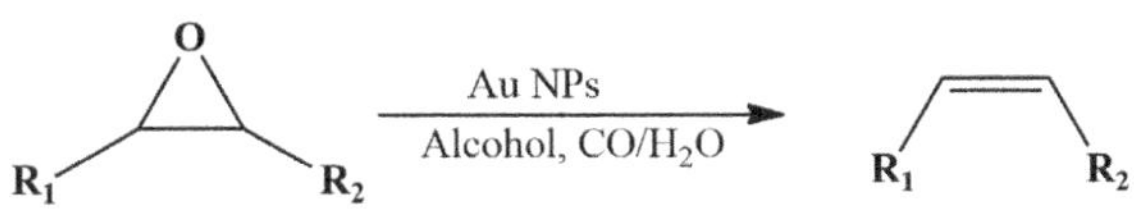

SCHEME 1.11 Synthesis of substituted but-2-ene.

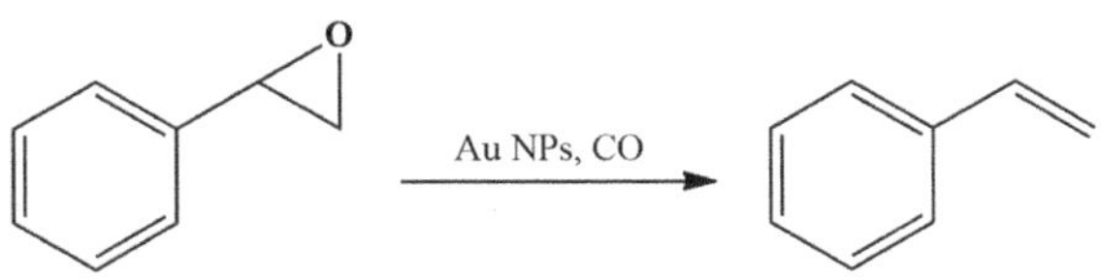

SCHEME 1.12 Synthesis of Styrene.

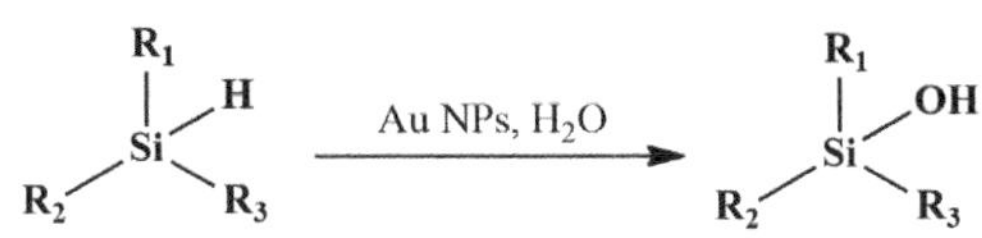

SCHEME 1.13 Synthesis of Silanols.

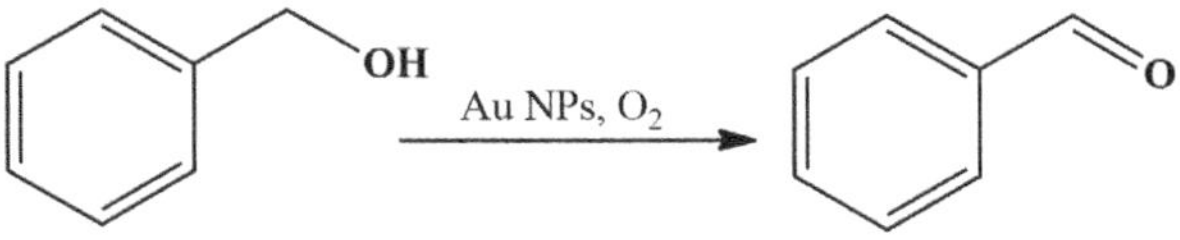

SCHEME 1.14 Selective oxidation of benzyl alcohol.

many organic transformations [98]. Other examples include biodiesel trans-esterification [99, 100], the elimination of harmful heavy metal ions from water [101] and artificial photosynthesis [102], Bromocresol green decomposition [103], vehicle gas exhaust purification [104]. Nanoparticles in catalysis and the importance of highly substituted pyridines as preferred medicinal scaffolds are in agreement with each other [105]. Catalyzing organic reactions with calcium oxide nanoparticles is efficient, non-explosive, environmentally safe and non-volatile while also being easy to handle.

It is evident from the literature that organic transformations with noble metal nanoparticles supported on metal oxides are an extremely important field of study. Numerous studies on metal oxides utilizing supported gold catalysts have been experimented over the last two decades, after the pioneering work of Haruta et al. [106, 107]. Previously, gold was thought to be catalytically inactive. But it was eventually inferred that gold can transform into an exceptionally active substance when broken down into nanoscale particles. Since then, catalysis with gold nanoparticles/supported gold nanocatalysts has become a very extensively studied topic in chemistry. The applications of gold nanoparticles as catalysis make their way into organic transformations such as selective oxidation of alcohols, CO oxidation, and selective reduction of nitro groups. There are various positive evaluations of gold catalysis available, including those by Huntings et al. and Arcadi [108, 109].

The size of the gold particles determines the catalytic activity of gold catalysts, which will be fully active when the particle size is within 8 nm in diameter [110, 111]. Development of feasible methods for generating supported gold catalysts with acceptable control over the particle size of gold and its stability still remains a challenge for scientists today [106, 110]. The AuNPs uniformly supported on substrates can be prepared via different synthetic methodologies like deposition-precipitation (DP) [112–165], co-precipitation (CP) [115–118] and impregnation (IP) [119–122] have been developed.

When using a large surface area support, it is possible for Au particles to scatter equally on the surface of the support. Nanofibers and nanotubes are good supports, as their surface area is very large. Zhu et al. embedded gold nanoparticles on TiO_2 (anatase) nanotubes and nanofibers using the conventional deposition-precipitation (DP) method. This results in the formation of small gold particles with high catalytic efficiency for the oxidation of CO [123, 124]. Furthermore, the nature of oxides is crucial because it affects the activities. Widely used catalyst supports are metal oxides such as SiO_2, TiO_2, Al_2O_3 and ZrO_2. Metal oxide supported gold catalysts find their application in a range of organic processes like water-gas shift reaction, CO oxidation, selective oxidation of alkenes, alcohols, and even alkanes and in the removal of NOx, VOCs from the atmosphere. These supports that are used can be categorized in two ways: Reproducible and irreproducible. The gold samples loaded on repeatable samples exhibit strong CO oxidation activity.

Moisture has an important role in low-temperature CO oxidation on supported gold catalysts, as it contributes to the development and regeneration of the surface active site [125, 126] Haruta et al. reported that this effect of the moisture is dependent on the catalyst support [125, 127]. On various types of supports, including insulating Al_2O_3 and SiO_2 as well as semiconducting TiO_2, increased activity was seen as moisture concentration increased. In spite of this, the exact mechanism of how moisture and catalytic supports work is still mostly unknown. In a recent coupled TG-FTIR study on Au/α-Fe_2O_3 catalysts for CO oxidation [128], it was demonstrated that small Au nanoparticles cannot directly activate the oxygen of the support lattice at low temperatures. Hence, lattice oxygen does not participate in the process; instead, the molecular oxygen species are responsible for the oxidation of CO at low temperature. However many studies have shown that molecular O_2 activation occurs primarily on the catalytic support [125, 129, 120]. As a result, these are most likely linked to the surface OH groups as well as the concentration and distribution of surface O-vacancies. On the contrary, the extensive studies on the photocatalytic water splitting reaction on TiO_2 surface [131] have shown that water molecules can either physically adsorb on TiO_2 surface or dissociate at oxygen vacancies, called defects, on the yielding surface OH groups. There might be a possibility that the water molecules can also have an interaction on the surface OH-groups on TiO_2 to enhance the adsorption and activation of molecular oxygen. This has been studied theoretically [132].

Other most-used metal nanoparticles such as catalysts for organic transformations are Palladium-based NPs, Iron oxide NPs, Zinc oxide NPs, Rhodium NPs, Aluminium NPs, Platinum NPs, and some bimetallic NPs. Bimetallic NPs include Ni-Pt, Au-Pd, and so forth.

In addition to their use as catalytic materials, palladium nanoparticles have several other applications, including hydrogen storage and sensing. Their synthesis has been extensively

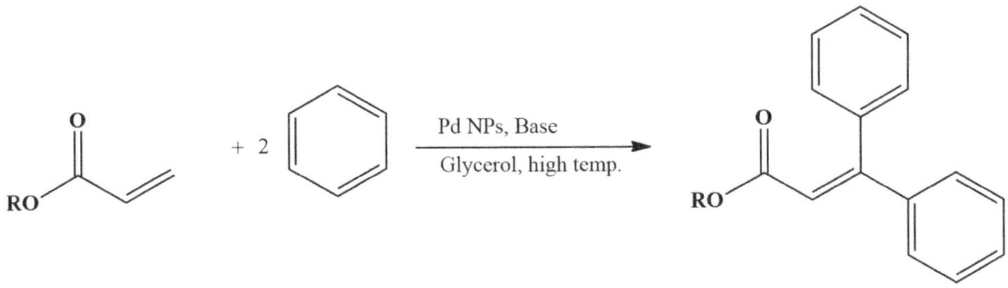

SCHEME 1.15 Synthesis of alkyl 3,3-diphenylacrylate.

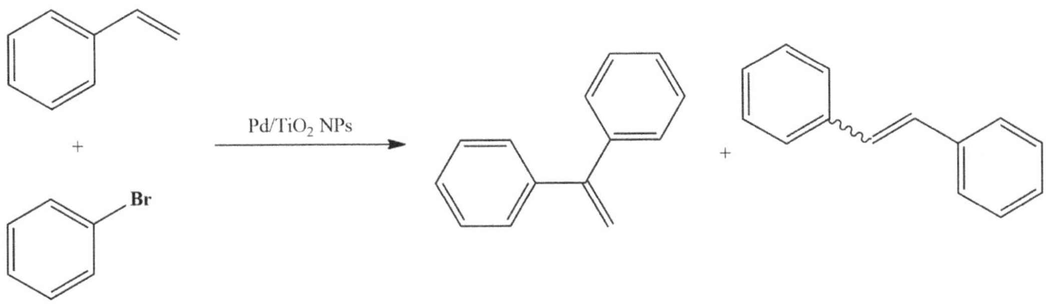

SCHEME 1.16 Reaction of Bromobenzene with styrene (A C-C Coupling reaction).

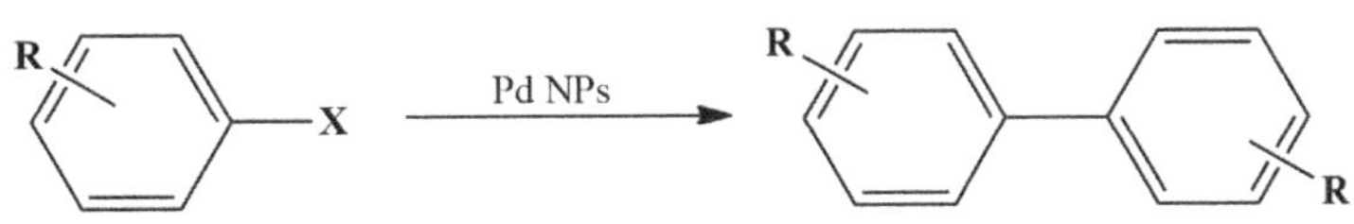

SCHEME 1.17 Synthesis of biaryls.

investigated, and interest in their qualities is developing as a result of that. Chemical and electrochemical synthesis of palladium nanoparticles employing a range of stabilizers, including as organic ligands, salt/surfactants, polymers and dendrimers, and their potential benefits in catalytic applications have been extensively investigated during the past several decades [133–137]. A few exemplar studies are represented below.

Iron and/or Iron oxide nanoparticles a great role in synthetic organic chemistry. Fe nanoparticles (usually referred to as) make their catalytic roles in many organic transformations. Alkene and alkyne hydrogenation catalysts have been developed using iron nanoparticles. For environmentally friendly alkene and alkyne hydrogenation processes, iron nanoparticles were produced. Schemes 1.18–1.20 represent few examples of organic transformations using iron nanoparticles [138–140].

1.5 RECENT DEVELOPMENTS IN ADVANCED NANOCATALYSTS

Catalysis by nanoparticles can have distinctive effects on the catalytic reaction. The detailed study of these distinctive features will give a better idea and understanding of nanocatalysts. Some of the characteristic properties of nanomaterials critically affect their synergistic action. Following are few characteristic properties of nanomaterials that are sorted: (i) quantities that are straightforwardly identified with bond length (like lattice parameter, density, binding energy); (ii) quantities that are

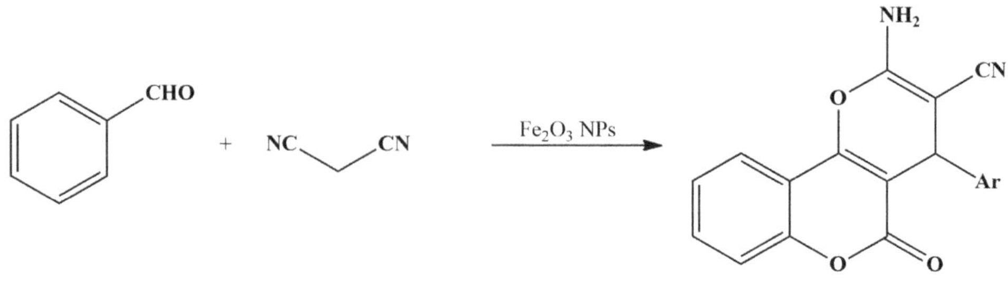

SCHEME 1.18 Synthesis of 2-amino-5-oxo-4-aryl-4,5-dihydropyrano[3, 2*c*]chromene-3-carbonitriles.

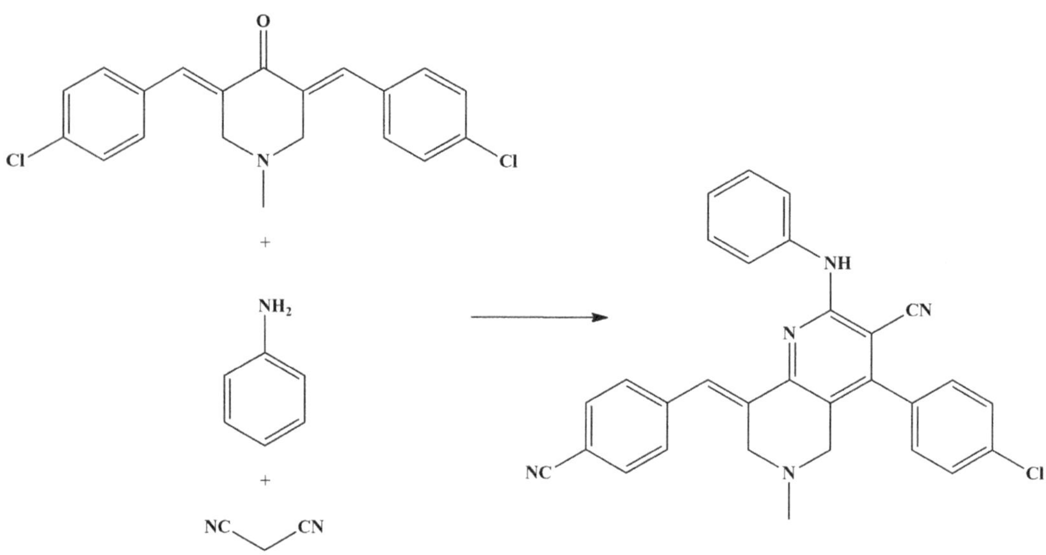

SCHEME 1.19 Synthesis of 1,6-naphthyridine analogues by alpha Fe_2O_3 NPs.

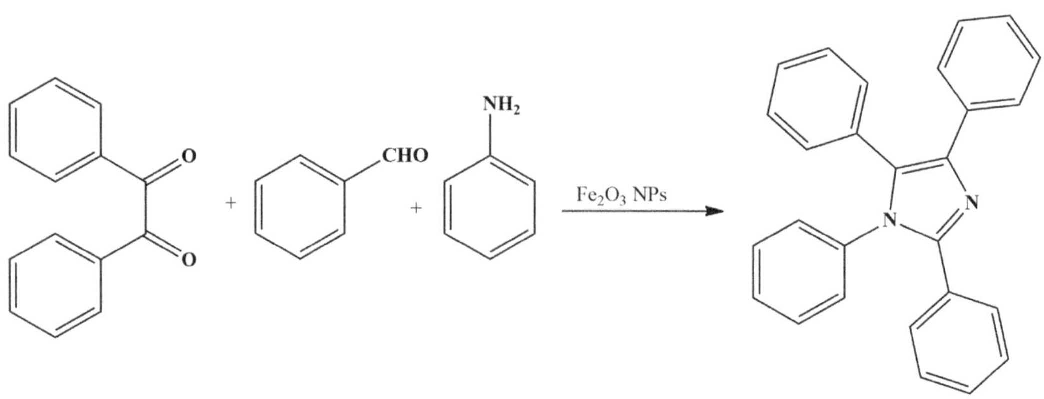

SCHEME 1.20 Synthesis of imidazoles.

dependent on cohesive energy; (iii) characteristics that vary with energy density, and (iv) qualities resulting from the interaction between energy density and atomic cohesive energy.

The addition of an interatomic interface causes the performance of a substance or a group of atoms to differ from that of a confined particle. Changing the relative quantity of the under-constituted surface atoms provides an additional chance to control the attributes of a nanocatalyst in comparison to its bulk equivalent. As a result, contributions from under-composed atoms and contributions from interatomic interfaces might be the first stage of consideration in overcoming any difficulties between individual atoms and the entire material in both chemical and physical forms. In terms of bond unwinding and its impacts on bond vitality, the effect of atomic coordination decrease is huge, and it reliably combines the display of a surface, a nanocatalyst, and an amorphous state. Nanoscale catalysts have recently been the subject of substantial academic and business exploratory considerations because of the multiple potential benefits that can arise from their implementation. The following are some of the most recent areas where the nanocatalyst applications are huge:

- Environmental protection
- Photocatalysis
- Drug delivery
- Production of biodiesel
- Water purification
- Fuel cells
- Solar cells
- Solid Rocket propellant

1.6 NANOSTRUCTURED CATALYSTS FOR GREENER AND SUSTAINABLE ORGANIC PROCESSES

Nanocatalysis is a rapidly expanding field that is a critical component of "sustainable technology and organic transformations" and is relevant to practically all sorts of catalytic organic transformations. Nanocatalysts that have found useful in catalysis include magnetic, graphene-based, mixed metal oxides, core-shell nanocatalysts, nano-supported catalysts and nanocatalysts have all been used in catalytic applications. Magnetic nanocatalysts are unique among reusable nanocatalysts because of their low preparation cost, exceptional activity, high selectivity, high stability, rapid recovery, and good recyclability.

The discipline of sustainable organic synthesis has recently encompassed a very good quantity of scientific creative discoveries. Many of these have been complemented by better and more efficient synthetic procedures that avoid, most importantly, the use of harmful reagents and reactants. Significant results are produced with the combination of organic reactions and the newer generation of nanocatalysts. These flexible semi heterogeneous nanocatalysts with a wide surface area are excellent replacements for traditional catalysts. Their physical properties like structure form, size and composition assist the reaction to achieve the highest catalytic activity, selectivity, and attaining stability as well. These notable advantages of nanocatalysts are dependent on the nanosize effect; catalytic efficiency generally increases as the size of nanostructures decreases. The surface free energy increases when the active site size is reduced to nanoscale levels. As a result the particles congregate into microscopic clusters and, hence, catalytic productivity is reduced. Furthermore, as the size of the catalysts reduces to nanoscale dimensions, isolation and recovery become more difficult; in most circumstances, separation using standard filtration is not an easy process. As a result, it is critical to incorporate appropriate support materials/nanomaterials when designing effective, reusable, and recyclable nanocatalysts. To date, these nanostructured materials are employed in various types of organic transformations as catalysts. Transformations include, but are not limited to, hydrogen-transfer reactions, chemo-selective oxidations, coupling reactions, oxidative aminations, asymmetric hydrogenations, C-H activations, Mannich reactions, Oxidative esterification, and so forth. For diverse catalytic applications, morphology dependent nanocatalysts, magnetic nanocomposites, graphene-supported nanocatalysts, hybrid nanostructured catalysts, integrated nanocatalysts, and core-shell nanocatalysts were used.

1.7 OUTLOOK

For production of heterocycles, green nanocatalysts have various advantages, especially in the industrial production. The advantages include short reaction times, high yields, use of cost-effective reagents, easy work-up technique, and very accurate reactions can be observed with good selectivity. Nanocatalysis can be used to synthesize a number of heterocycles that are extremely difficult to synthesize by conventional methods. Additionally, a variety of modifications can be made on the surface of the nanocatalyst to improve the catalytic efficiency. In most reactions, the spent catalyst may be easily removed from the reaction mixture and reused without affecting its catalytic activity. On the basis of nanoparticles, a wide range of novel processes for synthesizing diverse kinds of organic molecules, including organic functional group transformation, have been established. Notable development has been made using these nanocatalysts in terms of diversity of the organic reactions, activity, selectivity, and reusability. But still, the leaching of metal in nanocatalysts under harsh conditions or continuous flow reactions in flow reactors remains the major concern yet to be solved. To solve these challenges, new more resilient and improved multifunctional nanomaterials, as well as necessary techniques for the decorating of homogenous metals, organic ligands, or catalytic entities, must be designed and developed.

REFERENCES

1. *Norio Taniguchi, On the Basic Concept of Nanotechnology, Proceedings of the International Conference on Production Engineering*, Tokyo, Part 2 (Tokyo: JSPE), 18, 1074.
2. Naomi Lubick, Kellyn Betts, Silver socks have cloudy lining | Court bans widely used flame retardant, *Environmental Science & Technology* 42 (11), 3910 2008, https://doi.org/10.1021/es0871199.
3. Alan S. Edelstein, Robert C. Cammarata, Nanomaterials: Synthesis, *Properties, and Applications*, second ed. CRC Press, 1998.
4. Christopher G. Levins, Christian E. Schafmeister, The synthesis of curved and linear structures from a minimal set of monomers, *Journal of Organic Chemistry* 70, 9002, 2005, https://doi.org/10.1021/jo051639u.
5. Wei Liu, Catalyst Technology Development From Macro-, Micro- Down To Nano-Scale, China *Particuology*, 3, 383, 2005, https://doi.org/10.1142/S1672251505000801.
6. Andrew A. Herzing, Christopher J. Kiely, Albert F. Carley, Philip Landon, Graham J. Hutchings, Identification of Active Gold Nanoclusters on Iron Oxide Supports for CO Oxidation, *Science*, 321, 1331, 2008, DOI: 10.1126/science.1159639.
7. Beatriz Roldan Cuenya, Synthesis and catalytic properties of metal nanoparticles: Size, shape, support, composition, and oxidation state effects, *Thin Solid Films*, 518, 3127, 2010, https://doi.org/10.1016/j.tsf.2010.01.018.
8. Narayanan R, in “Shape-dependent Nanocatalysis and the effect of catalysis on the shape and size of colloidal metal nanoparticles” thesis submitted to Georgia Institute of Technology, 2005.
9. (a) Kin Hong Liew, Poh Lee Loh, Joon Ching Juan, Mohd Ambar Yarmo, Rahimi M. Yusop, QuadraPure-Supported Palladium Nanocatalysts for Microwave-Promoted Suzuki Cross-Coupling Reaction under Aerobic Condition, *The Scientific World Journal*, 2014 doi: dx.doi.org/10.1155/2014/796196. (b) Günter Schmid, Benedetto Corain, *European Journal of Inorganic Chemistry*, 3081, 2003, https://doi.org/10.1002/ejic.200300187.
10. Wenbo Hou, Nicole A. Dehm, Robert W.J. Scott, Alcohol oxidations in aqueous solutions using Au, Pd, and bimetallic AuPd nanoparticle catalysts, *Journal of Catalysis*, 253, 22, 2008, https://doi.org/10.1016/j.jcat.2007.10.025.

11. (a) Feng Shi, Man Kin Tse, Marga-Matina Pohl, Angelika Bruckner, Shengmao Zhang, Matthias Beller, *Angewandte Chemie International Edition*, 46, 8866, 2007, https://doi.org/10.1002/anie.200703418. (b) Maby Martínez, Rogelio Ocampo, Luz Amalia Rios, Alfonso Ramírez, Oscar Giraldo, *Journal of the Brazilian Chemical Society*, 22(12), 2322, 2011, https://doi.org/10.1590/S0103-5053201100 1200012.
12. (a) Zohreh Shobeiri, Mehrdad Pourayoubi, Akbar Heydari, Teresa Mancilla Percino, Marco A. Leyva Ramírez, Ultrasound assisted synthesis of $Cs_{2.5}H_{0.5}PW_{12}O_{40}$: An efficient nano-catalyst for preparation of -amino ketones via aza-Michael addition reactions, *Competes Rendus Chimie* 14, 597, 2011, https://doi.org/10.1016/j.crci.2011.03.002. (b) Lan Ma, Qingrong Peng, Dehua He, *Catalytic Behaviors of Amorphous Co-B Catalysts in Hydroformylation of 1-Octene, Catalysis Letters*, 130, 137, 2009, https://doi.org/10.1007/s10562-009-9839-8.
13. Javad Safari, Shiva Dehghan Khalili, Mehran Rezaei, Sayed Hossein Banitaba, Fereshteh Meshkani, Nanocrystalline magnesium oxide: a novel and efficient catalyst for facile synthesis of 2,4,5-trisubstituted imidazole derivatives, *Monatshefte für Chemie-Chemical Monthly*, 141, 1339, 2010, https://doi.org/10.1007/s00706-010-0397-y.
14. Selvaraj Mohana Roopan, Fazlur Rahman Nawaz Khan, ZnO nanoparticles in the synthesis of AB ring core of camptothecin, *Chemical Papers*, 64(6), 812, 2010, https://doi.org/10.2478/s11696-010-0058-y.
15. Madhumitha G, Selvaraj Mohana Roopan, Devastated crops: multifunctional efficacy for the production of nanoparticles, *Journal of Nanomaterials*, 2013, 1, 2013, https://doi.org/10.1155/2013/951858.
16. Sang-Bum Lee, Young In Park, Mi-Sook Dong, Young-Dae Gong, Identification of 2,3,6-trisubstituted quinoxaline derivatives as a Wnt2/β-catenin pathway inhibitor in non-small-cell lung cancer cell lines, *Bioorganic & Medicinal Chemistry Letters*, 20, 5900, 2010, DOI: 10.1016/j.bmcl.2010.07.088.
17. Nermien M. Sabry, Hany M. Mohamed, Essam Shawky A. E. H. Khattab, Shymaa S. Motlaq, Ahmed M. El-Agrody, Synthesis of 4*H*-chromene, coumarin, 12*H*-chromeno2,3-*d*]pyrimidine derivatives and some of their antimicrobial and cytotoxicity activities, *European Journal of Medicinal Chemistry*, 46 (2), 765, 2011, doi: 10.1016/j.ejmech.2010.12.015.
18. R. Raj, P. Singh, P. Singh, J. Gut, P.J. Rosenthal, and V. Kumar, "Azide-alkyne cycloaddition en route to 1H-1,2,3- triazole-tethered 7-chloroquinoline-isatin chimeras: synthesis and antimalarial evaluation," *European Journal of Medicinal Chemistry*, vol. 62, pp. 590–596, 2013.
19. M. M. Sit^onio, C.H. Carvalho Jr., I.A. Campos et al., "Antiinflammatory and anti-arthritic activities of 3, 4-dihydro-2, 2- dimethyl-2*H*-naphthol1,2-*b*]pyran-5, 6-dione (β-lapachone)," *Inflammation Research*, vol. 62, pp. 107–113, 2013.
20. Ramin Ghorbani-Vaghei, Seyedeh Mina Malaekehpoor, N- Bromosuccinimide as an efficient catalyst for the synthesis of indolo 2,3-*b*]quinolines, *Tetrahedron Letters*, 53, 4751, 2012, https://doi.org/10.1016/j.tetlet.2012.06.125.
21. Kurosh Rad-Moghadam, Seyyedeh Cobra Azimi, $Mg(BF_4)_2$ doped in BMIm]BF_4]: a homogeneous ionic liquid-catalyst for efficient synthesis of 1, 8-dioxo-octahydroxanthenes, decahydroacridines and 14-aryl-14 *H*-dibenzoa, j]xanthenes," *Journal of Molecular Catalysis A*, 363-364, 465, 2012, https://doi.org/10.1016/j.molcata.2012.07.026.
22. Tong-Shou Jin, Jin-Chong Xiao, Su-Juan Wang, Tong-Shuang Li, Ultrasound-assisted synthesis of 2-amino-2-chromenes with cetyl trimethylammonium bromide in aqueous media, *Ultrasonics Sonochemistry*, 11 (6), 393, 2004, https://doi.org/10.1016/j.ultsonch.2003.10.002.
23. Cheng-Liang Ni, Xiao-Hui Song, Hong Yan, Xiu-Qing Song, Ru-Gang Zhong, Improved synthesis of diethyl 2,6-dimethyl-4-aryl-4*H*-pyran- 3,5-dicarboxylate under ultrasound irradiation, *Ultrasonics Sonochemistry*, 17 (2), 367, 2010, https://doi.org/10.1016/j.ultsonch.2009.09.006.
24. Guo-Feng Chen, Hui-Ming Jia, Li-Yan Zhang, Bao-Hua Chen, Ji-Tai Li, An efficient synthesis of 2-substituted benzothiazoles in the presence of $FeCl_3$/Montmorillonite K-10 under ultrasound irradiation, *Ultrasonics Sonochemistry*, 20, 627, 2013, https://doi.org/10.1016/j.ultsonch.2012.09.010.
25. Aidin Lak, Mahyar Mazloumi, Matin Sadat Mohajerani, Saeid Zanganeh, Mohammad Reza Shayegh, Amir Kajbafvala, Hamed Arami, and Seyed Khatiboleslam Sadrnezhaad, Rapid formation of mono-dispersed hydroxyapatite nanorods with narrow-size distribution via Microwave Irradiation, *Journal of the American Ceramic Society*, vol. 91(11), 3580, 2008 DOI: 10.1111/j.1551-2916.2008.02690.x.
26. Matin Sadat Mohajerani, Mahyar Mazloumi, Aidin Lak, Amir Kajbafvala, Saeid Zanganeh, S. K. Sadrnezhaad, "Self-assembled zinc oxide nanostructures via a rapid microwave-assisted route, *Journal of Crystal Growth*, 310 (15) 3621, 2008, https://doi.org/10.1016/j.jcrysgro.2008.04.045.

27. Radha Narayanan, Synthesis of green nanocatalysts and industrially important green reactions, *Green Chemistry Letters and Reviews*, 5, 707, 2012, https://doi.org/10.1080/17518253.2012.700955.
28. Mohamed, R.M., McKinney, D.L., Sigmund, W.M., Enhanced nanocatalysts, *Materials Science and Engineering R*, 73, 1, 2012, doi:10.1016/j.mser.2011.09.001.
29. Mahyar Mazloumi, Nikta Shahcheraghi, Amir Kajbafvala, Saeid Zanganeh, Aidin Lak, Matin Sadat Mohajerani, S. K. Sadrnezhaad, 3D bundles of self-assembled lanthanum hydroxide nanorods via a rapid microwave-assisted route, *Journal of Alloys and Compounds*, 473, 1-2, 283–287, 2009.
30. Amir Kajbafvala, Hamed Ghorbani, Asieh Paravar, Joshua P. Samberg, Ehsan Kajbafvala, S.K. Sadrnezhaad, Effects of morphology on photocatalytic performance of zinc oxide nanostructures synthesized by rapid microwave irradiation methods, *Superlattices and Microstructures*, 51, 512, 2012, https://doi.org/10.1016/j.spmi.2012.01.015.
31. M.R. Bayati, Roya Molaei, Amir Kajbafvala, Saeid Zanganeh, H.R. Zargar, K. Janghorban, Investigation on hydrophilicity of micro-arc oxidized TiO_2 nano/micro-porous layers, *Electrochimica Acta*, 55, 20, 5786, 2010, https://doi.org/10.1016/j.electacta.2010.05.021.
32. Amir Kajbafvala, Joshua P. Samberg, Hamed Ghorbani, Ehsan Kajbafvala, and S. K. Sadrnezhaad, Effects of initial precursor and microwave irradiation on step-by-step synthesis of zinc oxide nano architectures, *Materials Letters*, 67, 342, 2012, https://doi.org/10.1016/j.matlet.2011.09.106.
33. Mahyar Mazloumi, Saeid Zanganeh, Amir Kajbafvala, Parisa Ghariniyat, shadi Taghavi Aidin Lak, Matin Sadat Mohajerani, S.K. Sadrnezhaad, Ultrasonic induced photoluminescence decay in sonochemically obtained cauliflower-like ZnO nanostructures with surface 1D nanoarrays, *Ultrasonics Sonochemistry*, 16 (1) 11, 2009, https://doi.org/10.1016/j.ultsonch.2008.05.011.
34. Saeid Zanganeh, Amir Kajbafval, Navid Zanganeh, Roya Molaei, M.R. Bayati, H.R. Zargar, S.K. Sadrnezhaad, Hydrothermal synthesis and characterization of TiO_2 nanostructures using LiOH as a solvent, *Advanced Powder Technology*, 22(3), 336, 2011, https://doi.org/10.1016/j.apt.2010.04.010.
35. Amir Kajbafvala, Saeid Zanganeh, Ehsan Kajbafvala, H.R. Zargar, M.R. Bayati, S.K. Sadrnezhaad, Microwave-assisted synthesis of narcis-like zinc oxide nanostructures, *Journal of Alloys and Compounds*, 497, 325, 2010, DOI:10.1016/j.jallcom.2010.03.057.
36. Amir Kajbafvala, Mohammad Reza Shayegh, Mahyar Mazloumi, Saeid Zanganeh, Aidin Lak, Matin Sadat Mohajerani, S. K. Sadrnezhaad, Nanostructure sword-like ZnO wires: rapid synthesis and characterization through a microwave-assisted route, *Journal of Alloys and Compounds*, 469, 293, 2009, https://doi.org/10.1016/j.jallcom.2008.01.093.
37. Aidin Lak, Mahyar Mazloumi, Matin Mohajerani, Amir Kajbafvala, Saeid Zanganeh, Hamed Arami, S.K. Sadrnezhaad, Self-assembly of dandelion-like hydroxyapatite nanostructures via hydrothermal method, *Journal of the American Ceramic Society*, 91, 3292, 2008, https://doi.org/10.1111/j.1551-2916.2008.02600.x.
38. Saeid Zanganeh, Amir Kajbafvala, Navid Zanganeh, Matin Sadat Mohajerani, Aidin Lak, M.R. Bayati, H.R. Zargar, S.K. Sadrnezhaad, Self-assembly of boehmite nanopetals to form 3D high surface area nanoarchitectures, *Applied Physics A*, 99, 317, 2010, https://doi.org/10.1007/s00339-009-5534-2.
39. Mahyar Mazloumi, Mehdi Attarchi, Aidin Lak, Matin Sadat Mohajerani, Amir Kajbafvala, Saeid Zanganeh, S.K. Sadrnezhaad, Boehmite nanopetals self-assembled to form rosette-like nanostructures, *Materials Letters*, 62, 4184, 2008, https://doi.org/10.1016/j.matlet.2008.06.025.
40. Saeid Zanganeh, Morteza Torabi, Amir Kajbafvala, Navid Zanganeh, M.R. Bayati, Roya Molaei, H.R. Zargar, S.K. Sadrnezhaad, CVD fabrication of carbon nanotubes on electrodeposited flower-like Fe nanostructures, *Journal of Alloys and Compounds*, 507, 494, 2010, https://doi.org/10.1016/j.jallcom.2010.07.216.
41. An-Hui Lu, E. L. Salabas, Ferdi Schuth, Magnetic Nanoparticles: Synthesis, Protection, Functionalization, and Application, *Angewandte Chemie International Edition*, 46, 1222 2007, https://doi.org/10.1002/anie.200602866.
42. Aharon Gedanken, Using sonochemistry for the fabrication of nanomaterials, *Ultrasonics Sonochemistry* 11, 47, 2004, https://doi.org/10.1016/j.ultsonch.2004.01.037.
43. C.N.R. Rao, S.R.C. Vivekchand, Kanishka Biswas, A. Govindaraj, Synthesis of inorganic nanomaterials, *Dalton Transactions*, 3728, 2007, https://doi.org/10.1039/B708342D.
44. Y. Mastai, A. Gedanken in: C.N.R. Rao, A. Müller, A.K. Cheetham (editors), *Chemistry of Nanomaterials*, Wiley-VCH, Weinheim, vol. 1, 113, 2004.

45. Devinder Mahajan, Elizabeth T Papish, Kaumudi Pandya, Sonolysis induced decomposition of metal carbonyls: kinetics and product characterization, *Ultrasonics Sonochemistry* 11, 385, 2004, https://doi.org/10.1016/j.ultsonch.2003.10.009.
46. Jongnam Park, Jin Joo, Soon Gu Kwon, Youngjin Jang, Taeghwan Hyeon, Synthesis of monodisperse spherical nanocrystals, *Angewandte Chemie International Edition*, 46, 4630, 2007, DOI: 10.1002/anie.200603148.
47. Didier Astruc, Feng Lu, Jaime Ruiz Aranzaes, Nanoparticles as recyclable catalysts: the frontier between homogeneous and heterogeneous catalysis, *Angewandte Chemie International Edition*, 44, 7852, 2005, doi: 10.1002/anie.200500766.
48. Cheng Pan, Katrin Pelzer, Karine Philippot, Bruno Chaudret, Fabrice Dassenoy, Pierre Lecante, Marie-José Casanove, *Journal of the American Chemical Society*, 123, 7584, 2001, https://doi.org/10.1021/ja003961m.
49. Aiken III, J.D., Finke, R.G., A Review of Modern Transition-Metal Nanoclusters: Their Synthesis, Characterization, and Applications in Catalysis. *Journal of Molecular Catalysis A: Chemical*, 145, 1, 1999, http://dx.doi.org/10.1016/S1381-1169(99)00098-9.
50. Robin J. White, Rafael Luque, Vitaliy L. Budarin, James H. Clark, Duncan J. Macquarrie, Supported metal nanoparticles on porous materials. *Methods and applications, Chemical Society Reviews*, 38, 481, 2009, https://doi.org/10.1039/B802654H.
51. Rahimizadeh M, Bakhtiarpoor Z, Eshghi H, Porde M, Rajabzadeh G. TiO_2 nanoparticles: an efficient heterogeneous catalyst for synthesis of bis(indolyl)methanes under solvent-free conditions, *Monatshefte für Chemie - Chemical Monthly* 140(12),1465, 2009, DOI:10.1007/s00706-009-0205-8.
52. Sodeh Sadjadi, Samaheh Sadjadi, Rahim Hekmatshoar, Ultrasound-promoted greener synthesis of benzoheterocycle derivatives catalyzed by nanocrystalline copper(II) oxide, *Ultrasonics Sonochemistry*, 17(5), 764, 2010, DOI: 10.1016/j.ultsonch.2010.01.017.
53. Wu Zhang, Qinglong Zeng, Xinming Zhang, Yujie Tian, Yun Yue, Yujun Guo, Zhenghua Wang, Ligand-Free CuO Nanospindle Catalyzed Arylation of Heterocycle C-H Bonds. *The Journal of Organic Chemistry*, 76(11), 4741, 2011, https://doi.org/10.1021/jo200452x.
54. Leila Moradi, Zeynab Ataei, Efficient and green pathway for one-pot synthesis of spirooxindoles in the presence of CuO nanoparticles, *Green Chemistry Letters and Reviews*, 10, 380, 2017, https://doi.org/10.1080/17518253.2017.1390611.
55. Mohd. Samim, N. K. Kaushik, Amarnath Maitra, Effect of size of copper nanoparticles on its catalytic behaviour in Ullman reaction, *Bulletin of Materials Science* 30, 535, 2007, https://doi.org/10.1007/s12034-007-0083-9.
56. Pedro Quaresma, Leonor Soares, Lívia Contar, Adelaide Miranda, Inês Osório, Patrícia A. Carvalho, Ricardo Franco, Eulália Pereira, Green photocatalytic synthesis of stable Au and Ag nanoparticles, *Green Chemistry* 11, 1889, 2009, https://doi.org/10.1039/B917203N.
57. Sandip Saha, Anjali Pal, Subrata Kundu, Soumen Basu, Tarasankar Pal, Photochemical Green Synthesis of Calcium-Alginate-Stabilized Ag and Au Nanoparticles and Their Catalytic Application to 4-Nitrophenol Reduction, *Langmuir*, 26(4), 2885, 2009, https://doi.org/10.1021/la902950x.
58. Guowu Zhan, Mingming Du, Jiale Huang, Qingbiao Li, Green synthesis of Au/TS-1 catalysts via two novel modes and their surprising performance for propylene epoxidation *Catalysis Communications*, 12(9), 830, 2011, https://doi.org/10.1016/j.catcom.2011.01.026.
59. Ranjeet A. Bapat, Tanay V. Chaubal, Chaitanya P. Joshi, Prachi R. Bapat, Hira Choudhury, Manisha Pandey, Bapi Gorain, Prashant Kesharwani, An overview of application of silver nanoparticles for biomaterials in dentistry, *Materials Science and Engineering*: C 91,881, 2018, https://doi.org/10.1016/j.msec.2018.05.069.
60. Sukumaran Prabhu, Eldho K. Poulose, Silver nanoparticles: mechanism of antimicrobial action, synthesis, medical applications, and toxicity effects. *International Nano Letters*, 2, 32, 2012, https://doi.org/10.1186/2228-5326-2-32.
61. Kholoud M. M. Abou El-Noura, Ala'a Eftaiha, Abdulrhman Al-Warthan, Reda A. A. Ammar, Synthesis and applications of silver nanoparticles, *Arabian Journal of Chemistry*, 3, 135, 2010, https://doi.org/10.1016/j.arabjc.2010.04.008.
62. Quang Huy Tran, Van Quy Nguyen, Anh-Tuan Le, Silver nanoparticles: synthesis, properties, toxicology, applications and perspectives. *Advances in Natural Sciences: Nanoscience and Nanotechnology* 4(3), 033001, 2013, https://doi.org/10.1088/2043-6262/4/3/033001.

63. Douglas Roberto Monteiro, Luiz Fernando Gorup, Aline Satie Takamiya, Adhemar Colla Ruvollo-Filho, Emerson Rodrigues de Camargo, Debora Barros Barbosa, The growing importance of materials that prevent microbial adhesion: antimicrobial effect of medical devices containing silver, *International Journal of Antimicrobial Agents*, 34(2), 103, 2009, https://doi.org/10.1016/j.ijantimicag.2009.01.017.
64. Maqusood Ahamed, Mohamad S. AlSalhi, M.K.J. Siddiqui, Silver nanoparticle applications and human health, *Clinica Chimica Acta* 411(23-24), 1841, 2010, https://doi.org/10.1016/j.cca.2010.08.016.
65. Jorge Garc´ıa-Barrasa, Josè Lopez-de-luzuriaga, Miguel Monge, Silver nanoparticles: synthesis through chemical methods in solution and biomedical applications. *Central European Journal of Chemistry*, 9(1), 7, 2011, https://doi.org/10.2478/s11532-010-0124-x.
66. Julia Fabrega, Samuel N. Luoma, Charles R. Tyler, Tamara S. Galloway, Jamie R. Lead, Silver nanoparticles: Behaviour and effects in the aquatic environment, *Environment International*, 37(2), 517, 2011, https://doi.org/10.1016/j.envint.2010.10.012.
67. Panagiotis Dallas, Virender K. Sharma, Radek Zboril, Silver polymeric nanocomposites as advanced antimicrobial agents: Classification, synthetic paths, applications, and perspectives, *Advances in Colloid and Interface Science*, 166(1-2), 119, 2011, https://doi.org/10.1016/j.cis.2011.05.008.
68. A.Y. Kim, Hee Seon Bae, Suhwan Park, Sungkyun Park, Kang Hyun Park, Silver nanoparticle catalyzed selective hydration of nitriles to amides in water under neutral conditions, *Catalysis Letters*, 141(5), 685, 2011, https://doi:10.1007/s10562-011-0561-y
69. Kiyotomi Kaneda, Takato Mitsudome, Tomoo Mizugaki, Koichiro Jitsukawa, Development of heterogeneous o\Olympic medal metal nanoparticle catalysts for environmentally benign molecular transformations based on the surface properties of hydrotalcite, *Molecules*, 15(12), 8988, 2010, https://doi.org/10.3390/molecules15128988.
70. Domna Iordanidou, Tryfon Zarganes-Tzitzikas, Constantinos G. Neochoritis, Alexander Dömling, Ioannis N. Lykakis, Application of silver nanoparticles in the multicomponent reaction domain: A combined catalytic reduction methodology to efficiently access potential hypertension or inflammation inhibitors, *ACS Omega*, 3(11), 16005, 2018, https://doi.org/10.1021/acsomega.8b02749.
71. Jayanta Saha, Arjuara Begum, Avik Mukherjee, Santosh Kumar, A novel green synthesis of silver nanoparticles and their catalytic action in reduction of Methylene Blue dye, *Sustainable Environment Research*, 27, 245, 2017, https://doi.org/10.1016/j.serj.2017.04.003.
72. Y. Zhou, C.Y. Wang, Y.R. Zhu, Z.Y. Chen, A novel ultraviolet irradiation technique for shape-controlled synthesis of gold nanoparticles at room temperature. *Chemistry of Materials*, 11, 2310, 1999, https://doi.org/10.1021/cm990315h.
73. Pengxiang Zhao, Na Li, Didier Astruc, State of the art in gold nanoparticle synthesis, *Coordination Chemistry Reviews*, 257–638, 2013, https://doi.org/10.1016/j.ccr.2012.09.002.
74. Zhiyang Zhang, Zhaopeng Chen, Fangbin Cheng, Yaowen Zhang, Lingxin Chen, Highly sensitive onsite detection of glucose in human urine with naked eye based on enzymatic-like reaction mediated etching of gold nanorods, *Biosensors and Bioelectronics*, 89, 932, 2017, https://doi.org/10.1016/j.bios.2016.09.090.
75. Zhenjiang Zhang, Jing Wang, Xin Nie, Tao Wen, Yinglu Ji, Xiaochun Wu, Yuliang Zhao, Chunying Chen, Near infrared laser induced targeted cancer therapy using thermoresponsive polymer encapsulated gold nanorods, *Journal of the American Chemical Society*, 136, 7317, 2014, https://doi.org/10.1021/ja412735p.
76. Yuanchao Zhang, Wendy Chu, Alireza Dibaji Foroushani, Hongbin Wang, Da Li, Jingquan Liu, Colin J Barrow, Xin Wang, Wenrong Yang, New gold nanostructures for sensor applications: a review, *Materials*, 7, 5169, 2014, DOI: 10.3390/ma7075169.
77. Yuan Zhang, Jun Qian, Dan Wang, Yalun Wang, Sailing He, Multifunctional gold nanorods with ultrahigh stability and tunability for in vivo fluorescence imaging, SERS detection, and photodynamic therapy, *Angewandte Chemie International Edition*, 52, 1148, 2013, https://doi.org/10.1002/anie.201207909.
78. Xiaoying Zhang, Gold nanoparticles: recent advances in the biomedical applications. Cell Biochemistry and Biophysics, 72, 771, 2015, https://doi.org/10.1007/s12013-015-0529-4.
79. Krishnendu Saha, Sarit S. Agasti, Chaekyu Kim, Xiaoning Li, Vincent M. Rotello, Gold Nanoparticles in Chemical and Biological Sensing, *Chemical Reviews*, 112(5), 2739, 2012, https://doi.org/10.1021/cr2001178.

80. Narges Elahi, Mehdi Kamali, Mohammad Hadi Baghersad, Recent biomedical applications of gold nanoparticles: A review, *Talanta*, 184, 537, 2018, https://doi.org/10.1016/j.talanta.2018.02.088.
81. G. Schmid, U. Simon, Gold nanoparticles: assembly and electrical properties in 1-3 dimensions. *Chemical Communications*, 697, 2005, https://doi.org/10.1039/B411696H.
82. Urmila Saxena, Pranab Goswami, Electrical and optical properties of gold nanoparticles: applications in gold nanoparticles-cholesterol oxidase integrated systems for cholesterol sensing. *Journal of Nanoparticle Research*, 14, 813, 2012, https://doi.org/10.1007/s11051-012-0813-9.
83. Chuan-Jian Zhong, Jin Luo, Derrick Mott, Mathew M. Maye, Nancy Kariuki, Lingyan Wang, Peter Njoki, Mark Schadt, Stephanie I-Im. Lim, Yan Lin, Gold-Based Nanoparticle Catalysts for Fuel Cell Reactions, *Nanotechnology in Catalysis*, 289, 2007, https://doi.org/10.1007/978-0-387-34688-5_14.
84. Vasant R. Choudhary, Anirban Dhar, Prabhas Jana, Rani Jha, Balu S. Uphade, A green process for chlorine-free benzaldehyde from the solvent-free oxidation of benzyl alcohol with molecular oxygen over a supported nano-size gold catalyst, *Green Chemistry*, 7(11), 768, 2005, https://doi.org/10.1039/B509003B.
85. Laise N.S. Pereira, Carlos E.S. Ribeiro, Aryane Tofanello, Jean C.S. Costa, Carla V.R. de Moura, Marco A.S. Garcia, Edmilson M. de Moura, Gold Supported on Strontium Surface-Enriched $CoFe_2O_4$ Nanoparticles: A Strategy for the Selective Oxidation of *Benzyl Alcohol, Journal of the Brazilian Chemical Society*, 30(6), 1317, 2019, https://doi.org/10.21577/0103-5053.20190030.
86. Vitaly Gitis, Rolf Beerthuis, N. Raveendran Shiju, Gadi Rothenberg, Organosilane oxidation by water catalyzed by large gold nanoparticles in a membrane reactor, *Catalysis Science & Technology*, 4, 2156, 2014, https://doi.org/10.1039/C3CY00506B.
87. Takato Mitsudome, Kiyotomi Kaneda, Gold nanoparticle catalysts for selective hydrogenations, *Green Chemistry*, 15, 2636, 2013, https://doi.org/10.1039/C3GC41360H.
88. Jianbo Zhao, Liming Ge, Haifeng Yuan, Yingfan Liu, Yanghai Gui, Baoding Zhang, Liming Zhou, Shaoming Fang, Heterogeneous gold catalysts for selective hydrogenation from nanoparticles to atomically precise nanoclusters. *Nanoscale*, 11, 11429, 2019, https://doi.org/10.1039/C9NR03182K.
89. Babak Karimi, Farhad Kabiri Esfahani, Unexpected golden Ullmann reaction catalyzed by Au nanoparticles supported on periodic mesoporous organosilica (PMO). *Chemical Communications*, 47, 10452, 2011, https://doi.org/10.1039/C1CC12566D.
90. Takato Mitsudome, Akifumi Noujima, Yusuke Mikami, Tomoo Mizugaki, Koichiro Jitsukawa, Kiyotomi Kaneda, Room-Temperature Deoxygenation of Epoxides with CO Catalyzed by Hydrotalcite-Supported Gold Nanoparticles in Water, *Chemistry – A European Journal*, 16(39),11818, 2010, https://doi.org/10.1002/chem.201001387.
91. Akifumi Noujima, Takato Mitsudome, Tomoo Mizugaki, Koichiro Jitsukawa, Kiyotomi Kaneda. Gold Nanoparticle-Catalyzed Environmentally Benign Deoxygenation of Epoxides to Alkenes, *Molecules*, 16(10), 8209, 2011, DOI: 10.3390/molecules16108209.
92. Søren Kegnaes, Jerrik Mielby, Uffe V. Mentzel, Claus H. Christensenc, Anders Riisager, Formation of imines by selective gold-catalysed aerobic oxidative coupling of alcohols and amines under ambient conditions, *Green Chemistry*, 12(8), 1437, 2010, https://doi.org/10.1039/C0GC00126K.
93. Rafael L. Oliveira, Pedro K. Kiyohara, Liane M. Rossi, Clean preparation of methyl esters in one-step oxidative esterification of primary alcohols catalyzed by supported gold nanoparticles, *Green Chemistry*, 11(9), 1366, 2009, https://doi.org/10.1039/B902499A.
94. Didier Astruc, Nanoparticles and Catalysis, Wiley, 1 2008, DOI:10.1002/9783527621323.
95. Masoud Zabeti Wan Mohd Ashri Wan Daud, Mohamed Kheireddine Aroua, Optimization of the activity of CaO/Al_2O_3 catalyst for biodiesel production using response surface methodology, *Applied Catalysis A General*, 366, 154, 2009, https://doi.org/10.1016/j.apcata.2009.06.047.
96. Ayhan Demirbas, Biodiesel from sunflower oil in supercritical methanol with calcium oxide, *Energy Conversion and Management*, 48, 937, 2007, https://doi.org/10.1016/j.enconman.2006.08.004.
97. M. López Granados, M.D. Zafra Poves, D. Martín Alonso, R. Mariscal, F. Cabello Galisteo, R. Moreno-Tost, J. Santamaría, J. L. G. Fierro, Biodiesel from sunflower oil by using activated calcium oxide, *Applied Catalysis B: Environmental*, 73, 317, 2007 https://doi.org/10.1016/j.apcatb.2006.12.017.
98. N. A. Oladoja, I. A. Ololade, S.E. Olaseni, V. O. Olatujoye, O. S. Jegede, A. O. Agunloye, Synthesis of nano calcium oxide from a gastropod shell and the performance evaluation for Cr (VI) removal from aqua system, *Industrial & Engineering Chemistry Research*, 51, 639, 2012, https://doi.org/10.1021/ie201189z.

99. Sandra Luz Martínez, Rubi Romero, José Carlos López, Amaya Romero, Víctor Sánchez Mendieta, Reyna Natividad, Preparation and characterization of CaO nanoparticles/NaX zeolite catalysts for the transesterification of sunflower oil, *Industrial & Engineering Chemistry Research*, 50, 2665, 2011, https://doi.org/10.1021/ie1006867.
100. Marinković Dalibor M., Stanković Miroslav V., Veličković Ana V., Avramović Jelena M., Cakić Milorad D., Veljković Vlada B., The synthesis of CaO loaded onto Al_2O_3 from calcium acetate and its applications in the transesterification of the sunflower oil, *Advanced Technologies*, 4(1), 26, 2015, DOI: 10.5937/savteh1501026M.
101. Anhua Cai, Xurong Xu, Haihua Pan, Jinhui Tao, Rui Liu, Ruikang Tang, Kilwon Cho, Direct synthesis of hollow vaterite nanospheres from amorphous calcium carbonate nanoparticles via phase transformation, *The Journal of Physical Chemistry C*, 114, 12948, 2010, https://doi.org/10.1021/jp801408k.
102. Mohammad Mahdi Najafpour, Sara Nayeri, Babak Pashaei, Nano-size amorphous calcium manganese oxide as an efficient and biomimetic water oxidizing catalyst for artificial photosynthesis: back to manganese, *Dalton Trans* 40, 9374, 2011, https://doi.org/10.1039/C1DT11048A.
103. Jejenija Osuntokun, Damian C. Onwudiwe, Eno E. Ebenso, Aqueous extract of broccoli mediated synthesis of CaO nanoparticles and its application in the photocatalytic degradation of bromocrescol green. *IET Nanobiotechnology*, 12(7), 888, 2018, DOI: 10.1049/iet-nbt.2017.0277.
104. J. Safaei-Ghomia, M. A. Ghasemzadeha, M. Mehrabib, Calcium oxide nanoparticles catalyzed one-step multicomponent synthesis of highly substituted pyridines in aqueous ethanol media, *Scientia Iranica*, 20(3), 549, 2013, https://doi.org/10.1016/j.scient.2012.12.037.
105. George E. Hoag, John B. Collins, Jennifer L. Holcomb, Jessica R. Hoag, Mallikarjuna N. Nadagouda, Rajender S. Varma, *Journal of Materials Chemistry*, 19(45), 8671, 2009, https://doi.org/10.1039/B909148C.
106. Haruta M., Tsubota S., Kobayashi T., Kageyama H., Genet M.J., Delmon B., Low-Temperature Oxidation of Co over Gold Supported on TiO_2, Alpha-Fe_2O_3, and Co_3O_4, *Journal of Catalysis*, 144, 175, 1993, https://doi.org/10.1006/jcat.1993.1322.
107. Haruta M., Yamada N., Kobayashi T., Iijima S., Gold catalyst prepared by coprecipitation for low-temperature oxidations of hydrogen and of carbon monoxide, *Journal of Catalysis*, 115, 301, 1989, https://doi.org/10.1016/0021-9517(89)90034-1.
108. (a) Antonio Arcadi, Alternative synthetic methods through new developments in catalysis by gold, *Chemical Reviews* 108, 3266, 2008, https://doi.org/10.1021/cr068435d. (b) A. Stephen K. Hashmi, Graham J. Hutchings, Gold catalysis, *Angewandte Chemie International Edition*, 45, 7896, 2006, https://doi.org/10.1002/anie.200602454.
109. M. Valden, X. Lai, D. W. Goodman, Onset of catalytic activity of gold clusters on titania with the appearance of nonmetallic properties, *Science* 281, 1647, 1998, DOI: 10.1126/science.281.5383.1647.
110. Masatake Haruta, Size- and support-dependency in the catalysis of gold, *Catalysis Today* 36, 153, 1997, https://doi.org/10.1016/S0920-5861(96)00208-8.
111. Rodolfo Zanella, Suzanne Giorgio, Chae-Ho Shin, Claude R Henry, Catherine Louis, Characterization and reactivity in CO oxidation of gold nanoparticles supported on TiO_2 prepared by deposition-precipitation with NaOH and urea, *Journal of Catalysis*, 222, 357, 2004, https://doi.org/10.1016/j.jcat.2003.11.005.
112. Rodolfo Zanella, Suzanne Giorgio, Claude R Henry, Catherine Louis, Alternative methods for the preparation of gold nanoparticles supported on TiO_2, *Journal Physical Chemistry B*, 106,7634, 2002, https://doi.org/10.1021/jp0144810.
113. J.M.C. Soares, Peter Morrall, Alison Crossley, Peter Harris, Michael Bowker, Catalytic and noncatalytic CO oxidation on Au/TiO_2 catalysts, *Journal of Catalysis*, 219, 17, 2003, https://doi.org/10.1016/S0021-9517(03)00194-5.
114. Xiao-Ying Wang, Shu-Ping Wang, Shu-Rong Wang, Ying-Qiang Zhao, Jing Huang, Shou-Min Zhang, Wei-Ping Huang, Shi-Hua Wu, The preparation of Au/CeO_2 catalysts and their activities for low-temperature CO oxidation, *Catalysis Letters*, 112, 115, 2006, https://doi.org/10.1007/s10562-006-0173-0.
115. Saleh Al-Sayari, Albert F. Carley, Stuart H. Taylor, Graham J. Hutchings, Au/ZnO and Au/Fe_2O_3 catalysts for CO oxidation at ambient temperature: comments on the effect of synthesis conditions

on the preparation of high activity catalysts prepared by coprecipitation, *Topics in Catalysis*, 44, 123, 2007, https://doi.org/10.1007/s11244-007-0285-9.

116. Benjamin E. Solsona, Tomas Garcia, Christopher Jones, Stuart H. Taylor, Albert F. Carley, Graham J. Hutchings, Supported gold catalysts for the total oxidation of alkanes and carbon monoxide, *Applied Catalysis A: General*, 312, 67, 2006, https://doi.org/10.1016/j.apcata.2006.06.016.
117. Mikhail Khoudiakov, Mool C. Gupta, Sarojini Deevi, Au/Fe_2O_3 nanocatalysts for CO oxidation: A comparative study of deposition-precipitation and coprecipitation techniques, *Applied Catalysis A: General*, 291, 151, 2005, https://doi.org/10.1016/j.apcata.2005.01.042.
118. A. M. Visco, F. Neri, G. Neri, A. Donato, C. Milone, S. Galvagno, X-ray photoelectron spectroscopy of Au/Fe_2O_3 catalysts, *Physical Chemistry Chemical Physics*, 1, 2869, 1999, https://doi.org/10.1039/A900838A.
119. H. H. Kim, S. Tsubota, M. Date, A. Ogata, S. Futamura, Catalyst regeneration and activity enhancement of Au/TiO_2 by atmospheric pressure nonthermal plasma, *Applied Catalysis A General*, 329, 93, 2007, https://doi.org/10.1016/j.apcata.2007.06.029.
120. Wen-Cui Li, Massimiliano Comotti, Ferdi Schüth, Highly reproducible syntheses of active Au/TiO_2 catalysts for CO oxidation by deposition-precipitation or impregnation, *Journal of Catalysis*, 237, 190, 2006, https://doi.org/10.1016/j.jcat.2005.11.006.
121. Shi-Hua Wu, Xiu-Cheng Zheng, Shu-Rong Wang, Dong-Zhan Han, Wei-Ping Huang, Shou-Min Zhang, TiO_2 supported nano-Au catalysts prepared via solvated metal atom impregnation for low-temperature CO oxidation, *Catalysis Letters*, 96, 49, 2004, https://doi.org/10.1023/B:CATL.0000034279.03771.c2.
122. Baolin Zhu, Kairong Li, Yunfeng Feng, Shoumin Zhang, Shihua Wu, Weiping Huang, Synthesis and catalytic performance of gold-loaded TiO_2 nanofibers, *Catalysis Letters*, 118, 55, 2007, https://doi.org/10.1007/s10562-007-9139-0.
123. Baolin Zhu, Qi Guo, Xueliang Huang, Shurong Wang, Shoumin Zhang, Shihua Wu, Weiping Huan, Characterization and catalytic performance of TiO_2 nanotubes-supported gold and copper particles, *Journal of Molecular Catalysis A: Chemical*, 249, 211, 2006, https://doi.org/10.1016/j.molcata.2006.01.013.
124. Geoffrey C Bond, David T Thompson, Gold-catalysed oxidation of carbon monoxide, *Gold Bulletin*. 33, 41, 2000, DOI:10.1007/BF03216579.
125. C.K. Costello, J.H. Yang, H. Y. Law, Y. Wang, J.N. Lin, L.D. Marks, M.C. Kung, H.H. Kung, On the potential role of hydroxyl groups in CO oxidation over Au/Al2O3, *Applied Catalysis A: General*, 243(1), 15, 2003, 10.1016/S0926-860X(02)00533-1
126. Masakazu Date, Mitsutaka Okumura, Susumu Tsubota, Masatake Haruta, Vital role of moisture in the catalytic activity of supported gold nanoparticles, *Angewandte Chemie International Edition*, 116(16), 2181, 2004, https://doi.org/10.1002/ange.200453796
127. Ziyi Zhong, James Highfield, Ming Lin, Jaclyn Teo, Yi-fan Han, Insights into the oxidation and decomposition of CO on Au/alpha-Fe_2O_3 and on alpha-Fe_2O_3 by coupled TG-FTIR, *Langmuir*, 24(16), 8576, 2008, https://doi.org/10.1021/la800395k
128. Masatake Haruta, Catalysis - Gold rush, Nature, 437, 1098, 2005, https://doi.org/10.1038/4371098a
129. Mayfair C. Kung, Robert J. Davis, Harold H. Kung, Understanding Au-catalyzed low-temperature CO oxidation, *Journal Physical Chemistry C*, 111(32), 11767, 2007, https://doi.org/10.1021/jp072102i.
130. Andrea Vittadini, A. Selloni, François P. Rotzinger, M. Grätzel, Structure and energetics of water adsorbed at TiO2 anatase (101) and (001) surfaces, *Physical Review Letters*, 81, 2954, 1998, https://doi.org/10.1103/PhysRevLett.81.2954.
131. L. M. Liu, B. McAllister, H. Q. Ye, P. Hu, Identifying an O_2 supply pathway in CO oxidation on Au/TiO_2(110): A density functional theory study on the intrinsic role of water, *Journal of the American Chemical Society*, 128, 4017, 2006, https://doi.org/10.1021/JA056801P.
132. Michael A. Henderson, William S. Epling, Charles H. F. Peden, Craig L. Perkins, Insights into photoexcited electron scavenging processes on TiO_2 obtained from studies of the reaction of O_2 with OH Groups adsorbed at electronic defects on TiO_2(110), *Journal of Physical Chemistry B*, 107(2), 534, 2003, https://doi.org/10.1021/jp0262113.

133. Mathieu Delample, Nicolas Villandier, Jean-Paul Douliez, Séverine Camy, Jean-Stéphane Condoret, Yannick Pouilloux, Joël Barrault, François Jérôme, Glycerol as a cheap, safe and sustainable solvent for the catalytic and regioselective β,β-diarylation of acrylates over palladium nanoparticles, *Green Chemistry*, 12(5), 804, 2010, https://doi.org/10.1039/B925021B.
134. Mhamed Lemhadri, Henri Doucet, Maurice Santelli, Alkynylation of aryl bromides with propargylamines catalyzed by a palladium-tetraphosphine complex, *Synthesis* 8, 1359, 2005, DOI: 10.1055/s-2005-865284.
135. Vincenzo Calò, Angelo Nacci, Antonio Monopoli, Antonio Fornaro, Luigia Sabbatini, Nicola Cioffi, Nicoletta Ditaranto, Heck reaction catalyzed by nanosized palladium on chitosan in ionic liquids, *Organometallics*, 23(22), 5154, 2004, https://doi.org/10.1021/om049586e.
136. Arash Ghorbani-Choghamarani, Masoomeh Norouzi, Suzuki, Stille and Heck cross-coupling reactions catalyzed by Fe_3O_4@PTA-Pd as a recyclable and efficient nanocatalyst in green solvents, *New Journal of Chemistry*, 40(7), 6299, 2016, https://doi.org/10.1039/C6NJ00088F.
137. Antonio Monopoli, Vincenzo Calo, Francesco Ciminale, Pietro Cotugno, Carlo Angelici, Nicola Cioffi, Angelo Nacci, Glucose as a clean and renewable reductant in the Pd- nanoparticle-catalyzed reductive homocoupling of bromo- and chloroarenes in water, *The Journal of Organic Chemistry*, 75(11), 3908, https://doi.org/10.1021/jo1005729
138. Mehdi Khoobi, Leila Ma'mani, Faezeh Rezazadeh, Zeinab Zareie, Alireza Foroumadi, Ali Ramazani, Abbas Shafiee, One-pot synthesis of 4*H*-benzo*b*]pyrans and dihydropyrano*c*]chromenes using inorganic-organic hybrid magnetic nanocatalyst in water, *Journal of Molecular Catalysis A*, 359, 74, 2012, https://doi.org/10.1016/j.molcata.2012.03.023
139. Shahnaz Rostamizadeh, Mohammad Azad, Nasrin Shadjou, and Mohammad Hasanzadeh, (α-Fe_2O_3)-MCM-41-SO_3H as a novel magnetic nanocatalyst for the synthesis of N-aryl-2-amino-1, 6-naphthyridine derivatives, *Catalysis Communications*, 25, 83, 2012, https://doi.org/10.1016/j.catcom.2012.04.013
140. Bahador Karami, Khalil Eskandari, Abdolmohammad Ghasemi, Facile and rapid synthesis of some novel polysubstituted imidazoles by employing magnetic Fe_3O_4 nanoparticles as a high efficient catalyst, *Turkish Journal of Chemistry*, 36, 601, 2012, https://doi.org/10.3906/kim-1112–49

2 Nanocatalysts in Oxidation and Sulfoxidation Reactions

Hanna S. Abbo, Tariq Shah, Divine M. Yufanyi, Nader G. Khaligh and Salam J.J. Titinchi

CONTENTS

2.1 Sulfur Redox Reactions 25
2.1.1 Sulfur Electrochemistry in Li–S Batteries 26
2.1.2 Liquid Sulfur-Redox Reaction 26
2.2 Molybdenum Catalyst Used for Sulfoxidation Reactions in Aqueous Medium 27
2.2.1 Characterization of Molybdenum Catalyst 28
2.3 TiO_2 Catalyst for Enantioselective Sulfoxidation 30
2.4 Selective Oxidation of Alkylsulfides 31
2.5 Applications of Nanocatalyst 34
2.5.1 Use in Environment and Especially in Wastewater Treatment 36
2.5.2 Applications of Nanocatalyst in Mechanical Industries 36
2.5.3 Applications of Nanocatalyst in Drugs Delivery 36
2.5.4 Energy Harvesting Applications 37
2.5.5 Materials Engineering Applications 37
2.6 Nickel Nanocatalyst in the Enhancement of Hydrogen Oxidation Reactions 38
2.6.1 Ni@h-BN Nanocatalyst Characterization 38
2.7 Organic and Inorganic Hybrid Nanocatalyst for Oxidation Reaction 40
2.7.1 Characterization of Ag-HAp Catalyst 41
2.8 Metal Organic Frameworks as Oxidation Catalyst 42
2.9 Metals as Heterogeneous Catalysts for Different Oxidation Reactions 44
2.9.1 The Conversion of Glucose to Gluconic Acid 44
2.9.2 The Oxidative Conversion of Hydrogen to Hydrogen Peroxide 45
2.9.3 The Oxidation of Alcohols 45
References 46

2.1 SULFUR REDOX REACTIONS

The electrochemistry of sulfur in many natural electrolytes is a regular heterogeneous interaction, and this unequivocally is a function of the physicochemical properties of the heterogeneous interfaces on which these reactions occur [1]. It is thus imperative understand these interfacial reactions of sulfur redox responses in a functioning cell. Indeed, oxygen and sulfur are in Group VI of the Periodic Table. Accordingly, they share some similar properties – for example, multi-electron transfers and, furthermore, multiphase advances as far as their interfacial redox practices. The electrochemistry of oxygen, which includes the reduction of oxygen as well as the development of oxygen, has been unequivocally examined as test responses in contemporary material science to assess the execution of nanostructure electrocatalysts [2].

DOI: 10.1201/9781003218708-3

2.1.1 Sulfur Electrochemistry in Li–S Batteries

If the ideas of energy science for the electrochemistry of oxygen are possibly inserted into the electrochemistry of sulfur, numerous innovative techniques may be suggested to diminish the over-potential and upgrade of the response energy in a functioning Li–S battery. Knowledge of this can help manage the cost of production for the judicious plan and upgrade of terminal/electrolyte interfaces in addition to advancing the commonsense uses of Li–S batteries. The update and advancement of the electrical and electronics industry is motivated by the expanding public necessity, engaging for energy storage systems of high capacity that is a crucial segment of electronic items and electric vehicles [3]. Batteries composed of lithium and sulfur are considered the leading next-generation innovations owing to the very high theoretical energy thickness of 2600 Wh kg^{-1} [4]. Unfortunately, the pragmatic use of the Li–S battery is thwarted by a series of challenges: (a) the tremendous fluctuation in volume of the sulfur cathode during lithiation/delithiation prompts the pulverization and cracking of electrodes; (b) sulfur insulation and its released items (Li_2S_2/Li_2S) prompt a high redox over potential and slow reaction energy; and (c) lithium polysulfides (LPSs) solvent intermediates in fluid electrolytes, disintegrate, diffuse, and decompose in electrolytes as well as at interfaces, prompting loss of dynamic materials and the destabilization of the interface. The obstacles mentioned above meet up to deliver present-day Li–S batteries having low Coulombic proficiency, lacking sulfur use, helpless cycling steadiness, and extreme electrode corrosion [3, 4]. As a rule, an ordinary Li–S battery comprises a sulfur cathode and a lithium metal anode, as well as a reasonable electrolyte in either the fluid or solid state [5, 6]. Redox electrochemical reactions of sulfur in aprotic fluid electrolytes (or polymeric gel electrolytes comprising a negligible part of fluid fraction of liquid solvents) incorporate confounded multiphase development, multistep charge-move/non transfer techniques.

2.1.2 Liquid Sulfur-Redox Reaction

Attributable to the extraordinary solubility of high-demand LPSs in fluid electrolytes, LPSs are readily broken down in natural electrolyte as soon as they are surrounded through preliminary sulfur electroreduction or Li_2S electro-oxidation. In this way, the mode of electrical contact changes with change in the conductive substrate from solid to solid and solid-fluid contact [7]. Traditional assessment considered that fluid LPSs would be better for electrical contact than the solid sulfur/Li_2S, since it (LPSs) can diffuse to empty conductive surfaces while solids cannot. Conductive species with a moderately low-surface-area have been genuinely utilized in earlier studies [8, 9]. The absolute surface area of conduction has never been a hurdle, but the issue derives from the steady contact loss among carbon and LPSs. Consequently, adsorption of a functioning LPS particle on the surface of the cathode can lead to the electron being transferred through the solid/fluid interface with electrochemical reactions possibly occurring. In such manner, a routine non-polar conductive surface, such as a carbon surface, is negative as far as LPS adsorption is concerned [10, 11]. The interface having a high attraction for dissolvable LPSs is basic for the immobilization of LPSs on the surface as well as the ensuing transfer of charge across the interfaces. Several research endeavors on the properties of solid/liquid interfaces through alteration of the strong surface using appropriate LPS attraction – for example, doping of a carbon matrix with heteroatoms and incorporation of inorganic nanomaterials into conductive frameworks – are known [12]. The binding energies of LPSs on host materials are not only improved by addition of dopants or decorators but, in addition they conceivably offer active sites for electrocatalytic change of soluble LPSs (Scheme 2.1a) [13, 14].

Notably, beginning active species and reduced product are profoundly solvable in the course of liquid–liquid transformation (higher-order LPSs to lower order LPSs during release and lower-order LPSs to higher-order LPSs conversely) in a functioning cell. The force that binds soluble

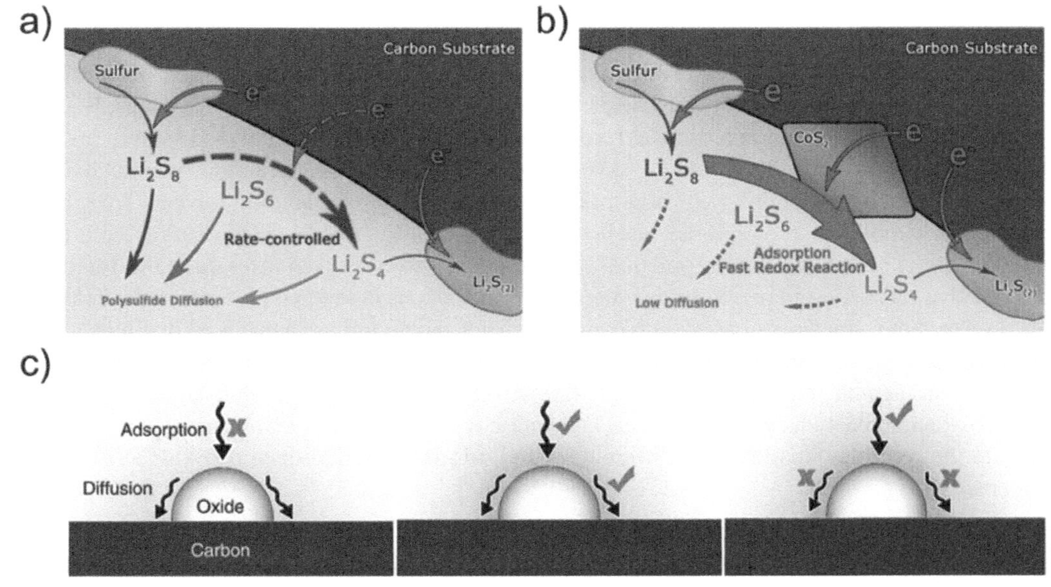

SCHEME 2.1 Schematic representation of the transformation of sulfur species on (a) nonpolar conductive carbon substrates and (b) polar CoS_2-designed conductive carbon substrates (Source Yuan 2013) [13]. (c) Scheme of the LPSs adsorption, dissemination on the outside of different nonconductive substrates (Source Yuan 2013 and 2019) [14].

LPSs to substrates, substrates/interfaces, ought to be moderate to fulfill both the adsorption of reactant and the desorption of product. Particularly for materials for electrical insulation, extra surface of LPS species on solid substrates is especially significant for their electrochemical transformation due to the transfer of electron on the surface (Scheme 1a) [14]. Appropriately, the harmony between adsorption on a surface and desorption of LPSs at receptive interfaces ought to be seriously considered. As a general rule, a solid binding strength is occurring because of significant transfer of charge among LPSs and the receptive interface. Such a transfer of charge initiates possibly LPS oxidation into sulfates/polythionates on substrates that are oxidizing, for example, metal oxides of high-valency [15] or LPS reduction into sulfides on materials that are reducing such as metals and metal compounds of low-valency [16]. Under these conditions, therefore, LPS deterioration totally changes the composition. Furthermore, the properties of the interface, as well as the binding energy, ought to be reexamined at the renovated surfaces/interfaces [17].

2.2 MOLYBDENUM CATALYST USED FOR SULFOXIDATION REACTIONS IN AQUEOUS MEDIUM

Sulfoxidation is perhaps the most vital reaction in organic synthetic reactions. Numerous sulfoxides as well as sulfones are utilized as adaptable intermediates for different natural conversions [18]. Furthermore, these compounds are confirmed as active drug elements for a few remedial medication molecules [19]. The sulfoxidation of natural sulfides is a straightforward technique specifically leading to sulfoxides or sulfones. A novel solvent that has certainly gained incredible consideration in the domain of "green" or practical chemistry, is water [20]. Several reports of self-catalyzed [21] in addition to sulfoxidation reactions that are catalyzed by metal in aqueous medium, are known. The drawbacks of these techniques are possibly high temperatures or potentially long reaction time, poor ability to recycle the catalyst, controlled amount of substrates, poorly oxidizable functional groups, essential promoters and experimental conditions. This requires the improvement of sulfoxidation conditions in aqueous medium.

Molybdenum compounds are prevalent in the domains of natural, inorganic, and organic chemistry [22]. A few natural reactions have used compounds of oxomolybdenum as oxidants. Recently, research focus has centered on responses resulting from oxidation that is catalyzed by systems based on molybdenum. In particular, MoO_2Cl_2 was synthesized and employed as a mild and effective catalyst in sulfoxidation reactions with high practical compatibilities in natural solvents [23]. Its use in the specific oxidization of sulfides has been accounted for by several prominent research teams. With the aim of conveying further practical "green" conventions for responses due to sulfoxidation, the objective was to design reactions that are applied in aqueous medium over other natural solvents. Since surfactant frameworks based on metals have been shown to have beneficial qualities to solubilize reactants in aqueous medium and simultaneously catalyze the reactions, in this current investigation it was demonstrated that a surfactant framework based on molybdenum can act as a catalyst for sulfoxidation reactions in water. Its reactant action was examined as far as the selectivity and the yield for the controlled oxidation of different natural sulfides with hydrogen peroxide as oxidant, is concerned.

2.2.1 Characterization of Molybdenum Catalyst

The molybdenum complex $(C_{19}H_{42}N)_2[MoO(O_2)_2(C_2O_4)]{\cdot}H_2O$, with cetyl-trimethyl-ammonium cation (CTA) as surfactant and a fundamental piece of the catalyst was prepared to examine sulfoxidation reactions in humid conditions. The complex was characterized by IR, NMR, ESI-MS and XRD methods. Characteristic sharp and solid peaks for (Mo=O) and (O–O) vibrational frequencies were found at 948 and 859 cm^{-1} in the IR range, respectively, while those for $(Mo–(O_2))$ moieties occurred at 648 and 583 cm^{-1} (Figure 2.1).

These frequencies are within acceptable range of comparative bonds [24]. The single crystal X-ray diffraction method was used to unambiguously confirm the structure of the complex. Slow evaporation of a solution of the complex in chloroform–ethanol (1: 1) mixture afforded single crystals appropriate for XRD examination (Figure 2.2a). The complex **1** crystalized with one molecule of water of crystallization and the asymmetric unit comprises a dianionic molybdenum segment, two CTA cations and a molecule of water (Figure 2.2b).

The oxo ligand in the coordination sphere occupies the axial positions alongside an oxygen molecule of the oxalate ligand, while four oxygen atoms of the peroxo groups and the other oxygen

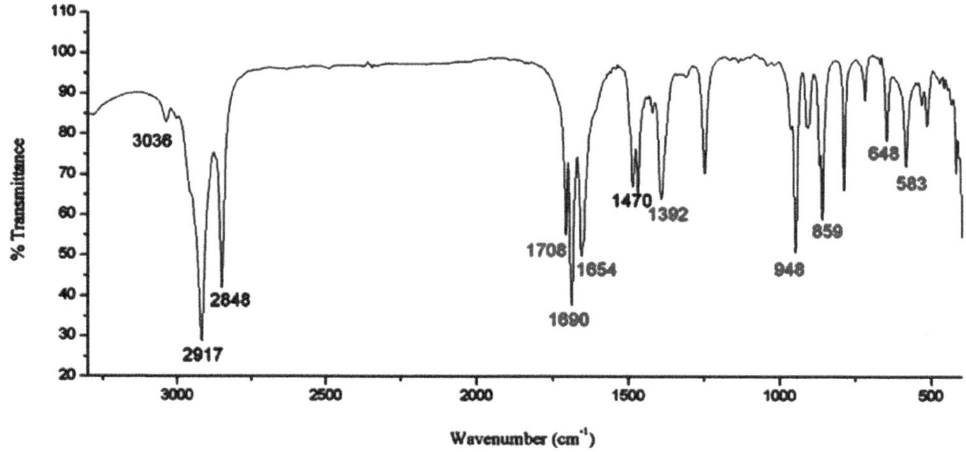

FIGURE 2.1 IR spectra of the complex (Source Holder 2013) [22].

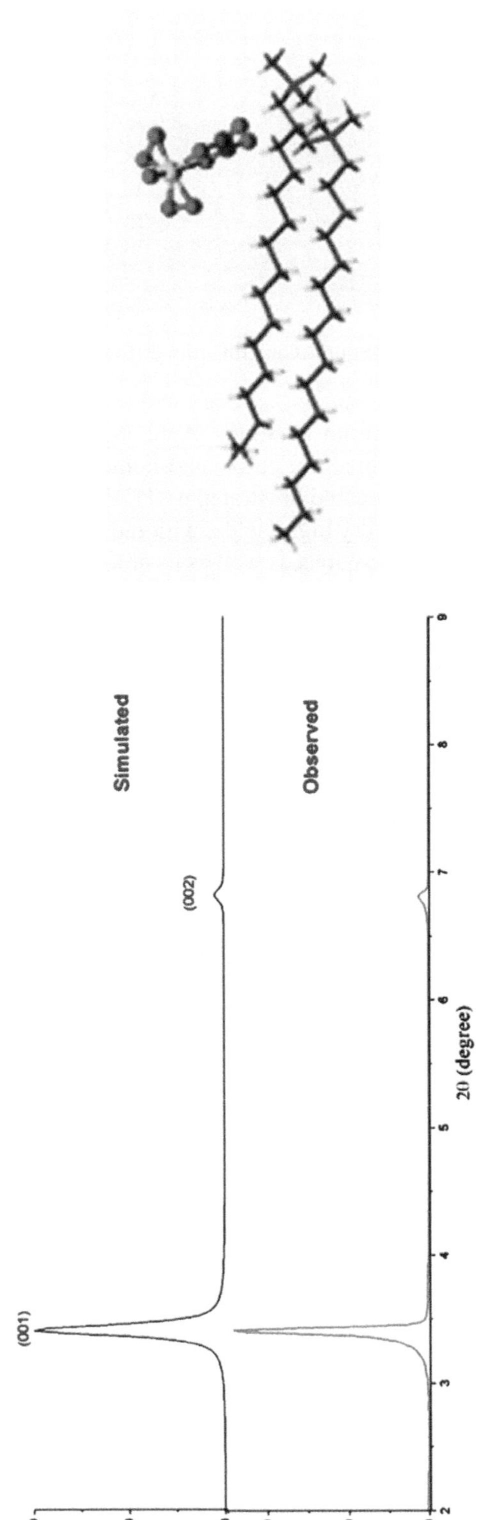

FIGURE 2.2 (a) Powder XRD pattern of the complex and (b) Crystal structure of the complex (Source Chakravarthy 2014) [25].

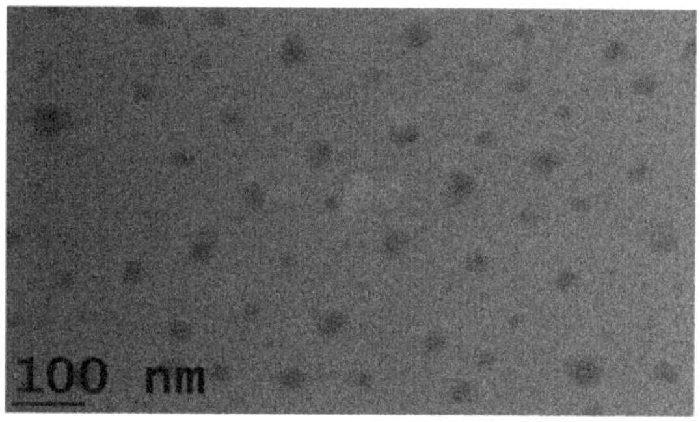

FIGURE 2.3 TEM image of the complex (Source Chakravarthy 2014) [25].

of the oxalate ligand occupy the equatorial positions. Agreement between the observed and the simulated powder XRD patterns confirmed the purity and homogeneity of the complex. The size and shapes of aggregates were determined and illustrated by TEM (Figure 2.3) [25]. Both sulfoxides and sulfones were obtained in great to very high yields, with the catalyst indicating high functional group similarity. The ease of use of this catalyst as well as its ability to be recycled in liquid medium indicates a more "green" and eco-friendly procedure toward sulfoxidations.

2.3 TIO$_2$ CATALYST FOR ENANTIOSELECTIVE SULFOXIDATION

Although different homogeneous enantioselective catalysts for sulfide oxidation have been successful, heterogeneous processes have been less explored [26–28]. The assembled titanium compound shown in Figure 2.4 has been effectively applied as catalyst for the carbonyl-ene reaction (Table 2.1). Observation of the light-yellow color in the fluid period of the reaction blend, under some trial conditions, proposed conceivable filtering of the heterogeneous active species into the arrangement stage during catalysis shown in Figure 2.4. The development of genuine heterogeneous titanium catalysts, which are self-supported, applied in the enantioselective oxidation of aryl and alkyl sulfides is shown here. The asymmetric sulfoxidation catalysts that were prepared were very stable, and in addition it has the possibility of reuse multiple times without loss of enantioselectivity and it is affordable compared to sulfoxides which are optically active up to >99.9 percent. SEM micrographs indicated that the solid particles were micrometer-size range (Figure 2.4), and were non-crystalline as shown by powder X-ray diffraction patterns.

Heterogeneous catalysts (2a–c) were evaluated for enantioselective sulfide oxidation sub-1a (Table 2.2), with CCl_4 selected as solvent after an initial assessment of the solubility. This room temperature reaction utilized 5 mol% of catalysts 2a–c. Oxidant effects on the reactivity and enantioselectivity of the catalyst was examined using thioanisole as substrate. Table 2.2 shows that the arrangement of decane using tert-butyl hydroperoxide (TBHP, 5–6 m in decane) was better than its liquid arrangement (70% arrangement in water) as far as enantioselectivity is concerned. In any case, the oxidation of sub-1a using a 30 percent aqueous solution of hydrogen peroxide (H_2O_2) as oxidant in the presence of 2a led to a severe reduction in the enantioselectivity of the reaction to 31.1 percent, despite the fact that the response could continue easily with complete change of sulfide (sub-1a). Partial catalyst deterioration in the presence of a huge amount of water in the reaction system, is most likely responsible for this situation. Making use of cumene hydroperoxide (CMHP, 80% in cumene) which is less sensitive and less receptive as oxidant rather than TBHP, led to the accomplishment of a very high enantioselectivity (99.2% ee) alongside a yield of 38 percent.

Especially in asymmetric sulfonation catalysis, the self-supporting heterogeneous TiO_2 probably can be reused for more than a month without important loss in activity and resistance to selectivity loss (>99.9°) as a result of its stability and ease to recycle. An enantioselective reaction with TiO_2

FIGURE 2.4 (a) SEM micrograph of self-supported TiO_2 catalyst, (b) assembled catalysts carbonyl-ene reactions, (c) self-supported TiO_2 catalyst [29].

catalyst is a good example of heterogeneous catalysis. The properties of self-supported catalysts – for example easy manufacturing, strong chiral chemically active sites in the solids, just as ease of recovery and straightforward reusing – are especially significant in emergent strategies for the combination of active compounds in the development of synthetic methods. The approach described here points out a feasible novel route for the development of chiral catalysts for asymmetric synthesis. Current effort includes research on various substrates, especially those with pharmaceutical importance, as well as further optimization of oxidants and reaction media [29].

2.4 SELECTIVE OXIDATION OF ALKYLSULFIDES

This study focused on the sustainability of the sulfoxidation of ethyl chloride ethyl sulfide (CEES) over strong solid catalysts. A few strong solid catalysts with various proportions of Brønsted to Lewis

TABLE 2.1
Enantioselective Catalysis of Assembled Catalysts Carbonyl-ene Reactions[a]

Entry	Catalyst	Solvent	4 Å MS[b]	Stirring	Temp [°C]	Time [h]	Yield [%][c]	ee [%][d]
1	**1a**	Toluene	-	+	RT	48	91	94
2	**1a**	Toluene	-	+	0	120	85	95
3	**1a**	Toluene	+	+	0	120	96	95
4	**1a**	Toluene	+	+	RT	120	93	92
5	**1a**	Free	-	+	0	120	65	92
6	**1a**	Toluene	-	-	0	120	75	94
7	**1b**	Toluene	-	+	RT	48	32	9
8	**1b**	Toluene	-	+	0	120	9	24
9	**1c**	Toluene	-	+	RT	30	99	96
10	**1c**	Toluene	-	+	0	96	95	91
11	**1c**	Toluene	+	+	RT	96	85	92
12	**1c**	Toluene	+	+	0	96	90	95
13	**1c**	Free	-	+	0	96	90	92
14	**1c**	Diethyl ether	-	+	RT	120	99	95
15	**1d**	Toluene	-	+	RT	96	99	95
16	**1d**	Toluene	-	+	0	96	99	98
17	**1d**	Toluene	-	+	-10	96	99	98
18	**1d**	Diethyl ether	-	+	RT	96	99	97

[a] The reaction was performed by using 3.0m of 6 with the ratio of 6/7/5=1:2:0.01 on a 1.25 mmol scale. [b] 15 mg of 4 MS (dried in vacuo atx400 8C for 8 h) was added. [c] Isolated yield. [d] The *ee* was determined by HPLC.

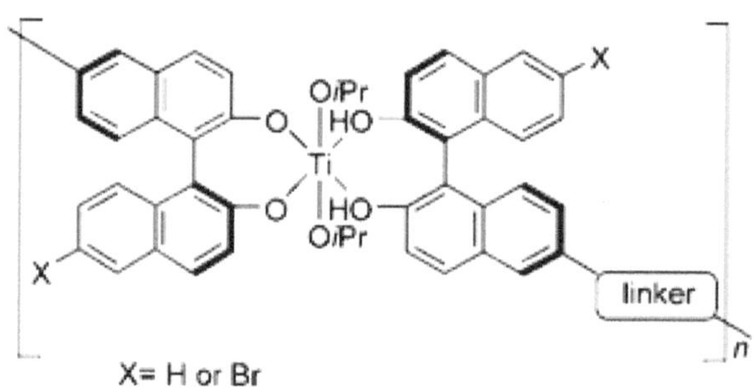

Structure of Catalysts **1a** - **1d**

TABLE 2.2
Enantioselective Oxidation of Sulfides Catalyzed by Self-Supported Catalysts

Entry	Catalysts	Substrates	oxidant	Yield [%]	ee [%]
1	**2a**	Sub-1a	TBHP	68.8	74.5
	2a	Sub-1a	TBHP	32.9	97.5
3	**2a**	Sub-1a	H_2O_2	32.8	31.1
4	**2a**	Sub-1a	CMHP	38.6	99.2
5	**2a**	Sub-1a	CMHP	35.7	87.7
6	**2b**	Sub-1a	CMHP	37.8	99.5
7	**2c**	Sub-1a	CMHP	37.1	98.7
8	**2a**	Sub-1b	CMHP	36.6	<99.9

TABLE 2.2 (Continued)
Enantioselective Oxidation of Sulfides Catalyzed by Self-Supported Catalysts

Entry	Catalysts	Substrates	oxidant	Yield [%]	ee [%]
9	**2b**	Sub-1b	CMHP	41.5	99.8
10	**2c**	Sub-1b	CMHP	41.5	99.1
11	**2a**	Sub-1c	CMHP	30.7	<99.9
12	**2b**	Sub-1c	CMHP	30.1	99.8
13	**2c**	Sub-1c	CMHP	31.0	<99.9
14	**2a**	Sub-1d	CMHP	41.0	98.6
15	**2b**	Sub-1d	CMHP	38.5	96.4
16	**2c**	Sub-1d	CMHP	44.9	97.0
17	**2a**	Sub-1e	CMHP	32.9	<99.9
18	**2b**	Sub-1e	CMHP	36.5	<99.9
19	**2c**	Sub-1e	CMHP	34.7	<99.9
20	**2b**	Sub-1f	CMHP	20.5	89.1
21	**2a**	Sub-1g	CMHP	36.3	75.5
22	**2a**	Sub-1a	CMHP	45.7	99.9

Substrates:
a: R=H, R'=Me;
b: R=4-Me, R=Me;
c: R=4-Br, R'=Me;
d: R=4-F, R=Me
e: R=3-Br, R'=Me;
f: R4-NO_2, R'=Me;
g: R-H, R'=et

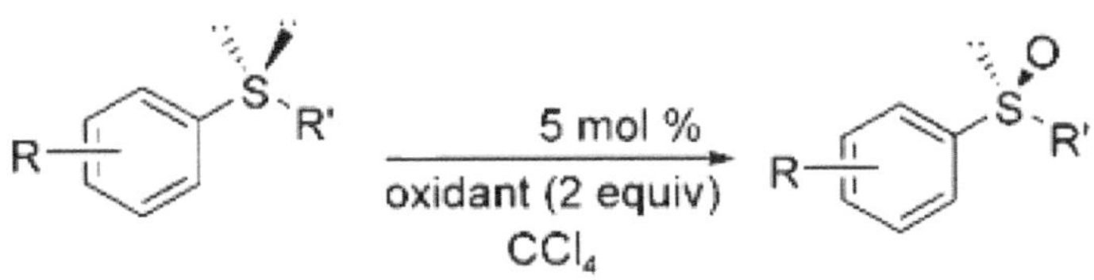

Structure of the catalyst form **2a** - **2c** and substrates.

TABLE 2.3
Solid Acid Catalyst Characterization

Materials	S_{BET} (m^2g^{-1})	Average Pore Width (Å)	Pore Volume Sites ($mmolg^{-1}$)	Total Acid Sites ($mmol\ g^{-1}$)	Brønsted/ Lewis Ratio	Brønsted Site ($mmol\ g^{-1}$)
Zr-P	347	59	0.51	1.53	0.37	0.41
ϒ-Al_2O_3 (nano-powder)	25	131	0.09	0.22	0.0	0.0
Activated Al_2O_3	136	75	0.25	0.59	∞	0.0
Amberlyst--15	42	305	0.32	2.35 (4.7°)	--	2.35

solid sites were picked to analyze the effectiveness of the catalyst as well as the impact of Brønsted or Lewis acids sites on the rates of the CEES sulfoxidation reaction. The characteristic physico-chemical properties of the strong solid catalyst are shown in Table 2.3. With regard to reactant implementation tests, the same equivalent amount of acid sites 5 mM for Amberlyst-15 was chosen. The precursor solution of tribromide/nitrate received a similar measure of strong acid catalyst in each trial, in order to contrast catalytic effectiveness. Figure 2.5 illustrates the rates of reaction of CEES oxidation utilizing the tribromide/nitrate framework with a different solid acid catalyst. The general sulfoxidation rate coefficients over the distinctive solid catalyst were determined utilizing

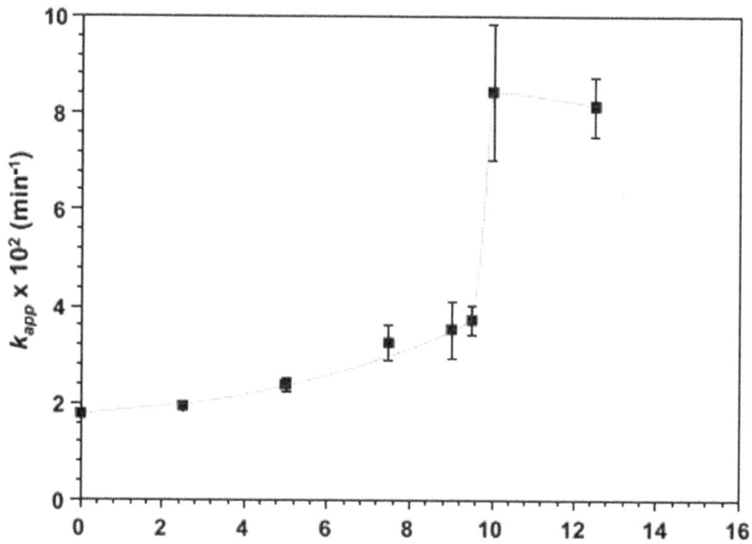

FIGURE 2.5 Effect of Brønsted acid concentration (Source Le 2019) [31].

first order kinetics. Amberlyst-15, comprising Brønsted acid sites, enhances the activity of the catalyst while the presence of Lewis acid sites in the structure of other solid catalysts, affected oxidation adversely. The γ-Al_2O_3 nanopowder with a reduced number of Lewis acid sites demonstrated an activity slightly higher than that of activated Al_2O_3 or Zr-P. Remarkably, Zr-P reactivity is very low, in spite of the fact that it possesses Brønsted acid sites. However, sulfoxidation catalytic activity were not improved in Lewis acid sites, due mainly to anions adsorbed onto the active sites and their transformation with little or no action, such as was previously reported for the halide anions adsorbed on aluminum oxides Lewis acid sites.

The number of acid sites (2.35 mmol g^{-1} for Amberlyst-15 tried here) reaction time of sulfoxidation for the reaction framework with the solid acid catalyst was assessed and the rate coefficients plotted as a function of acid sites is presented in Figure 2.5.

The reaction speed of sulfoxidations, related by different conditions of the reaction, showed that fast sulfoxidation can possibly be enhanced and it could be found in the addition of acid. The conclusion clarifies well that examination of other parameters is fundamental for a productive plan of another system for the catalytic reaction. To expand the application range to heterogeneous catalysis, the researcher achieved a novel solid active catalyst for the quick sulfoxidation reaction arrangement. Part of the acid sites was set up by the application of different solid active catalysts having various proportions of Brønsted to Lewis acid sites during the sulfoxidation in the arrangement of tribromide and nitrate (Figure 2.6). Brønsted acid (Amberlyst-15) displayed enhanced reaction performance compared to other solid catalysts, while Lewis acid or mixed acid (Zr-P) catalysts demonstrated an antagonistic impact on the time of the sulfoxidation reaction, primarily because of tribromide anion that is adsorbed [30, 31].

2.5 APPLICATIONS OF NANOCATALYST

Due to the unique properties of nanocatalysts, it can be used in variety of applications. Some of the vital applications are given below.

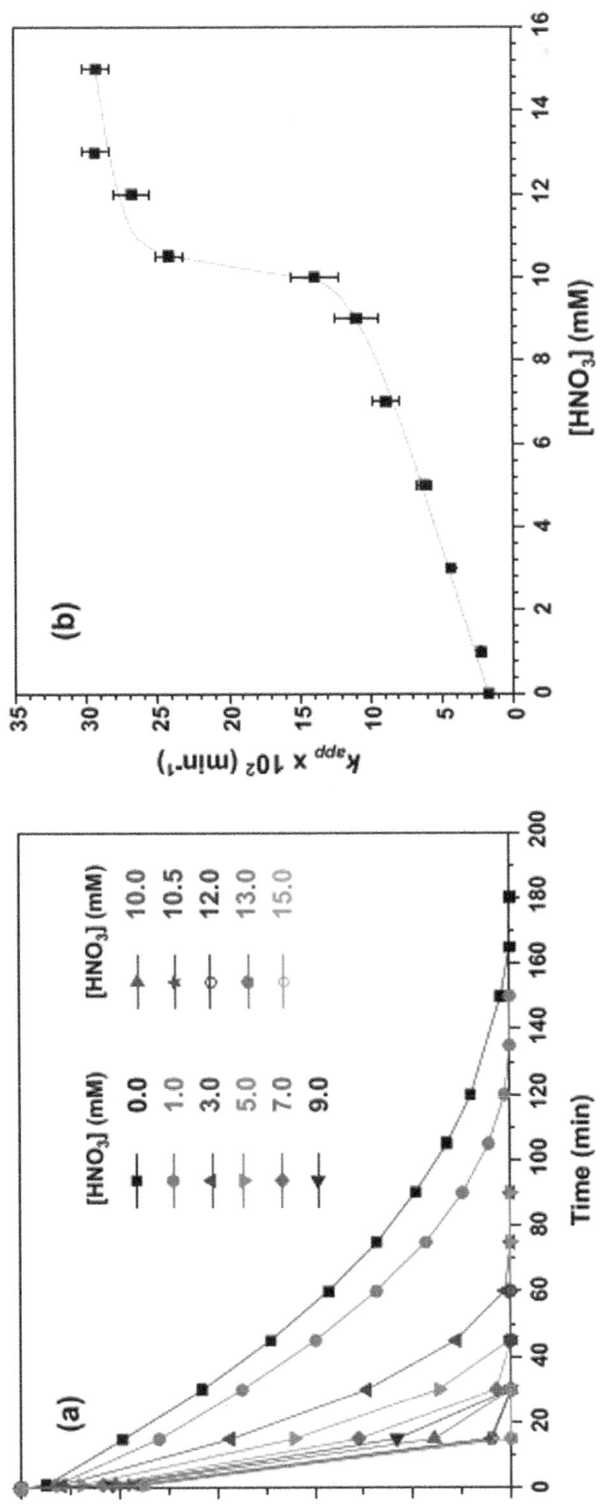

FIGURE 2.6 Acid concentration effect on oxidation of CEES (Source Le 2019) [31].

2.5.1 Use in Environment and Especially in Wastewater Treatment

The large surface area of the designed NPs in household and mechanical applications prompts the arrival of these materials into the environment. An evaluation of the environmental hazards of these NPs requires us to understand the usefulness, reactivity, eco-harmfulness and how persistent they are. Design and applications in different materials can increase the amount of these NPs in ground water and soil, which presents the most critical exposure avenue for the evaluation of environmental hazards. The large surface-to-mass proportion explains why regular NPs take responsibility for an important part in the solid/water pollutants that can be absorbed on the surfaces of NPs, co-encouraged during the arrangement of normal NPs that had pollutants adsorbed on their surface. Connection of impurities with NPs is a function of the qualities of the NPs, such as dimensions, synthesis, morphology, sponginess, collection/disaggregation and total construction. Luminophores are unpredictable in the environment and are shielded from ecological oxygen when doped inside the silica structure. The vast majority of natural uses of nanotechnology are divided into three classes:

(1) Environmental green chemistry and the prevention of pollution;
(2) Materials remediation due to soiling with unsafe constituents, and
(3) Sensors for ecological phases.

The removal of heavy metals, such as Hg, Pb, Th, Cd and arsenic, from normal water has attracted extensive consideration on account of their antagonistic consequences for ecological and human health. Super-paramagnetic iron oxide NPs are a powerful material that sorbs these poisonous delicate metals. NP-induced photodegradation is a likewise exceptionally normal practice, and numerous nanomaterials have found application due to this. Rogozea et al. [32] utilized NiO/ZnO NPs altered silica pair in the design due to photodegradation. The high surface area of NPs because of small size (< 10 nm), worked with the productive photo catalytic degradation reaction. A similar research team has described the combination of various NPs and revealed their optical, and fluorescence applications in degradation processes [33–35].

2.5.2 Applications of Nanocatalyst in Mechanical Industries

The excellent mechanical properties of NPs revealed through their young modulus, strain and stress properties, have enabled these to find many applications in mechanical industries, predominantly in coating , oils, and applications in cement. In addition, this property may be very important to achieve more grounded nano-devices for diverse applications. Control of tribological properties at the nanoscale level can be achieved by insertion of NPs in the metal and polymer network to increase their mechanical strength. The rolling mode of NPs in the contact region that is greased could give extremely low contact and wear. These NPs afford improved properties of sliding and delamination, which could likewise impact in low contact and wear, and consequently increase oil impact [36]. Coating can prompt different precisely solid attributes, as it increases toughness and wear hardness. Al, TiO_2 as well as carbon-based NPs have effectively been proven to have beneficial mechanical properties in coatings [37].

2.5.3 Applications of Nanocatalyst in Drugs Delivery

Over recent years these nanocatalysts have been of significant importance in medical applications as biodegradable NPs in drug delivery devices [38]. Research in drug delivery devices has witnessed the use of diverse polymers, since they can successfully convey the medications to the targeted site of therapeutic activity, while limiting the side effects. Pharmacologically active medications can be released in a controlled manner to the exact active site to attain the therapeutically optimum degree.

Utilization of liposomes as a potential drug transporter rather than traditional methods is due to their exceptional potential benefits, which include the ability to protect drugs from degradation, focus on the active site and decrease in the lethal feature and opposite secondary effects. In any case, constructive work on drugs coated with liposomes has been limited due to inborn problems of medical origin, such as squat epitome productivity, quick water spillage in the item of blood segments and exceptionally poor storage. Polymeric NPs present some basic benefits compared to these materials such as liposomes. For instance, NPs help to expand the ratability of medications or issues and has advantageous measured medication discharge properties. The majority of metallic as well as semiconductor NPs have a massive probability for malignant growth determination and treatment by virtue of their surface plasmon reverberation (SPR) upgraded light dispersing and retention. Au NPs productively convert the solid retained light into limited warmth and this can be used for the specific laser photograph warm treatment of malignant growth [39]. Ag NPs are progressively being utilized in wound dressings, catheters and different items because of their antimicrobial action. Antimicrobial properties are incredibly essential in material, medication, water sterilization and food packaging. Thus, the antimicrobial qualities of inorganic NPs add more importance to this significant perspective, when contrasted with natural mixtures, which are generally poisonous to the organic frameworks. Functionalization of these NPs with different groups enables them to specifically survive attack by microbial species. NPs such as TiO_2, ZnO, $BiVO_4$ as well as those based on Cu and Ni, have found such applications due to their appropriate antibacterial efficacies [40, 41].

2.5.4 Energy Harvesting Applications

Current research has cautioned us on the constraints and shortage of petroleum derivatives in the near future because of their non-renewable nature. In this way, researchers are shifting exploration methodologies toward the production of sustainable energy sources from well-accessible sources at reduced cost. NPs have been found to present the greatest possibility for this application given their large surface area, optical conductivity and reactant nature. NPs are widely used in photocatalytic applications in particular, to produce energy from photoelectrochemical (PEC) and electrochemical water splitting [42]. Adjacent to the splitting of water, electrochemical CO_2 decreases to fill antecedents, and solar cells and piezoelectric generators likewise present advanced alternatives to produce energy [43]. NPs are additionally employed in energy stockpiling applications to save the energy into various structures at the nano level. Nano-generators are made that can change the mechanical energy into power by utilizing piezoelectricity, which is an original way to deal with energy creation [44].

2.5.5 Materials Engineering Applications

Crystalline materials in the nanoscale give exceptionally exciting substances to material science. The assembly of NPs results in physicochemical properties that initiate outstanding mechanical, optical, electrical, and imaging properties that are incredibly searched for in specific applications in the clinical, commercial, and biological areas [45]. The focus of NPs is on the design and preparation of organic molecules in the size range < 100 nm, with interesting and novel practical properties. Documentation of the expected advantages of nanotechnology has been achieved by numerous producers at different levels; bulk-production of several attractive items are being mass-delivered such as microelectronics, aviation and also drug industries. Comparatively, health wellness items form the biggest class of nanotechnology customer items, trailed by the electronic and PC class just as home and nursery classification. Future transformations in numerous enterprises such as food packaging and pressing are seen to be based on nanotechnology. The domains of bio-photonics and material science have recently seen a surge in interest due to resonant energy transfer (RET) framework which includes naturally colored particles and noble metal NPs [46]. It is gradually becoming increasingly normal to have NPs in economically accessible items. Noble metals such as Au and Ag display several colors in the visible region on plasmon resonance because of aggregate motions of the electrons at the surface of NPs. Factors such as the size, shape, the inter-particle distance, and the dielectric property of the surrounding environment of NPs determines the resonance frequency of the solid. Metal NPs with plasmon-absorbance properties have been exploited for wide applications and uses, including substance sensors and biosensors [47, 48].

2.6 NICKEL NANOCATALYST IN THE ENHANCEMENT OF HYDROGEN OXIDATION REACTIONS

Currently, there is much interest in primary electrolytes for example alkaline polymer electrolytic fuels cells (APEFc) synthesized on alkaline polymer. These fuel cells, (APEFc), in comparison to proton exchange membrane fuels cells (PEMFc), present more advantages [49–54]. Nevertheless, the kinetic process of hydrogen oxidation reaction (HOR) takes place faster in acid electrolyte in comparison to APEFc which is slow in alkaline media. For instance, the best catalyst for HOR is the Pt nanocatalyst, while the HOR magnitude in alkaline media is two orders compared to acid media [55, 56]. These two challenges require urgent solutions in APEFc technology. Solutions to these challenges include discovery of an appropriate pathway for the kinetics of HOR in alkaline media as well as replacement of the platinum group metals with high performance non-metal nanocatalyst [57–60]. CrO-decorated Ni nanoparticles for the retention of Ni metallic activity as well as for the suppression of oxidation have been reported [61]. Results obtained indicate that there exists a correlation between the hydrogen binding energy on the surface of catalyst HOR activity. The synthesis of a Ni nanocatalyst on nitrogen that is doped with carbon nanotubes, which is highly performant, has been reported [62]. Modulation of the d-orbital of Ni atoms results in a decreased adsorption on the surface of the nanocatalyst. Also, a reduction of hydrogen binding energy in alkaline media as a result of high resistance to surface oxidation, and the HOR, was observed [63, 64]. Nano-reactors such as 2D materials and metal surfaces, for example graphene and hexagonal boron nitride, have recently been selected. Adsorption on the surfaces of metals and 2d layers can be intercalated by small molecules such as CO, H_2 and O_2. The strength of adsorption of molecules on the surface of metals reduces. Coverage of the metal surfaces enhances the metal catalyzed reaction besides modulation of the surface activity of the 2D layer. Adsorption is reduced by the 2D over layer and the possible modulation of Ni surfaces by h-BN core shells through the interaction of the small molecules H, O, OH, to cover Ni nanoparticles [65–74]. Of late, hydrogen oxidations have been broadly carried out with the use of nickel nonmetal catalysts. Easy oxidation of the Ni surfaces and decreased activity of Ni for HOR reactions are two shortcomings. This oxidation on the surfaces of nickel in air and electrolytes can be prevented through the use of dependable methods [75]. The development of inexpensive and performant catalysts for hydrogen oxidation reactions (HOR) in fuel cells, like alkaline membranes, is still a challenge. The preparation of a unique Ni@h-BN core shell nanocatalyst comprising encapsulated nanoparticles in a layer of h-BN shell has been reported by Gao et al. Results obtained were based on the assumption that HOR was improved by h-BN in comparison to bare Ni nanoparticles. Stronger binding with the Ni surface which results in a reduction of adsorption on the nickel could possibly lead to the attainment of optimal activity.

2.6.1 NI@H-BN NANOCATALYST CHARACTERIZATION

Ni@h-BN nanocatalyst has been prepared through a simple method, making use of carbon supported by boric acid and nickel nitrate solutions. The products, in diverse ratios, were annealed in NH_3 at 700°C. The particles were uniform in composition with sizes in the 10 to 15 nm range as shown on

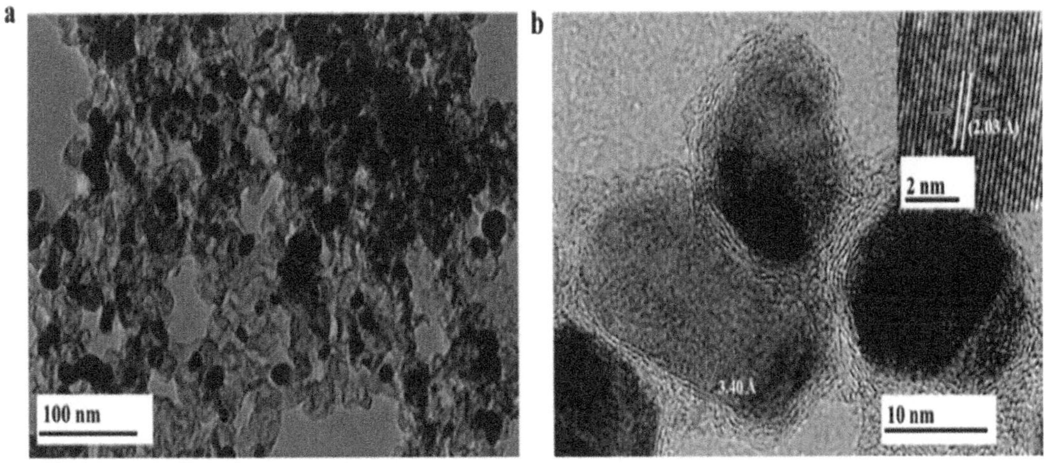

FIGURE 2.7 TEM image of Ni_3@(h-BN)1/C-700NH_3, (b) HRTEM image of Ni_3@(h-BN)1/C-700NH_3 calcinated in air, (b) lattice structure of the Ni particles (Source Gao 2017) [75].

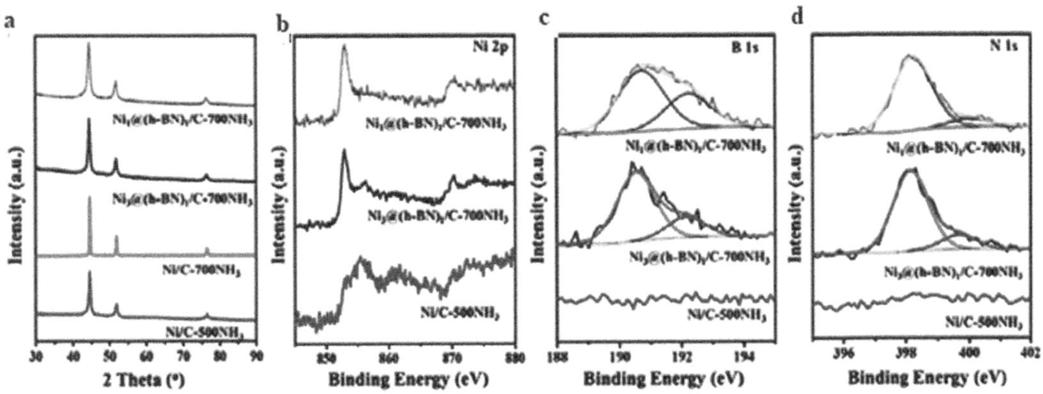

FIGURE 2.8 (a)XRD patterns of the Ni/C and Ni@h-BN/C samples treated in NH_3 at 500 or 700°C. XPS Ni 2p (b), B 1s (c), and N 1s (d) spectra of the Ni/C-500NH_3, Ni3@(h-BN)1/C-700NH_3, and Ni1@(h-BN)1/C-700NH_3 samples (Source Gao 2017) [75].

the TEM images (Figure 2.7) of the samples Ni/C-500 and Ni@h-BN/C-700. High resolution TEM revealed that a majority of the nanoparticles Ni@h-BN/C were covered with graphitic overlayer. Calcination of the samples in air at 500 °C followed by heating in NH_3 for 2 hr avoided interference by the carbon support. Calcinated samples of metal nanoparticles encapsulated with small layers of graphitic shells possess a spacing of 3.40 Å per interplaner distance in h-BN structure (002). The spacing of the Ni core layer is 2.03 Å. High resolution TEM confirms that Ni@h-BN core shells are also formed.

Peaks assigned to the Ni metal phase were also identified in the X-ray diffraction patterns (Figure 2.8) of diverse samples. The average crystallite sizes for four samples of Ni particles determined by the Debye-Scherrer equation were found to be 12.0, 38.1, 12.1 and 8.6 nm respectively. The nickel nanoparticles were aged up to 700 °C in NH_3 in the sample Ni/C-700 NH_3. Analysis of TEM and XRD results led to the assumption that treatment in NH_3 at high temperature, of nickel nanoparticles with the BN-shell, enabled these particles to remain extremely dispersed on the carbon support.

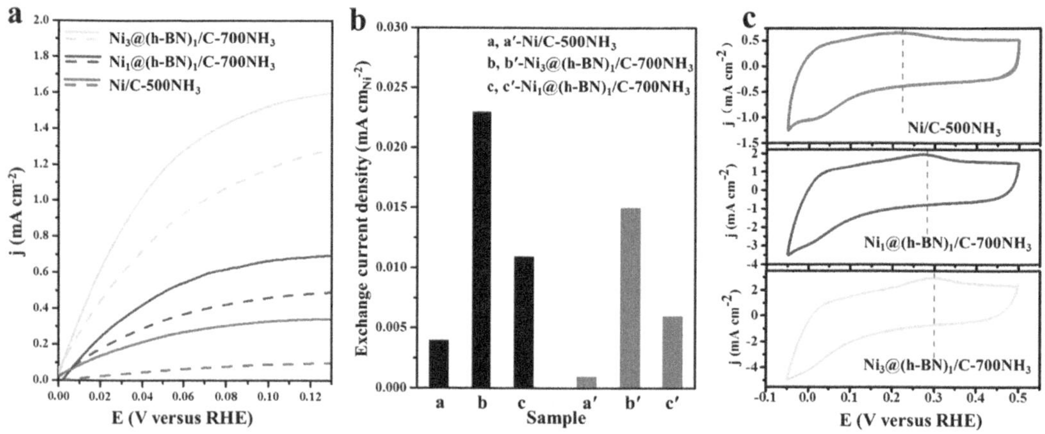

FIGURE 2.9 (a) Polarization curves from the Ni/C-500NH$_3$, Ni3@(h-BN)1/C-700NH$_3$, and Ni1@(h-BN)1/C-700NH$_3$ catalysts in H_2-saturated 0.1 M NaOH solution at 5 mV s^{-1} with a rotation rate of 2500 rpm (Source Gao 2017) [75].

HOR plots for reactions using the nanocatalysts Ni/C-500NH3, Ni$_3$@(h-BN)1/C-700NH$_3$, and Ni1@(h-BN)1/C-700NH$_3$, performed in a 1 M solution of NaOH saturated with H_2 at a rotation rate of 2500 rpm, are presented in Figure 2.9.

Sample Ni3@(h-BN)1/C-700NH$_3$ was found to be more performant than the others. The relation between overpotential and the current kinetic density was established using the Butler-Volmer equation.

A non-noble metal HOR catalyst consisting of ultrathin h-BN shells and nickel cores has been reported. Bare Ni nanoparticles in alkaline medium were less active toward HOR in comparison to those covered with a few layers of h-BN shells. Ni3@(h-BN)1/C-700NH$_3$ demonstrated a nanocatalyst density approximately six times more than that of Ni/C-500NH$_3$. The major conclusion from an analysis of experimental and theoretical study reports is that the HOR activity can be enhanced by h-BN core shells. APEFc technology conspicuously makes use of the Ni core-shell material [75].

2.7 ORGANIC AND INORGANIC HYBRID NANOCATALYST FOR OXIDATION REACTION

A new catalyst obtained by the immobilization of silver nanoparticles on functionalized magnetic hydroxyapatite (HAp) has been reported. Interest in the use of hydroxyapatite materials stems from their unique properties with potential applications in the domain of catalysis, protein and drug delivery and antimicrobial agents. The thermal and chemical stability of magnetic iron oxide nanoparticles renders it the most appropriate candidate for catalysts support. Ultrasonic irradiation technique can be employed in the synthesis of Hap nanoparticles which can later be used to prepare transition metal catalysts including silver nanoparticles. Coordination polymers based on Ag as well as AgNPs with magnetic HAp as support have previously been employed in the oxidative conversion of primary amines to hydroxyl amines. The catalytic oxidative conversion of primary amines has been used to explore the catalytic activity of Ag-HAp. Use was made of phenylethyl amine in the analysis of the activity of Ag-HAp nanocatalyst and the optimal reaction conditions. The catalyst HAp demonstrated efficiency in the partial oxidation of amines and it was used many times without loss of activity. Magnetic separation technique was employed to separate the nanostructure after its use as a productive heterogeneous nanocatalyst. The partial oxidative

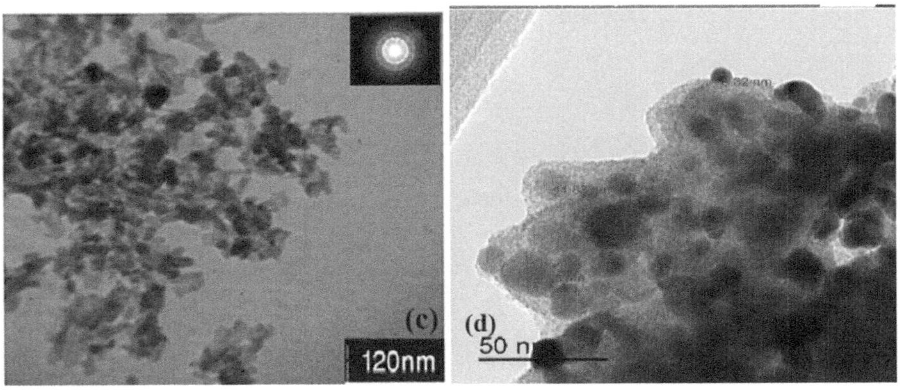

FIGURE 2.10 TEM and HRTEM micrographs of Ag-HAp catalyst (Source Bahadorikhalili 2020) [76].

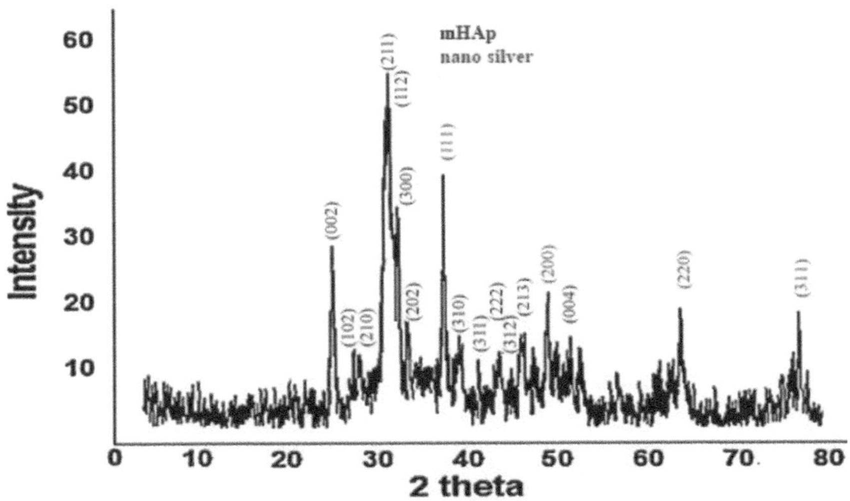

FIGURE 2.11 The XRD pattern of Ag-HAp catalyst (Source Bahadorikhalili 2020) [76].

conversion of essential amines to related N–monoalkylated hydroxylamines made use of the catalyst Ag-Hap same as the reduction of nitro compounds [76].

2.7.1 Characterization of Ag-HAp Catalyst

Microscopic techniques such as SEM and TEM were used to determine the morphology and average sizes of the nanoparticles that were prepared. While the Ag nanoparticles were found (Figure 2.10) to be spherical in shape with an average size of 200 nm, HAp NPs were uniform in shape but with sizes dispersed in the range 60–70 nm.

Peaks obtained in the powder XRD diffraction pattern (Figure 2.11) of the prepared catalyst indicated they were crystalline and were matched to the structure of Ag-HAp nanoparticles.

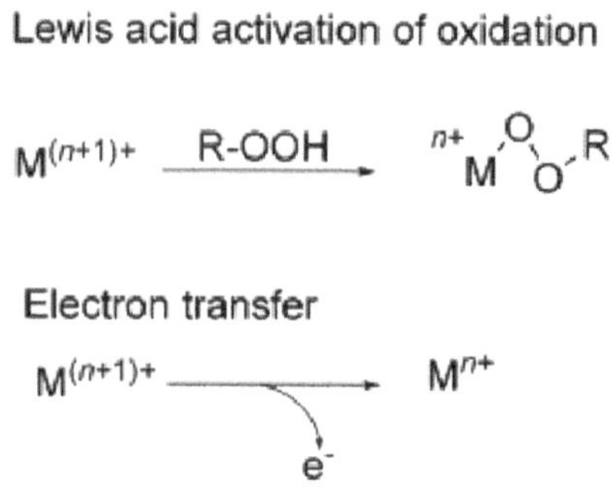

SCHEME 2.2 Probable mechanisms of oxidation concerning metal ions (Source Dhakshinamoorthy 2016) [77].

SCHEME 2.3 Mechanism the epoxidation of alkenes (Source Dhakshinamoorthy 2016) [77].

2.8 METAL ORGANIC FRAMEWORKS AS OXIDATION CATALYST

Catalytic oxidation reactions have been carried out with transition metals as the most reliable catalysts. These reactions catalyzed by the transition metal ion can proceed in one of two ways: through changes in oxidations states or through the action of the transition metal as Lewis acids (Scheme 2.2) [77].

MOFs can be employed as green catalysts in the oxidative conversion of organic compounds such as alkanes, cycloalkanes and alcohols. Epoxidation of alkenes makes use of hyperoxides as the oxidizing agent, while that of alkanes is molecular oxygen through an autoxidation mechanism where molecular oxygen reacts with a centered carbon radical to form a peroxyl radical chain reaction. The activation of oxygen, in high oxidation state, can be achieved by direct electron transfer as well as through the use of initiators or aldehydes [77–79]. Peroxide activation has been achieved by the use of metal ions behaving as Lewis acids; besides, the electron deficiency of peroxides can be increased by the presence of metal ions in zeolites and other catalysts. The mechanism for the epoxidation of alkenes using aluminosilicate and Ti zeolites is presented in Scheme 2.3.

Serving as heterogeneous catalysts, MOFs offer a few possible reaction sites that include metal ions with exchangeable coordination sites, which are not connected with linkers, diverse ligands with added active sites, or incorporated inside the empty spaces of the active species (Figure 2.12). MOFs with these types of active sites have found application as catalysts for oxidation reactions, with varying degrees of efficiency attained based upon the type of oxidation. Systems in which analogs of metal complexes, which serve as homogenous catalysts, and are connected to organic ligands are amongst the most performant MOF-based strong catalysts presently known for the epoxidation of alkenes [80–82]. Catalysts based on MOFs with pores comprising small noble metal nanoparticles that act as active sites, while the MOF plays a passive role in restricting them and blocking their growth, have been found to be the most performant MOF-based catalysts for the oxidation of alcohols [83–85].

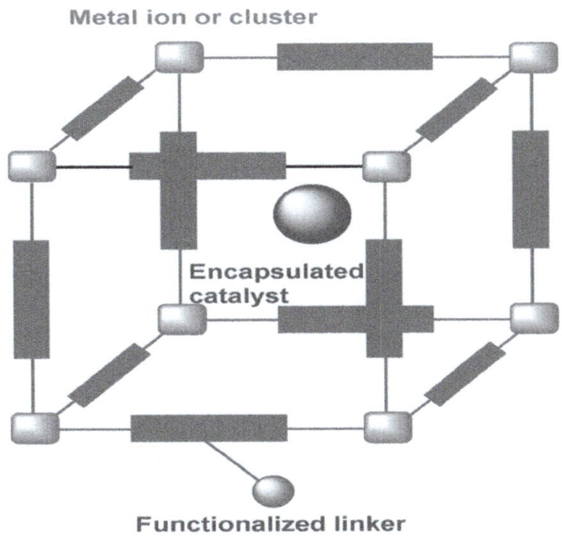

FIGURE 2.12 Oxidation sites locations in MOF (Source Dhakshinamoorthy 2016) [77].

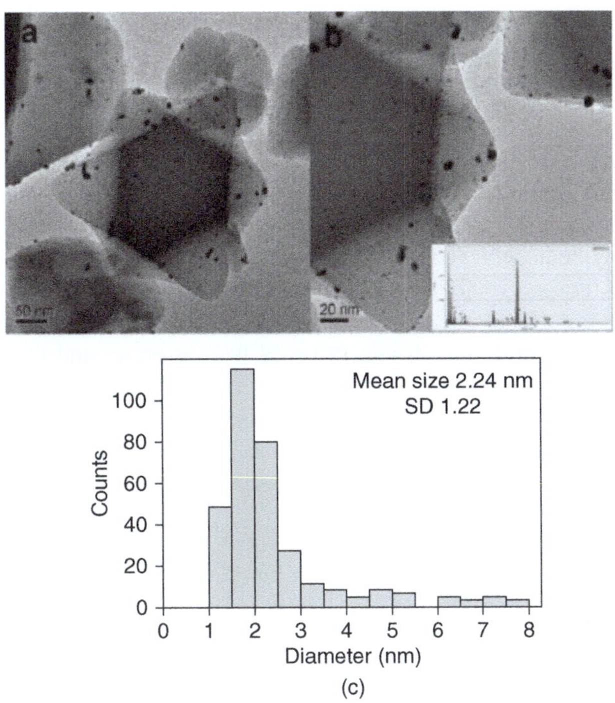

FIGURE 2.13 TEM images of Au nanoparticles for the catalytic reaction (Source Odoh 2016) [86].

These materials will be more interesting than the MOF catalysts that were mentioned before. TEM image (Figure 2.13) shows that Au nanoparticles are present in dispersed form on the surface. Materials that possess an enormous amount of single sites in a highly empty pore-volume system can be developed by combining the catalyst functionality and effectiveness with a metal, while reducing the number of segments of the catalysts. With all the above in consideration, the design and synthesis of MOFs in which the metal sites act as synergistic sites is possible in a single step without post-synthetic modifications [85].

MOFs are particularly applicable as strong catalysts in liquid-phase reactions, which is the best suitable condition for the production of compounds and fine chemicals, in contrast to that for petrochemicals. Oxidation reactions are most often utilized in the production of fine chemicals. The particular suitability of MOFs in catalyzing these reactions is a function of the characteristic type of sites that can be integrated in these materials, containing metal particles, metallic ions coordinated to the linkers, and noble metals trapped inside the MOFs. The autoxidation of hydrocarbons can be activated by MOFs with base metals; co-catalysts or the use of peroxides as oxidizing agents is often required for aerobic oxidation. MOFs with metal complexes connected to the linkers and making use of peroxides as oxidant is exclusively known presently for highly selective (cyclo) alkene epoxidation. It is still challenging to produce MOFs that are deficient in metal complexes yet capable of catalyzing epoxidation with molecular oxygen. This situation requires research attention, given the advantages of the catalyst. Alcohol oxidation in which oxygen is consumed can presently be effectively achieved by the use of MOFs comprising noble metals. Moreover, the upgrade of MOFs with base metals wherein the node acts as active sites for oxidation in which liquid oxygen is consumed is yet to be accomplished. Consequently, despite noteworthy progress within a short time coupled with successes previously accomplished, many targets are left unexplored in the domain of MOF oxidation catalysts.

2.9 METALS AS HETEROGENEOUS CATALYSTS FOR DIFFERENT OXIDATION REACTIONS

2.9.1 The Conversion of Glucose to Gluconic Acid

The oxidative conversion of glucose to gluconic acid under aerobic conditions (Figure 2.14) has received much research interest owing to its varied applications in diverse domains such as water-soluble purging, food additives and beverage bottle cleanser [87]. Such oxidative conversion of glucose had previously been achieved by means of biochemical pathways characterized by its difficulties: several steps, non-recyclability, and expensive nature [88]. The development of a catalytic technique is probably an effective pathway for the large-scale conversion of glucose to gluconic acid. Heavy metals like bismuth doped with Pt or Pd were used around the 1970s, but they were found to be unstable, poorly selective, and had a low rate of transformation. On the other hand, the use of bismuth on palladium or Pt/Pd on carbon support led to higher selectivity and dependability and very good rate of transformation, displaying the limitations of the significant metal support – less poisoning, for example, type of catalyst and the role of bismuth support are still being challenged [88].

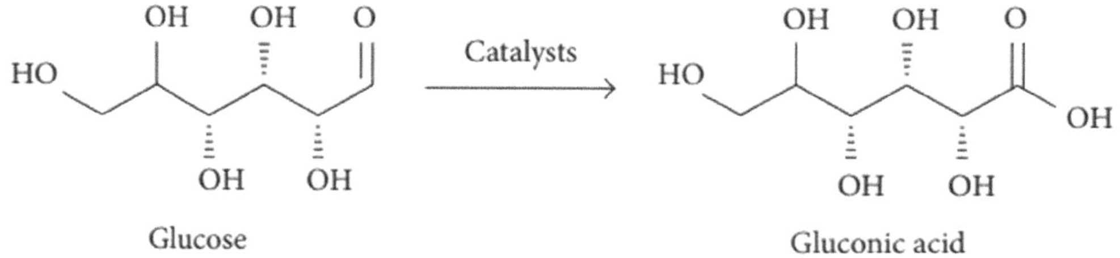

FIGURE 2.14 Glucose conversion to gluconic acid (Source Rahman 2014) [88].

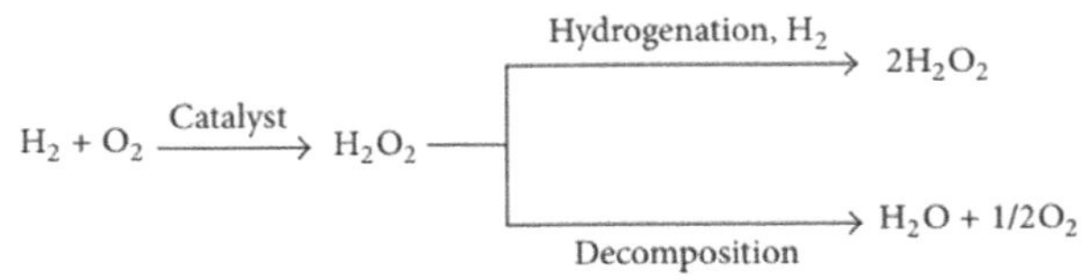

SCHEME 2.4 Formation through Hydrogenation of H_2O_2 and its decomposition (Source 2002) [89].

2.9.2 The Oxidative Conversion of Hydrogen to Hydrogen Peroxide

H_2O_2, an important industrial chemical, was previously basically used as a solid oxidant in different oxidation reactions, as disinfectants and as bleaching agents. It is considered as a green oxidizing agent given that water is its only side-product. Nowadays, health and climate assurance considerations have led to increased research interest on green synthesis and catalysis. Industrial preparation of H_2O_2 from H_2 and O_2 for the past 90 years, has made use of supported Pd catalysts. Nevertheless, the instability of the product (H_2O_2) obtained is exhibited through deterioration at low-temperature or hydrogenation to water (Scheme 2.4) [89].

Edwards et al. [90] reported the utilization of Au-catalysts blended using co-precipitation resulting in very low rate of conversion of H_2O_2. Furthermore, it was observed that use of Au-Pd catalysts by impregnation improved the formation of H_2O_2. Analysis of five different catalyst supports precisely Al_2O_3, Fe_2O_3, TiO_2, SiO_2, and carbon showed that Au-Pd catalyst with carbon-support (Au-Pd/C) gave a higher conversion. Reports by Song et al. [91] indicated that activated carbon treated $KMnO_4$ in acidic solution led to an increase in the production of H_2O_2 (78%) from hydroxylamine as a result of the production of surface active quinoid species during oxidation reactions.

2.9.3 The Oxidation of Alcohols

The oxidative conversion of alcohols to aldehydes or ketones is an important organic synthetic reaction. Several organic and fine chemicals are obtained generally from ketones, particularly, acetone [92]. The use of compounds such as chromium (VI) reagents, dimethyl sulfoxide, permanganates, periodates, or N-chlorosuccinimide in stoichiometric amounts in synthesis renders the process expensive and toxic. Homogeneous catalysts such as Pd, Cu, and Ru have been found to be specific catalysts for the oxidation of alcohols. A high oxygen pressure, organic solvent are requirements for homogeneous catalysis leading to cost and environmental problems [93]. Biodegradation considerations have led to a sustained search for new and inoffensive reactants to the ecosystem for alcohol oxidation. A higher selectivity toward aldehyde was achieved by Prati and Porta [94] during the oxidation of essential alcohols using Au/C catalysts. Similarly, a porous Si/Sn bimetallic catalyst was prepared by Endud and Wong [95] through a post-synthetic modification of rice husk ash as Si precursor, and $SnCl_2$ as a source of tin. The use of MCM-48 modified with tin and TBHP as oxidant resulted in greater selectivity toward the aldehyde or the ketone in the case of the oxidation of benzyl alcohols. The reactant activity of gold doped with (5–30% Ag) has been explored by Chaki et al. [96] in alcohol oxidation under aerobic conditions. They observed an increase in the reactant activity of Au at <10% Ag. Recently, gold nanoparticles (AuNP) have been shown by Kidwai and Bhardwaj [97] to be extremely active in the oxidation of alcohol using hydrogen peroxide as oxidant. Increase in surface area of the AuNPs resulted in a corresponding increase in activity. Furthermore, reactions that employ gold catalysts are free of synthetic hazardous and toxic solvents, producing water as the side product. This technique is an incredible promise toward the development of sustainable green chemistry [98,99].

REFERENCES

1. Q. Zhao, J. Zheng, L. Archer, *ACS Energy Lett.* **2018**, *3*, 2104.
2. (a) C. Tang, M.-M. Titirici, Q. Zhang, *J. Energy Chem.* 2017, *26*, 1077; (b) C. Tang, H.-F. Wang, Q. Zhang, *Acc. Chem. Res.* 2018, *51*, 881; (c) C. Tang, Q. Zhang, *Adv. Mater.* 2017, *29*, 1604103; (d) H. Jin, C. Guo, X. Liu, J. Liu, A. Vasileff, Y. Jiao, Y. Zheng, S.-Z. Qiao, *Chem. Rev.* **2018**, *118*, 6337; (e) D. Yang, L. Zhang, X. Yan, X. Yao, *Small Methods* **2017**, *1*, 1700209; (f) Z. Yang, Z. Yao, G. Li, G. Fang, H. Nie, Z. Liu, X. Zhou, X. A. Chen, S. Huang, *ACS Nano* **2012**, *6*, 205.
3. (a) P. G. Bruce, S. A. Freunberger, L. J. Hardwick, J.-M. Tarascon, *Nat. Mater.* **2012**, *11*, 19; (b) W. Li, J. Liu, D. Zhao, *Nat. Rev. Mater.* **2016**, *1*, 16023; (c) S.-H. Chung, C.-H. Chang, A. Manthiram, *Adv. Funct. Mater.* **2018**, *28*, 1801188; (d) H. Yuan, L. Kong, T. Li, Q. Zhang, *Chin. Chem. Lett.* **2017**, *28*, 2180; (e) X.-Q. Zhang, X.-B. Cheng, Q. Zhang, *Adv. Mater. Interfaces* **2018**, *5*, 1701097; (f) X. Zhang, X. Cheng, Q. Zhang, *J. Energy Chem.* **2016**, *25*, 967; (g) J. Cui, T.-G. Zhan, K.-D. Zhang, D. Chen, *Chin. Chem. Lett.* **2017**, *28*, 2171; (h) L. Wang, Z. Zhou, X. Yan, F. Hou, L. Wen, W. Luo, J. Liang, S. X. Dou, *Energy Storage Mater.* **2018**, *14*, 22; (i) R. Zhang, X. Chen, X. Shen, X.-Q. Zhang, X.-R. Chen, X.-B. Cheng, C. Yan, C.-Z. Zhao, Q. Zhang, *Joule* **2018**, *2*, 764; (j) X.-Q. Zhang, C.-Z. Zhao, J.-Q. Huang, Q. Zhang, *Engineering* **2018**, *4*, 831.
4. (a) Q. Pang, X. Liang, C. Y. Kwok, L. F. Nazar, *Nat. Energy* **2016**, *1*, 16132; (b) S. Xin, Z. Chang, X. Zhang, Y.-G. Guo, *Natl. Sci. Rev.* **2017**, *4*, 54; (c) X.-B. Cheng, R. Zhang, C.-Z. Zhao, Q. Zhang, *Chem. Rev.* **2017**, *117*, 10403; (d) X. Shen, H. Liu, X.-B. Cheng, C. Yan, J.-Q. Huang, *Energy Storage Mater.* **2018**, *12*, 161; (e) Z. Cheng, H. Pan, H. Zhong, Z. Xiao, X. Li, R. Wang, *Adv. Funct. Mater.* **2018**, *28*, 1707597; (f) W. Guo, Y. Fu, *Energy Environ. Mater.* **2018**, *1*, 20; (g) Q. Pang, A. Shyamsunder, B. Narayanan, C. Y. Kwok, L. A. Curtiss, L. F. Nazar, *Nat. Energy* **2018**, *3*, 783; (h) Z. Li, B. Y. Guan, J. Zhang, X. W. Lou, *Joule* 2017, *1* , 576.
5. (a) H.-J. Peng, J.-Q. Huang, Q. Zhang, *Chem. Soc. Rev.* **2017**, *46*, 5237; (b) R. Fang, S. Zhao, Z. Sun, W. Wang, H.-M. Cheng, F. Li, *Adv. Mater.* **2017**, *29*, 1606823; (c) X. Chen, T.-Z. Hou, B. Li, C. Yan, L. Zhu, C. Guan, X.-B. Cheng, H.-J. Peng, J.-Q. Huang, Q. Zhang, *Energy Storage Mater.* **2017**, *8*, 194; (d) X. Li, X. Sun, *Adv. Funct. Mater.* **2018**, *28*, 1801323; (e) G. Li, Z. Chen, J. Lu, *Chem* **2018**, *4*, 3 H. Yuan, J.-Q. Huang, H.-J. Peng, M.-M. Titirici, R. Xiang, R. Chen, Q. Liu, Q. Zhang, *Adv. Energy Mater.* **2018**, *8*, 1802107.
6. (a) B. Liu, R. Fang, D. Xie, W. Zhang, H. Huang, Y. Xia, X. Wang, X. Xia, J. Tu, *Energy Environ. Mater.* 2018, *1*, 196; (b) J. Balach, J. Linnemann, T. Jaumann, L. Giebeler, *J. Mater. Chem. A* 2018, *6*, 23127; (c) D. Zheng, G. Wang, D. Liu, J. Si, T. Ding, D. Qu, X. Yang, D. Qu, *Adv. Mater. Technol.* **2018**, *3*, 1700233; (d) J. Park, S.-H. Yu, Y.-E. Sung, *Nano Today* **2018**, *18*, 35; (e) Y. Zhu, S. Wang, Z. Miao, Y. Liu, S.-L. Chou, *Small* **2018**, *14*, 1801987.
7. N. Xu, T. Qian, X. Liu, J. Liu, Y. Chen, C. Yang, *Nano Lett.* **2017**, *17*, 538.
8. (a) J. R. Akridge, Y. V. Mikhaylik, N. White, *Solid State Ionics* **2004**, *175*, 243; (b) Y. V. Mikhaylik, J. R. Akridge, *J. Electrochem. Soc.* **2004**, *151* , A1969.
9. Nagao, Motohiro, Akitoshi Hayashi, and Masahiro Tatsumisago. "Electrochemical properties of all-solid-state Li/S batteries with Li2S-P2S5 solid electrolytes." In *ECS Meeting Abstracts*, no. 1, p. 86. IOP Publishing, **2009**.
10. X. Ji, K. T. Lee, L. F. Nazar, *Nat. Mater.* **2009**, *8*, 500.
11. H.-J. Peng, Q. Zhang, *Angew. Chem., Int. Ed.* **2015**, *54* , 11018.
12. N. Wang, Z. Xu, X. Xu, T. Liao, B. Tang, Z. Bai, S. Dou, *ACS Appl. Mater. Interfaces* **2018**, *10*, 13573.
13. Yuan, Z., Peng, H. J., Hou, T. Z., Huang, J. Q., Chen, C. M., Wang, D. W. and Zhang, Q. (**2016**). Powering lithium–sulfur battery performance by propelling polysulfide redox at sulfiphilic hosts. *Nano letters*, *16*(1), 519–527.
14. (a) Yuan, H., Peng, H. J., Huang, J. Q. and Zhang, Q. Advanced Materials Interfaces, **2019,** 6(4), p. 1802046. (b) J. Sun, Y. Sun, M. Pasta, G. Zhou, Y. Li, W. Liu, F. Xiong, Y. Cui, *Adv. Mater.* 2016, *28*, 9797; (c) X. Chen, H.-J. Peng, R. Zhang, T.-Z. Hou, J.-Q. Huang, B. Li, Q. Zhang, *ACS Energy Lett.* **2017**, *2*, 795.
15. Q. F. Zhang, Y. P. Wang, Z. W. Seh, Z. H. Fu, R. F. Zhang, Y. Cui, *Nano Lett.* **2015**, *15*, 3780; b) X. Liang, C. Y. Kwok, F. Lodi–Marzano, Q. Pang, M. Cuisinier, H. Huang, C. J. Hart, D. Houtarde, K. Kaup, H. Sommer, T. Brezesinski, J. Janek, L. F. Nazar, *Adv. Energy Mater.* **2016**, *6*, 1501636.
16. M. Zhao, H.-J. Peng, Z.-W. Zhang, B.-Q. Li, X. Chen, J. Xie, X. Chen, J.-Y. Wei, Q. Zhang, J.-Q. Huang, *Angew. Chem., Int. Ed.* **2019**, *58*, 1812062.

17. H. Yuan, H. Peng, J. Huang, Q. Zhang. *Adv. Mater. Interfaces*. **2019**. *6*, 1970025.
18. E. N. Prelizhaeva Russ. *Chem. Rev.,* 2000, *69*, 367.
19. S. Caron, R. W. Dugger, S. G. Ruggeri, J. A. Ragan and D. H. B. Ripin. *Chem. Rev.,* **2006**, *106*, 294
20. M.–O. Simon and C.-J. Li, *Chem. Soc. Rev.*, **2012**, *41*, 1415.
21. M. Nagao, A. Hayashi, M. Tatsumisago. *Electrochim. Acta,* **2011**, *56*, 6055–6059.
22. A. Holder *Molybdenum: Its Biological and Coordination Chemistry and Industrial Applications*, Nova Science Publishers, New York, **2013**
23. H. Yuan, H.-J. Peng, J.-Q. Huang, Q. Zhang. *Adv. Mater Interfaces* **2019**, *6*,1802046
24. C.-J. Li, L. Chen, *Chem. Soc. Rev.*, **2006,** *35*, 68.
25. R. Chakravarthy, V. Ramkumar. D. Chand. *Green Chemistry* **2014**, 16, 2190–2196
26. (a) D. J. Weix, J. A. Ellman, *Org. Lett.* **2003**, *5*, 1317–1320; (b) G. Santoni, G. Licini, D. Rehder, *Chem. Eur. J.* **2003**, 9, 4700–4708; (c) J. T. Sun, C. J. Zhu, Z. Y. Dai. M. H. Yang, Y. Pan, H. W. Hu, *J. Org. Chem.* **2004**, 69, 8500–8503; (d) B. Pelotier, M. S. Anson, I. B. Campbell, S. J. F. Macdonald, G. Prime, R. F. W. Jackson, *Synlett* **2003**, 1055–1060
27. V. Thakur, A. Sudalai, *Tetrahedron: Asymmetry* **2003**, *14,* 407–410
28. B. H. Lipshutz, Y. Shin, *Tetrahedron Lett.* **2000**, 41, 9515–9521
29. X. Wang, X. Wang, H. Guo, Z. Wang, K. Ding. *Chem.–A Europ J.* **2005**, *11,* 4078–4088.
30. U. Miotti, G. Modena, L. Sedea Mechanism of formation of alkyl aryl sulphoxides by oxidation of alkyl aryl sulphides with bromine *J. Chem. Soc. B*, **1970**, 802–805
31. N H. Le, Y.-h Han, H. Jung, J. Cho, *J. Hazardous Mater*, **2019**, 379, 120830.
32. E. Rogozea, N. Olteanu, A. Petcu, C. Lazar, A. Meghea, M. Mihaly, Extension of optical properties of ZnO/SiO2 materials induced by incorporation of Au or NiO nanoparticles, *Opt. Mater.* **2016**, *56*, 45–48.
33. I. Khan, K. Saeed, I. Khan. *Arab. J. Chem.*, **2019**. *12*, 908–931.
34. E. Rogozea, A. Petcu, N. Olteanu, C. Lazar, D. Cadar, M. Mihaly, M., *Mater. Sci. Semicond. Process* **2017**, 57, 1–11.
35. K. Saeed, I. Khan, *Instrum. Sci. Technol.* **2016**. 44, 435–444.
36. D. Guo, G. Xie, J. Luo. *J. Phys. D. Appl. Phys.* **2014**. *47*, 13001.
37. S. Mallakpour, F. Sirous, *Prog. Org. Coatings*. **2015**. *85*, 138–145.
38. J. Zhang,M. Saltzman. *Chem. Eng. Prog.* **2013**.*109*, 25–30.
39. K. Prashant, H. El Ivan, M. El-Sayed. "Au nanoparticles target cancer." *Nano Today,* **2007**, *2*, 18–29.
40. O. Akhavan, R. Azimirad, S. Safa, E. Hasani, *J. Mater. Chem.* **2011**, *21*, 9634.
41. M., Masoud, M. Zarch, M. Darroudi, K. Sayyadi, S. Keshavarz, J. Sayyadi, A. Fallah, H. Maleki. *Applied Sciences,* **2020**, *10*, 7533.
42. V. Avasare, Z. Zhang, D. Avasare, I. Khan, A. Qurashi, *Int. J. Energy Res.* **2015,** *39*, 1714–1719.
43. X. Fang, J. Liu, V. Gupta, *Nanoscale*, **2013**. *5*, 1716.
44. Z. Wang, X. Pan, Y. He, Y. Hu, H. Gu, Y. Wang, Z. Wang, X. Pan, Y. He, Y. Hu, H. Gu, Y. Wang. *Adv. Mater. Sci. Eng* **2015**. s1–21.
45. H. Dong, B. Wen, R. Melnik, *Sci. Rep.* **2014**. *4*, 7037.
46. Y. Lei, W. Huang, M. Zhao, Y. Chai, R. Yuan, Y. Zhuo, *Anal. Chem.* **2015**. *87*, 7787–7794.
47. N. Khlebtsov, L. Dykman, L., *Plasmonic Nanopart.* **2010**. 37–85.
48. S. Unser, I. Bruzas, J. He, L. Sagle, L., *Sensors*, **2015**, *15*, 15684–15716.
49. C. Bianchini, P. K. Shen, *Chem. Rev.*, **2009**, *109*, 4183–4206.
50. S. Gu, R. Cai, T. Luo, Z. Chen, M. Sun, Y. Liu, G. He, Y. Yan, *Angew. Chem., Int. Ed.*, **2009**, *48*, 6499–6502.
51. S. Lu, J. Pan, A. Huang, L. Zhuang and J. Lu, *Proc. Natl. Acad. Sci. USA*, **2008**, *105*, 20611–20614.
52. J. Varcoe, R. Slade, *Fuel Cells*, 2005, *5*, 187–200.
53. G. Li, J. Pan, S. Lu, Y. Li, A. Huang, L. Zhuang, J. Lu, *Adv. Funct. Mater.*, **2010**, *20*, 312–319.
54. L. Gao, Y. Wang, H. Li, Q. Li, N. Ta, L. Zhuang, Q. Fu, X. Bao. *Chemical science*, **2017**, *8*, 5728–5734.
55. Y.-J. Wang, J. Qiao, R. Baker, J. Zhang, *Chem. Soc. Rev.*, **2013**, *42,* 5768–5787.
56. J. Durst, A. Siebel, C. Simon, F. Hasche, J. Herranz, H. A. Gasteiger, *Energy Environ. Sci.*, **2014**, *7*, 2255–2260.
57. W. Sheng, H. A. Gasteiger, Y. Shao-Horn, *J. Electrochem. Soc.*, **2010**, *157*, B1529–B1536.
58. T. Kenjo, *J. Electrochem. Soc.*, **1985**, *132*, 383–386.
59. K. Mund, G. Richter, F. von Sturm, *J. Electrochem. Soc.*, **1977**, *124*, 1–6.

60. W. Sheng, A. P. Bivens, M. Myint, Z. Zhuang, R. V. Forest, Q. Fang, J. G. Chen, Y. Yan, *Energy Environ. Sci.*, **2014**, *7*, 1719–1724.
61. H. Liu, Y. Liu, Y. Li, Z. Tang, H. Jiang, *J. Phys. Chem. C.* **2010**, *114*, 13362–13369.
62. Z. Zhuang, S. Giles, J. Zheng, G.Jenness, S. Caratzoulas, D. Vlachos, Y. Yan. *Nature commun.* **2016**, *7*, 1–8.
63. K. Asazawa, K. Yamada, H. Tanaka, A. Oka, M. Taniguchi T. Kobayashi, *Angew. Chem., Int. Ed.*, **2007**, *46*, 8024–8027; J. K. Nørskov, T. Bligaard, A. Logadottir, J. R. Kitchin, J. G. Chen, S. Pandelov, U. Stimming, *J. Electrochem. Soc.*, **2005**, *152*, J23–J26.
64. Q. Fu, X. Bao, *Chin. J. Catal.*, **2015**, *36,* 517–519.
65. Q. Fu, X. Bao, *Chem. Soc. Rev.*, **2017**, *46*, 1842–1874.
66. D. Deng, K. S. Novoselov, Q. Fu, N. Zheng, Z. Tian, X. Bao, *Nat. Nanotechnol.*, **2016,** *11,* 218–230.
67. Grånäs, M. Andersen, M. A. Arman, T. Gerber, B. Hammer, J. Schnadt, J. N. Andersen, T. Michely, J. Knudsen, *J. Phys. Chem. C*, **2013**, *117*, 16438–16447.
68. L. Jin, Q. Fu, A. Dong, Y. Ning, Z. Wang, H. Bluhm, X. Bao, *J. Phys. Chem. C*, **2014**, *118*, 12391–12398.
69. R. Larciprete, S. Ulstrup, P. Lacovig, M. Dalmiglio, M. Bianchi, F. Mazzola, L. Hornekær, F. Orlando, A. Baraldi, P. Hofmann, S. Lizzit, *ACS Nano*, **2012**, *6*, 9551–9558.
70. P. Sutter, J. T. Sadowski, E. A. Sutter, *J. Am. Chem. Soc.*, **2010**, *132*, 8175–8179.
71. M. Wei, Q. Fu, H. Wu, A. Dong, X. Bao, *Top. Catal.*, **2016,** *59*, 543–549.
72. T. Brugger, H. Ma, M. Iannuzzi, S. Berner, A. Winkler, J. Hutter, J. Osterwalder, T. Greber, *Angew. Chem., Int. Ed.*, **2010**, *49*, 6120–6124.
73. R. Mu, Q. Fu, L. Jin, L. Yu, G. Fang, D. Tan, X. Bao, *Angew. Chem., Int. Ed.*, **2012**, *51*, 4856–4859.
74. Z. Zhuang, S. A. Giles, J. Zheng, G. R. Jenness, S. Caratzoulas, D. G. Vlachos and Y. Yan, *Nat. Commun.*, **2016**, *7*, 10141.
75. Gao, L., Wang, Y., Li, H., Li, Q., Ta, N., Zhuang, L. and Bao, X. *Chemical science*, **2017**, *8*(8), 5728–5734.
76. Bahadorikhalili, S., Arshadi, H., Afrouzandeh, Z. and Ma'mani, L., *New Journal of Chemistry*, **2020,** *44*(21), 8840–8848.
77. Dhakshinamoorthy, A., Asiri, A.M. and Garcia, H., *Chemistry–A European Journal*, **2016**, 22(24), 8012–8024.
78. A. Corma, H. Garcia, *Chem. Rev.* **2003**, *103*, 4307–4366.
79. J. H. Clark, *Green Chem.* **1999**, *1*, 1–8.
80. I. V. Berezin, E. T. Denisov, *The Oxidation of Cyclohexane*, Pergamon, New York, **1996.**
81. K. Chen, P. Zhang, Y. Wang, H. Li. *Green Chem.,* **2014,** *16*, 5, 2344–2374.
82. E. Roduner, W. Kaim, B. Sarkar, V. Urlacher, J. Pleiss, R. Gläser, W. Einicke, G.Sprenger, U. Beifuß, E. Klemm, C. Liebner, H. Hieronymus. *Chem.Cat.Chem.*, **2013**, *5*, 82–112.
83. S. O. Odoh, C. J. Cramer, D. G. Truhlar, L. Gagliardi, Chem. Rev. **2015,** 115, 6051–6111.
84. T. Ishida, M. Nagaoka, T. Akita, M. Haruta. " *Chem.–A Europ J.,* **2008**, *14*, 8456–8460.
85. Gumus, Y. Karataş, M. Gülcan. *Catal. Sci. Tech.,* **2020**,*10*, 4990–4999.
86. H. Liu, Y. Liu, Y. Li, Z. Tang, H. Jiang, *J. Phys. Chem. C* **2010**, *114*, 13362–13369.
87. T. Ishida, H. Watanabe, T. Bebeko, T. Akita, M. Haruta, *Appl. Catal. A: Gen.* **2010**, *377*, 42–46.
88. Ali, M., Rahman, M., Sarkar, S. M. and Hamid, S. B. A., *Journal of Nanomaterials*, **2014**.
89. P. Landon, P. J. Collier, A. J. Papworth, C. J. Kiely, G. J. Hutchings, *Chem. Commun.*, **2002**, 2058–2059.
90. K. Edwards, A. Thomas, B. E. Solsona, P. Landon, A. F. Carley, G. J. Hutchings, *Catal. Today*, **2007**, *122*, 397–402.
91. W. Song, Y. Li, X. Guo, J. Li, X. Huang, W. Shen, *J Mol. Catal. A: Chem.* **2010**, *328*, 53–59.
92. T. Hayashi, K. Tanaka, M. Haruta, *J Catal*, **1998**, *178*, 566–575.
93. Y.-H. Kim, S.-K. Hwang, J. W. Kim, Y.-S. Lee, *Ind. Engin. Chem. Res.*, **2014**, *53*, 12548–12552.
94. Prati, F. Porta, Appl. *Catal A: Gen*, **2005**, *291*, 199–203.
95. S. Endud, K.-L. Wong, *Microporous Mesoporous Mater.*, **2007**, *101*, 256–263.
96. N. K. Chaki, H. Tsunoyama, Y. Negishi, H. Sakurai, T. Tsukuda, J. Phys. *Chem. C*, **2007**, *111*, 4885–4888,
97. Kidwai, S. Bhardwaj, *Appl. Catal. A: Gen*, **2010**, *387*, 1–4.
98. C. Y. Ma, J. Cheng, H. L. Wang et al., *Catal. Today*, **2010**, *158*, 246–251.
99. M. Ali, M. Rahman, S. Sarkar, S. Hamid. *J. Nanomaterials*, **2014**, *1*, 192038.

3 Correlation of Surface Properties and Catalytic Activity of Metal Aluminophosphates

C.J. Binish and A.V. Vijayasankar

CONTENTS

3.1 Introduction to Alumina-Based Porous Materials 49
3.1.1 Aluminophosphates (AlPO) 50
3.2 Synthesis of Alumina-based Materials 51
3.2.1 Preparation of Amorphous Mesoporous Metal Aluminophosphates 52
3.3 Characterization of Amorphous Mesoporous Aluminophosphates 53
3.3.1 Powder X-Ray Diffraction 53
3.3.2 X-Ray Diffraction Pattern of Aluminophosphates 53
3.3 Fourier Transform Infrared Spectroscopy (FTIR) 53
3.3.4 BET-Specific Surface Area 55
3.3.5 Temperature Programmed Desorption 55
3.3.6 X-Ray Photoelectron Spectroscopy 55
3.3.7 Raman Spectra 56
3.4 Correlation of Surface Properties on Catalytic Activity of Aluminophosphates 56
References 59

3.1 INTRODUCTION TO ALUMINA-BASED POROUS MATERIALS

The modification of surface properties of materials by altering physical forms, incorporation of metals or metal oxides with diverse surface energies, has received much attention in material research due to its wide scope in industrial applications.

A material is said to be porous when its cavities, channels or interstices on the surface are deeper than they are wide. On the basis of pore size, porous materials [1] are classified as; Microporous materials with pore size less than 2 nm, Mesoporous materials with pore size in between 2 to 50 nm and Macroporous materials with pore size above 50 nm. Porous nature of solid plays a significant role in determining surface properties and control of porosity of materials is of great importance in a variety of applications.

A large number of inorganic materials, such as oxides, clays, zeolites, phosphates are porous in nature, and most of them are used as catalysts and adsorbents in various industrial processes. Traditionally used materials like alumina [2], silica and its modified forms, like zeolites [3], aluminophosphates [4], metal-organic framework (MOF) [5] are found to be promising materials due to its porous nature.

Aluminum oxide, commonly known as alumina, and its modified form have gained significant interest as adsorbents [6] in catalytic [7] applications. Alumina-based compounds have contributed a large number of materials for different applications, due to their flexible surface and textural

DOI: 10.1201/9781003218708-4

properties. Alumina exhibits many unique properties, which include being abundant in nature, non-toxic, resistant to acid and alkali, thermally stable, simple and ecofriendly preparation protocol, tunable surface acidity/basicity, efficient adsorbent capacity and controllable porosity.

Alumina exists in different crystalline forms dependent upon the phase transition temperatures – among those, Alpha (α), Beta (β) and Gamma (γ) are the major ones [8]. The α- Alumina, also termed nano alumina, has a corundum structure, where Al^{3+} center possesses an octahedral structure with oxygen atoms. α- Alumina possesses a low specific surface area, resistance to heat and is chemically inert. β- Alumina is hexagonal, and the unit cell contains two alumina spinel-based blocks. γ-Alumina is also known as nano alumina, an activated form of alumina which exhibits higher dispersion and purity. Due to high specific surface area, inert nature, resistance to high temperature, conductivity, mechanical strength, adsorption capacity, nano structural alumina is proved as potential material as catalyst support and adsorbent.

Alumina materials and its composites can be physically modified into different forms like film [9], beads [10] and membranes [11]. Alumina materials modified as membranes are used as support or layer due to their major impressive characteristics, like resistance in high transmembrane pressures, chemical inertness, hardness, porous texture, hydrophilic nature and covalent bonding.

Alumina films gained significant interest than other modified forms of alumina ike beads and membranes due to its wide range of applications in catalysis [12], metal removal [13], separation of impurities [14] in water treatment, corrosion-resistant protective layers [15], materials for mechanical protection [16] and as sensors [17]. The major merits of alumina films are their easiness of preparation, flexibility in tuning the surface charge and porous properties; they are stable in high temperatures and resistant to chemicals.

Physico chemical properties of alumina films can be modified by the deposition process. The nature of resultant material depends on the material and method followed for deposition. Alumina films incorporated with metal [18, 19] and metal oxides [20], have specific advantages over other materials. Generally, the deposition of atoms over film is nonequilibrium in nature. Surface texture, morphology, conductivity, and optical properties of composite films are highly influenced by the type of materials used and methodology followed for depositing.

Alumina frameworks incorporated with silicates generated a new family of potential materials known as zeolites [21]. Aluminum atoms in zeolites are always tetrahedrally coordinated with a negatively charged framework. Crystalline aluminosilicates or zeolites are classified based on the pore size, They are (a) Small-pore zeolites (8-ring structure with diameter of 0.30–0.45 nm), (b) Medium-pore zeolites (ten–ring structure with diameter of 0.45–0.60 nm), (c) Large-pore zeolites(twelve ring structure with diameter of 0.6–0.8 nm) and (d) Extra-large pore zeolites (fourteen-ring structure with diameter of 0.7–1.0 nm). Zeolites have already proven to be potential materials in the fields of sustainable energy generation [22], energy storage [23], catalysis [24], carbon dioxide absorption or removal [25], air-pollution control [26], and water treatment, mainly in heavy metal removal [27]. Zeolites possess unique merits such as large surface area, tunable three-dimensional framework, and tunable pore size, acidity/basicity, which enables zeolites to be good catalysts.

3.1.1 Aluminophosphates (AlPO)

Crystalline Aluminophosphates are microporous materials with a three-dimensional framework structure, similar to zeolites build of alternating AlO_4 and PO_4 tetrahedron. Aluminophosphate framework is neutral, surface is mildly hydrophilic and their pore volume varies from 0.3–0.8 nm [28].

Aluminophosphate exhibits high specific surface area, microporosity, high thermal and hydrothermal stability, flexible surface acid /base properties have made them a distinct class of potential materials deployed as catalysts [29] and adsorbents [30]. As per industrial requirement, the physico chemical properties like type of metal phosphate, size of particles, porous nature and texture of the materials plays a significant role.

Researchers made several attempts to synthesize aluminophosphate materials with a larger pore size by using various organic templates, which lead to the generation of mesoporous aluminophosphates [31]. Further,, the incorporation of metals, mainly transition metals, into aluminophosphate framework results in the formation of metal aluminophosphates. Amorphous mesoporous aluminophosphates materials have six coordinated (octahedral) and four coordinated (tetrahedral) aluminum environments. $AlPO_4$-Al_2O_3 systems are built up of [PO_4] tetrahedral and [AlO_4], [AlO_5], and [AlO_6] polyhedral units. These modifications in the structure and surface nature of the materials broadened the scope of aluminophosphates in redox catalysis. P/Al molar ratio determines the surface acidity or basicity of the materials, which influence the catalytic activity of aluminophosphate materials. These materials exhibited efficient catalytic activity in redox reactions compared to crystalline materials [33].

3.2 SYNTHESIS OF ALUMINA-BASED MATERIALS

Generally, alumina occurs as the minerals corundum, gibbsite, diaspore and bauxite. Alumina can be synthesized by either physical or chemical methods. Alumina was prepared by physical methods like ball milling, Laser Ablation Synthesis in Solution method, flame spray and so forth. Precipitation, coprecipitation, hydrothermal method, combustion method and sol-gel are the chemical methods followed for the synthesis of alumina-based materials. The physico-chemical properties of alumina materials are determined by the synthetic route, pre- and post-treatments followed for the preparation of the material.

Ball milling, otherwise known as solid-state synthesis, is one of the most important processes used to synthesize alumina materials with smaller particle size [34]. Generally, the ball milling method follows the top-down approach. The principle involved in the milling process is the collision between the tiny rigid balls, which creates friction that results in reduction of particle size of reactants, increased temperature, pressure, and internal energy. The size of synthesized particles is determined by the quality of dispersion, milling time, size of rigid balls and rotational speed of balls. Studies indicate that the size of Alumina nanoparticles can be reduced to 1.4 μm from the original size of 70 μm. Also, the results have proved that the particle size of commercial zeolite powder can be reduced up to 0.2 to 0.3 μm from 45 μm by employing the ball-milling technique.

Laser Ablation Synthesis in Solution method is considered a top-down physical approach for the generation of nanoparticles [35]. Nanoparticles are prepared with a laser beam hitting a solid target in a liquid and condensation of the plasma plume. Since the ablation is occurring in a liquid and air, the environment allows for creating a plume with stronger confinement. The Laser Ablation Synthesis in Solution method is followed to produce more refined and smaller nanoparticles. Using the laser ablation method, alumina is synthesized with particle size lower than 100 nm. Research has proved that alumina-based films can also be synthesized using this method.

Precipitation is a separation technique to segregate or settle down, or sediment, a solid material from a solution. These solid materials are prepared by a reaction with the help of a precipitating agent. The size of the particle of the precipitate is determined by nature, concentration, molar ratio, temperature and rate of mixing of reactants.

Generally, precipitates are formed by either nucleation or growth of particles. A stable solid formed on the solid surfaces by a combination of ions, atoms or molecules are known as nuclei. A nucleus will grow if they are exposed to the ions or atoms that lead to bigger particles. If nucleation continues without particle growth it leads to a precipitate with a wide quantity of small particles. Slow addition of precipitating agents results in super-saturation, which improves the rate of nucleation.

Coprecipitation is a process in which soluble compounds are carried out of solution by a precipitate [36]. Coprecipitations are of four different types: (1) Surface adsorption happens for precipitates with larger surface areas. (2) Mixed-crystal formation happens when one of the ions in the crystal lattice is substituted by another ion. (3) Occlusion is a precipitation process by which a contaminant is trapped into a growing crystal. (4) Mechanical entrapment is a type of mechanism in which a little amount of solution is trapped in between two growing crystals which are closer to each other.

The hydrothermal method is used to synthesize nanoparticles at low temperatures. Hydrothermal synthesis is usually conducted in autoclaves by mixing a metal precursor along with water at high temperatures. The major benefits of the hydrothermal method include low calcination temperature, low agglomeration level and high phase purity. The reaction kinetics and the surface properties of the resultant nanomaterials are highly influenced by the reaction conditions, such as pH, temperature and pressure.

The combustion method is one of the cost-efficient methods for synthesizing highly pure nanoparticles. The combustion method is a self-sustained redox reaction which requires an oxidizer and a fuel. Generally, metal nitrates are used as a source of metals, and an organic material like glycine or urea acts as fuel. Material synthesized by the combustion method possesses a high surface area due to lower agglomeration. Solution combustion synthesis is reported to obtain different forms of alumina like films and membranes. Particle size of alumina nanomaterials can be reduced to 2.0 - 2.5 nm by the combustion method [37].

The sol-gel process generates a gel from a sol that is an integrated network of colloidal particles. The metal sources used for sol-gel process are either metal alkoxides or soluble metal halides of respective metal, which are converted to sol and gel by hydrolysis and polycondensation reactions.

The sol will be converted to a gel, which is a liquid phase of continuous inorganic network that will be subjected to calcination, which favors further polycondensation reaction. The porous nature and surface properties of synthesized products are influenced by the pH, sol composition, nature of solvents, stirring time and heat treatments followed during preparation. Researchers have proved that the particle size of alumina nanoparticles can be reduced to 20–30 nm [38] size and zeolite particle size to 70–80 nm [39] by using the sol-gel method. Many other techniques [40] have been utilized to synthesize nanostructures of alumina, including plasma anodization, spray pyrolysis, chemical vapor deposition, sputtering, electron beam evaporation, and so forth.

Alumina composite films [41], one of the most promising modified forms of alumina, can be synthesized by the sol-gel process. Aluminum alkoxide [42] or aluminum inorganic salt [43] can be used as starting reagents. There are reports on gel films synthesized by the aqueous sol-gel process. Materials synthesized from aqueous sol are not as pure as those from alkoxide. Amorphous films prepared using polymeric sols as precursors exhibited porosity in the range of 25–37 percent with a pore diameter in the range of 2.0–2.7 nm. Alumina films prepared by the sol-gel method have well-defined pore-size distribution, high porosity, large specific surface area, uniformity, homogeneity and high thermal stability.

3.2.1 Preparation of Amorphous Mesoporous Metal Aluminophosphates

Metal salt is used along with a precipitating agent to prepare oxides of metals. Alumina is prepared by precipitating aluminum hydroxide from aluminum salt, using aqueous ammonia as a precipitating agent [32]. Drying or calcination of aluminum hydroxide leads to the formation of alumina. Pure aluminophosphates was precipitated from aluminum salt and orthophosphoric acid using aqueous ammonia as a precipitating agent. Dried precipitate results in the formation of aluminophosphates. Amorphous Metal aluminophosphate were synthesized by methods like precipitation, coprecipitation, sol-gel and hydrothermal methods. Metals are incorporated into the framework of $AlPO_4$ by coprecipitation of metal salt along with aluminophosphate. Aluminum salt, orthophosphoric acid and transition metal salt are dissolved in water, and aqueous ammonia is added to the hot solution accompanied by constant stirring. The pH is maintained at alkaline level, and the precipitate formed is filtered and dried. The dried precipitate is calcined in a muffle furnace. The addition of precipitation agents, type of metal used, pH, molar ratio, calcination temperature post- and pre-treatments determine the crystallinity, particle size and porous properties of the metal aluminophosphate material.

3.3 CHARACTERIZATION OF AMORPHOUS MESOPOROUS ALUMINOPHOSPHATES

3.3.1 Powder X-Ray Diffraction

Powder x-ray diffraction (PXRD) technique is one of the most extensive characterization techniques used for the determination of crystallinity of particles. PXRD analysis provides details regarding the crystallinity, the phase, lattice parameters and particle size of a powdered solid material. The phase determination and composition of crystal or non-crystalline materials can be performed by locating the position and intensity of the peaks with standard reference patterns available at the database known as International Centre for Diffraction. The average particle size of the materials can be determined by the Debye -Scherrer equation using the broadening of the most intense peak available in the XRD spectrum. The general pXRD peaks observed for alumina materials are given in Table 3.1.

3.3.2 X-Ray Diffraction Pattern of Aluminophosphates

PXRD pattern of the aluminophosphates and metal incorporated aluminophosphates, calcined at 500°C exhibited noncrystalline nature of the materials. A broad band observed in the range $2\theta = 15^{\circ}$–30° range is characteristic of amorphous aluminophosphate samples. Iron incorporated aluminophosphates exhibited sharp diffraction peaks at 2θ values approximately at 23, 26.9 and 38.5 at d spacing values of 3.8, 3.3 and 2.3Å respectively, indicating the presence of iron species in amorphous aluminum phosphate matrix [44]. For γ-Al2O3 film deposited on $SrTiO_3$ substrate[45], two Prominent Peaks are observed at angles (2θ) 45.51 and 46.51.

3.3 Fourier Transform Infrared Spectroscopy (FTIR)

Fourier-transformed infrared spectroscopy (FTIR) is used to determine the composition and functional groups present in a compound by analyzing the absorption of electromagnetic radiation of wavelength 4000–400 cm^{-1}. A molecule becomes IR active if its dipole moment changes by absorbing IR radiation. The position and intensity of FTIR bands are attributed to the nature of bonds, different functional groups and the strength of the bond, which conveys the information related to molecular structure.

TABLE 3.1
pXRD Peaks Observed for Different Phases of Alumina

Phase	Crystal Structure	XRD Peaks($2\theta^{\circ}$)
α	Rhombohedral	26, 32, 43, 57, 68
γ	Tetragonal, hexagonal, cubic	46, 67, 37
η	Cubic spinel	66
κ	Orthorhombic	34, 67
θ	Monoclinic	67, 46, 36, 31
δ	Orthorhombic, Tetragonal	67, 45, 37
χ	Cubic, hexagonal	67, 42, 50

TABLE 3.2
FTIR Bands of Alumina

Absorption Peak (cm^{-1})	Mode	Related to
460-524	Stretching	Al-O
595,656	Stretching in Oh structure	Al-O
715	Stretching in Td structure	Al-O
3460	Symmetric bending	Al-O-H
1600-1650		O-H deformation of water
400-800		Al-O-Al
3460-3550	Stretching	O-H
1637	Stretching	Al-O-H
1332		Al-O

TABLE 3.3
FTIR Bands of Aluminium Phosphate

Absorption Peak (cm^{-1})	Mode	Related to
1100	Symmetric Stretching	P-O
1250	asymmetric stretching	P-O
480	symmetric bending	P-O
695	asymmetric bending	P-O
1123	symmetric stretching	$[PO_4^{3-}$
3404	stretching	O-H
1666		H_2O
1402	Bending	$-CH_2$
662, 611		Nano aluminium

The FTIR spectra obtained below 1000 cm^{-1} for alumina or alumina films is due to general characteristics of alumina; the absorption peaks in the range of 600 cm^{-1} is assigned to amorphous aluminum oxide [46]. The absorption peaks of alumina at (430, 472, 607, 649, 688, 860) cm^{-1} are attributed to the stretching vibrations mode of the Al-O bond. Absorption bands appear at (430, 484, 617, 638, 650, 698, 781) cm^{-1} that confirm the formation of Al-O bonds. All absorption peaks appearing at (487–780) cm^{-1} relate to the stretching vibrations mode of Al-O bonds [47]. The FTIR bands observed for alumina are given in Table 3.2.

Aluminophosphates and metal aluminophosphates exhibited absorption bands at 1100–1140 cm^{-1} are attributed to asymmetric vibrations of phosphate. The shoulder peaks in the range of 700–500 cm^{-1} is due to the symmetric stretching mode of P-O-P and bending mode of O-P-O bonds. Both aluminophosphates and metal-aluminophosphates exhibited OH stretching vibration region at 3600–3000 cm−1 are attributed to the surface hydroxyl groups associated with the metal, aluminum and phosphorus atoms. The weak absorption bands around 1500–1600 cm^{-1}, confirms the presence of water molecules and oxidized carbonate phase of the calcined samples. Metal aluminophosphates exhibited broad and weak absorption bands around 1623–1640 cm^{-1} indicated the presence of adsorbed water molecules which is a characteristic of mesoporous materials. The absorption peak at 691cm^{-1}, due to stretching vibrations of Al-O and P-O bonds. FTIR bands observed for aluminium phosphate are given in Table 3.3.

TABLE 3.4
Surface Area of Alumina at Different Temperatures

Temperature (°C)	Type of Alumina	Total Specific Surface Area(m^2/g)
700	γ	159
900	δ	114
1000	δ	141
1100	α	98
1500	α	25
1550	α	14

3.3.4 BET-Specific Surface Area

The Brunauer-Emmett-Teller (BET) technique is also used for the determination of the surface area of nanoscale materials. It is based on the principle of physical adsorption of a gas on a solid surface, and it was named by the initials of the surnames of its developers, Brunauer, Emmett and Teller [48].

The surface area of a material is determined by the synthesis method followed for the preparation of the material, specifically pre- and post-treatment methods. New surfaces are created during splitting up of a bigger particle into smaller particles, which results in the increase in surface area of the materials. Generation of new pores also create new surfaces, which lead to change in surface area of a material [49].

Evaluation of pore characteristics and dimensions by the BET method is performed by evaluating the known amount of unreactive gas that has undergone adsorption and desorption over a porous material. The temperature of the solid sample is kept constant, while the pressure or concentration of the adsorbing gas is increased, which leads to the adsorption of gas over the solid. Surface area of alumina at different temperatures are given in Table 3.4. The surface area of aluminophosphates can be improved by intervention of transition metal into the aluminophosphates [50].

3.3.5 Temperature Programmed Desorption

Temperature-programmed desorption (TPD) is used for evaluation of active sites or the nature of acidity or basicity of sites on surface of material [51]. An adsorbate like pyridine or ammonia is allowed to adsorb at low temperature over a porous material, and then it is heated with a temperature program $\beta(t) = dT/dt$ in which temperature T is a linear function of time t.

TPD can be used to elucidate the mechanisms of reactions including adsorption, surface reaction and desorption. The total number of acidic sites was determined using pyridine as the adsorbate. The reduction in mass due to desorption of pyridine from the acidic sites was determined as a function of total surface acidity. Metal aluminophosphates exhibited higher surface acidity compared to aluminophosphates.

3.3.6 X-Ray Photoelectron Spectroscopy

X-ray photoelectron spectroscopy (XPS) is one of the surface science techniques used to understand the environment of metal ions, chemical composition and chemical states of the elements [52]. It is based on the photoelectric effect; Monochromatic X-rays are irradiated over a sample which will be absorbed by a core or valence electron. The electron will be emitted if the binding energy is less than the energy of the incident photon. The binding energy of the emitted electrons can be determined.

The XPS spectra of Titanium aluminophosphates confirmed the presence of three different Titanium species with their binding energies separated by about 1 eV [53]. The first species with the lowest binding energy arising at $Ti2p_{3/2}$ = 457.8458.0 eV is characteristic of octahedral titanium species. The medium peak at $Ti2p_{3/2}$ = 458.8–459.0 eV is attributed to Ti species in some intermediate coordination environment. The high binding energy peak at $Ti2p_{3/2}$ = 460.0–460.2 eV belongs to isolated Ti^{4+} in tetrahedral position.

The binding energy of Mg 1s in magnesium aluminophosphate was found to be 1305 eV, which is slightly higher than that of MgO [54]. Mg incorporated in aluminophosphates possesses a greater tendency to draw electrons in comparison with MgO. The Al 2p, P 2p and O 1s signals for the metal aluminophosphate, shifted toward higher binding energy than that of aluminophosphates. The same is observed in case of zinc aluminophosphates [55].

3.3.7 Raman Spectra

Raman spectroscopy is a scattering technique used to determine and provide valuable information on the structure, chemical composition and surface functional groups for crystalline and disordered phases [56]. In Raman spectra, the interaction of a laser beam with the sample molecules generates a scattered light, which is produced because of the inelastic collision between incident radiation and vibrating molecules. The incident light, which undergoes inelastic scattering that has a different frequency is used to generate the Raman spectrum. The scattering constitutes Rayleigh scattering if the frequency of incident radiation has a frequency equal to most of the scattered radiations. Stokes lines will appear in the Raman spectrum, if the frequency of incident radiation is higher than frequency of scattered radiation and an anti-Stokes line will appear if the frequency of incident radiation is lower than frequency of scattered radiation.

The FT-Raman spectra of the aluminophosphates shows high frequency bands between 2980 and 2800 cm^2, which is attributed to the organic templates used for the preparation of aluminophosphates. The bands exhibited at low-frequency regions, below 1400 cm^2 is used to differentiate between different aluminophosphate structures. The stretching bands at 1100, 1000, 500, and 250 cm^2 confirm the pore size and ring size of aluminophosphates.

3.4 CORRELATION OF SURFACE PROPERTIES ON CATALYTIC ACTIVITY OF ALUMINOPHOSPHATES

Catalytic materials are associated with one or more different kinds of active sites which, to be effective in organic, are acidic, basic and redox sites. In recent years, the aluminophosphates and metal aluminophosphates gained significant attention due to their promising nature as catalysts and adsorbents. The flexible surface properties, texture, chemical composition and thermal stability play a major role in determining the catalytic activity of aluminophosphates. Metal aluminophosphates were found to be selective in organic transformations with much higher rates and selectivity than conventional solid acids. Amorphous aluminophosphates and metal aluminophosphates were used as catalysts in alkylation [57], transesterification [58], dehydrogenation [59], aldol condensation [60], alkylation [61], isomerization [62], cracking [63], and oxidation [64] of several organic substrates reactions. The high catalytic activity of aluminophosphates [44] and metal aluminophosphates was attributed to the synergistic effect of strong basic and weak acid and base sites with suitable strengths. It is observed that P/Al molar ratio and transition metal loading over aluminophosphates determines the acidic-basic property of the material. Amorphous aluminophosphates incorporated with II group elements, III group elements and transition metals were used as efficient catalysts in industrially important organic transformations [65].

The nature and number of surface-active sites of catalysts depends upon the precursors, synthesis methods and porous nature of the material. Aluminum ions are considered as Lewis acid or Bronsted basic sites, and act as electron pair acceptors. The hydroxyl ions present in the alumina act as electron donors, which is otherwise called Lewis base. Two neighboring hydroxyl ions present on the surface of alumina undergoes dehydration and form a strained oxygen bridge that act as Lewis acid sites [66].

The optimization of preparation methods can influence the nucleation rate and crystal growth that decide the shape and size of the pores. Stirring plays a vital role in the size of the crystal. Large crystals can be obtained with the help of a nucleation suppressor. However, the morphology of the materials is generally modified by insertion, incorporation or addition of metals or transition metals to frame work by altering the pH.

Pore structure has a significant role in the catalytic efficiency of a catalyst [67]. The pore dimensions determine the shape selective properties of the catalyst. The required pore size, pore volume and pore shape of catalytic material can reduce the problems related to shape selectivity and molecular traffic control selectivity. However, synthesis conditions have a great influence on the nature of the porous structure of catalytic material mainly in mesoporous material. The molar ratio of phosphorus to aluminum is a key parameter for tuning the acidic or basic property, which largely influences the catalytic performance of the aluminophosphate materials [68][69]. However, it is impossible to synthesize porous aluminophosphate with P/Al ratio close to or above 1:1.

Transesterification is one of the organic reactions used by large numbers of organic chemical industries for preparation of esters [70] [32]. Transesterification is a process of conversion of one ester to another ester by interchanging an alkoxy moiety. The transesterification reaction between diethyl malonate and benzyl alcohol is shown in Scheme 3.1. Biodiesel production by conversion of fats available in oils is another important application of transesterification. Numerous acid catalysts including traditional homogeneous catalysts were employed to increase the yield of transester products, but the speed for reaction was very slow, which led to economic limitations. The transesterification reaction between diethyl malonate and benzyl alcohol is shown in Scheme 1.

SCHEME 3.1: TRANSESTERIFICATION OF DIETHYL MALONATE WITH BENZAL ALCOHOL

Alumina-based heterogeneous catalysts [71] were found to be efficient in terms of yield and selectivity of products for transesterification reactions. Amorphous aluminophosphates and metal aluminophosphates have been found to be catalytically active for the liquid phase transesterification reaction. The yield and selectivity of the products are greatly influenced by the structural properties of alcohols and the textural properties of catalysts. The highest activity of metal aluminophosphate is due to its mesoporous nature with uniform pore size distribution, larger surface area and higher surface acidity.

Amorphous mesoporous iron aluminophosphates [72] were reported as efficient catalysts in the transesterification reaction of esters with primary, secondary and aryl alcohols. The high specific surface area, narrow and broad pore size distribution are attributed to catalytic activity of Amorphous mesoporous iron aluminophosphates in the transesterification reaction. Mesoporous iron aluminophosphates materials with pore size range of 9–14 nm prepared by the simple coprecipitation method with an iron loading of 0.012–0.1 mol-generated high yields of dibenzyl malonate

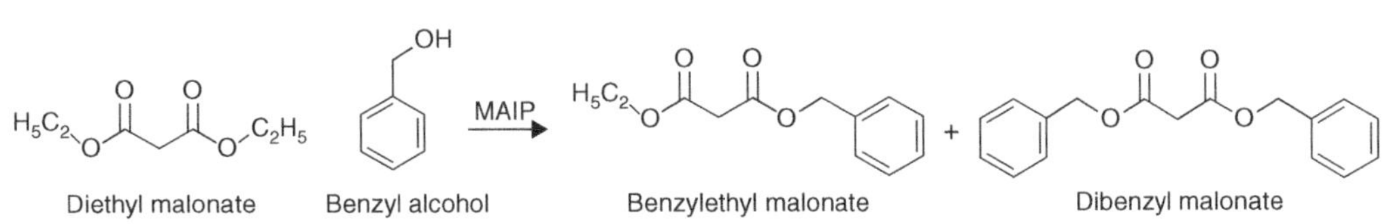

SCHEME 3.1 Transesterification of Diethyl Malonate with Benzyl Alcohol.

with high selectivity. The addition of precipitating agent and metal loading plays a significant role in the formation of pores on the surface of the material. Amorphous iron aluminophosphates with high specific area mesoporous pores can be synthesized by coprecipitation and urea hydrolysis [73].

The molar ratio of Al: Fe: P plays a major role in the surface and textural properties of the material. The presence of iron reduces the particle size and crystallinity of pure aluminophosphate. The catalytic activity of the iron aluminophosphates depends upon the amount of iron in the catalyst. The formation of hydrated alumina and polycondensed phosphates hinders the catalytic activity of the material [70]. The amount of iron acts as a control for designing the surface properties and catalytic properties of amorphous iron aluminophosphates.

In the transesterification reaction between diethyl malonate and benzyl alcohol, iron aluminophosphates exhibited better catalytic activity with more yield and higher selectivity. The formation of monosubstituted ester was due to the surface acidity of the catalysts. The excess alcohol undergoes reaction with monobenzyl substituted esters to form dibenzyl esters. The mesoporous nature of iron aluminophosphates with uniform pore size distribution was responsible for the formation of dibenzyl ester without transition state or product shape selectivity restrictions. For the formation of dibenzyl substituted esters, surface acid sites of intermediate strength was enough compared to formation of monobenzyl ester. The inexpensive nature, ease of preparation, and reusability of the catalyst are the advantages of amorphous mesoporous iron aluminophosphates.

Amide moiety is observed in different naturally occurring compounds as well as in synthesis of many pharmaceutically important compounds [74]. Synthesis of amide from its derivatives catalysed by different systems has been reported for several years. The application of amides in organic, biological, and materials chemistry mandates the design of more ecofriendly protocols and efficient catalysts for amide synthesis. The main drawbacks of reported methods were usage of non-ecofriendly chemicals and catalyst, disposal of acids or bases generated as byproducts or waste, and energy wastage due to the duration of reaction time.

Synthesis of amide by direct transformation of ester to amide, Scheme 3.2 is rarely reported in literature. Solvent-free reactions over solid acid catalysts are highly significant due to high selectivity, conversion and flexible reaction conditions.

SCHEME 3.2: GENERAL CHEMICAL EQUATION OF THE ESTER-AMIDE EXCHANGE REACTION

Metal aluminophosphates exhibited high catalytic efficiency in synthesis of amides. Iron and copper aluminophosphates exhibited 100 percent selectivity for benzanilide, the former being more active than the later [73]. Cobalt, nickel and vanadium aluminophosphates generated a small percentage of benzyl alcohol as well as toluene along with amides due to interchange and reduction reaction of methyl benzoate. The presence of acid sites on the surface of metal aluminophosphates improves the yield of amides. Lewis acid sites facilitate the generation of ester carbocation. The amine and ester attach to each other by bonding lone pairs of electrons and carbocation. The reaction intermediate becomes detached from the reaction by the release of an alcohol. The mesoporous surface

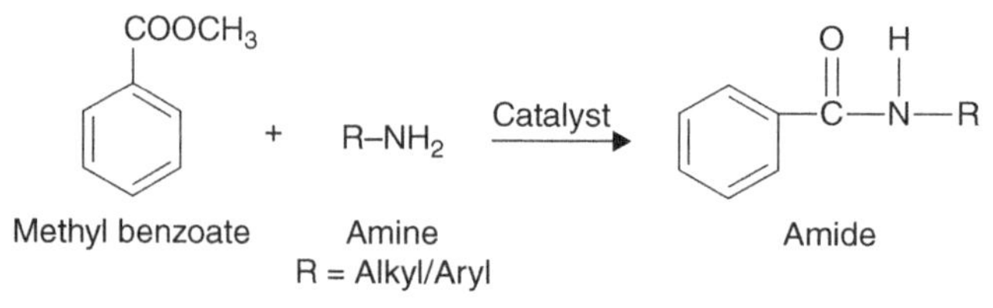

SCHEME 3.2 General Chemical Equation of the Ester-Amide Exchange Reaction.

and surface sites with intermediate acid strength were among the major factors that improved the catalytic activity of iron aluminophosphates in amidation reaction.

The catalyst was applicable for a variety of amines with high-yield desired products. It was observed that the yield of amides decreased with the increase in the number of carbon atoms and number of branches of alkyl amines. In case of aryl amines, the position of substituents in the ring and the nature of substituents, electron releasing or electron withdrawing property play a significant role in the amidation reaction. The electronic deficient groups present on the ring also improve the yield of the reaction.

The presence of nitro groups in ortho position and para position of amines exhibited lower yields due to low nucleophilicity and steric hindrance. However, the electron-withdrawing substituents in the meta position exhibited a high yield of corresponding amides when compared to substituents attached to ortho and para positions. Electronic transitions in aromatic amines also play a significant role in amidation reactions.

The formation and removal of corresponding azeotropic alcohol was not observed in presence of metal aluminophosphates as catalysts. The formation of other side reactions and its products was not observed over iron aluminophosphate catalysts. The insoluble nature of metal aluminophosphates in organic solvents supports the easy removal of catalyst from the reaction mixture. In some cases, the speed of reaction is low due to the heterogeneous character of the reaction.

Aluminophosphate and metal aluminophosphates gained prominence in the catalysis and adsorption processes due to their environmental and economic benefits. A heterogeneous catalyst like aluminophosphates has advantages of minimal catalyst loading, resistance to moisture and its reusability [73] when compared to conventional Lewis acids or bases as catalysts. Aluminophosphates gained significance as potential catalysts due to its correlation of catalytic activity with surface properties like surface acidity or basicity, strength of acidic sites, basic sites and redox sites, specific surface area and pore structure. Solid Lewis acid catalysts are highly oxophilic and capable of forming bonds with oxygen donor ligands. Aluminophosphates possess unique properties, such as tunable surface acidity and the possibility to modify surface properties by altering the chemical environment by loading metal into the framework of aluminophosphates. Amorphous metal aluminophosphates are easy to prepare, are nontoxic, inexpensive, and environmentally acceptable. Hence, they are employed as catalysts or adsorbents for the synthesis of complex molecules and highly efficient from the viewpoint of green chemistry.

REFERENCES

1. H. Zheng, F. Gao, and V. Valtchev, "Nanosized inorganic porous materials: Fabrication, modification and application," *J. Mater. Chem. A*, vol. 4, no. 43, pp. 16756–16770, 2016, doi: 10.1039/c6ta04684c.
2. R. M. Belekar and S. J. Dhoble, "Activated Alumina Granules with nanoscale porosity for water defluoridation," *Nano-Structures and Nano-Objects*, vol. 16, pp. 322–328, 2018, doi: 10.1016/j.nanoso.2018.09.007.
3. Y. Li, L. Li, and J. Yu, "Applications of Zeolites in Sustainable Chemistry," *Chem*, vol. 3, no. 6, pp. 928–949, 2017, doi: 10.1016/j.chempr.2017.10.009.
4. C. Li, S. Wu, G. Yu, X. Yang, G. Liu, and W. Zhang, "Removal of low-concentration ammonia from ambient air by aluminophosphates," *Chem. Res. Chinese Univ.*, vol. 34, no. 3, pp. 480–484, 2018, doi: 10.1007/s40242-018-7281-4.
5. M. Wen, G. Li, H. Liu, J. Chen, T. An, and H. Yamashita, "Metal-organic framework-based nanomaterials for adsorption and photocatalytic degradation of gaseous pollutants: Recent progress and challenges," *Environ. Sci. Nano*, vol. 6, no. 4, pp. 1006–1025, 2019, doi: 10.1039/c8en01167b.
6. T. P. M. Chu et al., "Synthesis, characterization, and modification of alumina nanoparticles for cationic dye removal," *Materials* (Basel), vol. 12, no. 3, pp. 1–15, 2019, doi: 10.3390/ma12030450.
7. H. R. Shaterian, K. Azizi, and N. Fahimi, "Phosphoric acid supported on alumina: A useful and effective heterogeneous catalyst in the preparation of α-amidoalkyl-β-naphthols, α-carbamato-alkyl-β–naphthols, and 2-arylbenzothiazoles," *Arab. J. Chem.*, vol. 10, pp. S42–S55, 2017, doi: 10.1016/j.arabjc.2012.07.006.

8. S. Lamouri et al., "Control of the γ-alumina to α-alumina phase transformation for an optimized alumina densification," *Bol. la Soc. Esp. Ceram. y Vidr.*, vol. 56, no. 2, pp. 47–54, 2017, doi: 10.1016/j.bsecv.2016.10.001.
9. Y. Kobayashi, T. Ishizaka, and Y. Kurokawa, "Preparation of alumina films by the sol-gel method," *J. Mater. Sci.*, vol. 40, no. 2, pp. 263–283, 2005, doi: 10.1007/s10853-005-6080-8.
10. J. Kim et al., "Bead-shaped mesoporous alumina adsorbents for adsorption of ammonia," *Materials (Basel)*, vol. 13, no. 6, 2020, doi: 10.3390/ma13061375.
11. H. K. Shahzad, M. A. Hussein, F. Patel, N. Al-Aqeeli, M. A. Atieh, and T. Laoui, "Synthesis and characterization of alumina-CNT membrane for cadmium removal from aqueous solution," *Ceram. Int.*, vol. 44, no. 14, pp. 17189–17198, 2018, doi: 10.1016/j.ceramint.2018.06.175.
12. A. V. Nartova, A. V. Bukhtiyarov, R. I. Kvon, and V. I. Bukhtiyarov, "The model thin film alumina catalyst support suitable for catalysis-oriented surface science studies," *Appl. Surf. Sci.*, vol. 349, pp. 310–318, 2015, doi: 10.1016/j.apsusc.2015.04.177.
13. T. C. Prathna, D. N. Sitompul, S. K. Sharma, and M. Kennedy, "Synthesis, characterization and performance of iron oxide/alumina-based nanoadsorbents for simultaneous arsenic and fluoride removal," *Desalin. Water Treat.*, vol. 104, no. April, pp. 121–134, 2018, doi: 10.5004/dwt.2018.21960.
14. B. Barik, P. S. Nayak, L. S. K. Achary, A. Kumar, and P. Dash, "Synthesis of alumina-based cross-linked chitosan-HPMC biocomposite film: An efficient and user-friendly adsorbent for multipurpose water purification," *New J. Chem.*, vol. 44, no. 2, pp. 322–337, 2019, doi: 10.1039/c9nj03945g.
15. M. Bouzbib, A. Pogonyi, T. Kolonits, Á. Vida, Z. Dankházi, and K. Sinkó, "Sol–gel alumina coating on quartz substrate for environmental protection," *J. Sol-Gel Sci. Technol.*, vol. 93, no. 2, pp. 262–272, 2020, doi: 10.1007/s10971-019-05193-y.
16. M. Nofz, *Alumina thin films*. 2018.
17. L. Juhász and J. Mizsei, "A simple humidity sensor with thin film porous alumina and integrated heating," *Procedia Eng.*, vol. 5, pp. 701–704, 2010, doi: 10.1016/j.proeng.2010.09.206.
18. A. Dandapat, D. Jana, and G. De, "Pd nanoparticles supported mesoporous γ-Al2O3 film as a reusable catalyst for reduction of toxic CrVI to Cr III in aqueous solution," *Appl. Catal. A Gen.*, vol. 396, no. 1–2, pp. 34–39, 2011, doi: 10.1016/j.apcata.2011.01.032.
19. Dandapat, D. Jana, and G. De, "Synthesis of thick mesoporous γ-alumina films, loading of Pt nanoparticles, and use of the composite film as a reusable catalyst," *ACS Appl. Mater. Interfaces*, vol. 1, no. 4, pp. 833–840, 2009, doi: 10.1021/am800241x.
20. V. I. Mikhaylov, E. F. Krivoshapkina, A. L. Trigub, V. V. Stalugin, and P. V. Krivoshapkin, "Detection and Adsorption of Cr(VI) ions by Mesoporous Fe-Alumina Films," *ACS Sustain. Chem. Eng.*, vol. 6, no. 7, pp. 9283–9292, 2018, doi: 10.1021/acssuschemeng.8b01598.
21. M. Moshoeshoe, M. Silas Nadiye-Tabbiruka, and V. Obuseng, "A Review of the Chemistry, Structure, Properties and Applications of Zeolites," *Am. J. Mater. Sci.*, vol. 2017, no. 5, pp. 196–221, 2017, doi: 10.5923/j.materials.20170705.12.
22. N. Kim, H. Park, N. Yoon, and J. K. Lee, "Zeolite-Templated Mesoporous Silicon Particles for Advanced Lithium-Ion Battery Anodes," 2018, doi: 10.1021/acsnano.8b01129.
23. J. Xu, X. Xiao, S. Zeng, M. Cai, and M. W. Verbrugge, "Multifunctional Lithium-Ion-Exchanged Zeolite-Coated Separator for Lithium-Ion Batteries," *ACS Appl. Energy Mater.*, vol. 1, pp. 7237–7243, 2018, doi: 10.1021/acsaem.8b01716.
24. O. Paper, "Catalytic Applications of Zeolites in Chemical Industry," pp. 888–895, 2009, doi: 10.1007/s11244-009-9226-0.
25. Megías-sayago, R. Bingre, L. Huang, G. Lutzweiler, and P. Granger, "CO 2 Adsorption Capacities in Zeolites and Layered Double Hydroxide Materials," vol. 7, no. August, pp. 1–10, 2019, doi: 10.3389/fchem.2019.00551.
26. M. Cazorla and M. W. Grutzeck, "Indoor air pollution control: Formaldehyde adsorption by zeolite rich materials," no. March 2018, 2020.
27. N. Elboughdiri, "The use of natural zeolite to remove heavy metals Cu (II), Pb (II) and Cd (II), from industrial wastewater The use of natural zeolite to remove heavy metals Cu (II), Pb (II) and Cd (II), from industrial wastewater," no. Ii, 2020, doi: 10.1080/23311916.2020.1782623.
28. M. Tiemann and M. Fröba, "Mesoporous aluminophosphates from a single-source precursor," pp. 406–407, 2002.

29. M. García et al., "Synthesis and Characterization of Aluminophosphates Type - 5 and 36 Doubly Modified with Si and Zn and Its Catalytic Application in the Reaction of Methanol to Hydrocarbons (MTH)," *Top. Catal.*, no. 0123456789, 2020, doi: 10.1007/s11244-020-01266-3.
30. Q. Liu, C. Ocean, A. E. Garcia-Bennett, and N. Hedin, "Aluminophosphates for CO 2 Separation," pp. 91–97, 2011, doi: 10.1002/cssc.201000256.
31. Z. Xiaofeng, L. I. N. Shen, C. Xinqing, C. Jiebo, Y. Liuyi, and L. U. O. Minghong, "Preparation of mesoporous aluminophosphate using poly (amido amine) as template," vol. 2, no. 4, pp. 419–421, 2007, doi: 10.1007/s11458-007-0079-4.
32. A. V. Vijayasankar, N. Mahadevaiah, Y. S. Bhat, and N. Nagaraju, "Mesoporous aluminophosphate materials: influence of method of preparation and iron loading on textural properties and catalytic activity," pp. 369–378, 2011, doi: 10.1007/s10934-010-9387-z.
33. V. Vijayasankar, S. Deepa, B. R. Venugopal, and N. Nagaraju, "Amorphous Mesoporous Iron Aluminophosphate Catalyst for the Synthesis of 1, 5-Benzodiazepines," *Chinese J. Catal.*, vol. 31, no. 11–12, pp. 1321–1327, 2010, doi: 10.1016/S1872-2067(10)60120-9.
34. N. Panigrahi, R. Chaini, A. Nayak, and P. Chandra Mishra, "Impact of milling time and method on particle size and surface morphology during nano particle synthesis from α-Al2O3," *Mater. Today Proc.*, vol. 5, no. 9, pp. 20727–20735, 2018, doi: 10.1016/j.matpr.2018.06.457.
35. R. A. Ismail, S. A. Zaidan, and R. M. Kadhim, "Preparation and characterization of aluminum oxide nanoparticles by laser ablation in liquid as passivating and anti-reflection coating for silicon photodiodes," *Appl. Nanosci.*, vol. 7, no. 7, pp. 477–487, 2017, doi: 10.1007/s13204-017-0580-0.
36. A. S. Jbara, Z. Othaman, A. A. Ati, and M. A. Saeed, "Characterization of γ- Al_2O_3 nanopowders synthesized by co-precipitation method," *Mater. Chem. Phys.*, vol. 188, pp. 24–29, 2017, doi: 10.1016/j.matchemphys.2016.12.015.
37. F. Deganello and A. K. Tyagi, "Solution combustion synthesis, energy and environment: Best parameters for better materials," *Prog. Cryst. Growth Charact. Mater.*, vol. 64, no. 2, pp. 23–61, 2018, doi: 10.1016/j.pcrysgrow.2018.03.001.
38. F. Mirjalili, M. Hasmaliza, and L. C. Abdullah, "Size-controlled synthesis of nano a -alumina particles through the sol–gel method," *Ceram. Int.*, vol. 36, no. 4, pp. 1253–1257, 2010, doi: 10.1016/j.ceramint.2010.01.009.
39. N. S. Ahmedzeki and B. A. Al-tabbakh, "Synthesis and Characterization of Nanocrystalline Zeolite Y Synthesis and Characterization of Nanocrystalline Zeolite Y," March, 2016.
40. S. Said, S. Mikhail, and M. Riad, "Recent processes for the production of alumina nano-particles," *Mater. Sci. Energy Technol.*, vol. 3, pp. 344–363, 2020, doi: 10.1016/j.mset.2020.02.001.
41. S. M. Riyadh, K. D. Khalil, and A. H. Bashal, "Structural properties and catalytic activity of binary poly (Vinyl alcohol)/Al2O3 nanocomposite film for synthesis of thiazoles," *Catalysts*, vol. 10, no. 1, 2020, doi: 10.3390/catal10010100.
42. Y. Yasuda and T. Morita, "*Journal of Asian Ceramic Societies*," *Integr. Med. Res.*, vol. 3, no. 1, pp. 139–143, 2015, doi: 10.1016/j.jascer.2014.12.004.
43. S. Popovic and N. Vdovic, "Chemical and microstructural properties of Al-oxide phases obtained from AlCl 3 solutions in alkaline medium," vol. 59, pp. 12–19, 1999.
44. A. V. Vijayasankar and S. Govindaraju, "Role of incorporated transition metal on surface properties and catalytic activity of mesoporous vanadium aluminophosphates in the synthesis of tetrahydroquinolin-5-(1H)-ones," *Chem. Data Collect.*, vol. 28, p. 100419, 2020, doi: 10.1016/j.cdc.2020.100419.
45. G. Balakrishnan, R. V. Babu, K. S. Shin, and J. I. Song, "Growth of highly oriented γ- And α-Al2O3 thin films by pulsed laser deposition," *Opt. Laser Technol.*, vol. 56, pp. 317–321, 2014, doi: 10.1016/j.optlastec.2013.08.014.
46. S. A. Naayi, A. I. Hassan, and E. T. Salim, "FTIR and X-ray diffraction analysis of Al2O3 nanostructured thin film prepared at low temperature using spray pyrolysis method," *Int. J. Nanoelectron. Mater.*, vol. 11, no. Special Issue BOND21, pp. 1–6, 2018.
47. K. Djebaili, Z. Mekhalif, A. Boumaza, and A. Djelloul, "Xps, Ftir,Edx,Xrd Analysis of Al2O3 Scales Grown on Pm2000 Alloy," *J. Spectrosc.*, vol. 2015, pp. 1–16, 2013.
48. S. Brunauer, P. H. Emmett and E. Teller, vol. 60, no. 35, 1977.
49. P. Sinha, A. Datar, C. Jeong, X. Deng, Y. G. Chung, and L. C. Lin, "Surface Area Determination of Porous Materials Using the Brunauer-Emmett-Teller (BET) Method: Limitations and Improvements," *J. Phys. Chem. C*, vol. 123, no. 33, pp. 20195–20209, 2019, doi: 10.1021/acs.jpcc.9b02116.

50. P. K. Kiyohara, H. Souza Santos, A. C. Vieira Coelho, and P. De Souza Santos, "Structure, surface area and morphology of aluminas from thermal decomposition of Al(OH)(CH3COO)2 crystals," *An. Acad. Bras. Cienc.*, vol. 72, no. 4, pp. 470–495, 2000, doi: 10.1590/s0001-37652000000400003.
51. K. T. Wang, S. Nachimuthu, and J. C. Jiang, "Temperature-programmed desorption studies of NH_3 and H_2O on the $RuO_2(110)$ surface: Effects of adsorbate diffusion," *Phys. Chem. Chem. Phys.*, vol. 20, no. 37, pp. 24201–24209, 2018, doi: 10.1039/c8cp02568a.
52. H. Bluhm, "X-ray photoelectron spectroscopy (XPS) for in situ characterization of thin film growth," *Situ Charact. Thin Film Growth*, pp. 75–98, 2011, doi: 10.1533/9780857094957.2.75.
53. R. Aiello et al., "XANES and XPS studies of titanium aluminophosphate molecular sieves Chemical Composition a," pp. 125–133, 2002.
54. H. Wang, Y. Wang, W. Liu, H. Cai, J. Lv, and J. Liu, "Amorphous magnesium substituted mesoporous aluminophosphate: An acid-base sites synergistic catalysis for transesterification of diethyl carbonate and dimethyl carbonate in fixed-bed reactor," *Microporous Mesoporous Mater.*, vol. 292, no. September 2019, p. 109757, 2020, doi: 10.1016/j.micromeso.2019.109757.
55. L. Sun, J. R. Deng, and Z. S. Chao, "Catalysis over zinc-incorporated berlinite (ZnAlPO4) of the methoxycarbonylation of 1,6-hexanediamine with dimethyl carbonate to form dimethylhexane-1,6-dicarbamate," *Chem. Cent. J.*, vol. 1, no. 1, pp. 1–9, 2007, doi: 10.1186/1752-153X-1-27.
56. S. Mourdikoudis, R. M. Pallares, and N. T. K. Thanh, "Characterization techniques for nanoparticles: Comparison and complementarity upon studying nanoparticle properties," *Nanoscale*, vol. 10, no. 27, pp. 12871–12934, 2018, doi: 10.1039/c8nr02278j.
57. H. Hentit, K. Bachari, M. S. Ouali, M. Womes, B. Benaichouba, and J. C. Jumas, "Alkylation of benzene and other aromatics by benzyl chloride over iron-containing aluminophosphate molecular sieves," *J. Mol. Catal. A Chem.*, vol. 275, no. 1–2, pp. 158–166, 2007, doi: 10.1016/j.molcata.2007.05.032.
58. J. Shi et al., "Amorphous mesoporous aluminophosphate as highly efficient heterogeneous catalysts for transesterification of diethyl carbonate with dimethyl carbonate," *Catal. Commun.*, vol. 12, no. 8, pp. 721–725, 2011, doi: 10.1016/j.catcom.2011.01.002.
59. Q. Gao, Y. Huang, and R. Xu, "Selective oxidative dehydrogenation of isobutane over unidimensional aluminophosphate molecular sieves (AlPO4-5, AlPO4-41, AlPO4-25)," *Chem. Commun.*, no. 16, pp. 1905–1906, 1996, doi: 10.1039/cc9960001905.
60. A. Hamza and N. Nagaraju, "Amorphous metal-aluminophosphate catalysts for aldol condensation of n-heptanal and benzaldehyde to jasminaldehyde," *Cuihua Xuebao/Chinese J. Catal.*, vol. 36, no. 2, pp. 209–215, 2015, doi: 10.1016/S1872-2067(14)60206-0.
61. P. Sreenivasulu, N. Viswanadham, T. Sharma, and B. Sreedhar, "Synthesis of orderly nanoporous aluminophosphate and zirconium phosphate materials and their catalytic applications," *Chem. Commun.*, vol. 50, no. 47, pp. 6232–6235, 2014, doi: 10.1039/c4cc02614d.
62. L. H. Gielgens, I. H. E. Veenstra, V. Ponec, M. J. Haanepen, and J. H. C. van Hooff, "Selective isomerisation of n-butene by crystalline aluminophosphates," *Catal. Letters*, vol. 32, no. 1–2, pp. 195–203, 1995, doi: 10.1007/BF00806114.
63. L. Feng, X. Qi, Z. Li, Y. Zhu, and X. Li, "Synthesis and characterization of magnesium-substituted aluminophosphate molecular sieves (MgAPO-11) and their kinetic study of catalytic cracking of n-hexane," *Cuihua Xuebao / Chinese J. Catal.*, vol. 30, no. 4, pp. 340–346, 2009, doi: 10.1016/s1872-2067(08)60101-1.
64. P. Selvam and A. Sakthivel, "Selective Catalytic Oxidation over Ordered Nanoporous Metallo-Aluminophosphates," *Liq. Phase Oxid. via Heterog. Catal. Org. Synth. Ind. Appl.*, pp. 95–125, 2013, doi: 10.1002/9781118356760.ch3.
65. G. Liu et al., "Thermally stable amorphous mesoporous aluminophosphates with controllable P/Al ratio: Synthesis, characterization, and catalytic performance for selective O-methylation of catechol," *J. Phys. Chem. B*, vol. 110, no. 34, pp. 16953–16960, 2006, doi: 10.1021/jp062824u.
66. X. Jin, P. Man, and J. Blanchard, "Preparation of amorphous silica-aluminas with enhanced acidic properties and spectroscopic identification of their acid sites," 2019.
67. J. C. Groen, L. A. A. Peffer, and J. Pérez-Ramírez, "Pore size determination in modified micro- and mesoporous materials. Pitfalls and limitations in gas adsorption data analysis," *Microporous Mesoporous Mater.*, vol. 60, no. 1–3, pp. 1–17, 2003, doi: 10.1016/S1387-1811(03)00339-1.

68. M. Hartmann and L. Kevan, "Transition-Metal Ions in Aluminophosphate and Silicoaluminophosphate Molecular Sieves: Location, Interaction with Adsorbates and Catalytic Properties," *Chem. Rev.*, vol. 99, no. 3, pp. 635–663, 1999, doi: 10.1021/cr9600971.
69. S. T. Wilson, "S. T. Wilson UOP, Research and Development, Tarrytown Technical Center, Tarrytown, NY," *Stud. Surf. Sci. Catal.*, vol. 58, pp. 137–151, 1991.
70. A. V. Vijayasankar, C. U. Aniz, and N. Nagaraju, "Surface properties and the catalytic activity of amorphous iron aluminophosphates: Effect of fe loading," *J. Korean Chem. Soc.*, vol. 54, no. 1, pp. 131–136, 2010, doi: 10.5012/jkcs.2010.54.01.131.
71. V. V. A. and N. N. Hamza Annath, "Iranian Journal of Catalysis," *Iran. J. Catal.*, vol. 28, no. 2, pp. 13–21, 2021.
72. A. V Vijayasankar and N. Nagaraju, "Comptes Rendus Chimie Preparation and characterisation of amorphous mesoporous aluminophosphate and metal aluminophosphate as an efficient heterogeneous catalyst for transesterification reaction," *Comptes rendus - Chim.*, vol. 14, no. 12, pp. 1109–1116, 2011, doi: 10.1016/j.crci.2011.09.013.
73. V. Vijayasankar,N. Nagaraju, H. Kathyayini and Harikrishna Tumma, *Mesoporous Iron Aluminophosphate: An Efficient Catalyst for One Pot Synthesis of Amides by Ester-Amide Exchange Reaction.* John Wiley, 2012.
74. P. W. Seavill and J. D. Wilden, "The preparation and applications of amides using electrosynthesis," *Green Chem.*, vol. 22, no. 22, pp. 7737–7759, 2020, doi: 10.1039/d0gc02976a.

4 Carbon Supported Noble Metal Nanocatalysts for Liquid Phase Oxidation Reactions

Mabuatsela V. Maphoru

CONTENTS

4.1 Overview of Carbon Materials as Catalyst Supports 65
4.1.1 Liquid Phase Oxidation of Organic Compounds on Carbon-Supported Noble Metal Catalysts 67
4.1.2 Platinum and Palladium Nanocatalysts 68
4.1.3 Ruthenium Nanocatalysts 81
4.1.4 Gold and Silver Nanocatalysts 86

4.1 OVERVIEW OF CARBON MATERIALS AS CATALYST SUPPORTS

For many centuries, carbon remained one of the most studied elements in the world due to its broad applications in various fields of science and engineering. Its versatility owes to its orbital hybridizations, sp, sp^2 and sp^3 [1], which enable carbon to form different kinds of orientations and spatial dimensions, 3D, 2D, 1D and 0D, through networks of various chemical bonds [2, 3]. The formation of carbon allotropes, graphite and diamond [4], which are very important in the synthesis of many carbon materials, originate from these various bonds of carbon. Activated carbon [5], graphite [6], carbon black [7], graphene [8], carbon nanotubes [9], carbon nanospheres [10], carbon nanofibers [11], carbon nanowires [12], fullerenes [13], carbon nanodiamonds [14] and other nanostructured carbon materials [15] were made through the manipulation of carbon (Figure 4.1). Other than carbon-carbon bonds, the ability of carbon to bind to many other elements creates an opportunity for the formation of a wide range of industrially important chemical materials with interesting physicochemical properties. Carbon materials possess high electrical conductivity [16], thermal [17] and chemical stabilities [18], mechanical strength [19] and specific surface area [20], making them good candidates in catalysis [21, 22], sensors [19], energy production and storage [11, 23], water treatment [24], environmental sciences [25], nanomedicine [26] and many other scientific and engineering applications.

In catalysis, it very common for carbon materials to be used as catalysts in their own right [27] or as catalyst supports for metal and metal oxides nanoparticles (NPs) [28–31]. It is for these reasons that carbon materials remain one of the most studied materials in the field of catalysis. In heterogeneous catalysis, catalytically active metals are supported on a cheap support material to minimize the cost of the transition metal and also to reduce the chance of NPs agglomeration and leaching [32–34]. It is for this purpose that carbon materials are used in the tailoring of supported catalysts. Some of the key properties of carbon supports are, amongst others, their high chemical stability [18], high surface area and porosity for efficient adsorption of reactants and desorption of products and their diffusion [35, 36] high thermal stability [17], electron conductivity [16] and good mechanical strength [19].

DOI: 10.1201/9781003218708-5

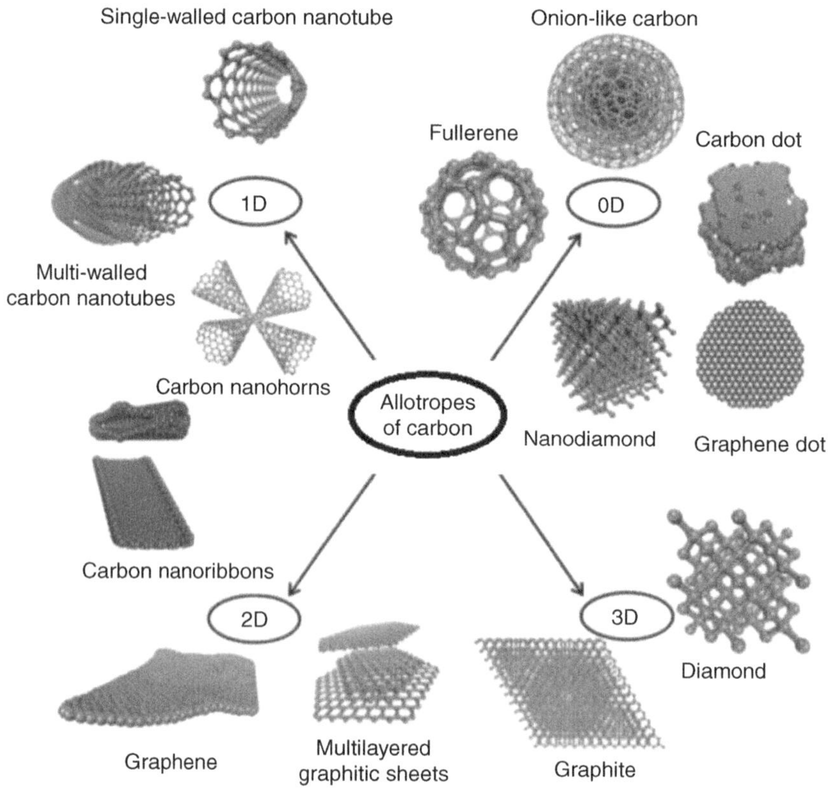

FIGURE 4.1 Carbon allotropes.

Source: Verma, 2021 [18].

It is of utmost importance that metal NPs be strongly anchored and homogeneously dispersed to yield uniform catalytic properties throughout the surface of the supported catalyst [37, 38]. However, due to the hydrophobic nature of carbon [39, 40], poor interactions between the support and the transition metal are often experienced [41]. Poor support-metal interactions are responsible for poor surface wetting in polar solvents [41] posing difficulties in loading metal NPs on carbon surfaces, leading to particle agglomeration, which reduce the catalytic activities of the supported catalyst [42, 43]. Many researchers have reported the use of chemical oxidants such as, HNO_3 [44], H_2SO_4 [45], $KMnO_4$ [46], H_2O_2 [47], ozone [48], and many others, as surface modifiers to increase the hydrophilicity of the carbon materials by modifying their surface structure to increase their interaction with hydrophilic solvents, reactants, catalysts substrates and metal NPs [42–43]. Sulfonation of carbon materials is also used for this purpose. Carbon materials were doped with heteroatoms such as phosphorus and nitrogen, which proved to play a very significant role on the catalytic properties and performance of the carbon-supported catalysts [49, 50]. The dopant replaced carbon atoms in the lattice structure of carbon material to improve the metal-support interaction and electronic properties of the catalyst [51]. In addition, carbon support materials also play a very significant role in improving activity and selectivity of the catalysts [52]. Carbon is a convenient support, since recovering metals from its surface can be achieved with ease by its combustion in air at low temperatures [53].

4.1.1 Liquid Phase Oxidation of Organic Compounds on Carbon-Supported Noble Metal Catalysts

The use of noble metal NPs – , Pt, Pd, Ru, Au, Ag and Rh – as oxidation catalysts has contributed significantly in the field of heterogeneous catalysis and green chemistry [54–59]. They are often used in combination with clean oxidants such as air, oxygen and peroxides in the catalytic oxidation reactions [60–72]. They have the ability to retain their metallic states after noticeable numbers of catalytic reactions run, therefore making them good oxidation catalysts that are recyclable [73–75]. Amongst their important applications in catalytic oxidation, are the following: catalytic converters for the oxidation of automobile exhaust [76], photocatalytic removal of organic pollutants and pathogens in water [77], as catalysts in electrocatalytic cells [78], oxidation catalysts in the production of fine chemicals and synthesis of pharmaceutical drugs [79, 80].

It is now scientifically proven that the physicochemical properties of bulk noble metals are different from those of their NPs [81]. Variation in their physical and chemical environment, structure, shape and electronic and mechanical properties brings about their differences in their catalytic performances in terms of activity and selectivity [81]. Furthermore, noble metal NPs have high surface area to volume ratios, which increases a number of catalytic sites available for the catalytic oxidation reaction, thus enhancing their catalytic activities in comparison with their bulk metal counterparts [82]. With the continuous scientific development in nanoscience, it is now possible to design and synthesize metal NPs with a size of less than 10 nm and different shapes [83].

Noble metal NPs have high surface energy, which facilitates their agglomeration and sintering [84]. Often, noble metal NPs are anchored on support materials with high surface area to increase their stability and dispersion, and also to increase their surface area [85]. This is also done to reduce the cost of noble metals required to prepare a catalyst. Carbon supports such as, activated carbon [86, 87], carbon black [88], graphene [77], carbon nanofibers [89], carbon nanotubes [90] and other nanostructured carbon materials are used for this purpose. In addition, their low combustion temperature in air allows for ease of recovery of the noble metals from the catalyst [53].

Supported noble metal NPs are not an exception to catalyst deactivation and, like any other metals, they undergo over-oxidation, agglomeration, leaching and surface poisoning by undesired byproducts and contaminants [91–94]. Thus, preparing a highly stable heterogeneous noble metal catalyst with high dispersion of the active components is an ongoing challenge. Noble metals are supported on carbon supports through chemical and physical methods, which include impregnation [95], electroless deposition [92, 96], ultra-sonication [97], microwave-assisted loading [92, 98] and many others [99]. Catalyst preparation methods have a significant influence on the final structure of the catalysts, that is, shape, size, metal dispersion, surface area and metal-support interaction [82, 92, 100, 101], which affect the performance of the catalyst [102, 103].

Reducing catalyst deactivation by using catalyst support is not always sufficient. Addition of a second inactive metal, promoter, to form a bimetallic catalyst, is often considered to improve the structure of the catalyst [31, 70, 72, 92, 104, 105]. These metals do not catalyze reactions, but instead they improve the characteristics of catalysts so as to meet the requirements for a given transformation such as selectivity, activity and stability [31, 70, 72, 92, 104–106]. Furthermore, their presence prolongs the lifetime of the catalyst by protecting them from various catalyst deactivations [92, 93, 107, 108]. In the oxidation of 2-methyl-1-naphthol, it was observed that Bi improves metal dispersion of carbon supported Ag-Bi, Pd-Bi and Pt-Bi catalysts, and these catalysts were found to be more active than unpromoted Pt/AC catalyst [31, 92]. It was also observed that Pb and Sb promoted Pd catalysts, Pd-Sb/C and Pd-Pb/C offer high catalytic performance with high resistance to CO poisoning in direct formic acid fuel cells [104, 110]. Other than metal promoters, the use of bimetallic catalysts with two active nobles is common in catalytic oxidation reactions [111, 112]. In principle, the use of bimetallic catalyst, where the second metal is either a co-catalyst or a promoter, requires a lot of understanding of their sizes, shapes and their interaction, as to whether the metals are segregated, associated or form a core-shell [92, 105, 112, 113]. The structure of the catalyst needs to be taken into consideration since it plays a very important role in the activity of the catalyst. However, a clear understanding of the catalytic activity and selectivity originating from these bimetallic catalyst systems is still lacking.

Other than the catalyst, the type of an oxidant used and its concentration, the concentration of the starting material, the solvent and temperature play a very significant role in the outcome of an oxidation reaction and also on the structural condition of the used catalyst [60, 80, 114–117]. The addition of alkali and alkaline-earth metals, such as Na, K and Mg, in various catalytic systems, also improves the outcome of the reaction [73, 80, 112, 118]. In this chapter, various catalytic oxidation reactions on carbon-supported noble metals will be discussed in detail.

4.1.2 Platinum and Palladium Nanocatalysts

Catalytic transformation of glycerol, a low-cost chemical compound from the biodiesel industry, is one of the most important reactions in synthetic organic chemistry. Fine chemical compounds, such as glyceric acid, and dihydroxyacetone with various industrial applications, are produced from the oxidation of glycerol on supported noble metal catalysts [119, 120]. Duan and co-workers [72] synthesized Pt-Sb/CNT with various Pt and Sb loadings by impregnation methods for the oxidation of glycerol (GLY) with oxygen at 60°C. Addition of catalysts promoters, antimony in this case, are known to improve the selectivities of the catalyst in the oxidation reaction [70, 92, 104, 110]. Different metal loading yielded Pt-Sb/CNT catalysts with different particle sizes. Oxidation of GLY can undergo different reaction routes, where route 1 involves the oxidation of a secondary hydroxyl group to form dihydroxyacetone (DHA) and route 2 involves the oxidation of one of the primary hydroxyl groups to form glyceraldehyde (GLY), followed by the overoxidation of an aldehyde to glyceric acid (GLA) (Scheme 4.1). Interestingly, oxidation of GLY on unpromoted Pt/CNT followed route 2, where the yield and selectivity of GLY was increasing with time due to the conversion of GLY to GLA. When Pt-Sb/CNT was used, oxidation of GLY followed route 1 with high selectivities of DHA (Figures 4.2b,c). Selectivity of DHA decreases with time due to parallel the formation of GLYA and other over-oxidation products. Furthermore, conversion of glycerol and selectivity to DHA increases with the size of NPs on Pt-Sb/CNT catalysts (Figure 4.2a). The authors concluded that the difference in the oxidation route for glycerol on promoted and unpromoted Pt/CNT catalysts is related to the geometric structures of the catalysts.

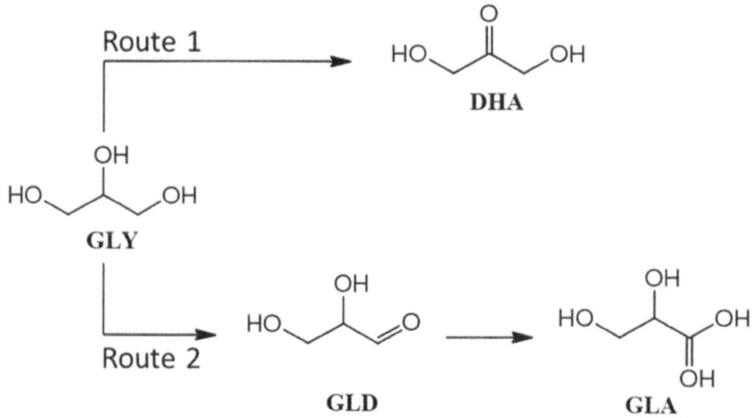

SCHEME 4.1 Oxidation of glycerol on precious metal catalysts. [Reaction conditions: GLY (15 mL, 0.3 M), Pt (2.7 wt%), catalyst (0.088 g), O_2 pressure (3 bar), stirring speed (400 rpm), 60°C, 4 h].

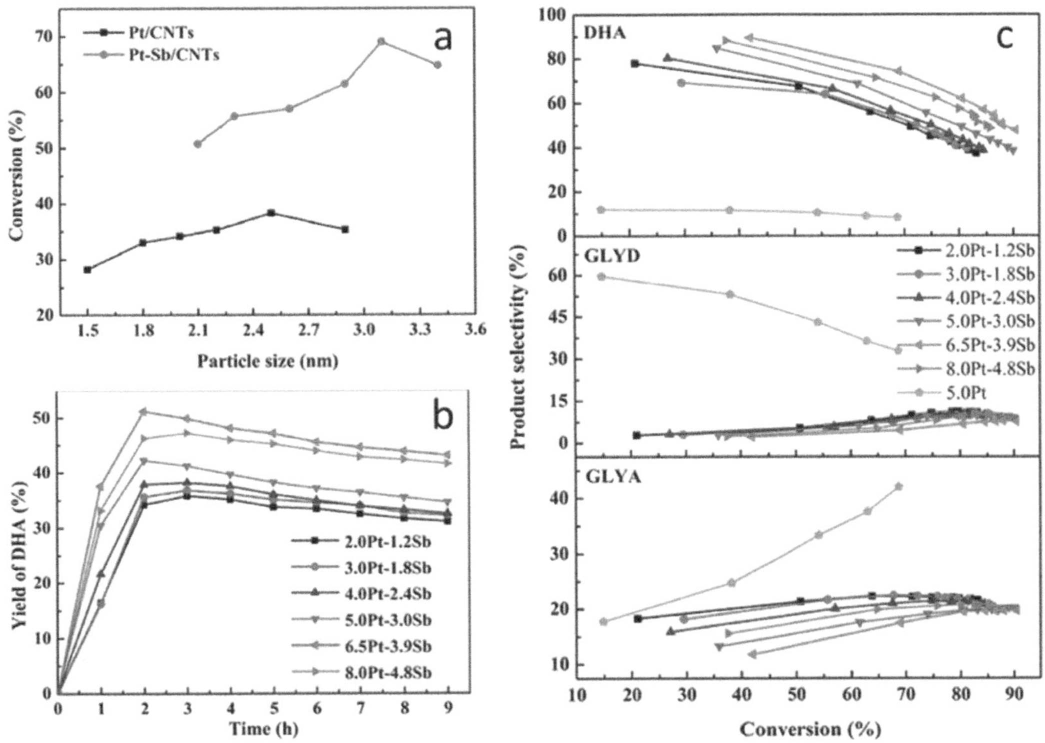

FIGURE 4.2 (a) Conversion of GLY against NPs size of Pt catalysts for 2 h; (a) DHA yield versus reaction time; (b) Selectivities of the products against the conversion of glycerol on Pt catalysts (c). [Reaction conditions: Aq. GLY solution (30 mL, 0.1 gmL^{-1}), GLY/Pt molar ratio (890), O_2 flow rate (150 mLmin^{-1}), 60°C].

Source: Duan, 2018 [72].

To confirm the influence of the geometric structures of the catalyst on their activities, Duan and co-workers [72] carried out density-functional theory (DFT), studied by creating Pt(111) surfaces and SbO-Pt(111) to understand the promoting effect of SbO_x (Figure 4.3). SbO and metallic Pt were chosen since XPS analysis indicated that Pt is in its metallic state while Sb is in its oxide form [72]. It was observed that monometallic Pt(111) adsorbs the oxygen atom of the primary alcohol on its surface while SbO-Pt(111) has also the secondary hydroxyl group being adsorbed on the SbO particles. The adsorption energy of glycerol on SbO-Pt(111), 0.52 eV, is higher than the adsorption energy of 0.40 eV obtained from a clean Pt(111) surface. In addition, the bond length of an O-H group of the secondary carbon on SbO-Pt(111) surfaces (0.991 Å) is longer than for the O-H obtained on Pt(111) (0.982 Å) and on free glycerol (0.978 Å). This served as a confirmation that the activation of a secondary hydroxyl group is preferred on SbO-Pt(111), which support the observation that the authors made on this reaction. DHA is used as a building blocks for degradable polymers and as a tanning agent in the cosmetics industry [121].

Tang and co-workers [122] also oxidized GLY to GLD and GLA on cobalt and nitrogen doped Pt encapsulated-porous carbon using oxygen as an oxidant. The incorporation of an additional metal to Pt is often used to protect Pt from over-oxidation, reduce the surface for formation of side products, minimize poisoning and increase Pt activity [31, 70, 72, 92, 104, 105]. The catalyst, Pt@Co-NC, was synthesized by pyrolysis of Pt NPs encapsulated in ZIF-67 composites (Figure 4.4a). Pt was not observed on XRD due to its low particle sizes while face centered cubic Co^0 was observed together with a diffraction peak at 2θ of 26° for carbon [122]. The presence of Co^0 (4f: 70.9 and 74.23 eV)

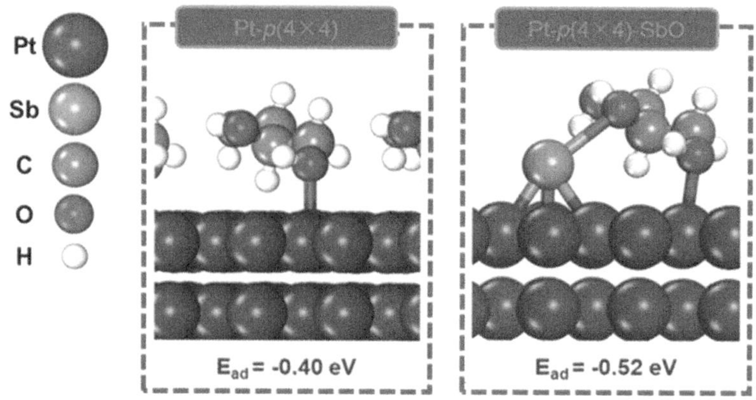

FIGURE 4.3 DFT studies for the lowest energy configurations of GLY on Pt(111) and SbO-Pt(111) surfaces.

Source: Duan, 2018 [18].

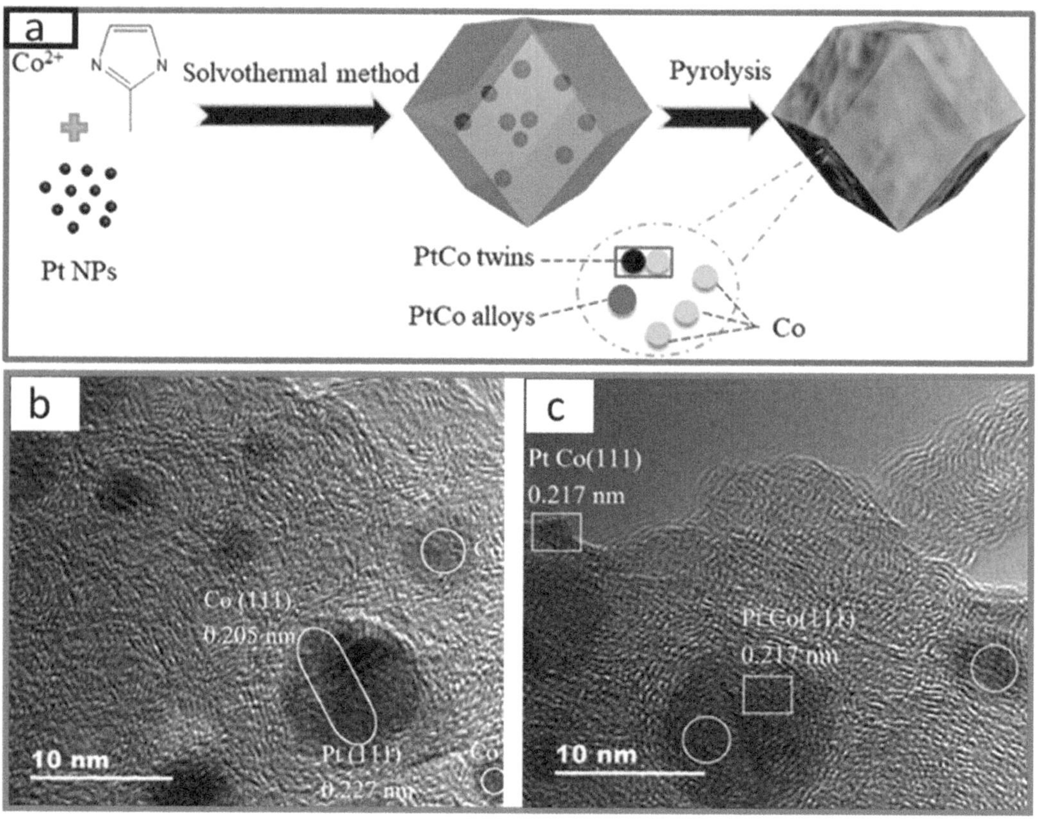

FIGURE 4.4 Synthetic route (a) and TEM images of Pt@Co-NC. [Rectangle: PtCo alloy, curved rectangle: PtCo twins, circle: Co elemental].

Source: Tang, 2020 [122].

and Pt^0 (4f: 70.9 and 74.23 eV) on the catalyst surface were also confirmed by the XPS data of the catalyst.

Lattice fringes on transmission electron microscopy (TEM) images showed that there were Co NPS, PtCo alloys and twinned PtCo NPs on the carbon surface of Pt@Co-NC (Figure 4.4b and c). The formation of PtCo alloys and twinned PtCo indicate that there is a good interaction between Pt and Co. The conversion of 61 percent GLY with selectivities of 54 percent and 37 percent for GLY and GLA were obtained, respectively. The byproducts were also present with the combined selectivity of 9 percent. The re-used catalyst maintained glycerol conversion of 61 percent and its structural stability. The preservation of its structure is related to close interaction of Pt and Co, which might not permit the movement of Pt NPs, thus avoiding metal agglomeration and leaching. The presence of mesopores on the catalyst is expected to confine Pt metals, therefore minimizing the possibility of particle agglomeration. Other Pt@Co-NC catalysts prepared by different methods, polyol reduction, galvanic replacement, and sol gel, gave conversions of 34 percent, 40 percent and 46 percent of glycerol, respectively. There are combined factors that owe to these differences and one of them is the structural differences that emerged due to different preparation methods. Pt@Co-NC prepared by Tang and co-workers [122] has Pt NPS encapsulated prior the preparation of Co-NC support while, for other methods, Pt NPs are deposited on the synthesized Co-CN supports. In addition, Tang and co-workers used a pyrolysis temperature of 800°C during the preparation of Pt@Co-NC, which produced PtCo alloys and twinned PtCo NPs and reduced thickness of organic framework to increase the catalytic active sites. Such does not apply to Pt@Co-NC catalysts prepared by polyol reduction, galvanic replacement, and sol gel methods, due to the absence of the pyrolysis step.

Meng and co-workers [123] compared the effect of iron oxides, Fe_3O_4 and FeO, and their loadings on oxidation of GLY with oxygen over Pt-Fe_3O_4/rGO and Pt-FeO/rGO catalysts. In all catalysts, Pt loading was kept at ~4.5 wt% while the amounts of Fe_3O_4 were varied to make a Fe/Pt ratio of 3.3, 8.5 and 17. TEM and EDX mapping images showed that Pt and iron oxides interact differently on the catalysts (Figure 4.5). Pt-Fe_3O_4/rGO-3.5 has individual Pt and Fe_3O_4 particles completely segregated while Pt-FeO/rGO has associated Pt and Fe_3O_4 NPs. The catalyst, Pt Fe_3O_4/rGO-13 has Pt NPs that lies next to Fe_3O_4, while Pt-Fe_3O_4/rGO-8.5 has Pt and Fe_3O_4 that are partially superimposed on each other.

Meng and co-workers [123] observed that Pt-Fe_3O_4/rGO-8.5 was the most active catalysts with TOF of 2677 h^{-1} and a selectivity of 50.2 percent GLA with GLY conversion of 29.5 percent at 60°C after 2 h. Catalysts without Fe_3O_4, Pt/rGO possess low catalytic activities, which proves that

TABLE 4.1
Oxidation of GLY on Pt@Co-NC Catalysts

				Product sel. (%)				
Entry	**Method**	**Used times**	**Conv. (%)**	**GLA**	**GLD**	**OXA**	**DHA**	**others**
1	This work	1	61	37	54	3.2	2.4	3.4
2	This work	2	61	37	53	4.9	2.3	2.8
3	This work	3	61	37	54	4.1	2.2	2.7
4	Polyol red.	1	34	26	69	3.0	0.5	1.5
5	Galvanic	1	40	44	47	2.0	1.0	6.0
6	Sol gel	1	46	44	50	2.4	1.9	1.7

Reaction conditions: GLY (15 mL, 0.3 M), Pt (2.7 wt%), catalyst (0.088 g), O_2 pressure (3 bar), stirring speed (400 rpm), 60°C, 4 h.

Source: Tang 2020 [122].

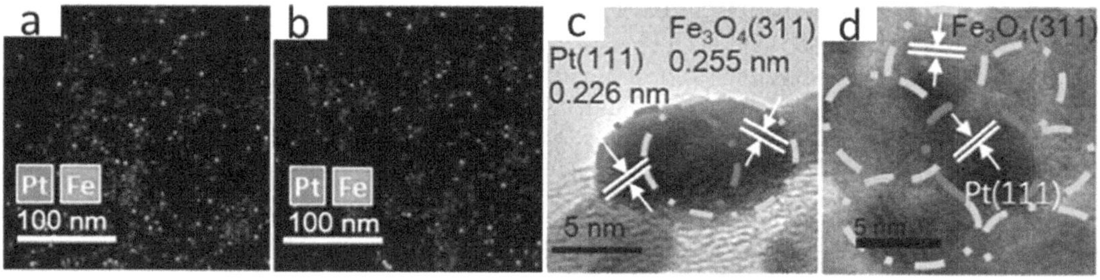

FIGURE 4.5 EDX mapping of FeO-Pt/rGO (a) and Pt-Fe_3O_4/rGO-3.3 (b), and TEM images of Pt-Fe_3O_4/rGO-8.5 (c) and Pt-Fe_3O_4/rGO-17 (d).

Source: Meng 2020 [123].

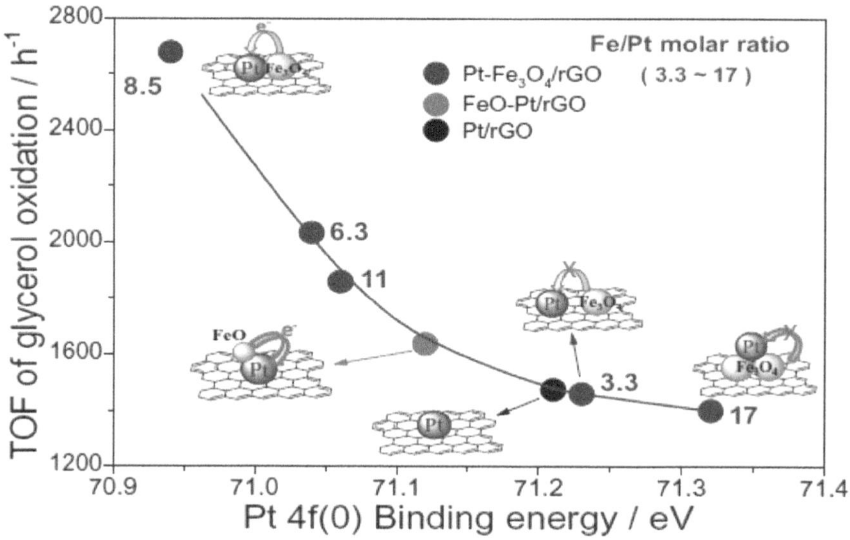

FIGURE 4.6 TOF of GLY for various Pt catalysts different against the binding energies of Pt $4f_{7/2}$ electron.

Source: Meng 2020 [123].

the presence of iron oxides has an influence on the catalytic performance of Pt catalysts in the oxidation of GLY. Differences in catalytic activities of Pt-Fe_3O_4/rGO with different Fe_3O_4 loadings are due to different interactions of Pt and Fe_3O_4 on their surfaces. The XPS results of the most active catalyst, Pt-Fe_3O_4/rGO-8.5, showed the negative shift in the binding energy of Pt $4f_{7/2}$, where 70.94 eV was obtained instead of 71.21 eV obtained for Pt/rGO (reference). Correlation of Pt $4f_{7/2}$ with TOFs obtained in the oxidation of GLY showed that activities of the catalysts increase with decreasing binding energies of Pt $4f_{7/2}$ electron (Figure 4.6). This negative shift means that Fe_3O_4 donate electrons to Pt thus increasing the electron density on Pt surface therefore facilitating the oxidation of GLY.

Selective oxidation of alcohol to epoxides, ethers, aldehydes, ketones and acids, is one of the important chemical reactions in the fine chemical industry. There is an exhausting amount of research on the catalytic oxidation of alcohols using supported noble metal catalysts [124, 125]. Yang and co-workers [111] carried out an oxidation of cinnamyl alcohol to cinnamaldehyde on bimetallic AuPt/CNR (CNR: carbon nanoreactor) using oxygen as an oxidant. XRD, SEM (Figure 4.7a) and TEM (Figure 4.7b) analysis showed that Au and Pt were not alloyed but were well dispersed on CNR with particle sizes of 6.1 nm and 1.9 nm, respectively. Comparing AuPt/CNR with monometallic Pt/CNR

TABLE 4.2
Catalyst Composition and Catalytic Results of Pt-based Catalysts on Oxidation of GLY

Entry	Catalyst	Fe/Pt ratio	Conv. (%)	Product sel. (%)			TOF (h^{-1})
				GLAD	GLA	DHA	
1	Pt/rGO	0	15.8	60.8	20.9	16.6	1474
2	FeO-Pt/rGO	3.3	13.3	64.	18.4	14.8	1637
3	Pt-Fe_3O_4/rGO-3.3	3.3	15.3	58.4	18.9	15.1	1460
4	Pt-Fe_3O_4/rGO-8.5	8.5	29.5	50.2	27.3	12.8	2677
5	Pt-Fe_3O_4/rGO-17	17	21.6	52.6	23.3	15.1	1400
6	Pt/Fe_3O_4	50	13.0	56.5	21.3	17.6	1150
7	Fe_3O_4/rGO	-	0.5	0	0	0	-

Reaction conditions: GLY (3 mmol), Pt/GLY molar ratio (1: 1000), H_2O (30 mL), O_2 flow (150 sccm), 60°C, 2 h.

Source: Meng 2020 [123].

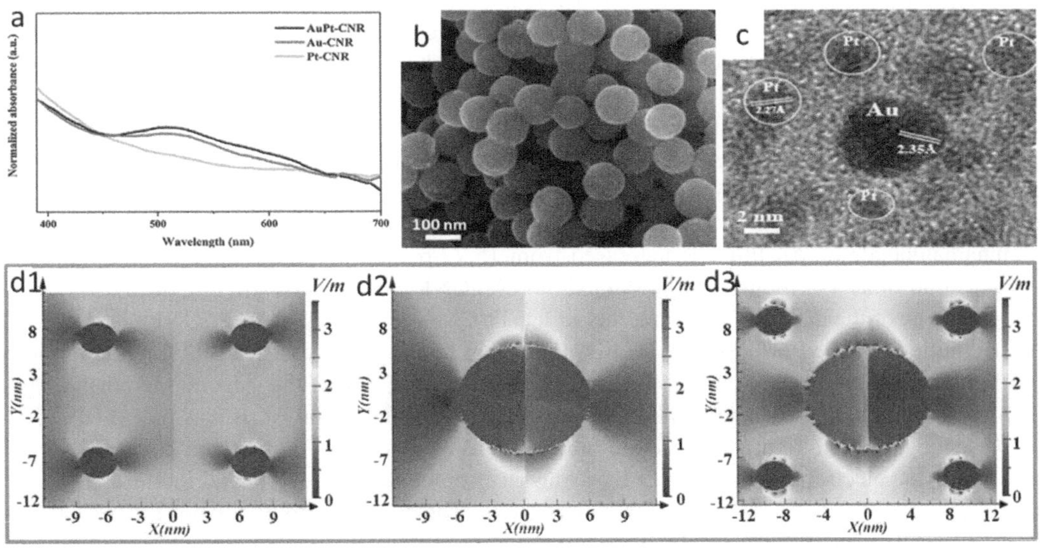

FIGURE 4.7 SEM (a) and TEM (b)images of Au-Pt-CNR catalyst, UV–Vis light absorption spectra of Pt-CNR, Au-CNR and Au-Pt-CNR catalysts (c), (and FDTD simulated spatial enhancement of the local electric field induced by LSPR effects under 550 nm light irradiation for P/CNR (d), Au/CNR (e), and Au-Pt/CNR (f).

Source: Yang 2020 [111].

and Au/CNR on their ability to absorb a visible light, a broad absorption bands that lies between 500 and 550 nm were observed for Pt/CNR, Au/CNR and AuPt/CNR (Figure 4.7c). This is due to LSPR effect of Au. It is clear that Pt/CNR does not absorb any lights in the UV-vis region since it contains small Pt NPS and small Pt NPs do not possess any plasmonic properties with insignificant optical adsorption. The absorption band for AuPt/CNR is more intense than that of Au/CNR (Figure 4.7c), which is due to synergic effects of Pt when it is at the close proximity with Au. Finite-difference time-domain (FDTD) simulations were conducted on the model that consist of one Au NP surrounded by four Pt NPs supported on carbon (based on TEM results in Figure 4.7b) to study the enhancement of local electric field by LSPR effects. The FDTD simulations studies showed

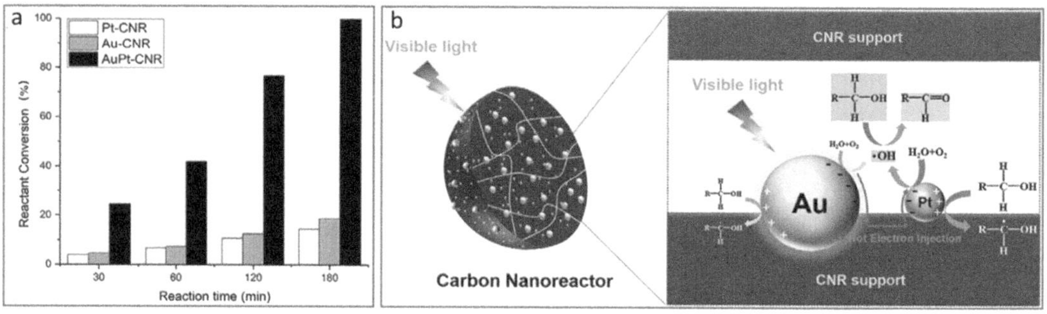

FIGURE 4.8 Conversion of cinnamyl alcohol on different noble metal catalysts against time (a) and proposed reaction mechanisms using hot electrons generated by local electric effect between Au and Pt on Au-Pt/CNR catalyst (b). [Reaction conditions: Aq. cinnamyl alcohol (0.01 M), O_2 pressure (2 bar), UV-vis light, catalysts (2 mM)].

Source: Yang 2020 [111].

that the electric field was induced on the external surface of the NPs for both Pt (Figure 4.7d) and Au NPs (Figure 4.7e) and this phenomenon was not observed on the core of NPs. The bimetallic catalyst, AuPt/CNR (Figure 4.7f), possess high electric field due to the incorporation of Pt and Au that improved its ability to absorb light as compared to Au/CNR. Electric field is responsible for the generation of hot electrons and therefore, in this case, the electrons will be generated on the outer surfaces of the NPs between Au and Pt.

When Yang and co-workers [111] oxidized cinnamyl alcohol on their catalysts in the dark environment, the conversions of cinnamyl alcohol were very low (<1 5.77%). However, when the reactions were carried out in the presence of visible light, the conversion of polar cinnamyl alcohol to cinnamaldehyde on Au-Pt/CNR increased from 15.77 percent to 76.77 percent under the same reaction conditions. For monometallic Pt/CNR and Au/CNR, conversions still remained low (<13%, 2h). Activities of the catalysts followed this order: Pt/CNR < Au/CNR < Au-Pt/CNR (Figure 4.8a). Low activities obtained on Pt/CNR are due to its low LSPR effect that lead to the generation of small number of hot electrons on its surface to allow the activation of the substrate by visible light. High catalytic activities were obtained on bimetallic Au-Pt/CNR catalyst. It was proposed that, in the presence of visible light, Au generate hot electrons on its surface by surface plasmonic resonance and those electrons are sent to Pt by using electron conducting CNR and then trapped by Pt NPs surrounding Au NP thus creating an electron sink. The electron sink facilitates the oxidation of the substrate. The mechanism for the oxidation of cinnamyl aldehyde is shown in Figure 4.8b. Low catalytic activities were obtained for the oxidation of a non-polar substrate, cyclohexanol (<15%, 3h), since hot electrons have low interaction with non-polar substrates.

Göksu and co-workers [126] oxidized various benzyl alcohols (BnOH) to benzaldehydes on Pt-Ni/SWCNT in the presence of KOH with oxygen at atmospheric pressures. The particle size of 2.10 ± 0.70 nm was obtained. XPS indicated that Pt is present as Pt^0 with Pt $4f_{7/2}$ of 71.5, and Pt^{2+} (Pt $4f_{7/2}$ = 72.7 eV), for typical PtO and $Pt(OH)_2$, with metallic Pt dominating. XPS confirmed that Ni was present as metallic Ni^0 and Ni^{2+}. A negative shift in the binding energy of Pt $4f_{7/2}$ from that of pure Pt (71.0 eV) was observed, which indicates that Pt is electron rich due to the donation of electrons from Ni during their interaction. In the catalytic oxidation of BnOH, there were no activities observed in the absence of base additives. Low yields of benzaldehydes, 8.8 ± 2.9% and 5.4 ± 4.6%, were obtained when K_2CO_3 and $NaHCO_3$ were used as bases in toluene at 80°C for 2 h, respectively. However, benzaldehyde yield of about 80% was obtained with KOH under the same reaction conditions. An increase in the amount of KOH from 1.0 mmol to 1.5 mmol increased the yield of benzaldehyde to >99%. Running this reaction for a longer period of time, 4 h, reduced the yield of benzaldehyde to 93%. This

TABLE 4.3
Oxidation of BnOHs to their Corresponding Benzaldehydes on PtNi@SWCNT Catalysts

Entry	Substrate	Conv. (%)	Sel. (%)	Yield (%)
1	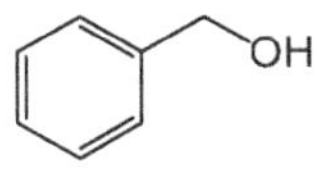	>98	100	>98 ± 1.1
2	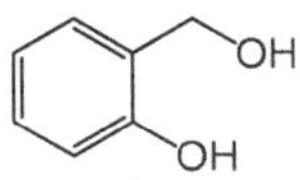	>98	100	>98 ± 1.3
3	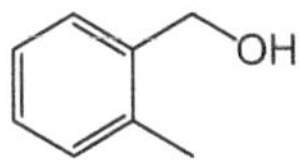	87	100	87 ± 3.7
4	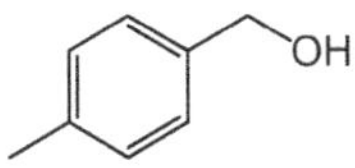	>97	100	>97 ± 1.9
5	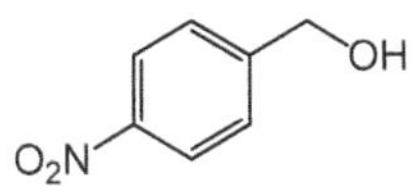	>97	100	>98 ± 1.1
6	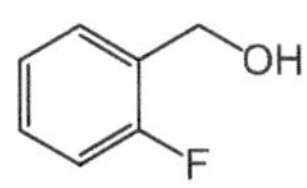	76	100	76 ± 4.6
7	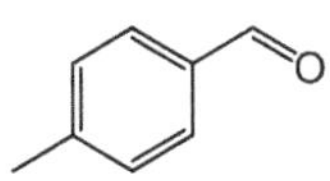	>97	100	>97 ± 2.2

Reaction conditions: BnOH (1.0 mmol), KOH (1.5 mmol), catalyst (2.0 mg with 5% wt metal content), toluene (3.0 ml), T (80°C), t (2 h), continuous flow of O_2.

Source: Göksu 2020 [126].

decrease in benzaldehyde yield is obviously due to over-oxidation of benzaldehyde to benzoic acid at elongated reaction times. When substituted BnOHs were tested, high yields of benzaldehydes were obtained from their corresponding substituted BnOHs (Table 4.3). Low yields of benzaldehydes were obtained for ortho substituted BnOHs due to steric effects on the substrates, which hinders the oxidation of the hydroxyl group of the BnOH (Table 4.3, entries 3 and 4, 6 and 7).

Macro aromatic frameworks are central building blocks of many pharmaceutical drugs and fine chemicals with various applications as ligands in enantioselective catalysis, organic dyes and cosmetics [126, 127]. Their synthesis can be achieved through the formation of carbon-carbon by coupling reactions. Maphoru and co-workers [92] carried out the oxidative coupling of 2-methyl-1-naphthol (2MN) on Pt/AC, Pt-Bi/AC and Pt-Sb/AC catalysts with H_2O_2 as an oxidant (Scheme 4.2).

The catalysts were prepared by electroless deposition and microwave-assisted methods with loadings of 5 percent for each metal. XRD analysis confirmed that all catalysts contained face-centered cubic Pt with $BiPO_4$ for Pt-Bi/AC (The AC support contained a substantial amount of PO_4^{3-}). For Pt-Sb catalysts, Sb diffraction peaks were absent on the XRD pattern of ED Pt-Sb catalyst while PtSb alloy diffraction peaks were observed for MW Pt-Sb. TEM images showed that ED Pt-Bi (Figure 4.9c), MW Pt-Bi (Figure 4.9d) and MW Pt-Sb (Figure 4.9f) have small

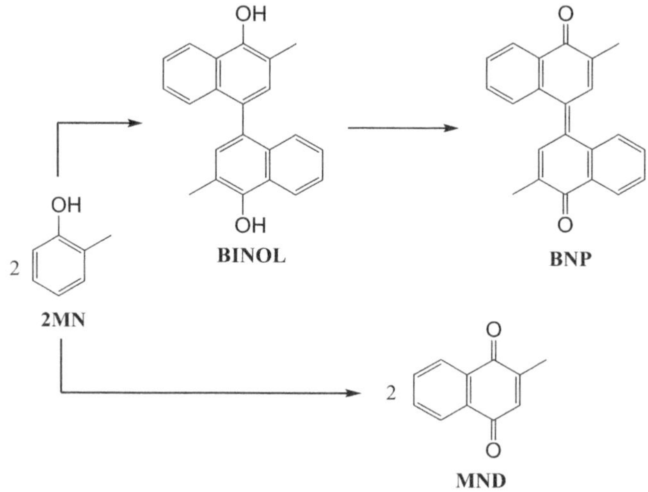

SCHEME 4.2 Oxidative coupling of 2-methyl-1-naphthol on Pt-Bi ad Pt-Sb catalysts [Reaction conditions: 2MN (6.34 mmol), catalyst (0.0401g), MeOH (25 mL), H_2O_2 (30%, 3.1 mL), 40 min].

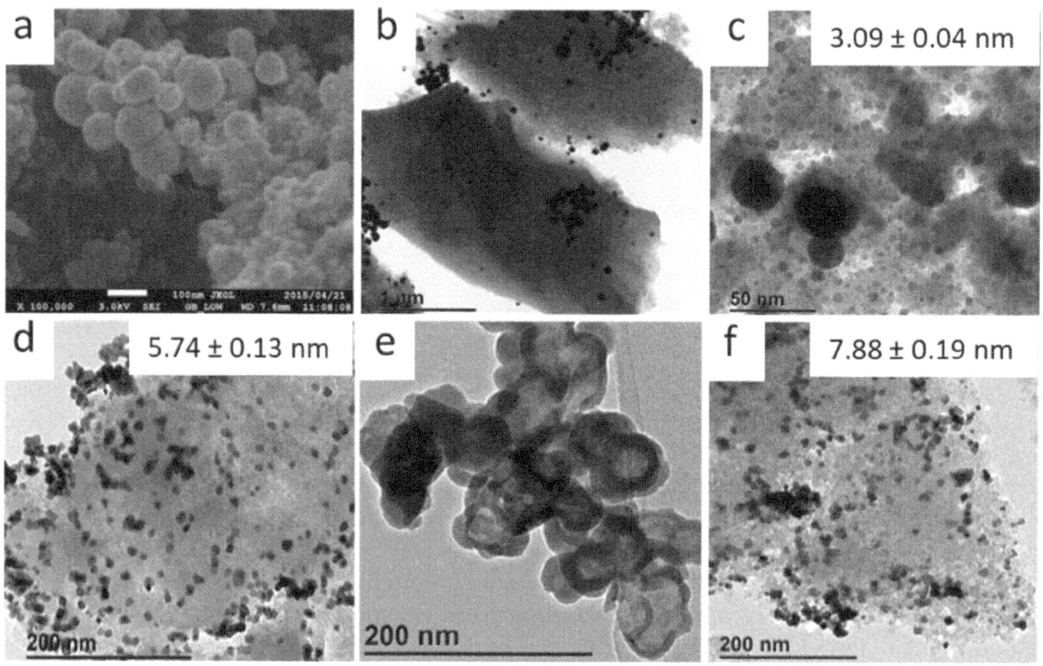

FIGURE 4.9 SEM image of Pt/AC(a) and TEM images of Pt/AC (b), ED Pt-Bi(c), MW Pt-Bi(d), ED Pt-Sb(e), MW Pt-Sb(f) (Pt, Bi and Sb %wt ~5% by ICP).

Source: Maphoru 2017 [92].

NPs with the average particle sizes of <8.0 nm while particles on Pt/AC are agglomerated (Figures 4.9a,b). However, Pt-Sb nanochains instead of nanoparticles were obtained for ED Pt-Sb catalyst (Figure 4.9e). Low particle sizes on promoted Pt catalysts owes to the addition of Bi or Sb promoters, which are known to enhance the metal dispersion by reducing the sizes of noble metal active ensembles [107, 108, 117].

TABLE 4.4
Oxidative Coupling of 2-methyl-1-naphthol on Promoted Pt Catalysts

				Product sel. (%)		
Entry	**Catalyst**	**T (°C)**	**Conv. (%)**	**BINOL**	**BNP**	**MND**
1	ED Pt	r.t.	28.9	67.1	30.6	-
2	ED Pt-Bi	r.t.	100	9	-	-
3	MW Pt-Bi	r.t.	86.0	75.1	24.3	-
4	ED Pt-Sb	r.t.	9.60	93.8	-	-
5	MW Pt-Sb	r.t.	75.2	85.6	12.3	
6	ED Pt	60	100	11.4	85.6	2.90
7	ED Pt-Bi	60	100	-	83.0	16.2
8	MW Pt-Bi	60	100	5.7	82.9	11.0
9	ED Pt-Sb	60	12.2	89.1	-	-
10	MW Pt-Sb	60	100	20.1	71.6	8.20

Reaction conditions: 2MN (6.34 mmol), catalyst (0.0401g), MeOH (25 mL), H_2O_2 (30%, 3.1 mL), 40 min.

Source: Maphoru 2017 [92].

In the coupling of 2MN, the intermediate product, 3,3'-dimethyl-1,1'-binaphthalenyl-4,4'-diol (BINOL), was obtained with high selectivities at room temperature (r.t.) while 3,3'-dimethyl-1,1'-binaphthalenylidene-4,4'-dione (BNP) was a major product with high conversions of 2MN in MeOH at 60°C (Table 4.4, entries 1–5). It is worth noting that ED Pt/AC possess low activity than all catalysts, except for ED Pt-Sb (Table 4.4, entries 1 and 6). Large agglomerated particles on ED Pt/AC reduce surface to volume ratio and number of active sites available for the reaction, which reduce its catalytic activity. Low catalytic activities obtained for VD Pt-Sb is due to low exposure of Pt (Table 4.4, entries 4 and 9), which was confirmed by the absence of Pt-fringes on HRTEM-SAED [92] of Pt-Sb nanochains (Figure 4.8e) suggesting that the amorphous Sb species are covering crystalline Pt. The catalyst with low average particle size, ED Pt-Bi, possessed superior activities. The activity at r.t. follows this pattern: ED Pt-Bi > MW Pt-Bi >MW Pt-Sb > ED Pt > ED Pt-Sb.

5-Hydroxymethylfurfural (HMF) is a biomass-derived chemical compound that can be used for the production of various important macromolecular products such as functional polymers, biodegradable plastics and biofuels through different chemical reactions [128]. Chen and co-workers [129] oxidized 5-hydroxymethylfurfural (HMF) to 2,5-furandicarboxylic acid (FDCA) on Pd catalyst supported on highly porous nitrogen and phosphorus-codoped graphene (Pd/HPGS) (Scheme 3). FDCA is a biomass originated monomer that can be used for the polymerization reactions of polyethylene terephthalate, polybutylene terephthalate, and a wide range of important polymers [130]. Chen and co-workers [129] synthesized their graphene support by solid-phase pyrolysis method followed by supporting Pd NPs onto the support using formaldehyde as a reducing agent. The geometry and chemical structure of graphene have been highly explored due to the interesting physicochemical properties they possess: a high chemical stability, surface area, mechanical strength, electron transfer ability, good electrical and thermal conductivities [131]. Doping of nitrogen, oxygen and phosphorus on graphene was confirmed by XPS. TEM images showed particles with fringes that have d-spacings of 0.194 nm and 0.224 nm and they belong to metallic Pd (200) and (111) facets (Figure 4.10a). EDX mapping indicated that Pd particles are more adsorbed on the sites that are more concentrated with N, P and O heteroatoms, which serves as a proof that heteroatoms are important in the stabilization of noble metal NPs. Pd/HPGS catalyst was compared with Pd/rGO, Pd/CNT and Pd/AC with the same Pd metal loading of ~4.0% for the oxidation of HMF to FDCA (Scheme 4.3).

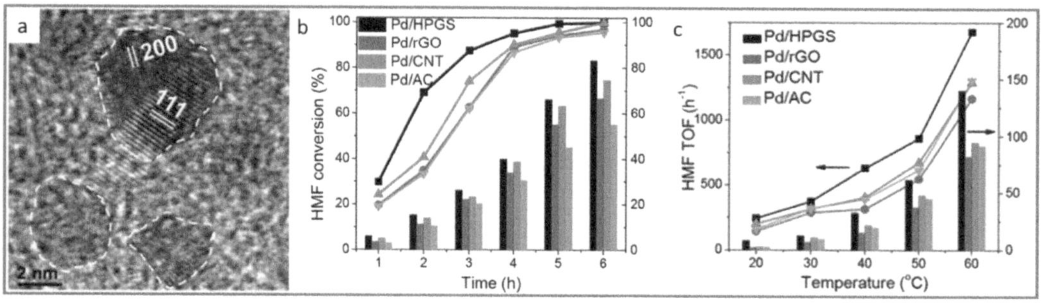

FIGURE 4.10 TEM image of Pd/HPGS (a), HMF conversion on carbon supported Pt catalysts against time (b), TOF of HMF on various carbon supported Pt catalysts as a function of time (c). [Reaction conditions: HMF/Pd (440 mol/mol), NaOH (1 equiv.), O_2 (500 sccm)].

Source: Chen 2017 [129].

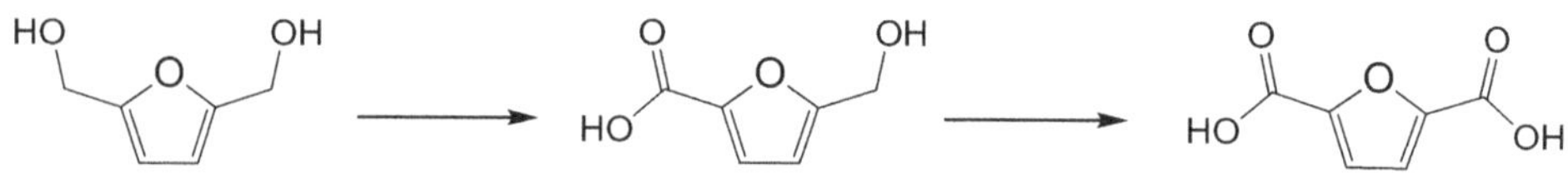

SCHEME 4.3 Oxidation of HMF to FDCA with noble metal catalysts. [Reaction conditions: HMF/Pd (440 mol/mol), NaOH (1 equiv.), O_2 (500 sccm)].

Pd/rGO and Pd/CNT were prepared by the same method that was used to prepare Pd/HPGS while Pd/AC was purchased from commercial suppliers. High conversion of HMF (Figure 4.10b) and TOF (Figure 4.10c) were obtained on Pd/HPGS as compared to other catalysts. Activities of these catalysts are related to their Pd^{2+} content, which were 79.6 percent, 33.3 percent, 46.8 percent and 61.5 percent (by XPS) for Pd/HPGS, Pd/rGO, Pd/CNT and Pd/AC, respectively. The transformation of Pd^0 and Pd^{2+} plays a very significant role in the catalytic cycle of HMF [129]. The role of surface Pd^{2+} is to accept electrons that are used to facilitate the removal of hydrogen on the germinal diol intermediate followed by the oxidation of the side chain of an alcohol to produce an aldehyde intermediate. The Pd^0 is an active metal and is responsible for activating oxygen molecules adsorbed on its surface. The authors concluded that nitrogen dopant heteroatoms on graphene are responsible for improving the Pd^{2+} ratio while P dopant plays a major role in decreasing the wall thickness and generating surface defects and micropores on the skeleton of the HPGS. The combination of these dopants allows superior interaction between HPGS support and Pd NPS therefore resulting in high catalytic activities.

German and co-workers [112] oxidized HMF to FDCA using bimetallic Pd-Au on mesoporous Sibunit carbon (Cp) in the presence of NaOH and oxygen as an oxidant at 60°C. It is clear from Table 4.5 that there is an insignificant difference in the conversion of HMF obtained on monometallic Ag, Au and Pd catalysts (entries 1, 2 and 3). High selectivity of FDCA (55%) was obtained on monometallic Pd catalysts while high selectivity of HFCA (5-hydroxymethyl-2-furancarboxylic acid), which is an intermediate product to FDCA, was obtained on monometallic Ag catalysts. Bimetallic Pd-Au/Cp brought an improvement on the selectivity of FDCA (60%) as compared to its monometallic catalysts counterparts (Table 4.5, entry 4). It is because the combination of two metals bring about high synergistic effects. Also, under the same reaction condition, Au possess a high potential in converting HFCA to FDCA, which assist Pd metal in this process. Not only metals affect the performance of the catalyst, but the heteroatoms on the carbon support also influence the outcome of the reaction due to surface changes brought by their presence [132]. German and co-workers [112] treated their carbon

TABLE 4.5
Catalytic Performances of Ag, Au, Pd and Pd-Au/C_p Catalysts for the Oxidation of HMF

Entry	Catalyst	Time, min	Conv. (%)	Product sel. (%)	
				HFCA	FDCA
1	Ag/Cp	120	98	98	2
2	Au/Cp	120	100	79	21
3	Pd/CP	120	99	45	55
4	Pd-Au/C_p	120	99	40	60
5	Pd-Au/C_p	15	83	69	31
6	Pd-Au/C_p-HNO_3	120	99	35	65
7	Pd-Au/C_p-HNO_3	15	57	57	43
8	Pd-Au/C_p-NH_4OH	120	99	30	70
9	Pd-Au/C_p- NH_4OH	15	38	38	62

Reaction conditions: HMF(0.15 M), H2O (0.15 L), NaOH/HMF (2 equiv.), O_2 pressure (3 atm), R (200), 60°C.

Source: German 2021 [112].

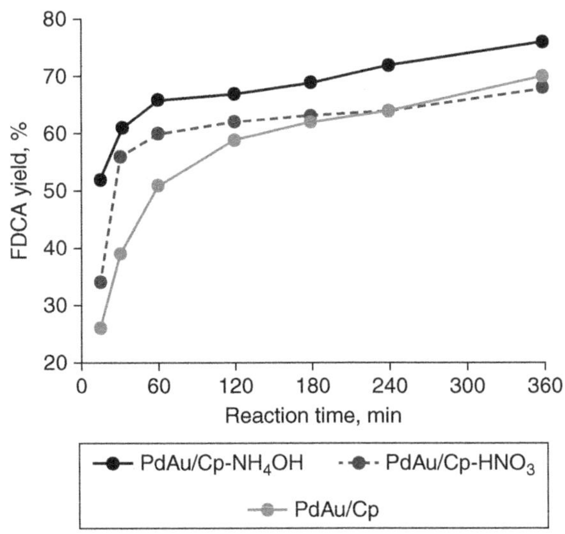

FIGURE 4.11 Influence of Sibunit support treatment for Pd-Au catalysts on the outcome of the oxidation of HMF. [Reaction conditions: HMF (0.15 M), H2O (0.15 L), NaOH/HMF (2 equiv.), O_2 pressure (3 atm), R (200), 60°C].

Source: German 2021 [112].

support with NH_4OH and HNO_3 separately. Pd-Au/Cp-NH_4OH gave high yield and selectivity of FDCA (70%) under the same reaction conditions with the same conversion obtained for Pd-Au/Cp and Pd-Au/Cp-HNO_3 (Figure 4.11, Table 4.5, entries 4–9). The same trend was observed when the reaction was ran for 15 min with these catalysts (Table 4.5, entries 5, 7 and 9). Pd-Au/Cp-NH_4OH has high concentration of nitrogen, which increase the concentration of Pd^{2+} and Au^+ species that are known to be responsible for the conversion of HMF and formation of FDCA. In addition, the presence of nitrogen renders the material basic thus promoting the formation of diacidic compound, FDCA [133]. Also, other authors indicated that N-heteroatoms improves the metal-support interaction, which protect the metal from agglomeration [134].

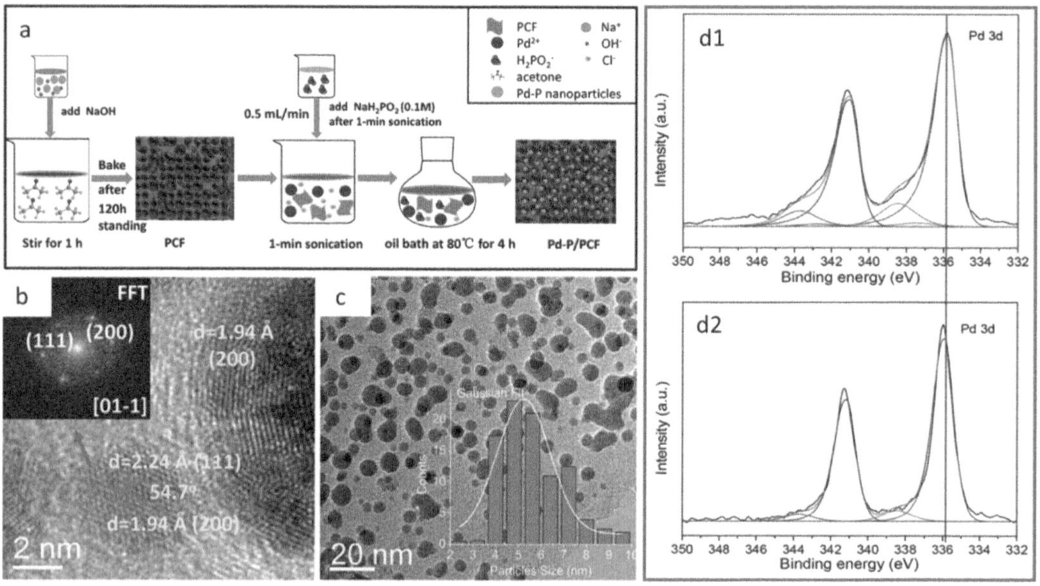

FIGURE 4.12 Synthetic route (a) and TEM images of Pd–P/PCF (b, c). Pd 3d peak shift on XPS spectra of as-prepared Pd/PCF (d) and used Pd/PCF catalysts (e).

Source: Guo 2018 [91].

Guo and co-workers [91] synthesized Pd-P nanoalloys supported on porous carbon frame (Figure 4.12a) as a catalyst for oxidation of BnOH to benzaldehyde. Pd particle agglomeration was observed on undoped Pd/PCF (reference catalyst) while Pd-P with the average particle size of 5.5 nm was obtained on Pd-P/PCF (Figure 4.12b). The d-spacings of 0.224 nm and 0.194 nm was obtained on the lattice fringes of Pd NPs, which are assigned to Pd(111) and Pd(200) facets, respectively (Figure 4.12c). EDX analysis confirmed the presence of Pd, P, C and O on the surface of Pd-P/PCF. XPS results showed that phosphorus is present as P^0 and P^v by P2p photoelectron emissions with binding energies of 130.2 eV and 133.5 eV on Pd-P/PCF, respectively. The ~0.2 eV shift from pure phosphorus (130.4 eV) explain that phosphorus accept electrons from Pd during the formation of Pd-P. Phosphorus doping was also confirmed by the positive shift of 0.2 eV in the binding energy of Pd^0 ($3d_{5/2}$ = 335.9 eV) [91]. This happens due to the reduction of electron density of electrons in the 3d orbital during the formation of Pd-P NPs. When Pd/PCF and Pd-P/PCF catalysts were tested for the oxidation of BnOH and high conversions of BnOH (Figure 4.13a) and selectivities (Figure 4.13b) to benzaldehyde were obtained on Pd-P/PCF as compared to Pd/PCF. The difference in their performance can be explained according to their difference in structural and chemical properties. Pd-P/PCF showed high metal stability with 3.35 percent Pd loss while 48.48 percent of Pd was lost on Pd/PCF. There was no shift in the binding energy of Pd electrons on recycled Pd-P/PCF relative to its fresh catalyst [91] while a shift was observed on used Pd/PCF catalyst (Figure 4.12e) from a fresh Pd/PCF (Figure 4.12d). The stability of Pd-P/PCF owes to the presence of phosphorus dopant. It was also observed by temperature programmed desorption (TPD) that the carbon support for Pd-P/PCF is more hydrophobic than Pd/PCF. A hydrophobic carbon is expected to promote the adsorption of hydrophobic BnOH followed by desorption of benzaldehyde therefore improving the overall reaction rate. The decrease in the electron density of the 3d orbital of Pd by phosphorus, facilitate desorption of benzaldehyde by weakening the Pd-alcohol intermediates bonds, which improves the activity of Pd-P/PCF.

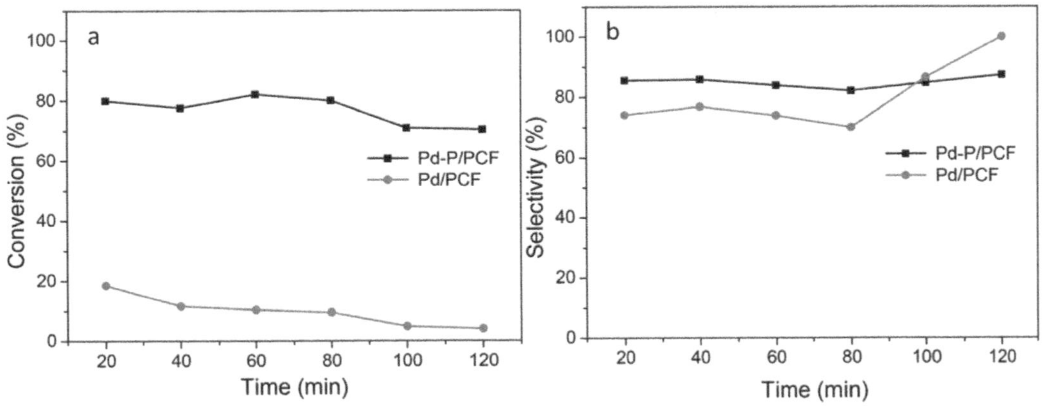

FIGURE 4.13 Conversion of BnOH (a) and selectivity of benzaldehyde on the Pd–P/PCF and Pd/PCF catalysts. [Reaction conditions: catalyst (10 mg), O_2 flow rate (5 mLmin^{-1}), substrate flow rate (0.05 mLmin^{-1}), 70°C, solvent free reaction.

Source: Guo 2018 [91].

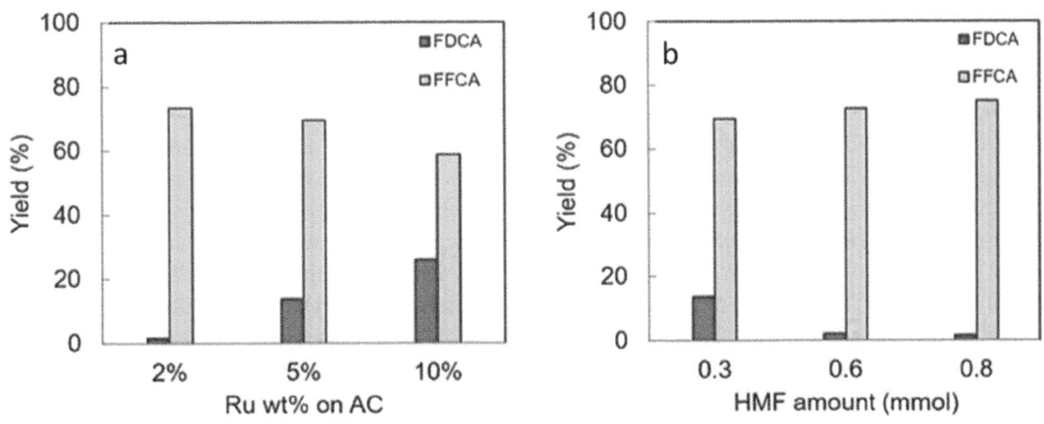

FIGURE 4.14 Influence of Ru loading (a) the amount of HMF (b) on the yields of FFCA and FDCA.

Source: Chen 2018 [118].

4.1.3 Ruthenium Nanocatalysts

Ruthenium based catalysts have been reported to have advantages such as high activity, cost effectiveness and environmental benignity, hence many researchers have studied these nanocatalysts in the catalytic transformation of chemical compounds [135, 136]. Chen and co-workers [118] oxidized HMF in water using activated carbon supported Ru catalyst (Ru/AC) with H_2O_2 as an oxidant. Four products, HFCA, DFF (2,5-diformylfuran), FFCA (5-formyl-2-furoic acid) and FDCA (Scheme 3), were formed where FFCA and FDCA were major products. It is clear that FFCA, which is an intermediate to FDCA, is a major product when Ru/AC with different Ru loadings (Figure 4.14a) or different HMF amounts (Figure 4.14b) are used. Furthermore, it was observed that the yield of FDCA increases with an increase in Ru loading (Figure 4.14a). This is understandable since Ru/AC catalyst with high Ru content contain sufficient amount of Ru to facilitate the rapid conversion of FFCA to FDCA in case there was metal poisoning that took place during the conversion of HMF to FFCA. High amount of HMF led to partial conversion of FFCA to FDCA, which resulted in low yields of FDCA. When $NaHCO_3$, Na_2CO_3 and NaOH were added to the reaction mixtures, $NaHCO_3$

TABLE 4.6
Stepwise Oxidation of HMF to FDCA on 5%Ru/AC in the Presence of Bases

	Step 1			Step 2			Yield (%)	
Entry	Base	t (h)	Catalyst reuse	Base	t (h)	Conv. (%)	FFCA	FDCA
1	$NaHCO_3$	1	yes		2	100	67	27
2	$NaHCO_3$	1	yes	$NaHCO_3$	2	100	68	31
3	$NaHCO_3$	1	yes	NaOH	2	100	23	63
4	$NaHCO_3$	1	yes	NaOH	4	100	14	77
5	$NaHCO_3$	1	yes	NaOH	5	100	11	82
6	$NaHCO_3$	1	fresh		2	100	53	44
7	$NaHCO_3$	1	fresh	NaOH	2	100	10	76

Reaction conditions: [5%Ru/AC, HMF (37.8 mg, 0.3 mmol), Na^+/HMF (2), volume of aq. HMF (14.5 mL), H_2O_2 (10%, dropping rate: of 1.6 mLh^{-1}), 1 bar, 785°C.

Source: Chen 2018 [118].

gave high yields of FFCA, 92 percent and 67 percent after 1 h and 3 h, respectively. The addition of a strong base, NaOH, to the mixture did not improve the yields of either FFCA or FDCA since it decomposes about 30 percent of HMF instead of promoting its oxidation. The use of the base in this reaction is to provide Na^+, which serves as a catalyst promoter that inhibit Ru poisoning [118]. To avoid a loss of HMF to decomposition, stepwise reactions were carried out. In the first step of the reaction where HMF was converted to FFCA, $NaHCO_3$ was added to the reaction mixture, while in the second step where FFCA was converted to FDCA, either NaOH or $NaHCO_3$ was added to the reaction mixture. High yields of FDCA were obtained when NaOH was used in step 2 (Table 4.6, entries 3 and 4). It was proposed that NaOH is necessary for providing OH^- to facilitate the conversion of FFCA to FDCA [118]. However, it is also clear that the catalyst undergo deactivation since in step 2, the recycled catalyst from step 1 gave a yield of 63 percent , which is lower than 76 percent that was obtained when the fresh catalyst was used in step 2. This was further confirmed by the results in entries 1 and 2 of Table 4.6.

Imines and their derivatives are well known key intermediates for the productions of nitrogen heterocycles, biologically active compounds and Schiff bases [137]. Zhang and co-workers [138] synthesized imines by oxidatively coupling alcohols and primary amines in the presence of 5%Ru/AC catalyst with oxygen at atmospheric pressure. A conversion of 87% was obtained for BnOH when coupled with aminobenzene to form N-benzylidenebenzylamine with a selectivity of > 99% in toluene after 15 h (Table 4.7, entry 1). Under the same reaction conditions, a high conversion of 97% for BnOH were obtained in toluene as compared to 92% and 69% that was obtained in benzotrifluoride and 1,4-dioxane, respectively. To determine the effect of substituents and their position on the outcome of the reaction, the authors reacted BnOH with a variety of aminobenzenes with electron-donating and electron-withdrawing groups that are ortho, meta or para to the amino group (Table 4.7). Ortho substituted amines were less reactive than meta and para substituted amines due to steric hindrance created by the substituent on the ortho position to NH_2 (Table 4.7, entries 2 and 3). Acyclicaliphatic amine coupled readily with BnOH to give high yields of imine (Table 4.7, entry 4). When aminobenzene was coupled with different substituted BnOHs, high yields of imines were realized (Table 4.7, entries 5 and 6). However, the coupling of acyclic aliphatic alcohol with aminobenzene gave low yield of imine (Table 4.7, entry 7). This is because the cross coupling of amines and alcohols requires that benzaldehyde be formed first, which is difficult to achieve on a non-activated aliphatic alcohols. About 63 percent of activity loss was observed on recycled 5%Ru/

TABLE 4.7
Oxidative of Coupling of Alcohol with Amines on 5%Ru/AC Catalysts

Entry	Alcohol	Amine	Imine	T (h)	Conv. (%)	Yield [1),2)] (%)
1		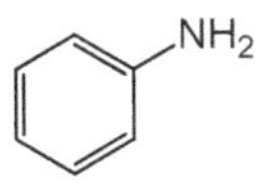	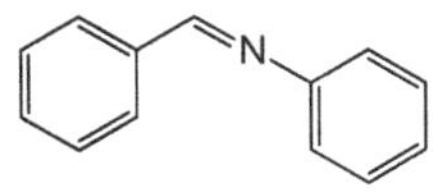	15	97	98 (81)
2	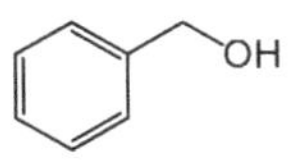	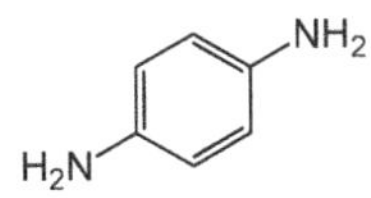	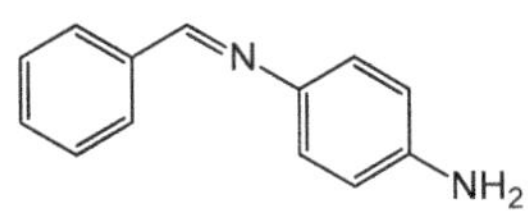	33	95	87(70)
3		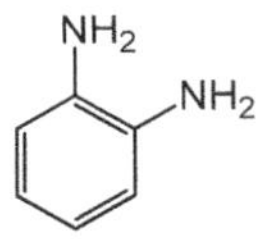	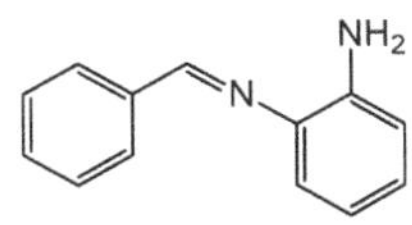	36	99	87(60)
4		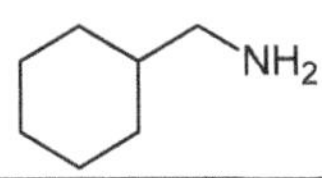	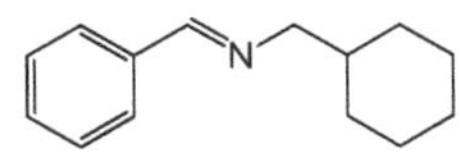	36	95	94(77)
5	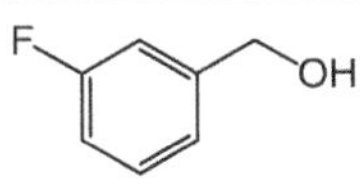	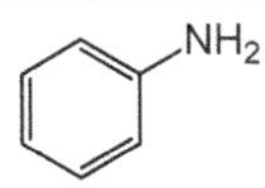	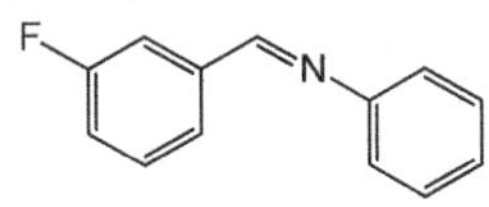	32	99	97(82)
6	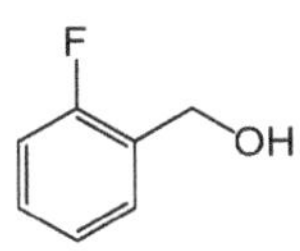		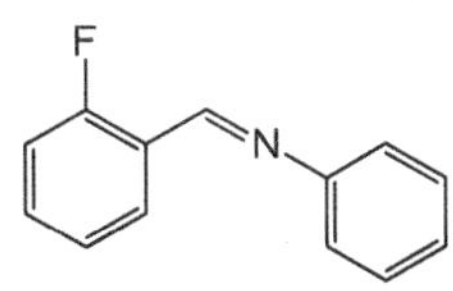	36	96	94(70)
7[3)]	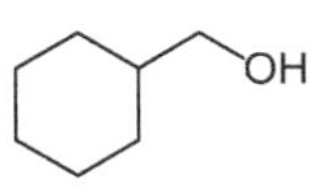		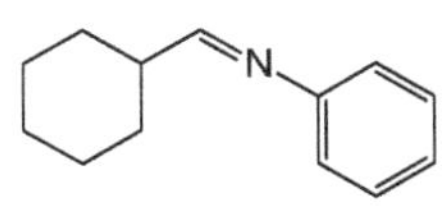	36	25	24

Reaction conditions: amine (0.55 mmol), alcohol (0.5 mmol), catalyst (5 mol%), toluene (5 mL), 90°C, oxygen balloon. 1)GC based yields, [2)] Isolated yields (in brackets), [3)] T= 120°C.

Source: Chen 2017 [138].

AC, which is associated to an increase in the particle size from 1.1 nm to 2.8 nm. Ru agglomeration was caused by weak metal-support interactions that often lead to sintering of metal NPs.

Sánchez-García and colleagues [139] oxidized BnOH to benzaldehyde using Ru supported on nitrogen-doped reduced graphene oxide (rGO) aerogels. The first N-doped rGO, GANH3, was prepared by reacting NH_3 with GO to reduce GO by reacting nitrogen with the oxygen atoms on its surface followed by thermal treatment to synthesize N-doped graphene with nitrogen bonded to the carbon lattice of rGO (Figure 4.15a). The second N-doped rGO, GADA, was prepared by coating GO with polydopamine to synthesis the amorphous carbon with nitrogen bonded to the amorphous carbon on the rGO sheet after thermal treatment (Figure 4.15a). XPS analysis showed that GADA has 30 percent of quaternary nitrogen while GANH3 has 15 percent of quaternary nitrogen. Ru/GANH3 and Ru/GADA were prepared by dry impregnation method. High I_G/I_D ratio (intensity of D-band to intensity of G-band of carbon on Raman spectrum) was

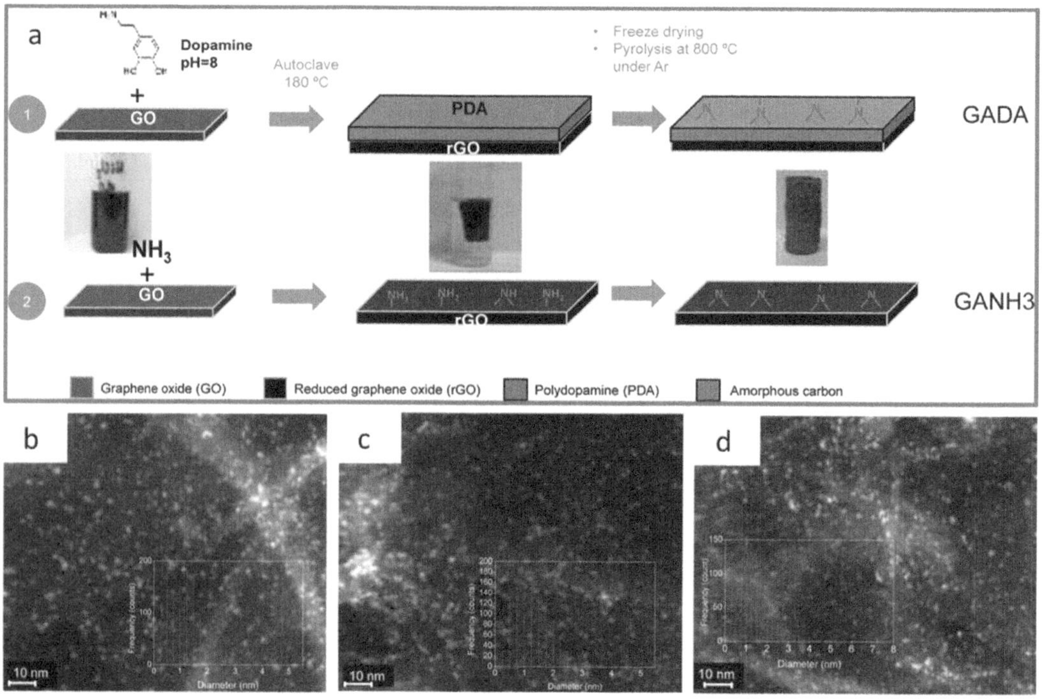

FIGURE 4.15 Synthesis route of GADA and GANH3 supports (a) and HAADF-STEM images of Ru/GA (b), Ru/GANH3 (c), Ru/GADA (d). (%w Ru: 5%).

Source: Sánchez-García 2020 [139].

obtained for GANH3 (1.08) and GADA (1.11) in comparison with undoped graphene aerogel (GA, 1.03). Therefore, N-doping created defects on GANH3 and GADA supports. GADA is more defective than GANH3 since nitrogen atoms on GADA are within the amorphous carbon coated on rGO while GANH3 does contain any amorphous carbon but contain nitrogen doped in the carbon lattice of rGO nanosheets only. High Ru dispersion were obtained on the catalysts (Figure 4.15b-d). High catalytic performance were observed on Ru/GADA catalyst than on Ru/GANH3, Ru/GA and Ru/Al_2O_3 (reference commercial catalyst) (Figure 4.16). This is because Ru/GADA has high Ru^0 content, 50 percent, in compassion with 41 percent, 45 percent, and 29 percent for Ru/GA, Ru/GANH3, Ru/Al_2O_3, respectively. Ru^0 is the active metal required to dehydrogenate the hydroxyl group on the alcohol to form the carbonyl group in this reaction. In addition, Ru/GADA has high hydrophobic surface with a contact angle of 113° with BnOH absorption of 11.9 mmol/g, which facilitate the adsorption of the benzene ring while the hydroxyl group on the BnOH is reacting with Ru. Furthermore, Ru/GADA has high quaternary nitrogen content, which is believed to enhance the oxidation activity of the catalyst [139]. Ru/Al_2O_3 catalyst performed poorly due to its low hydrophobicity (contact angle of 3°) and poor absorption of BnOH (1.5 mmol/g). Also, the 3D structure of Ru/Al_2O_3 also make some of its active sites to be inaccessible to the alcohol while the high portion of NPs on the surface of 2D sheet structure of reduced graphene oxide are accessible to the substrate. Furthermore, Ru/GADA and Ru/GANH3 are highly macroporous than Ru/Al_2O_3, which allows for an easy diffusion of the reactants and products, therefore enhancing their catalytic performances.

Salazar and co-workers [140] performed the oxidative esterification of HMF and MeOH to dimethyl2,5-furandicarboxylate (FDCM) using Co-Ru supported on N-doped carbon in the presence of K_2CO_3 and oxygen as an oxidant. To avoid the build-up of pressure on the cartridge of the continuous flow reactor, powder carbon support was replaced with carbon particles with irregular shape

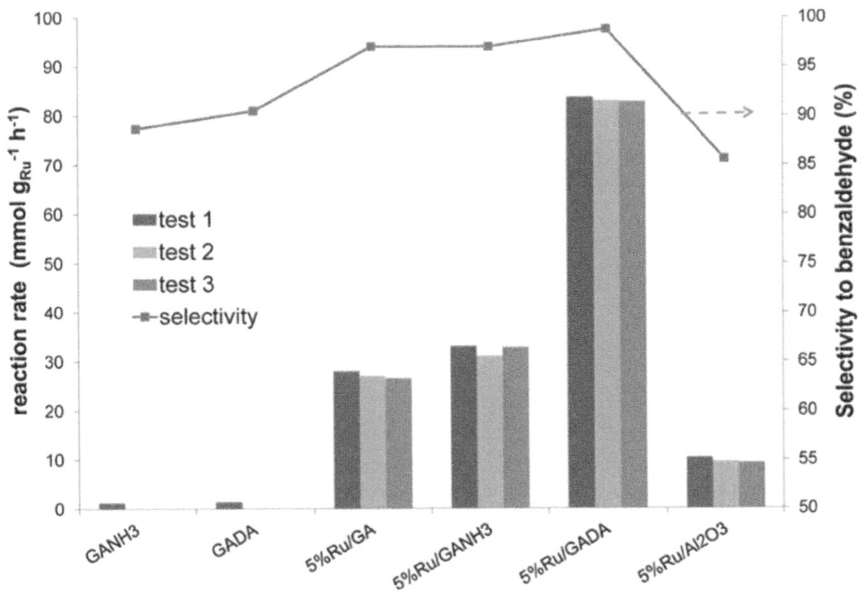

FIGURE 4.16 Selectivities of benzaldehyde and reaction rates obtained for the oxidation of BnOH. [Reaction conditions: O_2 pressure, (10 bar), 80°C, 6 h]. %).

Source: Sánchez-García 2020 [139].

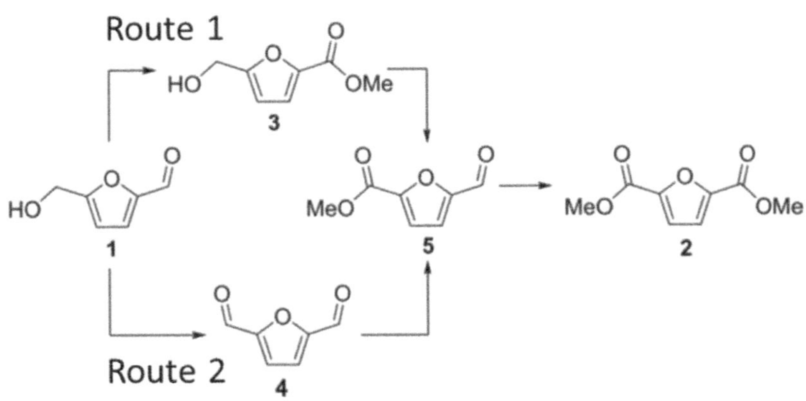

SCHEME 4.4 Esterification of HMF (1) to FDCM (2).

(C-i, -20 + 40 mesh) and cylindrical carbon pellets (C-p, ~0.82= nm). The authors proposed that their esters proceed through either route 1 or route 2, as shown in Scheme 4.

It is clear from the results shown in Table 4.8 that on C-p supported catalysts, the reaction proceeds through both routes 1 and 2 since significant selectivities of 3 and 4 were obtained with comparable conversion of the substrates (Table 4.8, entries 1 and 2). On C-i supported catalysts, route 1 dominates since 3 was obtained with high selectivity at high conversion of the substrate (Table 4.8, entry 5). Bimetallic catalysts, Co_xO_y-N+RuO_x-N@C-i possessed high catalytic activities than RuO_x-N+Co_xO_y-N@C-i and Co_xO_y-N+RuO_x-N@C-p catalysts. Superior catalytic activities on C-i supported catalysts is due to their high mesoporous structure, which allows a continuous adsorption of substrate and desorption of products. The difference in the catalytic activities of Co_xO_y-N+ RuO_x-N@C-i and RuO_x-N+Co_xO_y-N@C-i is related to their surface Co and Ru species (Table 4.9). The preparation of Co_xO_y-N+RuO_x-N@C-i involves incorporation of Ru to Co_xO_y-N@C-i template, which led to the migration of Co from the bulk

TABLE 4.8
Outcome of Oxidative Esterification of HMF with MeOH in a Continuous Flow Reactor

Entry	Catalyst	Conv. (%)	Product sel. (%)				
			2	3	4	5	Others
1	Co_xO_y-N@C-p	57	4	72	12	-	12
2	RuO_x-N@C-p	60	4	-	66	25	5
3	Co_xO_y-N+RuO_x-N@C-p	44	5	46	20	-	29
4	RuO_x-N+Co_xO_y-N@C-p	66	15	52	11	-	22
5	Co_xO_y-N@C-i	92	18	63	4	-	15
6	RuO_x-N@C-i	35	3	-	54	11	31
7	Co_xO_y-N+RuO_x-N@C-i	98	57	20	1	1	21
8	RuO_x-N+Co_xO_y-N@C-i	73	27	45	8	-	20

Reaction conditions: HMF (0.125 M), catalyst (600 mg), HMF solution, MeOH solution of 0.125 M HMF (0.2 $mLmin^{-1}$), K_2CO_3 (0.025 M), O_2 flow (1.4 $mLmin^{-1}$), 62°C, autogenous pressure, Residence time (10 min).

Source: Ref [140], Permission obtained from Elsevier. Salazar 2020 [139].

TABLE 4.9
Comparison of Elemental Analysis of Surface and Bulk Co and Ru

Entry	Catalyst	Co (wt%)		Ru (wt%)	
		Bulk Co	Surface Co	Bulk Ru	Surface Ru
1	Co_xO_y-N@C-i	2.23	2.78		
2	RuO_x-N@C-i			2.46	12.94
3	Co_xO_y-N+RuO_x-N@C-i	1.70	5.40	2.34	9.27
4	RuO_x-N+Co_xO_y-N@C-i	1.58	5.24	2.17	11.76

Source: Salazar 2020 [139].

catalyst into the catalyst surface and therefore enhancing its catalytic properties. However, during the preparation of Co_xO_y-N+RuO_x-N@C-i, where Co was incorporated on RuO_x-N@C-i template, Ru did not migrate from the bulk to the surface of the catalyst.

4.1.4 Gold and Silver Nanocatalysts

Gold has previously been deemed as being catalytically inactive as compared to other noble metals. However, with time, it was discovered that the activity of gold depends on its particle size, where small gold nanoparticles were found to be catalytically active [141–143]. Their electronic band structure is also different from bulk gold due quantum size effects [144]. Megías-Sayago and co-workers [142] studied the size dependece of Au/AC catalysts on the oxidation of HMF to FDCA in the presence of NaOH and oxygen (Scheme 3). Conversion of HMF remained constant for catalysts with average particle sizes ranging from 4 to 40 nm. High yields of FDCA were obtained on catalysts with Au particle sizes of 4 to16 nm while low yields were obtained on catalysts with particle sizes that are greater than 16 nm (Figure 4.17).

Glyceric acid (GLA) is a product obtained in the oxidation of glycerol and its application includes its use in the synthesis of polymers, as base material for functional surfactants and as a

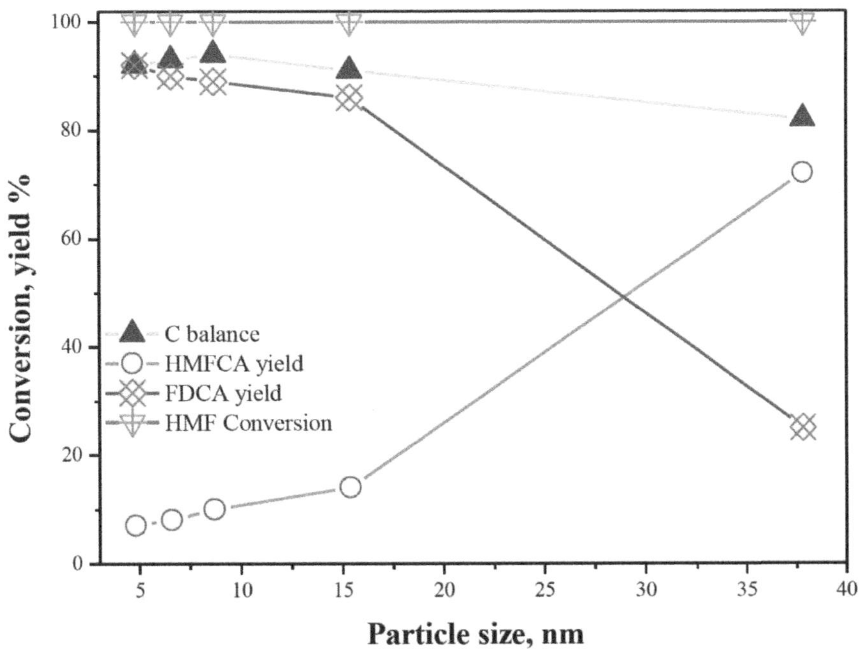

FIGURE 4.17 Catalytic outcome of the oxidation of HMF on Au/AC against Au particle sizes. (Reaction conditions: HMF: Au: NaOH molar ratio (1: 0.01: 2), pressure of O2 (10 bar), 70°C, 4 h, stirring rate (400 rpm).

Source: Megías-Sayago 2020 [142].

liver stimulant [145]. Murthy and Selvam [146] synthesized GLA from GLY using gold catalyst supported on ordered mesoporous carbon (OMC), CMK-3 and NCCR-56, and activated carbon (AC) with oxygen as an oxidizing agent (Scheme 1). Au/AC and Au/OMC were prepared by formation of an Au-sol followed by the deposition of the sol onto carbon support to form well dispersed face centered cubic Au NPs with narrow particle size distributions on carbon support. Oxidation of GLY on Au/AC catalysts with different particle sizes yielded GLA as a major product with tartronic acid, gycolic acid and oxalic acid as side products. The activity of the catalysts showed high dependency on particle sizes of Au NPs (Table 4.10). The conversion of GLY decreased with an increase in Au particle size. It is because large Au particles have low active surface sites due to the reduction of surface to volume ratio (reference). Both Au/CMK-3 and AU/NCCR-56, with TOFs of 96 h^{-1} and 99 h^{-1} respectively, showed superior activities than Au/AC (TOF 92 h^{-1}). High activities of Au/CMK-3 and AU/NCCR-56 owes to high porous structures of OMC. Au/OMC remained stable after reaction with an insignificant change on Au-particle size distribution while Au/AC underwent severe particle agglomeration that led to a decrease in catalyst activity on the recycled catalyst. GLY conversion on Au/AC was reduced from 88 percent to 48 percent due to reduction of surface to volume ratio of Au.

Quinones are biologically active compounds with anti-inflammatory, antifungal, antimicrobial, antitumor, molluscicidal, antiparasitic, leishmanicidal and trypanocidal activities [148–152]. Lots of drugs which are clinically approved or in trial contain quinone skeletons [148]. For example, atovaquone, in combination with proguanil is a recommended regimen for prophylaxisin areas of chloroquine resistance and for treatment of uncomplicated falciparum malaria [153]. Quinones also forms a huge family of vitamins, which can be used as antioxidants and as a prevention and treatment for illnesses such as osteoporosis and cardiovascular diseases [147]. Jawale and co-workers [80] carried out the oxidation of phenols to quinones and aminophenoxazinone on Au/CNT (CNT: carbon nanotubes) catalysts at room temperature with air as an oxidizing agent. The

TABLE 4.10
The Performance of as Prepared and Used Carbon Supported on Au Catalysts

Entry	Catalyst	PSD(nm)	Conv. (%)	Product sel. (%)					TOF
				GLA	TAT	GLO	OXA	others	
1	Fresh Au/CMK-3	3.9 ± 1.3	82	71	8	15	5	1	96
2	Used Au/CMK-3	4.1 ± 1.3	84	70	10	14	6	-	
3	Fresh Au/NCCR-56	3.1 ± 0.9	84	70	7	17	4	2	99
4	Used Au/NCCR-56	5.1 ± 1.8	84	71	6	16	5	2	-
5	Fresh Au/AC	4.1 ± 1.1	88	72	7	13	7	1	92
6	Used Au/AC	35 ± 11	48	76	3	11	10	-	-

Reaction conditions: GLY (0.3 M), GLY/Au (500 mol/mol), GLY/NaOH (4 mol/mol), H_2O (10 mL), pressure of O_2 (7 atm), 60°C, 5 h.

Source: Murthy 2019 [146].

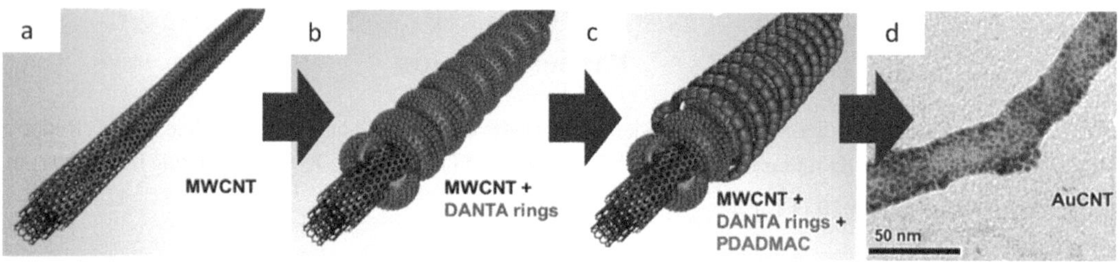

FIGURE 4.18 Depiction of the synthetic path of Au/CNT starting with MWCNT (a) followed by formation of polymerized nanorings from the addition of DANTA to MWCNT (b) and incorporation of cationic polymer, PDADMAC, to produce the second layer. (d) TEM image of Au/ACNT catalyst.

Source: Jawale 2014 [80].

catalyst was synthesized by using layer-by-layer approach where the surface of multiwalled carbon nanotubes (Figure 4.18a) were decorated with amphiphilic nitrilotriacetic-diyne lipid (DANTA). DANTA was then transformed to nanorings on CNT in water (Figure 4.18b). The resulting structure was polymerized under ultraviolet irradiation to produce a second layer of polycationic polymer (PDADMAC) (Figure 4.18c), in which gold NPs were incorporated to produce Au/CNT (Figure 4.18d). When a hydroquinone, 1,4-dihydroxyphenol, was oxidized for 12 h in THF and MeOH in the absence of either $NaHCO_3$ and K_2CO_3, the reaction did not occur. Addition of $NaHCO_3$ under the same reaction conditions yielded benzoquinone with yields of 41 percent and 50 percent in THF and MeOH, respectively. An influence of the solvent in the reaction was observed when a yield of 76 percent of benzoquinone was obtained when THF and MeOH were replaced with $CHCl_3/H_2O$ (3:1) in the absence of the base under the same reaction conditions. The reaction was further extended to the oxidation of substituted 1,4-dihydroxyphenols with Au/CNT in $CHCl_3/H_2O$ (3:1) with K_2CO_3 as an additive at r.t. (Table 4.11). Yields of above 95 percent were obtained for all substrates except for 2-chlorohydroquinone, which gave 2-chlorobenzoquinone with a yield of 70 percent, and 2-acetoxyhydroquinone, which did not react after 24 h. both chloro and acetoxy groups are electron withdrawing groups. Electron withdrawing groups have the ability to inhibit oxidation reaction by deactivating the phenol substrate [154]. Unsubstituted 1,2-dihydroxyphenol was found to be unreactive in contrast to reactive substituted phenols that yielded 1,2-benzoquinones with yields of 99 percent. When these reaction conditions were applied to 2-aminophenol substrate,

TABLE 4.11
Oxidation of Phenols on Au/CNT

Entry	Substrate	product	t (h)	Yield (%)
1	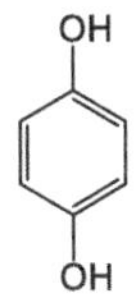	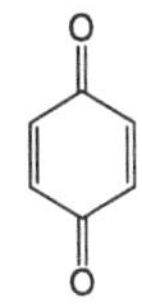	6	95
2	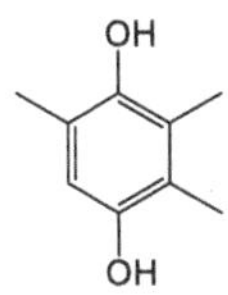	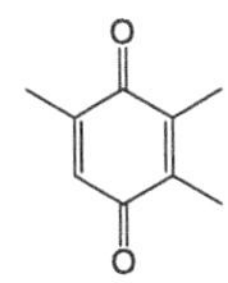	5	96
3	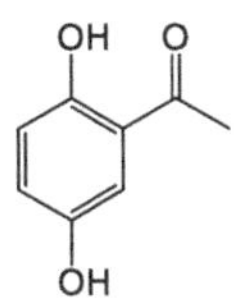	No reaction	24	-
4	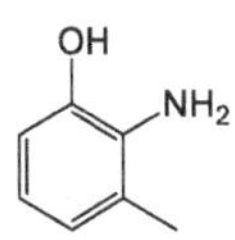	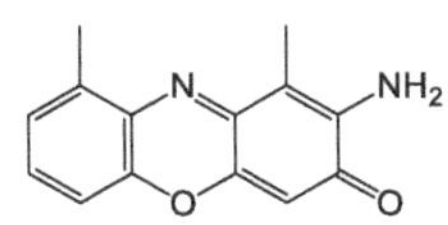	24	95

Reaction conditions: substrate (0.23 mmol), Au/CNT (0.13 mol %), K_2CO_3 (1 equiv.), $CHCl_3/H_2O$ 3:1 (2 mL), r.t., under air.

Source: Jawale 2014 [80].

2-aminophenoxazin-3-one, instead of aminoquinone, was obtained at a yield of 96% (Table 4.11, entry 3).

Gluconic acid, a product that can be produced from the oxidation of naturally abundant glucose, have many applications in pharmaceutical, food and textiles industries. It is currently mainly produced by biotechnological processes (fermentation). Lama and co-workers [155] carried out the oxidation of D-glucose to D-gluconic acid on Au/C with oxygen as an oxidant. The carbon supports were prepared from glucose (C-glucose) and glucosamine (C-glucosamine). Glucosamine was mainly used to prepare N-doped carbon support, N@C. H@C support was prepared by treating C-glucose with H_2 gas at 600°C while O@C support was prepared by a heat treatment of C-glucose in the presence of air at 400°C. Catalysts, Au/C, Au/O@C, Au/H@C were prepared from C-glucose while Au/N@C was prepared from C-glucosamine. Surface hydrophobicity of supports by adsorption of steam on carbon surface showed that O@C and N@C supports are very hydrophilic in comparison with untreated carbon and H@C. Hydrophilic surfaces promote interaction of adsorbed metals and this explains why Au/O@C (Figure 4.19c) and Au/N@C (Figure 4.19d) contained severe gold agglomerants while Au/H@C (Figure 4.19a) and Au/C (Figure 4.19 b) had relatively small particle sizes. Low catalytic activities were obtained on the hydrophilic catalysts with agglomerated metal particles, Au/O@C and Au/N@C (Figure 4.20). Agglomerated metals reduce the number of active sites for the reaction on the support material.

Azobenzenes are chromophores with a wide spectrum of applications such as indicators, food additives, organic colorants and cross-linkers in hydrogel to induce reversible photo-responsive phase transitions in response to polarized light [156, 157]. Gao and co-workers [158] conducted

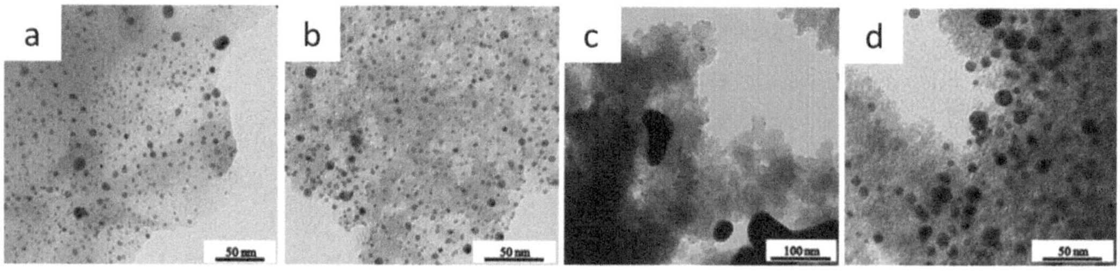

FIGURE 4.19 TEM images Au/H@C (a), Au@C (b), Au/O@C (c), and Au/N@C (d).

Source: Lama 2018 [155].

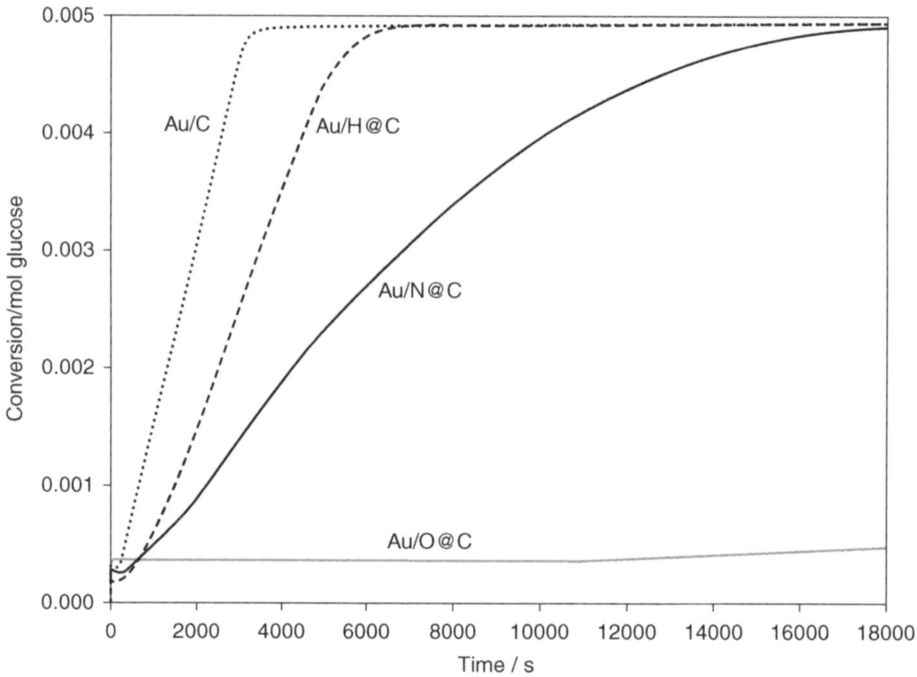

FIGURE 4.20 Conversion of glucose on Au@C, Au/Au@C, Au/O@C, and Au/N@C catalysts. [Reaction conditions: glucose (50 mL, 0.1 molL^{-1}), O_2 Flow (250 mLmin^{-1}), 45°C, magnetic stirring (800 rpm).].

Source: Lama 2018 [155].

oxidative coupling of anilines to azobenzenes with Au-Ag/C catalysts. Catalysts were prepared by pyrolysis of an octanuclear heterometallic Au(I)-Ag(I) cluster at elevated temperature (450°C to 600°C) in the presence of N_2 gas to yield Au-Ag/C. The presence of both Au and Ag on carbon matrices (ratio of 1:1) were confirmed by EDX analysis. TEM analysis indicated that catalysts prepared by using low pyrolysis temperature of 450°C, have high metal dispersion with the average particle size of 6.20 nm in contrast to sintered particles on the catalyst prepared with a high pyrolysis temperature of 600°C. Oxidative coupling of anilines were conducted by reacting aniline on Au-Ag/C catalyst with KOH in DMSO using air as an oxidant at 60°C for 24 h. High yields of 89 percent was obtained for Au-Ag/C catalyst prepared by pyrolysis temperature of 450°C, while yields of 75 percent, 73 percent and 59 percent were obtained for catalysts prepared by pyrolysis temperatures of 500°C, 550°C and 600°C. The difference in these catalytic results owes to their differences in Au-Ag particle sizes. Au-Ag/C prepared at 450°C (Figure 4.21a) has superior dispersion than catalysts prepared at 500°C (Figure 4.21b), 550°C (Figure 4.21c) and 600°C (Figure 4.21d). In addition, XPS

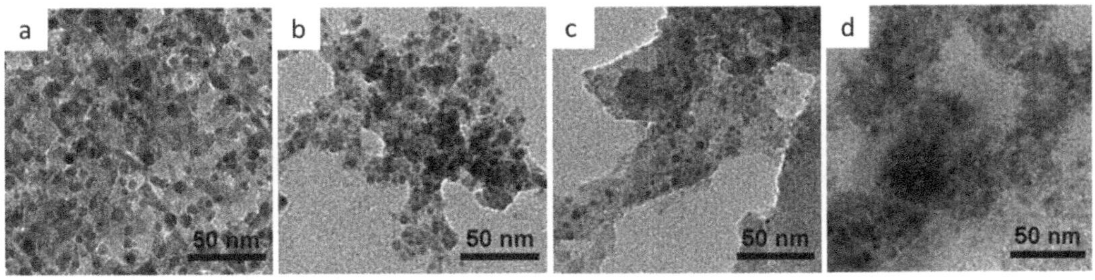

FIGURE 4.21 TEM images of Au-Ag/C catalysts prepared at pyrolysis temperatures of 450°C (a), 500°C (b), 550°C (c), and 600°C (d).

Source: Gao 2018 [158].

TABLE 4.12
Oxidative Coupling of Aniline to their Corresponding Anilines on Au-Ag/C Catalyst

Entry	Substrate	Product	Yield (%)
1	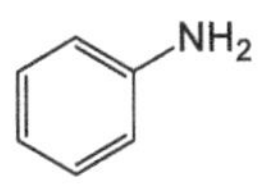	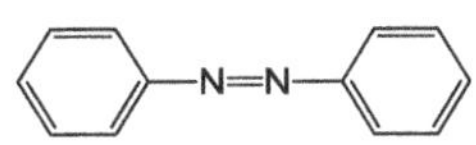	85
2	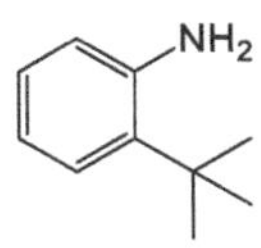	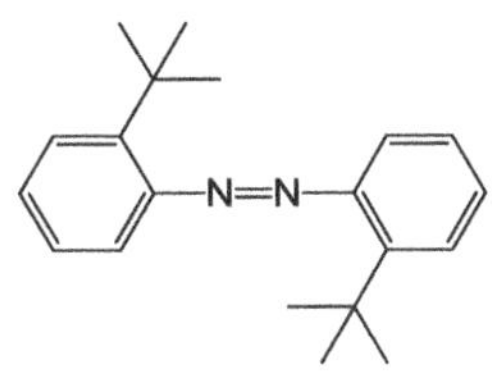	91
3	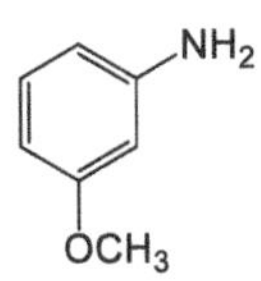	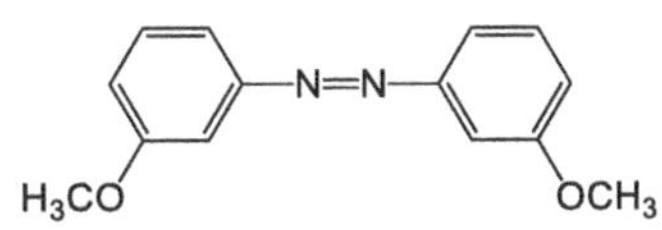	88
4	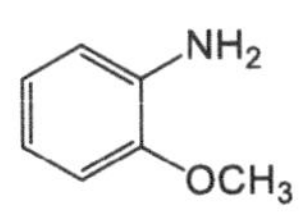	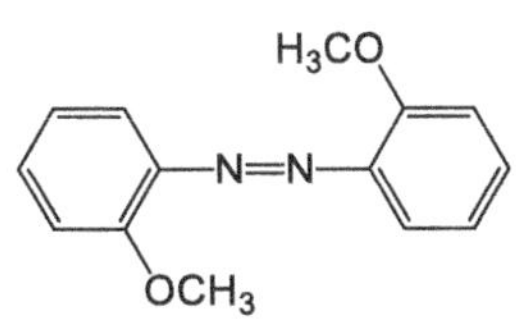	85

Reaction conditions: Aniline (2 mmol), AuAg/C-450 (2 mg), KOH (1 mmol, 0.056 g), DMSO (3 mL), 60 °C, 24 h, in air.

Source: Gao 2018 [158].

indicated that Au-Ag/C prepared at 450°C contained lots of doped phosphorus atoms on the carbon support. Doped heteroatoms promote high catalytic activities by generating active sites that allow spontaneous adsorption of gas oxidants on the catalyst surface. Au-Ag/C catalyst prepared at 450°C was tested in different bases, K_2CO_3, Na_2CO_3, $CsCO_3$, NaOH and Et_3N, where high yields were obtained in KOH. Yields of azobenzene dropped drastically in 1,4-dioxane, DMF, MeCN, toluene and water therefore demonstrating that the nature of the solvent has an influence on the outcome of this reaction. In addition, high yields of azobenzenes were obtained when substituted anilines were reacted on Au-Ag/C prepared at 450°C (Table 4.12).

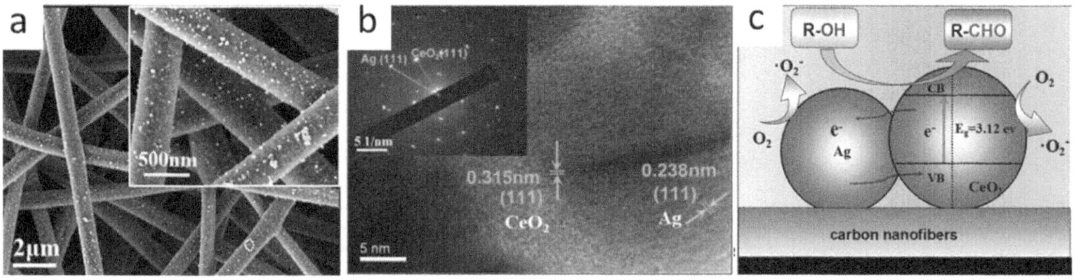

FIGURE 4.22 SEM (a) and TEM (b) images CeO_2/Ag@CNF catalysts. Illustration of proposed oxidation mechanism on CeO_2/Ag@CNF catalyst in the presence of UV-Vis light(c).

Source: Liu 2017 [158].

Liu and co-workers [159] carried out the photocatalytic oxidation of BnOH to benzaldehydes with CeO_2-Ag supported on carbon nanofibers (CNF). High particle dispersion with an average particle size of 19.26 nm and lattice space fringes of 0.238 nm for Ag (111) facets were obtained on CeO_2-Ag\CNF as shown by SEM (Figure 4.22a) and TEM (Figure 4.22b) images. XPS results also confirmed the presence of metallic Ag by binding energy peaks at 368.08 eV (Ag $3d_{5/2}$) and 374.09 eV (Ag $3d_{3/2}$). Ce^{4+} peaks with a binding energy of 898.4 eV (Ce $3d_{5/2}$) and 916.6eV (Ce $3d_{3/2}$) were also observed. A conversion of 96.25% with a selectivity of 69.04 percent for benzaldehyde was obtained for the oxidation of BnOH on CeO_2/Ag@CNF in MeCN with air as an oxidant at 90ºC under UV-Vis light. Low selectivity obtained for benzaldehyde is due to the formation of benzoic acid from the over-oxidation of benzaldehyde. When butyl alcohol was used as a substrate, a conversion of 87.50 percent was achieved with 45.05 percent selectivity of butanal. It is worth noting that butanoic acid was not formed as it was the case in the oxidation of BnOH. The authors proposed that silver NPs on the catalyst adsorb the incident photons under the irradiation light, which happen through the surface plasmon resonance (Figure 4.22c). The resulting electrons are transferred to CeO_2. Concurrently, the electron in the valence band (VB) of CeO_2 are excited to conduction band (CB) to create the electron/hole pairs. Oxygen (O_2) is reduced to superoxide by energy of the conduction band if its reduction-oxidation potential surpasses that of oxygen. The electron-hole pair created is used to facilitate the electron transfer process. The final products are then formed when the oxidized alcohol is moved to contact nanocatalysts.

REFERENCES

1. Hirsch, A. The Era of Carbon Allotropes. *Nat. Mater.*, 2010, 9, 868–871. https://doi.org/10.1038/nmat2885.
2. Shen, Y., Zhu, X., Zhu, L., Chen, B. Synergistic Effects of 2D Graphene Oxide Nanosheets and 1D Carbon Nanotubes in the Constructed 3D Carbon Aerogel for High Performance Pollutant Removal. *Chem. Eng. Prog.*, 2017, 314, 336–346. https://doi.org/10.1016/j.cej.2016.11.132.
3. Lv, F., Qin, M., Zhang, F., Yu, H., Gao, L., Lv, P., Wei, W., Feng, Y., Feng, W. High Cross-Plane Thermally Conductive Hierarchical Composite using Graphene-Coated Vertically Aligned Carbon Nanotubes/Graphite. *Carbon*, 2019, 149, 281–289. https://doi.org/10.1016/j.carbon.2019.04.043.
4. Zhang, W., Chai, C., Fan, Q., Song, Y., Yang, Y. Two Novel Superhard Carbon Allotropes with Honeycomb Structures, *J. Appl. Phys.*, 2019, 126, 145704. https://doi.org/10.1063/1.5120376.
5. Ji, T., Han, K., Teng, Z., Li, J., Wang, M., Zhang, J., Cao, Y., Qi. J. Synthesis of Activated Carbon Derived from Garlic Peel and its Electrochemical Properties. *Int. J. Electrochem. Sci.*, 2021, 16, 150653. https://doi.org/10.20964/2021.01.61.
6. Çakmak, G., Öztürk, T. Continuous Synthesis of Graphite with Tunable Interlayer Distance. Diam. Relat. Mater., 2019, 96, 134–139. 10.1016/J. https://doi.org/10.1016/j.diamond.2019.05.002.

7. Guo, X.-F., Kim, G.-J. Synthesis of Ultrafine Carbon Black by Pyrolysis of Polymers using a Direct Current Thermal Plasma Process. *Plasma Chem. Plasma Process*, 2010, 30, 75–90. https://doi.org/10.1007/S11090-009-9198-7.
8. Le Fevre, L.W., Cao, J., Kinloch, I.A., Forsyth, A.J., Dryfe, R.A.W. Systematic Comparison of Graphene Materials for Supercapacitor Electrodes. *Chemistryopen*, 2019, 8, 418–428. https://doi.org/10.1002/open.201900004.
9. Zhang, B., Piao, G., Zhang, J., Bu, C., Xie, H., Wu, B., Kobayashi, N. Synthesis of Carbon Nanotubes from Conventional Biomass-Based Gasification Gas. *Fuel Process. Technol.*, 2018, 180, 105–113. https://doi.org/10.1016/j.fuproc.2018.08.016.
10. Zhang, L., Peng, B., Wang, L., Guo, C., Wang, Q. Sustainable and High-Quality Synthesis of Carbon Nanospheres with Excellent Dispersibility via Synergistic External Pressure- and PSSMA Assisted Hydrothermal Carbonization. *Adv. Powder Technol.*, 2021, 32, 2449–2456. https://doi.org/10.1016/j.apt.2021.05.016.
11. Wu, F., Dong, R., Bai, Y., Li, Y., Chen, G., Wang, Z., Wu, C. Phosphorus-Doped Hard Carbon Nanofibers Prepared by Electrospinning as an Anode in Sodium-Ion Batteries. *ACS Appl. Mater. Interfaces*, 2018, 10, 21335–21342. https://doi.org/10.1021/acsami.8b05618.
12. Mao, J.-Y., Lin, F.-Y., Chu, H.-W., Harroun, S.G., Lai, J.Y., Lin, H.-J., Huang, C.-C. In Situ Synthesis of Core-Shell Carbon Nanowires as a Potent Targeted Anticoagulant. *J. Colloid And Interface Sci.*, 2019, 552 , 583–596. https://doi.org/10.1016/j.jcis.2019.05.086.
13. Dorel, R., Echavarren, A.M. From Palladium to Gold Catalysis for the Synthesis of Crushed Fullerenes and Acenes. *Acc. Chem. Res.*, 2019, 52, 1812–1823. https://doi.org/10.1021/acs.accounts.9b00227.
14. Ekimov, E.A., Kondrin, M.V., Lyapin, S.G., Grigoriev, Y.V., Razgulov, A.A., Krivobok, V.S., Gierlotka, S., Stelmakh S. High-Pressure Synthesis and Optical Properties of Nanodiamonds obtained from Halogenated Adamantanes. *Diam. Relat. Mater.*, 2020, 103 , 107718. https://doi.org/10.1016/j.diamond.2020.107718.
15. Verma, C., Quraishi, M.A., Ebenso, E.E., Hussain, C.M. Recent Advancements in Corrosion Inhibitor Systems through Carbon Allotropes: Past, Present, and Future. *Nano Select*, 2021, 1–20. https://doi.org/10.1002/nano.202100039.
16. Dong, X., Jin, H., Wang, R., Zhang, J., Feng, X., Yan, C., Chen, S., Wang, S., Wang, J., Lu, J. High Volumetric Capacitance, Ultralong Life Supercapacitors Enabled by Waxberry-Derived Hierarchical Porous Carbon Materials. *Adv. Energy Mater.*, 2018, 8, 1702695. https://doi.org/10.1002/aenm.201702695.
17. Mehrali, M., Latibari, S.T., Mehrali, M., Mahlia, T.M.I., Sadeghinezhad, E., Metselaar, H.S.C. Preparation of Nitrogen-Doped Graphene/Palmitic Acid Shape Stabilized Composite Phase Change Material with Remarkable Thermal Properties for Thermal Energy Storage. *Appl. Energy*, 2014, 135, 339–349. https://doi.org/10.1016/j.apenergy.2014.08.100.
18. Liu, Z., Zang, C., Ju, Z., Hu, D., Zhang, Y., Jiang, J., Liu, C. Consistent Preparation, Chemical Stability and Thermal Properties of a Shape-Stabilized Porous Carbon/Paraffin Phase Change Materials. *J. Clean. Prod.*, 2020, 247, 119565. https://doi.org/10.1016/j.jclepro.2019.119565.
19. Peng, X., Wu, K., Hu, Y., Zhuo, H., Chen, Z., Jing, S., Liu, Q., Chuanfu Liu, C., Linxin Zhong, L. A Mechanically Strong and Sensitive CNT/r-GO–CNF Carbon Aerogel for Piezoresistive Sensors. *J. Mater. Chem. A*, 2018, 6, 23550–23559. https://doi.org/10.1039/c8ta09322a.
20. Lin, X., Liang, Y., Lu, Z., Lou, H., Zhang, X., Liu, S., Zheng, B., Liu, R., Fu, R., Wu, D. Mechanochemistry: A Green, Activation-Free and Top-Down Strategy to High-Surface-Area Carbon Materials. *ACS Sustainable Chem. Eng.*, 2017, 5, 8535–8540. https://doi.org/10.1021/acssuschemeng.7b02462.
21. Pereira, L., Dias, P., Soares, O.S.G.P., Ramalho, P.S.F., Pereira, M.F.R., Alves, M.M. Synthesis, Characterization and Application of Magnetic Carbon Materials as Electron Shuttles for the Biological and Chemical Reduction of the Azo Dye Acid Orange 10. *Appl. Catal. B: Environ.*, 2017, 212 , 175–184. https://doi.org/10.1016/j.apcatb.2017.04.060.
22. Qiu, B., Xing, M., Zhang, J. Recent Advances in Three-Dimensional Graphene Based Materials for Catalysis Applications. *Chem. Soc. Rev.*, 2018, 47, 2165–2216. https://doi.org/10.1039/c7cs00904f.

23. Yu, D., Goh, K., Wang, H., Wei, L., Jiang, W., Zhang, Q., Dai, L., Chen, Y. Scalable Synthesis of Hierarchically Structured Carbon Nanotubes Graphene Fibres for Capacitive Energy Storage. *Nat. Nanotechnol.*, 2014, 9, 555–562. https://doi.org/10.1038/nnano.2014.93.
24. Jain, M., Khan, S.A., Pandey, A., Pant, K.K. Zior, Z.M., Blaskovich, M.A.T. Instructive Analysis of Engineered Carbon Materials for Potential Application in Water and Wastewater Treatment. *Sci. Total Environ.*, 2021, 793, 148583. https://doi.org/10.1016/j.scitotenv.2021.148583.
25. Xue, Q., Chen, H., Li, Q., Yan, K., Besenbacher, F., Dong, M. Room-Temperature High-Sensitivity Detection of Ammonia Gas Using the Capacitance of Carbon/Silicon Heterojunctions. *Energy Environ. Sci.*, 2010, 3, 288. https://Doi.Org/10.1039/B925172n.
26. Zhao, Q., Lin, Y., Han, N., Li, X., Geng, H., Wang, X., Cui, Y., Wang, S. Mesoporous Carbon Nanomaterials in Drug Delivery and Biomedical Application. *Drug Deliv.*, 2017, 24(2), 94–107. https://doi.org/10.1080/10717544.2017.1399300.
27. Mateo, W., Lei, H., Villota, E., Qian, M., Zhao, Y., Huo, E., Zhang, Q., Lin, X., Wang, C., Huang, Z. Synthesis and Characterization of Sulfonated Activated Carbon as a Catalyst for Bio-Jet Fuel Production from Biomass and Waste Plastics. *Bioresour. Technol.*, 2020, 297, 122411. https://doi.org/10.1016/j.biortech.2019.122411.
28. Song, P., Lei, Y., Hu, X., Wang, C., Wang, J., Tang, Y. Rapid One-Step Synthesis of Carbon-Supported Platinum-Copper Nanoparticles with Enhanced Electrocatalytic Activity via Microwave-Assisted Heating. J. Colloid Interface Sci., 2020, 574, 421–429. https://doi.org/10.1016/j.jcis.2020.04.041.
29. Rajesh, D., Mahendiran, C., Suresh, C. The Promotional Effect of Ag in Pd-Ag/Carbon Nanotube-Graphene Electrocatalysts for Alcohol and Formic Acid Oxidation Reactions. Chemelectrochem, 2020, 7, 2629–2636. https://doi.org/10.1002/celc.202000642.
30. Sahoo, M.K., G. Rao, G.R. Enhanced Methanol Electro-Oxidation Activity of Pt/rGO Electrocatalyst Promoted by NbC/Mo_2C Phases. *Chemistryselect*, 2020, 5, 3805–3814. https://doi.org/10.1002/slct.202000170.
31. Maphoru, M.V., Heveling, J., Kesavan Pillai, S. Oxidative Coupling of 2-Methyl-1-Naphthol: A Comparison Between Bismuth-Promoted Pt, Pd And Ag Catalysts. *Chemistryselect*, 2018, 3, 6224–6231. https://doi.org/10.1002/slct.201801148.
32. Megías-Sayago, C., Santos J.L., Ammari, F., Chenouf, M., Ivanova, S. Centeno, M.A., Odriozola, J.A. Influence of Gold Particle Size in Au/C Catalysts for Base-Free Oxidation of Glucose. *Catal. Today*, 2018, 306,183–190. https://doi.org/10.1016/j.cattod.2017.01.007.
33. Verga, L.G., Russell, A.E., Skylaris, C.-K. Ethanol, O, and CO Adsorption on Pt Nanoparticles: Effects of Nanoparticle Size and Graphene Support. *Phys. Chem. Chem. Phys.*, 2018, 20 , 25918–25930. https://doi.org/10.1039/c8cp04798g.
34. Jackson, C., Smith, G.T., Inwood, D.W., Leach, A.S., Whalley, P.S., Callisti, M., Polcar, T., Russell, A.E., Levecque, P., Kramer, D. Electronic Metal-Support Interaction Enhanced Oxygen Reduction Activity and Stability of Boron Carbide Supported Platinum. *Nat. Commun.*, 2017, 8 , 15802. https://doi.org/10.1038/ncomms15802.
35. Yu, K., Zhu,H., Qi, H., Liang, C. High Surface Area Carbon Materials Derived from Corn Stalk Core as Electrode for Supercapacitor. *Diam. Rel. Mater.*, 2018, 88, 18–22. https://doi.org/10.1016/j.diamond.2018.06.018.
36. Sun, F., Wang, L., Peng, Y., Gao, J., Pi, X., Qu, Z,, Zhao, G., Qin, Y. Converting Biomass Waste into Microporous Carbon with Simultaneously High Surface Area and Carbon Purity as Advanced Electrochemical Energy Storage Materials. *Appl. Surf. Sci.*, 2018, 436, 486–494. https://doi.org/10.1016/j.apsusc.2017.12.067.
37. Sun, X., Li, Y., Li, M.-J. Highly Dispersed Palladium Nanoparticles on Carbon-Decorated Porous Nickel Electrode: An Effective Strategy to Boost Direct Ethanol Fuel Cell up to 202 Mw cm^{-2}. *ACS Sustainable Chem. Eng.* 2019, 7, 11186–11193. https://doi.org/10.1021/acssuschemeng.9b00355.
38. Liu, D., Li, X., Chen, S., Yan, H., Wang, C., Wu,C., Haleem, Y.A., Duan, S., Lu, J., Ge, B., Ajayan, P.M., Luo, Y., Jiang, J., Song, L. Atomically Dispersed Platinum Supported on Curved Carbon Supports for Efficient Electrocatalytic Hydrogen Evolution. *Nat. Energy*, 2019, 4, 512–518. https://doi.org/10.1038/s41560-019-0402-6.
39. Yang, S., Chen, L., Wang, C., Rana, M., Ma, P.-C. Surface Roughness Induced Superhydrophobicity of Graphene Foam for Oil-Water Separation. *J. Colloid Interf. Sci.*, 2017, 508 , 254–262. https://doi.org/10.1016/j.jcis.2017.08.061.

40. Ghosh, S., Polaki, S.R., Ajikumar, P.K., Krishna, N.G., Kamruddin, M.K. Aging Effects on Vertical Graphene Nanosheets and their Thermal Stability. *Indian J Phys*, 2018, 92(3), 337–342. https://doi.org/10.1007/s12648-017-1113-0.
41. Sezer, N., Koç, M. Oxidative Acid Treatment of Carbon Nanotubes. *Surf. Interfaces*, 2019, 14, 1–8. https://doi.org/10.1016/j.surfin.2018.11.001.
42. Chu, H., Wei, L., Cui, R., Wang, J., Li, Y. Carbon Nanotubes Combined with Inorganic Nanomaterials: Preparations and Applications. *Coord. Chem. Rev.*, 2010, 254 1117–1134. https://doi.org/10.1016/j.ccr.2010.02.009.
43. He, D, Zeng, C., Xu, C., Cheng, N., Li, H., Mu, S., Pan, M. Polyaniline-Functionalized Carbon Nanotube Supported Platinum Catalysts. *Langmuir*, 2011, 27, 5582–5588. https://doi.org/10.1021/la2003589.
44. Woodhead, A.L., De Souza, M.L., Church, J.S. An Investigation into the Surface Heterogeneity of Nitric Acid Oxidized Carbon Fiber. *Appl. Surf. Sci.*, 2017, 401, 79–88. https://doi.org/10.1016/j.apsusc.2016.12.218.
45. Cobb, J.L., Laidlaw, F.H.J., West, G., Wood, G., Newton, M.E., Beanland, R., .Macpherson, J.V. Assessment of Acid and Thermal Oxidation Treatments for Removing sp^2 Bonded Carbon from The Surface Of Boron Doped Diamond. *Carbon*, 2020, 167, 1–10. https://doi.org/10.1016/j.carbon.2020.04.095.
46. Huang, H., Zhou, J., Xie, M., Liu, H. Mechanistic Study on Graphene Oxidation by $KMnO_4$ in Solution Phase and Resultant Carbon-Carbon Unzipping. *J. Phys. Chem. C*, 2020, 124, 11165–11173. https://doi.org/10.1021/acs.jpcc.0c01314.
47. Zhang, H., Li, Y., Zhao, Y., Li, G., Zhang, F. Carbon Black Oxidized by Air Calcination for Enhanced H_2O_2 Generation and Effective Organics Degradation. *ACS Appl. Mater. Interfaces*, 2019, 11, 27846–27853. https://doi.org/10.1021/acsami.9b07765.
48. Luo, J., Liu, Y., Wei, H., Wang, B., Wu, K.-H., Zhang, B., Su, D.S. A Green and Economical Vapor-Assisted Ozone Treatment Process for Surface Functionalization of Carbon Nanotubes. *Green Chem.*, 2017, 19, 1052–1062. https://doi.org/10.1039/c6gc02806c.
49. Yang, H., Ko, Y., Lee, W., Züttel, A., Kim, W. Nitrogen-Doped Carbon Black Supported Pt-M (M = Pd, Fe, Ni) Alloy Catalysts for Oxygen Reduction Reaction in Proton Exchange Membrane Fuel Cell. *Mater. Today Energy*, 2019, 13, 374–381. https://doi.org/10.1016/j.mtener.2019.06.007
50. Li, J.-S., Wang, X.-R., Li, J.-Y., Zhang, S., Sha, J.-Q., Liu, G.-D., Tang, B. Pomegranate-like Molybdenum Phosphide@Phosphorus-Doped Carbon Nanospheres Coupled with Carbon Nanotubes for Efficient Hydrogen Evolution Reaction. *Carbon*, 2019, 139, 234–240. https://doi.org/10.1016/j.carbon.2018.06.058
51. Wang, L., Meng, T., Sun, J., Wu, S., Zhang, M., Wang, H., Zhang, Y. Development of Pd/Polyoxometalate/Nitrogen-Doping Hollow Carbon Spheres Tricomponent Nanohybrids: A Selective Electrochemical Sensor for Acetaminophen. *Anal. Chim. Acta*, 2019, 1047 28–35. https://doi.org/10.1016/j.aca.2018.09.042.
52. Shao, T., Zhang, Q., Li, J., He, S., Zhang, D., Zhou, X. AgPt Hollow Nanodendrites Based on N Doping Graphene Quantum Dots for Enhanced Methanol Electrooxidation. *J. Alloys Compd.*, 2021, 882, 160607. https://doi.org/10.1016/j.jallcom.2021.160607.
53. Auer, E., Freund, A., Pietsch, J., Tacke, T. Carbons as Supports for Industrial Precious Metal Catalysts. *Appl. Catal. A: Gen.*, 1998, 173, 259–271. https://doi.org/10.1016/s0926-860x(98)00184-7.
54. Yoshida, T., Murayama, T., Sakaguchi, N., Okumura, M., Ishida, T., Haruta, M. Carbon Monoxide Oxidation by Polyoxometalate-Supported Gold Nanoparticulate Catalysts: Activity, Stability, and Temperature-Dependent Activation Properties. *Angew. Chem. Int. Ed.*, 2018, 57, 1523–1527. https://doi.org/10.1002/anie.201710424.
55. Huang, H., Wei, Y., Shen, B., Zhang, Y., He, H., Jiang, Q., Yang, L., Nanjundan, A.K., Na, J., Xu, X., Zhu, J., Yamauchi, Y. Synthesis Of Multiple-Twinned Pd Nanoparticles Anchored on Graphitic Carbon Nanosheets for use as Highly-Active Multifunctional Electrocatalyst in Formic Acid and Methanol Oxidation Reactions. *Adv. Mater. Interfaces*, 2020, 7, 2000142. https://doi.org/10.1002/admi.202000142.
56. Mostashari, S.M., Dehkharghani, R.A., Taromi, F.A., Farsadrooh, M.A Straightforward One-Pot Synthesis of Pd-Ag Supported on Activated Carbon as a Robust Catalyst Toward Ethanol Electrooxidation. *Int. J. Hydrog. Energy*, 2021, 46, 9406–9416. https://doi.org/10.1016/j.ijhydene.2020.12.108.

57. Kuznetsova, L.I., Kuznetsova, N.I. Cyclohexane Oxidation with an O_2-H_2 Mixture in the Presence of a Two-Component Pt/C–Heteropoly Acid Catalyst and Ionic Liquids. *Kinet. Catal.*, 2017, 58(5), 522–532. https://doi.org/10.1134/s0023158417050147.
58. Cattaneo, S., Stucchi, M., Villa, A., Prati, L. Gold Catalysts for the Selective Oxidation of Biomass-Derived Products. *Chemcatchem*, 2019, 11, 309–323. https://doi.org/10.1002/cctc.201801243.
59. Aschwanden, L., Mallat, T., Maciejewski, M., Krumeich, F., Baiker, A. Development of a New Generation of Gold Catalysts for Amine Oxidation. *Chemcatchem*, 2010, 2, 666–673. https://doi.org/10.1002/cctc.201000092.
60. Maphoru, M.V., Heveling, J., Kesavan Pillai, S. Solvent and Temperature Effects on the Platinum-Catalyzed Oxidative Coupling of 1-Naphthols. *Eur. J. Org. Chem.*, 2016, 331–337 https://doi.org/10.1002/ejoc.201501280.
61. Liu, H., Liu, Y., Li, Y., Tang, Z., Jiang, H. Metal-Organic Framework Supported Gold Nanoparticles as a Highly Active Heterogeneous Catalyst for Aerobic Oxidation of Alcohols. *J. Phys. Chem. C*, 2010, 114, 13362–13369. https://doi.org/10.1021/jp105666f.
62. Sheldon, R.A. Recent Advances in Green Catalytic Oxidations of Alcohols in Aqueous Media. *Catal. Today*, 2015, 247, 4–13. https://doi.org/10.1016/j.cattod.2014.08.024.
63. Shokouhimehr, M., Yek, S.M.-G., Nasrollahzadeh, M., Kim, A., Varma, R.S. Palladium Nanocatalysts on Hydroxyapatite: Green Oxidation of Alcohols and Reduction of Nitroarenes in Water. *Appl. Sci.*, 2019, 9(19), 4183. https://doi.org/10.3390/app9194183.
64. Freakley, S.J., He, Q., Kiely, C.J., Hutchings, G.J. Gold Catalysis: A Reflection on Where We are Now. *Catal. Lett.*, 2015, 145, 71–79. https://doi.org/10.1007/s10562-014-1432-0.
65. Faroppa, M.L., Musci, J.J., Chiosso, M.E., Caggiano, C.G., Bideberripe, H.P., Fierro, J.L.G., Siri, G.J., Casella, M.L. Oxidation of Glycerol with H_2O_2 on Pb-promoted Pd/Γ-Al_2O_3 catalysts. *Chin. J. Catal.*, 2016, 37(11), Issue 11, 1982–1990. https://doi.org/10.1016/S1872-2067(16)62531-7.
66. Santonastaso, M., Freakley, S.J., Miedziak, P.J., Brett, G.L., Edwards, J.K., Hutchings, G.J. Oxidation of Benzyl Alcohol using In Situ Generated Hydrogen Peroxide. *Org. Proc. Res. Dev.*, 2014, 18(11), 1455–1460. https://doi.org/10.1021/op500195e.
67. Tiruvalam, R.C., Pritchard, J.C., Dimitratos, N., Lopez-Sanchez, J.A., Edwards, J.K., Carley, A.F., Hutchings, G.J., Kiely, C.J. Aberration Corrected Analytical Electron Microscopy Studies of Sol-Immobilized Au + Pd, Au{Pd} And Pd{Au} Catalysts Used for Benzyl Alcohol Oxidation and Hydrogen Peroxide Production. *Faraday Discuss.*, 2011, 152, 63–86. https://doi.org/10.1039/C1FD00020A.
68. Ma, C.Y.N., Cheng, J., Wang, H.L., Hu, Q., Tian, H., He, C., Hao, Z.P. Characteristics of Au/HMS Catalysts for Selective Oxidation of Benzyl Alcohol to Benzaldehyde. *Catal. Today*, 2010, 158(3–4), 246–251. https://doi.org/10.1016/j.cattod.2010.03.080.
69. Parmeggiani, C., Cardona, F. Transition Metal Based Catalysts in the Aerobic Oxidation of Alcohols. *Green Chem.*, 2012, 14, 547–564. https://doi.org/10.1039/C2GC16344F.
70. Langa, S., Nyamunda, B.C., Heveling, J. Antimony-Modified Platinum Catalysts for the Selective and Stable Oxidation of Cinnamyl Alcohol with Hydrogen Peroxide. *Catal. Lett.*, 2016, 146, 755–762. https://doi.org/10.1007/s10562-015-1689-y.
71. Tareq, S., Saiman, M.I., Yun Hin, T., Abdullah, A.H., Rashid, U. (2018). The Impact of Hydrogen Peroxide as an Oxidant for Solvent-Free Liquid Phase Oxidation of Benzyl Alcohol using Au-Pd Supported Carbon and Titanium Catalysts. *Bull. Chem. React. Eng. Catal.*, 13(2), 373–385. Doi:10.9767/bcrec.13.2.1204.373-385.
72. Duan, X., Zhang, Y., Pan, Y.M., Dong, H., Chen, B., Ma, Y., Qian, G., Zhou, X., Yang, J., Chen, D. SbO_x-Promoted Pt Nanoparticles Supported on CNTs as Catalysts for Base-Free Oxidation of Glycerol to Dihydroxyacetone. *Aiche J.*, 2018, 64(11), 3979–3987. https://doi.org/10.1002/aic.16217.
73. Jawale, D.V., Gravel, E., Villemin, E., Shah, N., Geertsen, V., Namboothiri, I.N.N., Doris, E. Co-Catalytic Oxidative Coupling of Primary Amines to Imines Using an Organic Nanotube-Gold Nanohybrid. *Chem. Commun.*, 2014, 50, 15251–15254. https://doi.org/10.1039/c4cc07951e.
74. Alhumaimess, M.S., Alsohaimi, I.H., Alshammari, H.M., Aldosari, O.F., Hassan, H.M.A. Synthesis of Gold and Palladium Nanoparticles Supported on Cuo/rGO using Imidazolium Ionic Liquid for CO Oxidation. *Res. Chem. Intermed.*, 2020, 46, 5499–5516. https://doi.org/10.1007/s11164-020-04274-w.
75. Qin,X., Zhang, L., Xu, G.-L., Zhu, S., Wang, Q., Gu, M., Zhang, X., Sun, C., Balbuena, P.B., Amine, K., Shao, M. The Role of Ru in Improving the Activity of Pd toward Hydrogen Evolution and

Oxidation Reactions in Alkaline Solutions. *ACS Catal.*, 2019, 9, 9614–9621. https://doi.org/10.1021/acscatal.9b01744.

76. Lu, Y., Zhang, Z., Lin, F., Wang, H., Yong Wang, Y. Single-Atom Automobile Exhaust Catalysts. *Chemnanomat*, 2020, 6 (12), 1659–1682. https://doi.org/10.1002/cnma.202000407.
77. Mangalam, J., Kumar, M., Sharma, M., Joshi, M. High Adsorptivity and Visible Light Assisted Photocatalytic Activity of Silver/Reduced Graphene Oxide (Ag/rGO) Nanocomposite for Wastewater Treatment. *Nano-Struct. Nano-Objects*, 2019, 17, 58–66. https://doi.org/10.1016/j.nanoso.2018.11.003.
78. Samad, S., Loh, K.S., Wong, W.Y., Lee, T.K., Sunarso, J., Chong, S.T., Daud, W.R.W. Carbon and Non-Carbon Support Materials for Platinum-Based Catalysts in Fuel Cells. *Int. J. Hydrog. Energy*, 2018, 43, 7823–7854. https://doi.org/10.1016/j.ijhydene.2018.02.154.
79. Chen, C., Li, X., Wang, L., Liang, T., Wang, L., Zhang, Y., Zhang, J. Highly Porous Nitrogen- and Phosphorus-Codoped Graphene: An Outstanding Support for Pd Catalysts to Oxidize 5-Hydroxymethylfurfural into 2,5-Furandicarboxylic Acid. *ACS Sustainable Chem. Eng.*, 2017, 5, 11300–11306. https://doi.org/10.1021/acssuschemeng.7b02049.
80. Jawale, D.V., Gravel, E., Geertsen, V., Li, H., Shah, N., Namboothiri, I.N.N., Doris, E. 2014. Aerobic Oxidation of Phenols and Related Compounds using Carbon Nanotube-Gold Nanohybrid Catalysts. *Chemcatchem*, 2014, 6, 719–723. https://doi.org/10.1002/cctc.201301069.
81. Mendoza-Pérez, R., Guisbiers, G. Bimetallic Pt–Pd Nano-Catalyst: Size, Shape and Composition Matter. *Nanotechnology*, 2019, 30, 305702. https://doi.org/10.1088/1361-6528/ab1759.
82. Suchomel, P., Kvitek, L., Prucek, R., Panacek, A., Halder, A., Vajda, S., Zboril, R. Simple Size-Controlled Synthesis of Au Nanoparticles and their Size Dependent Catalytic Activity. *Sci. Rep.*, 2018, 8, 4589. https://doi.org/10.1038/s41598-018-22976-5.
83. Alamgholiloo, H., Rostamnia, S., Hassankhani, A., Liu, X., Eftekhari, A., Hasanzadeh, A., Zhang, K., Karimi-Maleh, H.K., Khaksar, S., Varma, R.S., Shokouhimehr, M. Formation and Stabilization of Colloidal Ultra-Small Palladium Nanoparticles on Diamine-Modified Cr-MIL-101: Synergic Boost to Hydrogen Production From Formic Acid. *J. Colloid Interface Sci.*, 2020, 567, 126–135. https://doi.org/10.1016/j.jcis.2020.01.087.
84. Ciapina, E.G., Santos, S.F., Gonzalez, E.R. The Electrooxidation of Carbon Monoxide on Unsupported Pt Agglomerates. *J. Electroanal. Chem.*, 2010, 644(2), 132–143. https://doi.org/10.1016/j.jelechem.2009.09.022.
85. Meduri, K., Stauffer, C., Qian, W., Zietz, O., Barnum, A., Johnson, G.O., Fan, D., Ji, W., Zhang, C., Tratnyek, P., Jiao, J. Palladium and Gold Nanoparticles on Carbon Supports as Highly Efficient Catalysts for Effective Removal of Trichloroethylene. *J. Mater. Res. Technol.*, 2018, 33(16), 2404–2413. https://doi.org/10.1557/jmr.2018.212.
86. Maphoru, M.V., Heveling, J., Kesavan Pillai, S. Oxidation of 4-Methoxy-1-Naphthol on Promoted Platinum Catalysts. *Kinet. Catal.*, 2017, 58, 441–447. https://doi.org/10.1134/S0023158417040103.
87. Akbayrak, S., Özçifçi, Z., Tabak, A. Noble Metal Nanoparticles Supported on Activated Carbon: Highly Recyclable Catalysts in Hydrogen Generation From The Hydrolysis of Ammonia Borane. *J. Colloid Interface Sci.*, 2019, 546, 324–332. https://doi.org/10.1016/j.jcis.2019.03.070.
88. Yoshii, T., Nakatsuka, K., Mizobuchi, T., Kuwahara, Y., Itoi, H., Mori, K., Kyotani, T., Yamashita, H. Effects of Carbon Support Nanostructures on the Reactivity of a Ru Nanoparticle Catalyst in a Hydrogen Transfer Reaction. *Org. Process Res. Dev.*, 2018, 22(12), 1580–1585. https://doi.org/10.1021/acs.oprd.8b00207.
89. Wang, D., Villa, A., Su, D., Prof. Prati, L., Schlögl, R. Carbon-Supported Gold Nanocatalysts: Shape Effect in the Selective Glycerol Oxidation. *Chemcatchem*, 2013, 5(9), 717–2723. https://doi.org/10.1002/cctc.201200535.
90. Labulo, A.H., Omondi, B., Nyamori, V.O. Suzuki-Miyaura Reaction and Solvent free Oxidation of Benzyl Alcohol by Pd/Nitrogen-Doped CNTs Catalyst. *J. Mater. Sci.*, 2018, 53:15817–15836. https://doi.org/10.1007/s10853-018-2748-8.
91. Guo, W., Niu, S., Shi, W., Zhang, B., Yu, W., Xie, Y., Ji, X., Wu, Y., Su, D., Shao, L. Pd–P Nanoalloys Supported on a Porous Carbon Frame as an Efficient Catalyst for Benzyl Alcohol Oxidation. *Catal. Sci. Technol.*, 2018, 8, 2333–2339. https://doi.org/10.1039/c8cy00554k.
92. Maphoru, M.V., Kesavan Pillai, S., Heveling, J. Structure-Activity Relationships of Carbon-Supported Platinum-Bismuth and Platinum-Antimony Oxidation Catalysts. *J. Catal.*, 2017, 348, 47–58. http://dx.doi.org/10.1016/j.jcat.2017.02.003.

93. Besson, M., Gallezot, P. Selective Oxidation of Alcohols and Aldehydes on Metal Catalysts. *Catal. Today*, 2000, 57: 127–141. https://doi.org/10.1016/s0920-5861(99)00315-6.
94. Jaska, C.A., Manners, I. Heterogeneous or Homogeneous Catalysis? Mechanistic Studies of the Rhodium-Catalyzed Dehydrocoupling of Amine-Borane and Phosphine-Borane Adducts. *J. Am. Chem. Soc.*, 2004, 126, 9776–9785. https://doi.org/10.1021/ja0478431.
95. Kim, Y., Kim, J., Kim, D.H. Investigation on the Enhanced Catalytic Activity of a Ni-Promoted Pd/C Catalyst for Formic Acid Dehydrogenation: Effects of Preparation Methods and Ni/Pd Ratios. *RSC Adv.*, 2018, 8, 2441–2448. https://doi.org/10.1039/c7ra13150j.
96. Hosseinabadi, P., Javanbakht, M., Naji, L., Ghafarian-Zahmatkesh, H. Influence of Pt Nanoparticle Electroless Deposition Parameters of the Electrochemical Characteristics of Nafion-Based Catalyst-Coated Membranes. *Ind. Eng. Chem. Res.*, 2018, 57, 2, 434–445. https://doi.org/10.1021/acs.iecr.7b03647.
97. Song, J., Xiao, Z., Jiang, Y., Chang, A., Zeng, J. Surfactant-Free Room Temperature Synthesis of Pd_xPt_y/C Assisted by Ultra-Sonication as Highly Active and Stable Catalysts for Formic Acid Oxidation. *Int. J. Hydrog. Energy*, 2019, 44, 11655–11663. https://doi.org/10.1016/j.ijhydene.2019.03.169.
98. Wang, H.-W., Dong, R.-X., Chang, H.-Y., Liu, C.-L., Chen-Yang, Y,-W. Preparation and Catalytic Activity of Pt/C Materials via Microwave Irradiation. *Mater. Lett.*, 2007, 61, 830–833. https://doi.org/10.1016/j.matlet.2006.05.067.
99. Munnik, P., De Jongh, P.E., De Jong, K.P. Recent Developments in the Synthesis of Supported Catalysts. *Chem. Rev.*, 2015, 115(14), 6687–6718. https://doi.org/10.1021/cr500486u.
100. Truszkiewicz, E., Kowalczyk, K., Dębska, A., Wojda, D., Iwanek, E., Kępinski, L., Mierzwa, B. Methanation of CO On Ru/Graphitized-Carbon Catalysts: Effects of the Preparation Method and the Carbon Support Structure. *Int. J. Hydrog. Energy*, 2020, 45(56), 31985–31999. https://doi.org/10.1016/j.ijhydene.2020.08.235.
101. Amin, R.S.A.A . Elzatahry, A.A., El-Khatib, K.M., Youssef, M.E. Nanocatalysts prepared by Microwave and Impregnation Methods for Fuel Cell Application. *Int. J. Electrochem. Sci.*, 2011, 6, 4572.
102. Francesco, I.N., Fontaine-Vive, F., Antoniotti, S. Synergy in the Catalytic Activity of Bimetallic Nanoparticles and New Synthetic Methods for the Preparation of Fine Chemicals. *Chemcatchem*, 2014, 6, 2784–2791. https://doi.org/10.1002/cctc.201402252.
103. Wu, Z., Borretto, E., Medlock, J., Bonrath, W., Cravotto, G. Effects of Ultrasound and Microwaves on Selective Reduction: Catalyst Preparation and Reactions. *Chemcatchem*, 2014, 6, 2762–2783. https://doi.org/10.1002/cctc.201402221.
104. Yu, X., Pickup, P.G. Pb and Sb Modified Pt/C Catalysts for Direct Formic Acid Fuel Cells. *Electrochim. Acta*, 2010, 55, 7354–7361. https://doi.org/10.1016/j.electacta.2010.07.019.
105. Wanjala, B., Luo, J., Fang, B., Mott, D., Zhong, C.-J. Gold-Platinum Nanoparticles: Alloying and Phase Segregation. *J. Mater. Chem.*, 2011, 21, 4012–4020. https://doi.org/10.1039/c0jm02682d.
106. Moniz, S.J.A., Bhachu, D., Blackman, C.S., Cross, A.J., Elouali, S., Pugh, D., Cabrera, R.Q., Vallejos, S. A Novel Route to Pt-Bi_2O_3 Composite Thin Films and their Application in Photo-Reduction of Water. *Inorg. Chim. Acta*, 2012, 380, 328–335. https://doi.org/10.1016/j.ica.2011.09.029.
107. Roy, K., Artiglia, L., Xiao, Y, Varma, A, van Bokhoven, J.A. Role of Bismuth in the Stability of Pt-Bi Bimetallic Catalyst for Methane Mediated Deoxygenation of Guaiacol, on APXPS Study. *ACS Catal.* 2019, 9, 3694–3699. DOI: 10.1021/acscatal.8b04699.
108. Besson M, Lahmer F, Gallezot P, Fuertes P, Flèche G. Catalytic Oxidation Of Glucose on Bismuth-Promoted Palladium Catalysts. *J. Catal.*, 1995, 152, 116–121 https://doi.org/10.1006/jcat.1995.1065.
109. Cai, Jindi, Yiyin Huang, Yonglang Guo. "Bi-Modified Pd/C Catalyst via Irreversible Adsorption and Its Catalytic Activity for Ethanol Oxidation in Alkaline Medium." *Electrochim. Acta*, 2013, 99 22–29.
110. Yu, X., Pickup, P.G. Codeposited PtSb/C Catalysts for Direct Formic Acid Fuel Cells. *J. Power Sources*, 2011, 196, 7951–7956. https://doi.org/10.1016/j.jpowsour.2011.05.051.
111. Yang, W., Zhao, J., Tian, H., Wang, L., Wang, X., Ye, S., Liu, J., Huang, J. Solar-Driven Carbon Nanoreactor Coupling Gold and Platinum Nanocatalysts for Alcohol Oxidations. *Small*, 2020, 16, 2002236. https://doi.org/10.1002/smll.202002236.

112. German, D., Pakrieva, E., Kolobova, E., Carabineiro, S.A.C., Stucchi, M., Villa, A., Prati, L., Bogdanchikova, N., Corberán, V.C., Pestryakov, A. Oxidation of 5-Hydroxymethylfurfural on Supported Ag, Au, Pd And Bimetallic Pd-Au Catalysts: Effect of the Support. *Catalysts*, 2021, 11, 115. https://doi.org/10.3390/catal11010115.
113. Yin, Y.B., Chen, L., Heck, K.N., Zhang, Z., Wong, M.S. Toward Glucuronic Acid Through Oxidation of Methyl-Glucoside Using Pdau Catalysts. *Catal. Commun.*, 2020, 135. 105895 https://doi.org/10.1016/j.catcom.2019.105895.
114. Phungjit, U., Hunsom, M., Pruksathorn, K. Effect of Temperature for Platinum/Carbon Electrocatalyst Preparation on Hydrogen Evolution Reaction. *Eng. J.*, 2021, 25(4). https://doi.org/10.4186/ej.2021.25.4.105.
115. Altaee, H., Alshamsi, H.A. Selective Oxidation of Benzyl Alcohol by Reduced Graphene Oxide Supported Platinum Nanoparticles. *J. Phys. Conf. Ser.*, 2020, 1664. https://doi.org/10.1088/1742-6596/1664/1/012074.
116. Mallat, T., Baiker, A. Reactions in "Sacrificial" Solvents. *Catal. Sci. Technol.,* 2011, 1, 1572–1583. https://doi.org/10.1039/c1cy00207d.
117. Mallat, T., Bodnar, Z., Hug, P., Baiker, A. Selective Oxidation of Cinnamyl Alcohol to Cinnamaldehyde with Air over Bi-Pt/Alumina Catalysts. *J. Catal.*, 1995, 153, 131–143. https://doi.org/10.1006/jcat.1995.1115.
118. Chen, C.-T., Nguyen, C.V., Wang, Z.-Y., Bando, Y., Yamauchi, Y., Bazziz, M.T.S., Fatehmulla, A., Farooq, W.A., Yoshikawa, T., Masuda, T., Wu, K.C.-W. Hydrogen Peroxide Assisted Selective Oxidation of 5-Hydroxymethylfurfural in Water under Mild Conditions. *Chemcatchem* 2018, 10, 361–365. https://doi.org/10.1002/cctc.201701302.
119. Zhou, Y., Shen, Y., Xi, J., Luo, X. Selective Electro-Oxidation of Glycerol to Dihydroxyacetone by PtAg Skeletons. *ACS Appl. Mater. Interfaces*, 2019, 11(32), 28953–28959. https://doi.org/10.1021/acsami.9b09431.
120. Yan, H., Yao, S., Zhao, S., Liu, M., Zhang, W., Zhou, X., Zhang, G., Jin, X., Liu, Y., Feng, X., Chen, X., Chen, D., Yang, C. Insight into the Basic Strength-Dependent Catalytic Performance in Aqueous Phase Oxidation of Glycerol to Glyceric *Acid. Chem. Eng. Sci.*, 2021, 230, 116191. https://doi.org/10.1016/j.ces.2020.116191.
121. Ciriminna, R., Fidalgo, A., Ilharco, L.M., Pagliaro, M. Dihydroxyacetone: An Updated Insight into an Important Bioproduct. *Chemistryopen*, 7(3), 233–236. https://doi.org/10.1002/Open.201700201.
122. Tang, T., Wang, Y., Dong, W., Liu, C., Xu, C. Reusable and Active Pt@Co-NC Catalysts for Oxidation of Glycerol. *Renew. Energy*, 2020, 153, 472–479. https://doi.org/10.1016/j.renene.2020.02.029.
123. Meng, Y., Wang, H., Dai, Y., Zheng, J., Yu, H., Zhou, C., Yang, Y. Modulating the Electronic Property of Pt Nanocatalyst on rGO By Iron Oxides for Aerobic Oxidation of Glycerol. *Catal. Commun.*, 2020, 144, 106073. https://doi.org/10.1016/j.catcom.2020.106073.
124. Song, H., Liu, Z., Gai, H., Wang, Y., Qiao, L., Zhong, C., Yin, X., Xiao, M. Nitrogen-Doped Ordered Mesoporous Carbon Anchored Pd Nanoparticles for Solvent Free Selective Oxidation of Benzyl Alcohol to Benzaldehyde by Using O_2. *Front. Chem.*, 2019, 7, 458. https://doi.org/10.3389/fchem.2019.00458,
125. Torbina, V.V., Vodyankin, A.A., Ten, S., Mamontov, G.V., Salaev, M.A., Sobolev, V.I., Vodyankina, O.V. Ag-Based Catalysts in Heterogeneous Selective Oxidation of Alcohols: A Review. Catalysts, 2018, 8, 447. https://doi.org/10.3390/catal8100447.
126. Göksu, H., Cellat, K., Şen, F. Single-Walled Carbon Nanotube Supported PtNi Nanoparticles (PtNi@SWCNT) Catalyzed Oxidation of Benzyl Alcohols to the Benzaldehyde Derivatives in Oxygen Atmosphere. *Sci. Rep.*, 2020, 10:9656. https://doi.org/10.1038/s41598-020-66492-.
127. Takeya, T., Otsuka, T., Okamoto, I., Kotani, E. Semiconductor-Mediated Oxidative Dimerization of 1-Naphthols with Dioxygen and *o*-Demethylation of the Enol-Ethers by SnO_2 without Dioxygen. *Tetrahedron*, 2004, 60, 10681–10693. https://doi.org/10.1016/j.tet.2004.09.003.
128. Kucherov, F.A., Romashov, L.V., Galkin, K.I., Ananikov, V.P. Chemical Transformations of Biomass-Derived C6-Furanic Platform Chemicals for Sustainable Energy Research, Materials Science, and Synthetic Building Blocks ACS. *Sustain. Chem. Eng.*, 2018, 7 , 8064–8092 https://doi.org/10.1021/acssuschemeng.8b00971.

129. Chen, C., Li, X., Wang, L., Liang, T., Wang, L., Zhang, Y., Zhang, J. Highly Porous Nitrogen- and Phosphorus-Codoped Graphene: An Outstanding Support for Pd Catalysts to Oxidize 5-Hydroxymethylfurfural into 2,5-Furandicarboxylic Acid. *ACS Sustainable Chem. Eng.*, 2017, 5, 11300–11306. https://doi.org/10.1021/acssuschemeng.7b02049.
130. Joshi, A.S., Lawrence, J.G., Coleman, M.R. Effect Of Biaxial Orientation on Microstructure and Properties of Renewable Copolyesters of Poly(Ethylene Terephthalate) with 2,5- Furandicarboxylic Acid for Packaging Application . *ACS Appl. Polym. Mater.*, 2019, 1, 1798–1810. https://doi.org/10.1021/acsapm.9b00330.
131. Farjadian, F., Abbaspour, S., Sadatlu, M.A.A., Mirkiani, S., Ghasemi, A., Hoseini-Ghahfarokhi, M., Mozaffari, N., Karimi, M., Hamblin, M.R. Recent Developments in Graphene and Graphene Oxide: Properties, Synthesis, and Modifications: *A Review. Chemistryselect*, 2020, 5, 10200–10219. https://doi.org/10.1002/Slct.202002501.
132. Cao, Y., Mao, S., Li, M., Chen, Y., Wang, Y. Metal/Porous Carbon Composites for Heterogeneous Catalysis: Old Catalysts with Improved Performance Promoted by N-Doping. *ACS Cata.*, 2017, 7, 8090–8112. https://doi.org/10.1021/acscatal.7b02335.
133. Feng, Y., Jia, W., Yan, G., Zeng, X., Sperry, J., Xue, B., Sun, Y., Tang, X., Lei, T., Lin, L. Insights into the Active Sites and Catalytic Mechanism of Oxidative Esterification of 5-Hydroxymethylfurfural by Metal-Organic Frameworks-Derived N-Doped Carbon. *J. Catal.*, 2020, 381, 570–578. https://doi.org/10.1016/j.jcat.2019.11.029.
134. Campisi, S., Chan-Thaw, C.E., Villa, A. Understanding Heteroatom-Mediated Metal–Support Interactions in Functionalized Carbons: A Perspective Review. *Appl. Sci.* 2018, 8, 1–25. https://doi.org/10.3390/app8071159.
135. Kamiuchi, N., Mitsui, T., Muroyama, H., Matsui, T., Kikuchi, R., Eguchi, K. 2010. Catalytic Combustion of Ethyl Acetate and Nanostructural Changes of Ruthenium Catalysts Supported on Tin Oxide. *Appl. Catal. B: Env.* 97:120–126. https://doi.org/10.1016/j.apcatb.2010.03.031.
136. Huang, B., He, Y., Zhu, Y., Wang, Z., Cen, K. 2020. SO_2 Electrocatalytic Oxidation Properties of Pt-Ru/C Bimetallic Catalysts with Different Nanostructures. *Langmuir*, 36:3111–3118. https://doi.org/10.1021/acs.langmuir.9b03286.
137. Martin, S.F. Recent Applications of Imines as Key Intermediates in the Synthesis of Alkaloids and Novel Nitrogen Heterocycles. *Pure Appl. Chem.*, 2009, 81(2), 195–204. https://doi.org/10.1351/pac-con-08-07-03.
138. Zhang, Y., Lu, F., Zhang, H.-Y., Zhao, J. Activated Carbon Supported Ruthenium Nanoparticles Catalyzed Synthesis of Imines from Aerobic Oxidation of Alcohols with Amines. *Catal. Lett.*, 2017, 147, 20–28. https://doi.org/10.1007/s10562-016-1930-3.
139. Wu, G., Swaidan, R., Lia, D., Lia, N. Enhanced Methanol Electro-Oxidation Activity of PtRu Catalysts Supported on Heteroatom-Doped Carbon. *Electrochim. Acta*, 2008, 53(26), 7622–7629. https://doi.org/10.1016/j.electacta.2008.03.082.
140. Salazar, A., Linke, A., Eckelt, R., Quade, A., Kragl, U., Mejía, E. Oxidative Esterification of 5-Hydroxymethylfurfural under Flow Conditions using a Bimetallic Co/Ru Catalyst. *Chemcatchem*, 2020, 12, 1–9. https://doi.org/10.1002/cctc.202000205.
141. Hutchings, G.J. Catalysis by Gold. *Catal. Today*, 2005, 100(1–2), 55–61. https://doi.org/10.1016/j.cattod.2004.12.016.
142. Megías-Sayago, C., Lolli, A., Bonincontro, D., Penkova, A., Albonetti, S., Cavani, F., Odriozola, J.A., Ivanova, S. Effect of Gold Particles Size over Au/C Catalyst Selectivity in HMF Oxidation Reaction, *Chemcatchem*, 2020, 12, 1177–1183. https://doi.org/10.1002/cctc.201901742.
143. Ketchie, W.C., Fang, Y.-L., Wong, B.S., Murayama, M., .Davis, R.J. Influence of Gold Particle Size on the Aqueous-Phase Oxidation of Carbon Monoxide and Glycerol. *J. Catal.*, 2007, 250(1), 94–101. https://doi.org/10.1016/j.jcat.2007.06.001.
144. Sahoo, S.R., Ke, S.-C. Spin-Orbit Coupling Effects in Au 4f Core-Level Electronic Structures in Supported Low-Dimensional Gold Nanoparticles. *Nanomater.*, 2021, 11, 554. https://doi.org/10.3390/nano11020554.
145. Habe, H., Fukuoka, T., Kitamoto, D., Sakaki, K. Biotechnological Production of D-Glyceric Acid and its Application. *Appl. Microbiol. Biotechnol.*, 2009, 84, 445–452. https://doi.org/10.1007/s00253-009-2124-3.

146. Murthy, P.R., Selvam, P. The Enhanced Catalytic Performance and Stability of Ordered Mesoporous Carbon Supported Nano-Gold with High Structural Integrity for Glycerol Oxidation. *Chem. Rec.*, 2019, 19, 1913–1925. https://doi.org/10.1002/tcr.201800109.
147. El-Najjar, N., Gali-Muhtasib, H., Ketola, R.A., Vuorela, P., Urtti, A., Vuorela, H. The Chemical and Biological Activities of Quinones: Overview and Implications in Analytical Detection. *Phytochem. Rev.*, 2011, 353–370. https://doi.org/10.1007/s11101-011-9209-1.
148. Ganapaty, S., Thomas, P.S., Fotso, S., Laatsch, H. Antitermitic Quinones from Diospyros Sylvatica. *Phytochemistry*, 65(9), 2004, 1265–1271. https://doi.org/10.1016/j.phytochem.2004.03.011.
149. Ogata, T., Okamoto, I., Kotani, E., Takeya, T. Biomimetic Synthesis of the Dinaphthofuranquinone Violet-Quinone, Utilizing Oxidative Dimerization with the ZrO_2/O_2 System. *Tetrahedron*, 2004, 60, 3941–3948. https://doi.org/10.1016/j.tet.2004.03.038.
150. Takeya, T., Doi, H., Ogata, T., Okamoto, I., Kotani, E. Aerobic Oxidative Dimerization of 1-Naphthols to 2,2'-Binaphthoquinones Mediated by $SnCl_4$ and its Application to Natural Product Synthesis. *Tetrahedron*, 2004, 60, 9049–9060. https://doi.org/10.1016/j.tet.2004.07.073.
151. Kholdeeva, O.A., Zalomaeva, O.V., Shmakov, A.N., Melgunov, M.S., Sorokin, A.B. Oxidation of 2-Methyl-1-Naphthol with H_2O_2 over Mesoporous Ti-MMM-2 Catalyst. *J. Catal.*, 2005, 236, 62–68. https://doi.org/10.1016/j.jcat.2005.09.022.
152. Zalomaeva, O.V., Trukhan, N.N., Ivanchikova, I.D., Panchenko, A.A., Roduner, E., Talsi, E.P., Sorokin, A.B., Rogov, V.A., Kholdeeva, O.A. EPR Study on the Mechanism of H_2O_2-Based Oxidation of Alkylphenols over Titanium Single-Site Catalysts. *J. Mol. Catal. A: Chem.*, 2007, 277, 185–192. https://doi.org/10.1016/j.molcata.2007.07.047.
153. López, L.I.L., Flores, S.D.N., Belmares, S. Y.S., Galindo, A.S. Naphthoquinones: Biological Properties and Synthesis of Lawsone and Derivatives-a Structured Review. *Vitae-Revista De La Facultad De Química Farmacéutica*, 2014, 21(3), 248–258.
154. Maphoru, M.V., Heveling, J., Pillai, S.K. Oxidative Coupling Of 1-Naphthols Over Noble And Base Metal Catalysts. *Chempluschem*, 2014, 79(1), 99–106. https://doi.org/10.1002/cplu.201300307.
155. Lama, S.M.G., Schmidt, J., Malik, A., Walczak, R., Silva, D.V., Vçlkel, A., Oschatz, M. Modification of Salt-Templated Carbon Surface Chemistry for Efficient Oxidation of Glucose with Supported Gold Catalysts. *Chemcatchem*, 2018, 10, 2458–2465. https:/doi.org/10.1002/cctc.201800104.
156. Léonard, E., Mangin, F., Villette, C., Billamboza, M., Len, C. Azobenzenes and Catalysis. *Catal. Sci. Technol.*, 2016, 6, 379–398. https://doi.org/10.1039/c4cy01597e.
157. Fedele, C., Netti, P.A., Cavalli, S. Azobenzene-Based Polymers: Emerging Applications as Cell Culture Platforms. *Biomater. Sci.*, 2018, 6, 990–995. https://doi:10.1039/c8bm00019k.
158. Gao, B.-B., Zhang, M., Chen, X.-R., Zhu, D.-L., Yu, H., Zhang, W.-H., Lang, J.-P. Preparation of Carbon-Based AuAg Alloy Nanoparticles by using the Heterometallic $[Au_4Ag_4]$ Cluster for Efficient Oxidative Coupling of Anilines. *Dalton Trans.*, 2018, 47, 5780–5788. https://doi.org/10.1039/c8dt00695d.
159. Liu, Y., Jiang, G., Li, L., Chen, H., Huang, Q., Du, X., Tong, Z. Electrospun CeO_2/Ag@carbon Nanofiber Hybrids for Selective Oxidation of Alcohols. *Powder Technol.*, 2017, 305, 597–601. https://doi.org/10.1016/j.powtec.2016.10.042.

5 Metal Oxide Nanomaterials for Visible Light Photocatalysis

G. Reenamole

CONTENTS

5.1 Introduction ... 103
5.2 Visible Light Photocatalysis ... 104
5.3 Mechanism of Photocatalysis ... 105
5.4 Various Types of Visible Light-Absorbing Photocatalysts ... 107
5.4.1 Modified Single Materials ... 108
5.4.2 Heterojunctions ... 108
5.4.3 Z-Scheme Heterojunctions ... 110
5.4.4 Carbon-Semiconductor Composites ... 110
5.4.5 Plasmonic Materials ... 111
5.4.6 Sensitized Photocatalysts ... 111
5.5 TiO_2 based Visible Light Active Metal Oxide Photocatalysts ... 112
5.5.1 TiO_2 Structure, Properties and Electronic Processes ... 112
5.5.2 Advanced TiO_2 Nanomaterials for Visible-Light Induced Applications ... 114
5.6 Non TiO_2 Based Visible Light Photocatalysts ... 118
5.6.1 WO_3 Based Photocatalysts ... 118
5.6.2 Bismuth-Based Photocatalysts ... 118
5.6.3 MoS_2 Based Visible Light Active Photocatalysts ... 119
5.6.4 Ag_3PO_4 Based Nanocomposites ... 119
5.7 Applications of Metal Oxides for Visible Light-Induced Photocatalysis ... 119
5.7.1 Environmental Applications ... 119
5.7.2 Energy Applications ... 124
5.7.3 Synthetic Applications ... 125
5.8 Future Outlook and Conclusion ... 126
References ... 126

5.1 INTRODUCTION

The field of semiconductor photocatalysis has largely grown over the past few decades of our lives and continues its diverse applications. The use of metal oxides as photocatalysts has gained enormous interest among the scientific community, especially as a potential candidate for energy and environmental applications as they possess advantages such as environmental friendliness and self-regeneration. From the discovery in 1972 by Fujishima and Honda of photo splitting of water using TiO_2 metal oxides [1], its photochemical sterilization and self-cleaning capabilities were highly focused and extended those capabilities towards several outdoor applications. The most important features of such materials are continuous absorption bands, narrow and intense emission spectra, processability, high chemical stability, and surface functionality. Most metal oxides may be used as photocatalysts due to the promising arrangement of electronic structure, light absorption capabilities, and charge transport characteristics. While TiO_2 is an ecofriendly material with a wide range of applications in industry, it has recently gained attention as part of environmental rehabilitation efforts

DOI: 10.1201/9781003218708-6

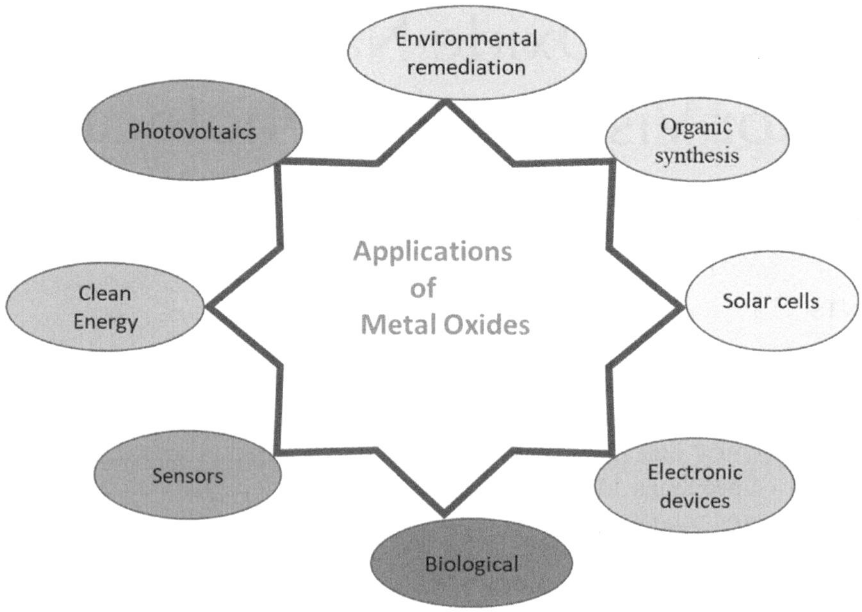

FIGURE 5.1 Applications of metal oxide nanomaterials.

in places plagued by chronic pollution. Products such as self-cleaning window films, air-purifying roofing tiles, air filters, and antifogging mirrors and so forth are some of the examples of commercialization of their photocatalytic properties – and, still, visible light applications are yet to be explored. At present, the use of air disinfection using a metal–oxide-based visible light photocatalyst is especially pertinent in connection with the increasing spread of COVID-19 coronavirus, which was first reported in Wuhan, China, and subsequently spread worldwide. Various applications of metal oxides are summarized in Figure 5.1. Furthermore, tunable surface characteristics and electronic band energy configuration of nanometer level metal oxides make them suitable to meet the requirements in commercial applications.

5.2 VISIBLE LIGHT PHOTOCATALYSIS

A photocatalyst is defined as a material that is proficient to absorb light, produce electron-hole pairs that allow chemical transformations of the participants and regenerate its chemical composition after each cycle of such interactions. Photocatalysis refers to the acceleration of a chemical reaction in the presence of light and photocatalyst, where the catalyst must be able to absorb quanta of suitable wavelengths depending on the band structure [2–4]. The essential part of the process is the action of ultraviolet (UV), visible (VIS) or infrared (IR) radiation so that it will initiate the chemical reaction catalyzed by a photocatalyst. Concepts for the use of photocatalysis should possess qualities such as high photoresponse towards UV and visible region of the electromagnetic spectrum, resistance to photo corrosion, inertness, low cost and low environmental effects and toxicity. However, visible light-mediated photocatalysis relies on the ability of photocatalysts to absorb low-energy visible light and engage in energy transfer (ET) processes with organic substrates. The looming energy crisis and fastest rising pollution issues triggered interest in the exploration of renewable energy sources, such as wind or solar light. In particular, exploiting the use of solar energy to perform advanced photooxidation and reduction processes and thereby degrade the toxic organic substances, has become the challenging area of research. Moreover, the photocatalytic process is found to be highly beneficial in terms of generating renewable and green energy carriers (e.g., H_2 and O_2)

from water splitting [5], use in decreasing pollution (e.g., degradation of organic contaminants via photocatalysis) [6], its use in energy conversion and materials storage (e.g., photocatalytic reduction of CO_2) [7–9] and use in synthetic fields where green chemistry is chosen as a major pathway to synthesis organic compounds [10]. Most commonly using catalysts are semiconductors with a characteristic electronic structure of filled valance band (VB) and empty conduction bands (CB). Electron-hole pairs are then generated as a result of the absorption, a photon with sufficient energy (greater than the band-gap energy) and thereafter migrated from the VB to the CB. Then, these electrons and holes migrate to the surface of photocatalysts and subsequently react with adsorbed electron acceptors and donors, respectively (See the details in section 6.3). Thus, an efficient semiconductor photocatalyst requires a band gap suitable for harvesting light, facile separation and transportation of charge carriers, and a proper valence band (VB) and conduction band (CB) edge potential for redox reaction being feasible. A large number of photocatalytic materials, including conventional semiconductors and recently emerging photoelectronic materials like quantum dots, plasmonic metal nanoparticles and 2D materials and so forth are increasingly studied for the purpose of effective light absorption, charge separation, and charge transfer.

5.3 MECHANISM OF PHOTOCATALYSIS

As discussed previously, the significant features of photocatalytic system are the desired band gap, suitable morphology, high surface area, stability and reusability. There are various metal oxides such as TiO_2, ZnO, MoO_3, CeO_2, ZrO_2, WO_3, Fe_2O_3 and SnO_2 which are used in semiconductor photocatalysis as they can act as sensitizers for the light-induced redox reactions due to their electronic structure, filled valence band and empty conduction band. When metal oxide is irradiated with light having energy *hυ* equal to or exceeding the band gap energy of the semiconductor, electrons from the valence band are excited to the conduction band, resulting in holes in the valence band [11] as shown in Figure 5.2. The photogenerated pair (e-/h+) is able to reduce and/or oxidize a compound adsorbed on the surface of photocatalyst since their lifetimes are long enough to participate in the redox reaction. The lifetime of this excited electron and the created positive hole should be long enough to enable their participation in reduction and oxidation reactions. In fact, the lack of a continuum of interband states in semiconductors assures an adequately extended lifetime for these photogenerated e–CB/h+VB pairs to initiate redox reactions on the catalyst surface. On the other

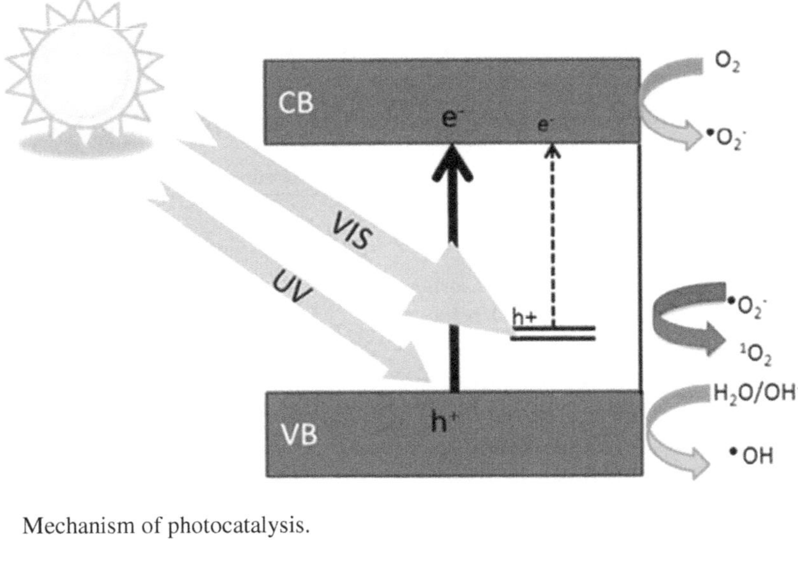

FIGURE 5.2 Mechanism of photocatalysis.

hand, both charges must migrate to the surface of the semiconductor particle to be available to the surrounding medium. However, these charge pairs can either recombine and discharge the input energy as heat, get trapped in metastable surface states, or can react with electron donors and electron acceptors adsorbed on the semiconductor surface [12]. The quantum efficiency of a photocatalytic reaction is expected to increase by increasing the lifetime of electron–hole pairs and the rate of the interfacial charge transfer process. A competition exists between the trapping and recombination of the photogenerated electron-hole pairs and that between the interfacial charge transfer and recombination of the trapped species. However, the net effect of these contests determines the overall quantum efficiency of the photocatalytic process.

In addition, the capacity of a photocatalyst to carry out certain reactions is determined by the relative positions of the catalyst and substrate's energy levels. If the adsorbed molecule's reduction potential is higher than that of the photoelectrons, it can be reduced, and if its potential is lower than that of the holes, it can be oxidized. The interesting characteristic feature of semiconducting metal oxides is the strong oxidizing power of their holes. The redox potential for photo-generated holes is +2.53 V versus the standard hydrogen electrode (SHE). They can react with water molecules in a one-electron oxidation step to produce highly active hydroxyl radicals (•OH), which can degrade the organic contaminants into CO_2, water and mineral acids [13]. The major drawbacks of TiO_2 based photocatalysts arise from the rapid charge recombination of the above-mentioned electron-hole pairs, thereby suppressing the quantum efficiency and the wide band gap of the material, which restricts light absorption to only the ultraviolet region (wavelength < 390 nm) and thus limiting the practical applications of metal oxide-based photocatalysts for solar light harvesting. The possible events in a simple semiconductor metal oxide photocatalysis and their lifetimes are shown in Table 5.1. TiOH represents the primary hydrated surface functionality of TiO_2, hVB+ is a valence band hole, eCB- is a conduction band electron and $Ti_{(III)}OH$ is the surface trapped conduction band electron. It is assumed that the hole-transfer occurs only through a surface-trapped hole species or through the hydroxyl radical. However, the rate of the decomposition of pollutants was found to be highly dependent on its adsorbed concentration, and which is an indirect indication of the hydroxyl radical concentration. The illuminated surface area of photocatalysts, light irradiance, reactants adsorption rate, electron-hole recombination rate, and various characteristics properties of the catalysts determine the pace of the photocatalytic process [14]. Studies show that there is a linear relationship between the photocatalytic rate and the amount of substrate adsorbed on the surface of the photocatalyst. The higher the surface area of the photocatalyst, the faster the rate of reaction of e- and h+ with the adsorbed substrates because of the larger number of substrates that surround the e-– h+ pairs [15]. Unlike the reaction of electrons and holes, it is difficult to evaluate the recombination

TABLE 5.1
Electronic Events and Lifetime on Light Absorption by TiO_2

Electronic Processes	Lifetime of the Species
a) Charge carrier generation	
$TiO2 + h\upsilon \rightarrow hVB+ + eCB^-$	10^{-15} (very fast)
b) OH• formation at the TiO2 surface	
$hVB+ + Ti^{IV}OH \rightarrow \{Ti^{IV}OH\bullet\} +$	10^{-9} (fast)
c) Charge trapping	
$eCB^- + T^{iIV}\ OH \rightarrow \{T^{iIII}\ OH\}$	10-10 (dynamic)
$eCB^- + TiIV \rightarrow Ti\ III$	10-8 (Irreversible)
d) Charge–carrier recombination	
$eCB^- + \{TiIVOH\bullet\}+ \rightarrow TiIVOH$	10-7 (slow)
$hVB+ + \{TiIIIOH\} \rightarrow TiIVOH$	10-9 (fast)

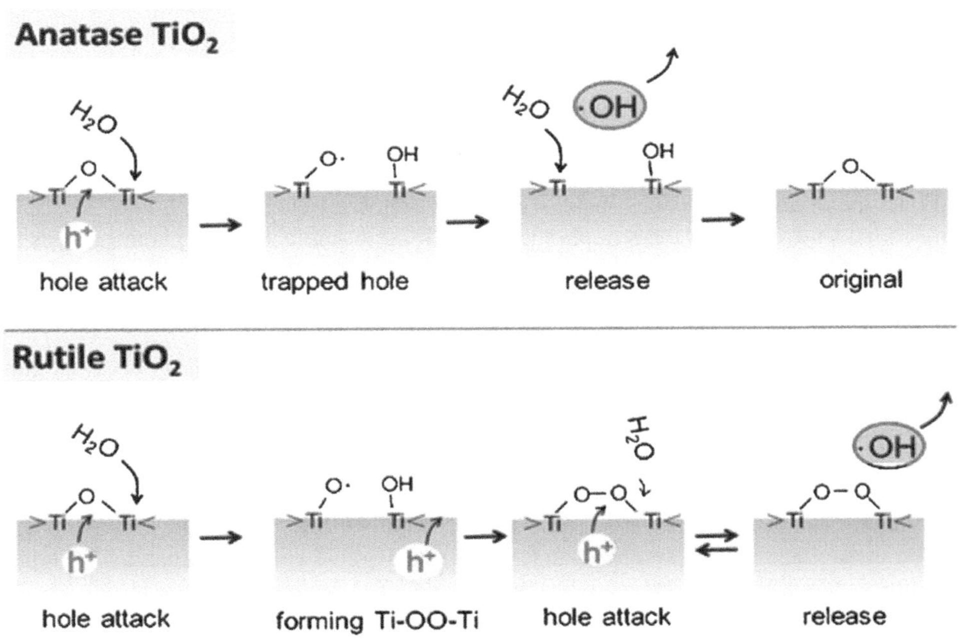

FIGURE 5.3 Scheme showing the mechanism of OH radical production with anatase and rutile. Reproduced with permission from Ref. [21] [Copyright 2014] American Chemical Society.

rate directly. It is assumed that recombination takes place at crystal defects [16–17]. The surface of the crystal can be considered a defective site and catalysts with higher surface area will have a higher recombination rate. But if the surface reaction predominates the recombination reaction, the catalyst with higher surface area is better and vice versa.

Band gaps narrowing up to 0.2 eV were reported for TiO_2 nanomaterials with 5–10 nm particle sizes, indicating that poorly crystalline nanoparticles and thin films had wider band gaps. For example, anatase (tetragonal), rutile (tetragonal), and brookite (orthorhombic) are three known polymorphs of TiO_2 and corresponding band gaps are 3.2, 3.0 and 3.4 eV, respectively [18]. Anatase TiO_2 is considered as the photocatalytically most active polymorph of TiO_2 due to superior mobility of its electron–hole pairs and improved surface hydroxyl density [19]. In contrast, the photocatalytic performance of rutile TiO_2 is not promising, and that of the brookite phase has not been systematically investigated [20]. Zhang et al. used two distinct probe molecules, coumarin and coumarin-3-carboxylic acid, to investigate the mechanism of OH radical generation in anatase and rutile photocatalysts [21]. Rutile TiO_2 was found to produce a smaller amount of OH radicals compared to anatase. The production of hydroxyl radicals on the anatase TiO_2 surface was shown to be responsible for the conversion of trapped holes. Whereas on rutile TiO_2 surfaces, Ti-peroxo (TiOO-Ti) formed by the combination of two trapped holes act as a catalyst to generate OH radicals from water (Figure 5.3]. Nevertheless, the design and fabrication of metal oxides, their modified oxides and nanocomposites and so forth with desired band gap, efficient Oh radical generation and corresponding optical properties are highly attractive nowadays and also in the near future.

5.4 VARIOUS TYPES OF VISIBLE LIGHT-ABSORBING PHOTOCATALYSTS

Though most semiconductors can be activated by ultraviolet light, effort to extend the activity towards visible region is imperative for their effective solar energy utilization. Because UV radiation accounts for just approximately 5 percent of the sun spectrum's energy, visible light accounts for 45 percent, and near-infrared (NIR) accounts for 50 percent. The mechanism of visible-light active photocatalysts' visible-light response is similar to that of UV-light active photocatalysts, with the exception that the photon energy required to initiate the photocatalytic cycle is lower, resulting in better selectivity. Several semiconductor materials possessing visible-light response like CdS, CdSe, WO_3, GaZnON, InP, Cu_2O, $BiVO_4$, Ag_2O, Bi_2MoO_6, Bi2WO_6, $RbPb_2Nb_3O_{10}$, and others have been reported for the effective detention of solar energy [22]. Among them, CdS is one of the most studied one. Important metal oxides such as TiO_2, ZnO, MoO_3, CeO_2, ZrO_2, WO_3, Fe_2O_3, SnO_2 and their mixed oxides are another class of photocatalysts used in semiconductor photocatalysis [23].

5.4.1 Modified Single Materials

In addition to the above semiconductors, a wide variety of photocatalyst materials exist either as single materials, heterojunctions, graphene-based photocatalysts, metal-organic framework (MOF) compounds, black phosphorus, BiOCl, $ZnFe_2O_4$ and so forth [24–27]. MOFs are known for their specific applications, including hydrogen generation and CO2 reduction, even though achieving visible light activity is still in its infancy. Besides, organic polymers have also been taking their place in the area of photocatalytic applications. Because of its lower band gap (2.8 eV), g-C3N4 is recognized as another suitable co-photocatalyst with a favorable conduction band (CB) edge position for fuel generation. Nevertheless, when compared to TiO_2, graphitic carbon nitride has a lower chemical stability and is more sensitive to hydroxyl radical based degradation [28]. Though graphitic carbon nitride was reported to be popular visible light active photocatalyst, its low chemical stability compared to TiO_2, high recombination rates, low conductivity and low solvent-accessible surface area and so forth resulted in their being poor candidates for many practical applications. While numerous researches have been done on developing new photocatalysts, it should be emphasized that many of the materials examined are unlikely to have practical use while showing promise in bench-scale tests. For example, material containing hazardous elements like CdS and $CsPbBr_3$ or uncommon and expensive components, as well as material that is brittle or chemically unstable, is unlikely to be practical, which likely adds to TiO_2 photocatalysis' popularity. The popularity of pure semiconductor material as visible light photocatalysts have been diminishing recently mainly due to the complications associated with achieving narrow bandgap with low recombination losses at a time. Strategies to improve their individual visible light performance are discussed in the following sections.

5.4.2 Heterojunctions

Another class of photocatalysts are semiconductor heterojunctions. They are combinations of metal–semiconductor or semiconductor–semiconductor which can enhance the visible light absorption without increasing the recombination. Because of the potential gradient of heterojunction between semiconductors, the speed of electron transport is increased by decreasing recombination sites of photoexcited charge carriers. To improve charge carrier separation and therefore prevent photogenerated charge-carrier recombination on photocatalysts, it is critical to develop and fabricate appropriate heterostructures. The band edge alignment is an essential consideration in heterojunction design for optimal photocatalytic activity [29]. Figure 5.4 is a schematic representation of various synergic combinations of materials which are listed as Type I, Type II and Type III. According to the literature, such a semiconductor combination can compensate for the individual materials' shortcomings by offering effective charge separation and increased photostability. The conduction band level of at least one of the semiconductors must be more negative than the other. This is an essential requirement for linking semiconductors. TiO_2 combined with CdS, for example, can be used for photocatalytic water splitting since TiO2 has a higher negative conduction band than EH_2/H_2O. A higher rate of hydrogen production was observed for coupled CdS–TiO_2 compared to

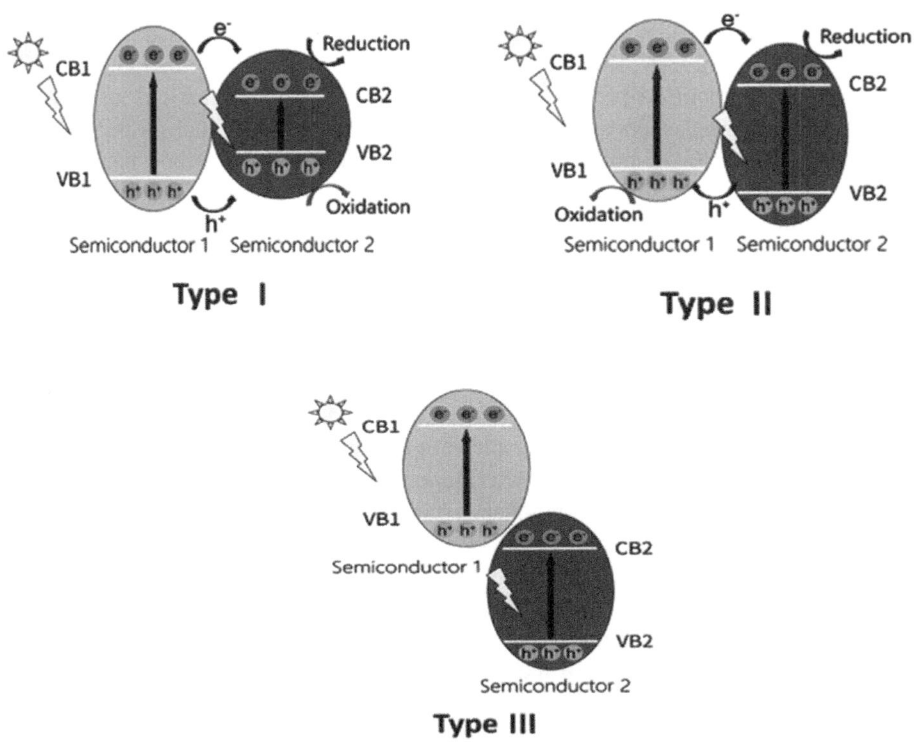

FIGURE 5.4 Schematic representation of Type I, Type II, and Type III heterojunctions.

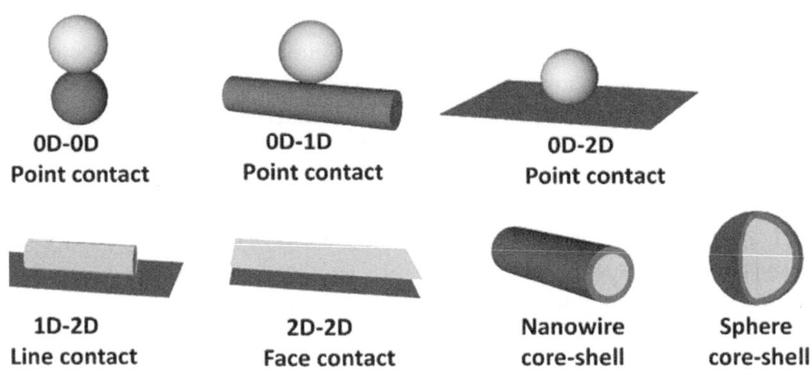

FIGURE 5.5 Schematic representation of heterojunctions based on their morphology. Reprinted with permission from Ref. [30] [Copyright 2020, AIP Publishers].

the pure under visible-light illumination as optical absorption of CdS–TiO_2 extends up to 520 nm after coupling.

Material morphology may also be used to classify heterojunctions, such as partial contact photocatalysts and core-shell photocatalysts [30]. Partial contact photocatalysts can have point contact when one of the photocatalysts is a nanoparticle, line contact for nanowire or nanoribbon/nanosheet catalysts, and face contact when both photocatalysts are two-dimensional, depending on the morphology of two semiconductors (Figure 5.5). In addition to energy levels and morphology, the lattice mismatch is a factor to consider when choosing materials to create a heterojunction, with materials with a low lattice mismatch being favored [31]. Various heterojunction photocatalysts such as CeO_2–AgI, g-C_3N_4/Bi2WO6, Ag_2O/Bi5O, In_2S_3/$CdIn_2S_4$ nanotubes, CdS/MoS_2, amorphous NiO/g-C_3N_4, ZnS/g-C_3N_4, g-C_3N_4/ZnIn2S_4, g-C_3N_4/ phosphorene, $ZnIn_2S_4$–In2O3, and p-CaFe2O_4/ n-ZnFe2O4 have been reported previously [32–39]. Moreover, combinations of semiconductor photocatalysts with other polymers and/or carbon-based nanostructures, such as chitosan, graphene, reduced graphene oxide (rGO), polyaniline and nitrogen-doped rGO, carbon nanotubes (CNTs), polypyrrole, as well as polyaniline, have also been studied [40–44]. MOF-based heterojunctions, such as zeolitic imidazolate framework-8 (ZIF-8) nanoclusters on g-C3N4 nanotubes, have also been studied in order to combine the photocatalytic activity of g-C3N4 with increased CO2 adsorption of a MOF. Finally, in heterogeneous photocatalysis, an appealing set of Janus nanostructures, that is, nanostructures having surfaces with two different characteristics, were discovered [45].

5.4.3 Z-Scheme Heterojunctions

By minimizing recombination and utilizing plasmonic effects, visible light absorption may also be achieved by adding particular types of heterojunctions, such as Z-scheme heterojunctions and plasmonic photocatalysts. This form of photocatalyst is very interesting for water-splitting processes, since it has a theoretical maximum solar energy conversion efficiency of 40 percent, which is greater than what single photocatalysts can accomplish (30%). In a Z-scheme heterojunction, electrons are transferred from one material's CB to the other material's valence band (VB) through an electron mediator, resulting in a greater difference between oxidation and reduction energy levels than in a heterojunction [46–47]. Reversible redox mediators in solution have been widely employed in liquid Z-scheme systems. Fe3+/Fe2+, IO 3 /I, and NO 3 /NO 2 are common electron mediators in liquid Z-scheme systems. Nonetheless, solid-state Z-scheme catalysts are of great interest due to light-harvesting problems in liquid systems and the possibility of backward reactions. Conductors can act as electron mediators in a solid-state Z-scheme photocatalyst. Au nanoparticles, Ag nanoparticles, rGO, and carbon nanotubes (CNTs) are commonly utilized for these purposes [48]. Furthermore, materials with band gap alignment, which resembles Type II heterojunctions, except in charge carrier transfer route, are also reported and are known as direct Z-scheme photocatalysts and their observed electron transfer was across the interface between two semiconductors. In a direct Z-scheme heterojunction, the electrons in the lower CBM material recombine with the holes in the higher VBM material, leading in a greater difference between the oxidation (lower VBM position) and reduction (higher CBM position) energy levels, resulting in increased redox capacity. Z-scheme photocatalysts with different material combinations have been reported to date are graphene bridged 3PO_4/Ag/$BiVO_4$, CdS/RGO/g-C3N_4, WO_3/ g-C3N4, Ag2Mo2O_7/MoS_2, g-C_3N_4/ SnS_2, WO_3/g-C_3N_4, TiO_2/$CuInS_2$, $SrTiO_3$:La, TiO_2/CdS, α-Fe_2O_3/g-C_3N_4, Rh/C/$BiVO_4$:Mo, black phosphorus/$BiVO_4$, g-C3N_4/Ag/MoS_2, g-C_3N_4/nanocarbon/ $ZnIn_2S_4$, and MnO_2/monolayer g-C_3N_4 with Mn vacancies and so forth [49–58]. In general, Z-scheme photocatalysts have piqued interest for a wide range of applications, including environmental cleaning and fuel production (water splitting as well as CO_2 reduction).

5.4.4 Carbon-Semiconductor Composites

The use of carbonaceous materials with semiconductors and composites has been shown to improve visible-light activity, with carbon materials acting as adsorbents, electron sinks to prevent charge carrier recombination, or photosensitizers and supports to generate a higher density of electron–hole pairs. Any enhancement in the activity of these system could solve the issues related to band-gap narrowing, improving the adsorption characteristics of the photogenerated charge carriers and prolonging their lives [59]. Various forms of carbon such as graphite, carbon black, and graphitized materials have long been used in heterogeneous catalysis in support of metal particles. Moreover, the development of carbon nanostructures like carbon nanotubes, fullerene, graphene, nanorods and so forth, opened up multiple horizons to enhance the practical applications of visible light active semiconductor photocatalysts [60–63].

5.4.5 Plasmonic Materials

By integrating metal nanoparticles with plasmonic effects, specifically resonant oscillations of the electrons inside a plasmonic nanoparticle, it is possible to increase the visible light sensitivity of the above-mentioned semiconductors. A material's plasmonic wavelength is determined by its dielectric function, as well as its size and shape. The photocatalytic activity of plasmonic nanoparticles may be increased by a variety of methods, including hot electron injection, localized electromagnetic field augmentation, and plasmon resonance energy transfer (PIRET) from metal to semiconductor via dipole–dipole interactions [64]. In the first scenario, the wide band-gap semiconductor is not stimulated by visible light, but electrons are transferred from the noble metal's surface plasmon. In the second scenario, excited electrons are transferred from the semiconductor to the noble metal. Because the Fermi levels of these noble metals are lower than those of semiconductors (SC), photogenerated electrons are effectively transferred from the conduction band of SC to metal. The electron–hole recombination rate is considerably reduced as a result of this electron trapping mechanism, resulting in greater photocatalytic reactions. Noble metals like Ag, Pt and Au nanoparticles are largely known for this purpose since they possess strong localized surface plasmon resonance (LSPR). In a photocatalytic water splitting experiment using an alcohol-water solution, for example, the amount of hydrogen generated was 138 mL for Au/TiO2, compared to 7 mL for pure TiO2. Furthermore, the plasmon-excited electrons decay time of Au/TiO2 NPs was 1.5 ns, which is better than the few picoseconds of an individual Au NP, because electron migrations from the Au NP surface to the TiO2 NP help to extend the decay time and effectively limit the recombination of electrons-holes in Au NP. Nevertheless, use of low-cost non-noble metals like Bismuth and use of more abundant materials have also been reported [65]. Furthermore, because their plasmonic wavelengths may be adjusted up to 1000 nm, Al and Cu have been recommended as ideal candidates for plasmonic improvement of photocatalysis [66–67]. Combination of different plasmonic materials with various nanostructures like Au/Ag/Cu_2O core-shell, nanorod or Au/CuSe/Pt were also possible to extend the visible light absorption. Applications of plasmonic photocatalysts include water splitting, decomposition of organic compounds, and photovoltaic devices.

5.4.6 Sensitized Photocatalysts

Here, a visible light absorbing dye or semiconductor quantum dots (QDs) is anchored on metal oxide or semiconductor in such a way that excited electrons are transported from dyes/QD to the semiconductor's conduction band. In the absence of semiconductors, certain dyes can even produce electrons by absorbing visible light. (Figure 5.6). Similar systems are usually used in visible-light induced hydrogen generation, dye-sensitized water splitting and Dye sensitized solar cells (DSSC) and so forth. One of the most serious problems with dye-sensitized solar cells is dye degradation. Dye regeneration using sacrificial agents or redox systems like EDTA and I3/I pair were generally used to overcome this [68]. Electron injections in dye sensitization take place on a femtosecond scale, whereas electron–hole pair recombination takes place on a nanosecond to millisecond period [69–70]. The dye-sensitized photocatalytic processes have the advantage of a quick electron injection into the semiconductor and a delayed reverse reaction [71]. A viable photocatalyst, under laser light at 532 nm and solar simulated light with or without a cut filter for producing hydrogen from water/methanol solutions has been developed recently by coupling $[Ru(bipy)3]^{2+}$ dye molecules and delaminated graphene oxide [72]. The suggested process is that when $[Ru(bipy)3]^{2+}$ is stimulated by light, it enters the triplet excited state, which has an energy of 1.06 eV in comparison to the identical reference electrode. As a result, electron injection from the $[Ru(bipy)3]^{2+}$ LUMO into the

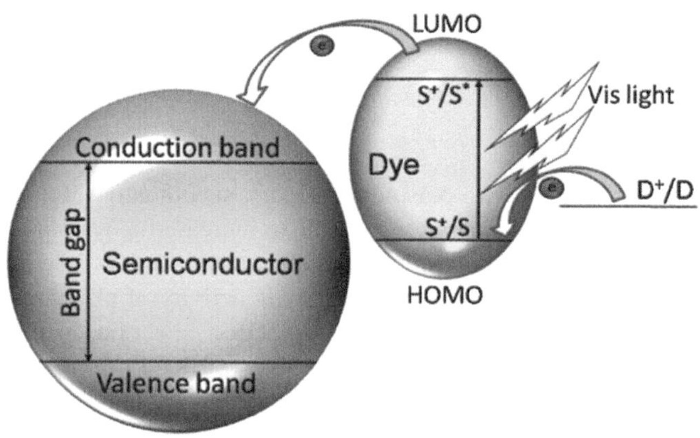

FIGURE 5.6 Mechanism of Dye sensitized visible light absorption.

CB of GO will be advantageous from a thermodynamic standpoint. Due to their capacity to collect hot electrons and produce numerous carriers, quantum dots (QDs) are effective sensitizers in solar cells. Sensitization using visible light absorbing QDs such PbS, InP, CdS, CdTe, CuInS2, Bi2S3, and CdSe has recently received a lot of interest for solar energy conversion [73]. Tuning the particle size of QDs is regarded as an efficient way to harvest the entire solar spectrum. For example, visible light absorption of CdSe QD is at 557, 543, and 505 nm for corresponding dots of particle sizes 2.3, 2.6, 3.0, and 3.7 nm respectively [74].

5.5 TIO_2 BASED VISIBLE LIGHT ACTIVE METAL OXIDE PHOTOCATALYSTS

5.5.1 TiO_2 Structure, Properties and Electronic Processes

TiO_2 is the most widely used n-type semiconductor photocatalyst mainly because it is inexpensive, chemically stable and biologically benign. Following the discovery of photo splitting of water using TiO_2 by Fujishima and Honda, there have been numerous significant developments in titanium dioxide during the last few decades [1]. Other significant innovations such as outstanding anti-fogging and self-cleaning abilities, attributed to the super hydrophilic properties of the photoexcited TiO_2 reported by Wang et al. and use of nano titanium dioxide as photoanodes in dye sensitized solar cell (DSSC), reported by Graetzel and O'Regan in 1991 and so forth has opened doors for the exploration of TiO_2 as a candidate in various potential applications [75–76]. Anatase, rutile, and brookite are the three polymorphs of TiO_2. When metastable anatase and brookite are calcined at temperatures exceeding 600°C, they change into the thermodynamically stable rutile phase. Their band gap is 3.2 eV, 3.0 eV and~3.2 eV for anatase, rutile and brookite respectively. Anatase and rutile are the main polymorphs and their key properties are summarized in Table 5.2. The various spatial configurations of TiO_6 octahedra help explain the crystal structure of TiO_2 polymorphs. Structurally, titanium (Ti^{4+}) atoms are co-ordinated to six oxygen (O2−) atoms, forming TiO_6 octahedra. The differences between the three crystal structures are the various degrees of distortion and 3-D assembly of the TiO6 octahedra. The octahedra in rutile share edges at (0 01) planes to form a tetragonal structure (Figure 5.7), while the octahedra in brookite share both edges and corners to form an orthorhombic structure [77]. Commercially known TiO_2, Degussa, is a mixture of both anatase and rutile and the effective passage of electrons from the conduction band of anatase to those of rutile TiO_2 resulted in higher photocatalytic activity of these biphasic TiO_2.

As described in section 6.3, mechanism of TiO_2 photocatalysis involves the generation of charge carriers like holes and electron as a result of photoexcitation. Charge carriers can be confined in the TiO2 lattice as Ti3+ and O defect sites, or they can recombine and dissipate energy. On a timeframe

TABLE 5.2
Structural and Physical properties of Anatase and Rutile

Property	Anatase	Rutile
Molecular weight (g/mol)	79.88	79.88
Crystal structure	Tetragonal	Tetragonal
Lattice constants (A) °	a =3.78	c = 9.52
	c = 2.96	a = 4.59
Density (g/cm3)	3.79	4.13
Ti O bond length (A) °	1.94 (4)	1.95 (4)
	1.97 (2)	1.98 (2)
Refractive index	2.55	2.75
Dielectric constant	31	114
Light absorption (nm)	<390	<415
Melting point (◦C)	1825	1825
Boiling Point (°C)	2500-3000	2500-3000

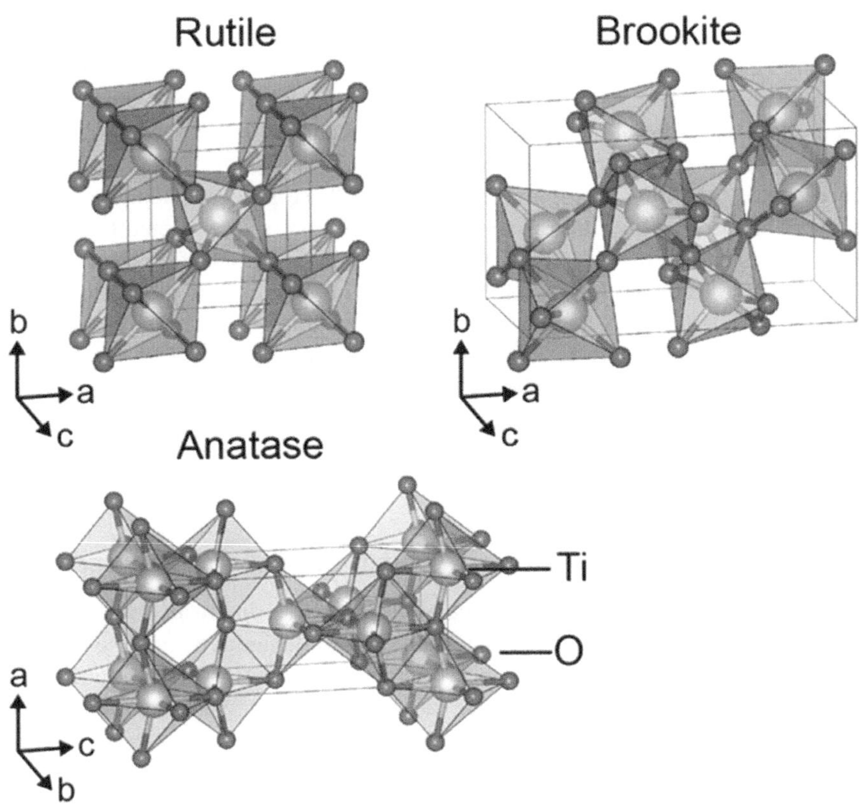

FIGURE 5.7 Crystal structure of various polymorphs of TiO_2. Reproduced with permission from Ref. [77] [Copyright 2017. Springer Nature Publishers].

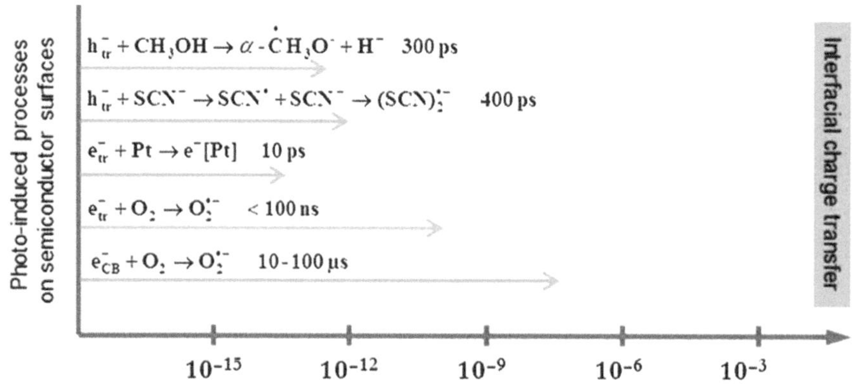

FIGURE 5.8 Schematic representation of interfacial charge transfer processes on TiO_2 surface.

of 30 ps, excited electrons were trapped as Ti^{3+} species, and around 90 percent or more of the photogenerated electrons recombined within 10 ns. The positive hole in the valence band can oxidize OH or water at the surface to create the hydroxyl radical (•OH), which is a highly strong organic pollutant oxidant. It is assumed that the hole-transfer occurs only through a surface trapped hole species or through the hydroxyl radicals and these hydroxyl radicals are the predominant oxidizing species in photo-activated TiO_2. The trapping and recombination of these electron hole pairs compete with one other, since they possess lifespan of picoseconds to nanoseconds. On the contrary, possibility of another race is expected between the interfacial charge transfer and recombination of the trapped species as shown in Figure 5.8, by which those process happens from millisecond to microsecond. Chemical species adsorbing on the surface are also expected to be efficient electron and hole scavengers. However, overall efficiency can be increased if one can increase the rate of interfacial charge transfer processes. After reacting with oxygen, the photo-excited electron in the conduction band is reduced to create superoxide radical anion ($O^{\bullet}-$) and hydroperoxide radical (•OOH) upon further reaction with H^+. These reactive oxygen species (ROS) are essential in the degradation of organic compounds. Moreover, H_2O_2 has also been implicated in the photocatalytic degradation of several inorganic and organic molecules [78]. As a result of the homolytic scission of hydrogen peroxide, extremely reactive •OH species are produced, or H_2O_2 acts as an electron acceptor.

5.5.2 Advanced TiO_2 Nanomaterials for Visible-Light Induced Applications

TiO_2 is one of the most widely studied photocatalytic materials due to its low cost and nontoxic nature, as well as its good chemical and thermal stabilities. Stoichiometric anatase titania has a relatively large band gap of 3.2 eV and it can absorb only UV light. The effective utilization of solar energy toward visible region is thus limited since the solar spectrum consists of only 4–5 percent UV light. Therefore, several strategies were developed to extend the absorption range of titanium dioxide into the visible-light region such as doping, sensitization, formation of heterojunctions, morphology tuning, generation of oxygen vacancies, surface modification and fabrication of composites with other materials. Light adsorption, charge recombination dynamics, surface structure, and charge transfer kinetics, among other things, can all benefit from modified TiO_2. Several innovative visible light active (VLA) photocatalytic materials and systems have emerged as promising materials as a result. Because of their huge surface area and aspect ratio, nanostructured metal oxides are reported to exhibit greater activity than bulk materials (ratio of length to width). The customized morphology, regulated porosity, vectorial charge transfer, and minimal recombination at grain boundaries of such structures all contribute to improved performance in photoinduced applications. Therefore, developing materials such as quantum dots (QDs), nanorods, nanotubes, nanowires, nanocubes and so forth has been one of the most popular approaches [79–82].

Another strategy is to develop carbonaceous-TiO_2 photocatalysts, mainly, using g-C_3N_4, fullerenes, carbon nanotubes and graphene oxide, where they can offer large specific surface area and high mobility of the charge carriers. For example, a research by Li and colleagues [83] employing a g-C3N4/TiO2 hybrid, which consisted of TiO2 microspheres wrapped in lamellar g-C3N4, demonstrated high photocatalytic efficiency for the inactivation of E. coli K-12 at visible light intensity of 30 mW/cm^2. Similarly, the g-C_3N_4/TNA (Titanium nanotube Array) membranes developed by Wang et al. via potentiostatic anodization have also proved their remarkable efficiency for complete E. coli inactivation and outstanding anti-fouling potential for water treatment under VL irradiation [84]. Nanocomposites made of properly integrated graphene with TiO_2 are found to be suitable photocatalysts with desired properties for visible light absorption. The use of graphene–TiO2 nanorod hybrid nanostructures in microcapacitors has recently been discovered, with the graphene rod-shaped TiO2 nanocomposite demonstrating a significant improvement in photocatalytic performance over the graphene spherical TiO_2 nanocomposite, which can be attributed to graphene's high electronic mobility and higher surface area of TiO_2 nanorods. Moreover, study on visible light activity of B-doped graphene/rod-shaped TiO_2 in oxidative photo-destruction of NOx gas [85] has enlightened the idea of using TiO_2–carbon nanocomposites in addressing environmental issues.

Another popular approach for extending the absorption response to longer wavelengths is doping. These methods alter the band structure of the photocatalyst band gap or introduce new defect states, causing the absorption edges to move toward the visible area. The dopant also functions as a recombination trap for e and h+, boosting TiO_2's photocatalytic and antibacterial activity while lowering its recombination rate. Common dopants used in TiO_2 are nitrogen, sulfur, carbon, phosphorus, and fluorine as non-metal dopants and Ag, Mg, Cu, Rb, Co, Mo, Au, Pt, Fe, Cr, V, and so forth, as metal dopants [86–90]. In addition, alkali metal (Li+, Na+, and K+)51 and chlorine doping has also been reported for g-C3N4. Carbon doped TiO_2, (C-TiO_2) surface as well as co-doped C-TiO_2 have been found to be useful for the treatment of drinking water. In the case of N as a dopant, it can be easily introduced in the TiO_2 structure due to its comparable atomic size with oxygen, small ionization energy and high stability. Extensive research on N-TiO_2 was devoted to understanding the underlying mechanism of their exceptional antibacterial and self-cleaning properties and it was concluded that, the bandgap narrowing of N-TiO_2 was responsible for the greater visible light absorption. Model pollutants like phenols, methylene blue, methyl orange (although dyes have significant absorption in the visible range), and rhodamine B, as well as many gaseous pollutants (e.g., volatile organic compounds, nitrogen oxides) have all been found to be efficiently destroyed by N doped TiO_2 photocatalyst. Besides, doping with F, C, S and so forth not only allows visible light absorption but also increases the lifetime of photogenerated charge carriers [91–92]. The development of localized energy states in the TiO_2 band gap owing to anion doping and/or oxygen vacancies was seen in an analysis of visible light activity of S doped TiO2 films produced employing a new sol–gel process based on the self-assembly methodology [93]. Thus, the considerable reduction in the cost of forming an O-vacancy is a significant benefit of hetero-atom doping. The O-vacancy level in anatase TiO_2 may function as a recombination center for holes and electrons, as well as the dopant (C, N, and S) itself. As a result, band-gap changes in TiO_2, such as the creation of localized mid-gap states and the narrowing of the band-gap, induce novel and unanticipated characteristics that may contribute more or less to visible-light absorption. Co-doped N-F-TiO_2 has also found potential in the visible light mediated degradation of a variety of pollutants in water, as they possess distinct N spin species in NF–TiO_2 [94].

Moussab Harb et al. performed a variety of selenium (Se) doping experiments at various substitutional sites for oxygen or titanium, interstitial sites, and combined substitutional and interstitial sites. Various structures such as Ti(1–2_x)O_2Se2x (Se^{4+} species), $TiO_{(2-x)}Se_x$ (Se^{2-} species), and $TiO_{(2-x)}$Se2x

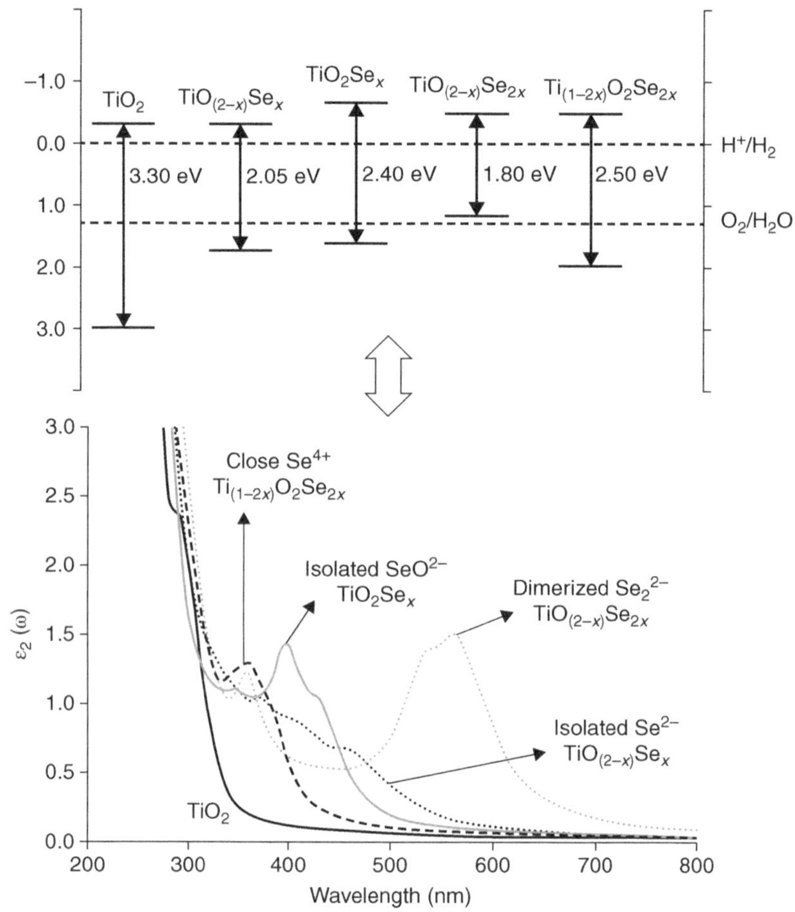

FIGURE 5.9 Effect of selenium doping in the visible light activation of anatase TiO2.

Reproduced with permission from Ref. [95] [Copyright 2013 American Chemical Society].

(Se_2^{2-} species) with visible-light optical absorption spectra due to a substantial decrease in the band gap were identified as shown in Figure 5.9 [95]. Transition metal modifications of TiO2, such as Cr, Co, V, and Fe, have been shown to extend TiO2's spectral sensitivity well into the visible range. Vibriofischeri, a biofilm-forming marine bacterium, was shown to be inactivated by Fe2O3-doped TiO2 NPs produced by ultrasonic-aided co-precipitation. Modifications of TiO_2 with transition metals such as Cr, Co, V and Fe have found to extend the spectral response of TiO_2 well into the visible region. Fe_2O_3-doped TiO_2 NPs synthesized via ultrasonic-aided co-precipitation is found to be another choice VLA TiO_2 for the inactivation of Vibriofischeri, a biofilm-forming marine bacterium.

Noble metals, such as Pt, Au, Pd, Rh, Ni, Cu, and Ag, have been shown to improve TiO_2 photocatalytic activity by reducing the chance of electron–hole recombination. Pd doped TiO_2 for the reduction of various organic compounds including acetophenone, and in membrane reactors, Au doped TiO_2 for photocatalytic water splitting of an alcohol–water solution, porous one-dimensional (1D) TiO_2 nanotubes (NTs) with Au synthesized by emulsion electrospinning followed by deposition–precipitation for environmental applications and so forth were some of the examples of plasmonic TiO_2 materials. Plasmon-induced visible light activity of TiO_2 was further demonstrated with platinum deposits on TiO_2 by Hwang et al. and with Ag modified TiO_2 by Seery et al. where both Pt and Ag trap the photo-generated electrons, and subsequently increase the photo-induced electron

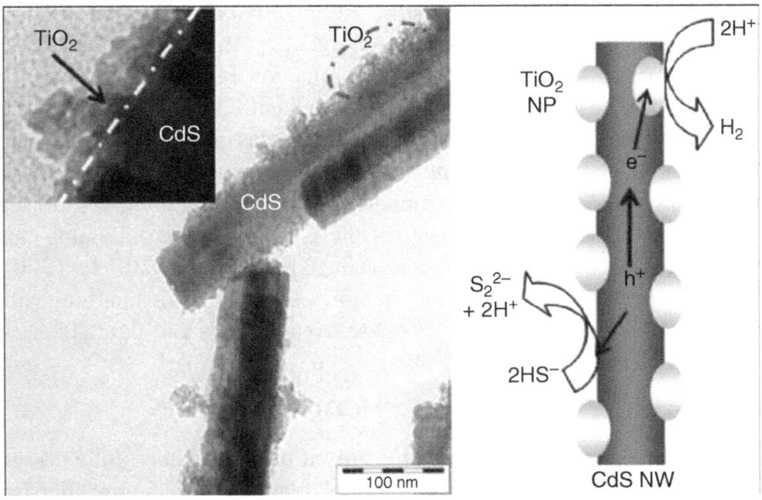

FIGURE 5.10 TEM and mechanistic image of the interface between CdS nanowires and TiO2 nanoparticles. TiO2 provide sites for collecting the photoelectrons generated from CdS nanowires, enabling thereby an efficient electron–hole separation. (Reprinted with permission from S. J. Jum, G. K. Hyun, A. J. Upendra, W. J. Ji, S. L. Jae, Int. J. Hydrogen Energy, 33 (2008) 5975 [Copyright 2008, Elsevier].

transfer rate at the interface [96–97]. Metal-nonmetal combination of dopants on TiO_2 was validated for many applications, including H_2 production, and it has been tabulated that N/Fe^{3+}, N/V^{5+}, N/Cr^{3+}, N/W^{5+}, N/Ce^{3+}, N/La^{3+}, N/Ga^{+} and C/V^{5+} combinations could be effective for the disinfection of wastewater, where the synergetic action of dopants led to increased oxygen vacancy production and band gap realignment, resulting in higher visible light activity [98–103]. Similar studies of co-doping and tri-doping with vanadium and/or chromium and nitrogen as non-metals on TiO_2 thin film under visible light also have yielded reasonable data on the enhanced visible light activity of TiO_2, where augmented activity was due to the band gap narrowing mechanism as discussed earlier. As a result, maximum visible photo-response observed was for tri-doped titania (V-Cr-N-TiO_2) for model dye, Methylene Blue (MB) degradation.

For the degradation of organic contaminants, the coupling of a large band gap semiconductor with a smaller one that may be triggered by visible light is of significant importance. The photostability of particles can be dramatically improved by blocking trap states by covering them with tiny coatings of a wide band gap material. CdS, for example, is an interesting material with an optimum 2.4 eV optimum band gap. However, in order to overcome the problems associated with photoanodic corrosion of CdS in aqueous environments it has been combined with a wide band gap semiconductor, such as TiO_2 (Figure 5.10), and the charge separation of photogenerated electrons and holes is enhanced as a result of this connection [104]. Mixed oxides of TiO_2 coupled with other semiconductors such as SiO_2, ZrO_2, ZnO, CuO, SnO, CdS, and so forth are known for their extended visible light absorption, but with a difference in mechanism with respect to each material with which it is coupled. For example, the enhanced photocatalytic activity of CuO/TiO_2 was attributed to photo-electron transfer from TiO_2's more negative conduction band (CB) to CuO, leaving the photogenerated hole on TiO2 to participate in oxidation reactions, whereas in the case of ZnO-TiO2 nano-composites synthesized using a solvo-thermal method, it was due to the formation of a hierarchical nanostructure with a large A-range of organic dyes such as MB, Congo red, Methyl Orange, and eosin red and so forth were tested effectively for degradation under visible light with α-Fe_2O_3/TiO_2 nanocomposite and excellent performance was reported to be attributed to the large surface area and charge separation [106].

However, the mechanism behind the visible light activity of sol-gel prepared WO_3–TiO_2 nanocomposites toward the degradation of imazapyr organic pollutant is demonstrated to be via sensitization. It is obvious that under visible light irradiation, photogenerated electrons from the valence band (VB) of WO_3 could be stimulated to the conduction band (CB), and holes from the VB of WO_3 could be transported to the VB of TiO_2. As a result, the holes in the TiO_2 VB accumulate. These positive holes and negative electrons combine with water molecules to create hydroxyl radicals and superoxide anions, which may efficiently eliminate organic pollutants in a short amount of time. In addition, it is worthy to mention that many of the gaseous phase pollutants can be destroyed with TiO_2 based P-type semiconducting nanocomposites. For example, $BiFeO_3$ (P–BFO) and TiO_2–$BiFeO_3$ (T/P–BFO) P-type nanocomposites are recognized for their outstanding visible light photocatalytic capabilities for the degradation of gas phase acetaldehyde and liquid phase phenol [107].

5.6 NON TIO_2 BASED VISIBLE LIGHT PHOTOCATALYSTS

The effective utilization of solar energy is possible only if one can successfully design and fabricate new material photocatalysts having suitable structural features with engineered band gap. Apart from modifying the popular metal oxides such as TiO_2 and ZnO, it is necessary to explore novel semiconductor materials that can absorb visible light. Few non-TiO_2 photocatalysts, such as WO_3, Ag_3PO_4, $BiVO_4$, and g-C_3N_4-based photocatalysts, have been reported to have appropriate energy band structures and high visible-light-driven efficiency.

5.6.1 WO_3 Based Photocatalysts

WO_3 is an n-type semiconductor having an electronic bandgap (Eg), which corresponds to the difference between VB and CB, where VB is produced by full O2p orbitals and the conduction band (CB) is generated by vacant W 5d orbitals. The ideal crystal phase of WO_3 is cubic, and the crystal phase changes with the degree of distortion from the ideal phase. WO_3 crystals are generally formed by corner and edge sharing of WO_6 octahedra. Though, monoclinic II (ε-WO_3), triclinic (δ-WO_3), monoclinic I (γ-WO_3), orthorhombic (β-WO_3), tetragonal (α-WO_3), and cubic WO_3 phases are known, only monoclinic WO_3 always shows the best photocatalytic efficiency. When compared to TiO_2, one benefit of WO_3 is that it can be irradiated by the visible solar spectrum's blue area. Furthermore, WO_3 is a viable option for wastewater treatment due to its exceptional stability in acidic conditions. Noble metals like platinum (Pt), palladium (Pd), Au, doped WO_3 shows better visible light efficiency compared to TiO_2. On the other side, the metal surface of Pt induces the photocatalyst is more hydro-phobic compared with a metal oxide surface. WO_3 *nanotubes* doped with Pt nanoparticles had a greater surface area and better visible-light-driven photocatalytic activity compared to Pt nanoparticle-loaded commercial WO_3. These results sparked a lot of interest in recent years, and they were thought to be the start of a WO_3 photocatalyst research boom.

5.6.2 Bismuth-Based Photocatalysts

With an optical band gap of 2.5 eV, bismuth tungstate (Bi_2WO_6) is one of the most effective Bi-containing photocatalysts. Low optical band-gap energy levels are generally caused by hybridization between Bi 2p and O 2s states [108]. Despite its promise, the use of bismuth tungstate for environmental purification requires more research before it can be used on a large basis. Bismuth ferrites, $BiFeO_3$, have optical band gaps ranging from 2.2 to 2.8 eV at ambient temperature, indicating that they are sensitive to visible light irradiation. The monoclinic -Bi_2O_4 nanorods became popular for water disinfection after demonstrating their efficiency against inactivating E. coli in 120 min of VL. Bismuth oxyhalide (BiOX) is another class of materials and the layered tetragonal structure of BiOX, which is made of $[Bi_2O_2]^{2+}$ slabs interleaved with double slabs of halogens, may contribute to the small band gap in VL photocatalysis. Bismuth oxybromide (BiOBr) nanosheets, hierarchical BiOI/BiOBr heterostructure, Bi_2O_4-decorated BiOBr (Bi_2O_4/BiOBr) nanosheets, BiOBr-xAgBr composite, BiOBr-AgBr, AgI/AgBr/BiOBr composites are some of the Bi based visible light active photocatalysts for water disinfection [109-112]. Narrow band gaped Bismuth vanadate ($BiVO_4$) and Ag^+, Er^{3+} and Y^+ (Er-Y-$BiVO_4$) doped ($BiVO_4$ are another widely explored materials for of E. coli disinfection up to 99.99 percent under UV or VL irradiation.

5.6.3 MoS_2 Based Visible Light Active Photocatalysts

Though MoS_2 is not an efficient photocatalysts itself due to its edge activity, it can be a sensitizer for other semiconductors to induce visible light activity. ZnO/MoS2 nanocomposite, ternary nanocomposite (CNTs-MoS_2-Ag), Cu film on the MoS2 nanofilm (Cu–MoS_2) and so forth are some of the recently developed VL-driven photocatalysts for water disinfection. The pyramid shaped MoS_2 and vertically aligned 2D MoS_2 nanofilms drew the attention of the scientific community since they outperformed almost all TiO_2-based photocatalysts in terms of antibacterial capability [113-115].

5.6.4 Ag_3PO_4 Based Nanocomposites

Ag_3PO_4 is a novel semiconducting photocatalysts which can achieve a quantum efficiency up to 90 percent at wavelengths longer than 420 nm. They're well-known for their superior aptitude for O_2 evolution from H_2O and organic dye degradation when exposed to visible light. Many techniques, including the construction of nanostructures, have been proposed to improve the efficiency of Ag_3PO_4 under visible-light irradiation. For example, single-crystalline Ag_3PO_4 rhombic dodecahedrons and cubes, hierarchical Ag_3PO_4 porous microcubes, Ag_3PO_4 tetrapod microcrystals and so forth are few structures with enhanced photocatalytic property. A facile methodology to controllably prepare Ag3PO4 crystals with various morphologies (including branched, tetrapod, nanorod-shaped, and triangular-prism-shaped Ag_3PO_4 crystals) using a solvent mixture of N,N dimethylformamide (DMF) and H_2O at room temperature has been reported (Figure 5.11) and it is noted that the branched Ag_3PO_4 sample having greater surface area contributes to the enhanced VLA Figure [116]. Ag_3PO_4 coupled with other semiconductors, carbon materials, or noble metals to improve the photocatalytic activity, such as Ag_3PO_4/Fe_3O_4, Ag_3PO_4/TiO_2, Ag_3PO_4/BiOCl, Ag3PO4/AgX (X = Cl, Br, I), Ag3PO4/SnO2, Ag3PO4/reduced graphite oxide sheets, Ag3PO4/ carbon quantum dots, and Ag3PO4/Ag composites and so forth and their heterostructures are recently emerged new VLA nanocomposite materials. Monoclinic Ag_3VO_4 has also been the subject of much research due to its narrow bandgap energy of 2.05–2.34 eV and its unique band structure. Nickel oxide (NiO) is a highly efficient P-type oxides, with a wide bandgap (3.6–4.0 eV) and visible-light induced antibacterial performance of NiO/Ag_3VO_4 nanocomposites obtained through ultrasonic-assisted synthesis is studied extensively against S. aureus, Streptococcus mutants, Proteus vulgaris and E. coli as potential disinfectant.

5.7 APPLICATIONS OF METAL OXIDES FOR VISIBLE LIGHT-INDUCED PHOTOCATALYSIS

5.7.1 Environmental Applications

(a) Water Treatment and Disinfection: The metal oxides-based visible light mediated photocatalysis has several applications in the fields of energy and the environment, including hazardous pollutant breakdown, wastewater treatment, fuel production, self-cleaning windows, antifogging coatings, and bacterial species disinfection. Various forms of pollution of water and air affect the living and working conditions and pose serious health problems. Various industries such as textiles,

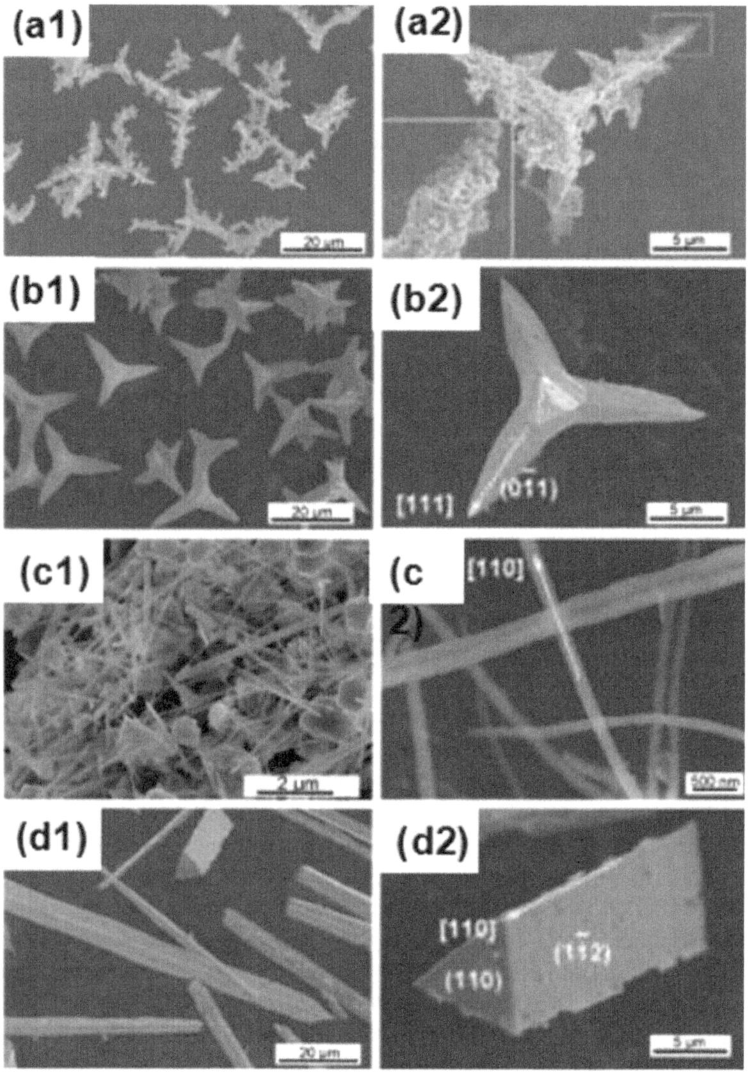

FIGURE 5.11 SEM images of (a) branched, (b) tetrapod, (c) nanorod-shaped, and (d) triangular-prism-shaped Ag3PO4 crystals. Reproduced with the permission from Ref. [116] [Copyright 2013, RSC publishers].

leather manufacturers, automobile industries, and others, emit harmful organic pollutants that contain extremely carcinogenic nitroaromatic dyes that are employed as coloring agents and end up in water bodies (Figure 5.12). Gases, including oxides of carbon and Sulphur, are also harmful to human health when their concentration percentage is increased in the atmosphere. Removing harmful contaminants from water bodies continues to be a major issue in our environmental system. Photocatalytic degradation of cyanotoxins, particularly the hepatotoxin microcystin-LR, has also been done using visible light active TiO_2 photocatalysts (MC-LR). MC-LR is a very hazardous pollutant of emerging concern that is often detected in surface waters, with N–TiO_2 showing the greatest MC-LR degradation. Under visible light irradiation, several phenoxyacid herbicides were photo catalytically converted using Fe-, N-doped anatase, and rutile TiO_2, as well as undoped anatase and rutile TiO_2. Photocatalytic nanoparticles of Fe_3O_4, TiO_2, gold (Au), ZnO and Cupric oxide (CuO) are experimentally proved to inhibit microbes like S. aureus, which is a biofilm forming bacteria [117].

FIGURE 5.12 Pictorial representation of environmental pollution (Picture courtesy: www.iamarsalan.com/how-can-we-reduce-air-and-water-pollution).

Not only do most metal oxides, such as ZnO, TiO_2, Al_2O_3, and others, have photocatalytic properties, but their surfaces are also excellent adsorbents of organic pollutants; 2,4-dichlorophenoxyacetic acid (2,4-D) is a common pesticide that can be detected in surface and ground water as a result of agricultural runoff. Under visible light, an Ag/TiO_2 photocatalyst hydrothermally produced with template-assisted techniques efficiently decomposed 2,4-D. Recently, the effectiveness of TiO_2's photocatalytic disinfection for drug-resistant microorganisms in water has been observed. Consequently, Panasonic has even created a commercial water sterilizer that uses TiO_2 and zeolite catalysts to ensure that underdeveloped countries have clean drinking water [118]. Solar pasteurization and solar disinfection (SODIS) are still emerging as cost-effective solar-based methods.

One of the advantages with this rising photocatalytic technology is its efficiency in removing heavy metals from water [119]. Pharmaceuticals and antibiotics expelled from pharmaceutical factories, hospital wastewater and so forth could be eliminated using photocatalysts such as Ag_3PO_4/Ag/$BiVO_4$ Z-scheme photocatalysts, doped porous carbon nitride, AgI/$Bi12O_{17}Cl_2$, Ag_2Mo2O_7/MoS_2 Z-scheme 1D/2D photocatalysts, α-$NiMoO_4$ nanorods/Ag nanoparticle plasmonic photocatalysts, doped ZnO and TiO2 composites with chitosan, Zn0.9Fe0.1S/Ni foam [120-121]. Some photocatalysts such as CeO_2—AgI [122], cyclodextrin decorated TiO_2 spheres, Ag_2O/Bi_5O_7I, and reduced graphene oxide (rGO) supported on silica and zirconia and so forth [123] have proven their effectiveness in degrading emerging pollutants like endocrine disruptors, bisphenol A and microplastics, and so forth. Photocatalytic water-treatment systems are generally regarded as unfeasible in the water-treatment sector, as traditional ozone-based methods have a considerably greater energy efficiency. However, the Purifies Photo-Cat, employing UV lamps, a popular photocatalytic water-treatment system, has been marketed. Lately, concentrating solar reactors brought attention. CPC, compound parabolic collector, has been the most studied photoreactor for wastewater and solar water treatment. A better variant, consisting of static mixers coated with TiO_2 or Fe_2O_3 in a borosilicate tube of a compound parabolic collector (CPC), has also been investigated for the breakdown of the antibiotic oxytetracycline. A variety of CPC pilot scale reactor images can be found at Plataforma Solar de Almería Website (www.psa.es/en/instalaciones/aguas.php).

In the context of commercialization, a cost-effective method, raceway pond reactors (RPRs) have recently been suggested [124] as solar photo-Fenton reactors in order to degrade low concentration pollutants. Similar experiments utilizing solar TiO_2 for photocatalytic degradation of chlorpyrifos in a natural farm runoff at a pH of circumneutral have demonstrated that in a slurry arrangement, 80 percent of the pollutant may be removed. Table 5.3 shows some instances of emerging developments in photocatalytic reactors for environmental applications, highlighting the techniques utilized, reactor types, and major findings obtained.

(b) Air purification*:* Metal oxide-based visible light photocatalysis might be used for air purification in addition to water treatment. NOx, SOx, CO, H2S, and volatile organic compounds are among the pollutants of concern (VOCs). VOCs are dangerous air pollutants that can be released into the atmosphere by a number of industrial operations. Based on a series of investigations, it has been demonstrated that metal oxide photocatalysis for air purification may be both economically and technologically competitive. For example, under visible light irradiation in an H2–O2 environment, a bifunctional photocatalyst made of nitrogen-doped and platinum-modified TiO2 (Pt/TiO2xNx) was shown to be efficient for the breakdown of benzene and other persistent VOCs. The design of gas-phase photocatalytic reactors with immobilized catalysts, arranged as conjugated plate; annular, monolith, packed bed; foam packed bed; fluidized bed and microfluidic photocatalytic reactors for air pollution control are moving fast toward real-scale applications. Photocatalytic self-cleaning

TABLE 5.3
Examples of New Trends of Photocatalytic Reactors for Environmental Applications

Process	Reactor Features	Major Result	Ref
Cr(VI) reduction under simulated solar light (41 WUV·m2) TiO_2 photocatalysis, in presence of different sacrificial agents	Micro–meso structured-reactor (NETmix) constructed with cellulose acetate sheet coated with a TiO2 thin film. Illuminated surface per unit of volume inside the reactor of 470 m2 m-3 and 1.36 g TiO2 per liter of liquid inside the reactor.	Fast and complete Cr(VI) reduction: Maximum reduction rate was attained using tartaric acid and 3.96% of photonic efficiency was achieved.	Ref: 125
TiO2 photocatalysis for n-decane oxidation at gas phase under simulated solar irradiation	Micro–meso-structured-reactor constructed with cellulose acetate sheet coated with a TiO2 thin film with equivalent of 1.95 g of TiO_2 per litre of air and illuminated surface per unit of volume inside the reactor of 349 m2 m–3	No catalyst deactivation was observed after 72 h of continuous use	Ref: 126
Photochemical reduction of Cr(VI) with UV tubular fluorescent lamps (Xelux 8 W, 16 mm diameter, 2.5 mW cm–2) in the presence of ethanol	Spiral-shaped reactor (SSR), only UV photolysis.	The designed reactor was effective when applied to real wastewater, showing a total Cr(VI) reduction of 51.8% and photonic efficiency of 2.52% and its configuration is suitable for scale-up	Ref: 127
Photocatalytic degradation of hydroxybutanedioic acid.	TiO2-coated optical fiber photoreactor.	30% degradation after 5-h irradiation in a multioptical fiber reactor	Ref: 128
Photocatalytic degradation of malic acid.	TiO2-coated optical fiber photoreactor.	75% maleic acid removal and 21% TOC removal after 20 h	Ref: 129

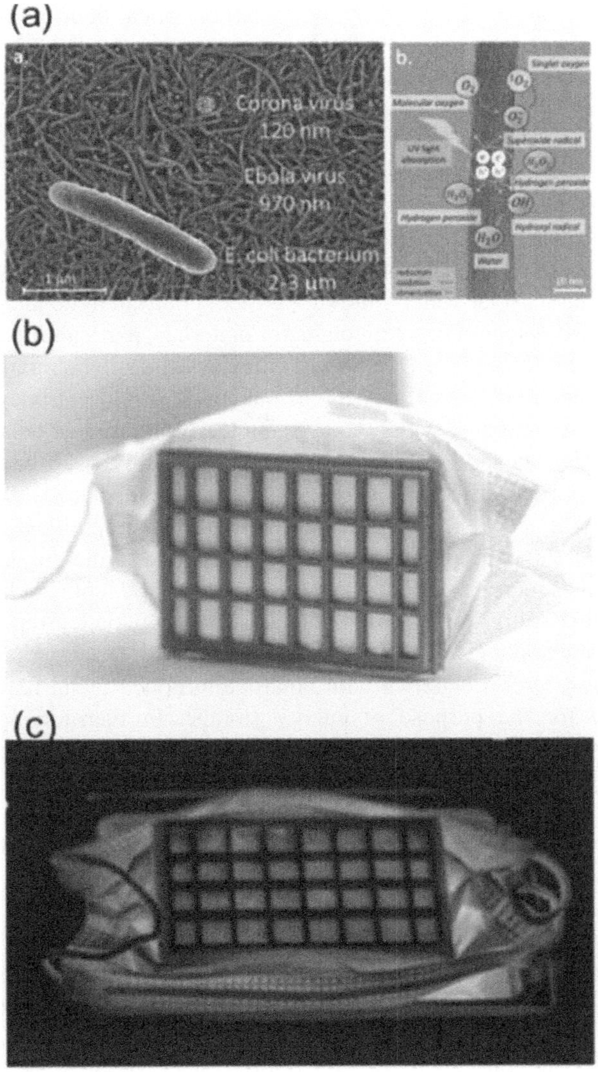

FIGURE 5.13 Photocatalytic air filters based on TiO2 nanowires: (a) photocatalytic action for the disinfection of microbial species; (b) photograph of photocatalytic mask-based filter; (c) mask disinfection under 365 nm. Reproduced with permission from Ref. [131].

and antimicrobial materials can be utilized in both indoor and outdoor environments, where they can begin photocatalytic reactions under natural solar light (outdoors) or LED irradiation (indoors) to clean their surfaces and/or treat air pollutants on a continual basis. The industrial world has given significant attention to this potential technology in terms of commercialization. Photocatalytic reactors for the direct cleaning of industrial exhausts, and novel digital printing method for the coating of tiles with visible light active Ag-TiO_2, solar photoreactors for CO_2 reduction and so forth are few known examples [130].

New applications are also emerging. In this regard, photocatalytic mask filter based on TiO_2 nanowires, a protection against airborne viruses including COVID-19 and could be able to photocatalytically sanitized under UV irradiation, photocatalytic metal-organic framework (MOF) based personal mask [131], which can be self-disinfected under visible light and so forth are some of the noteworthy inventions (Figure 5.13). Another pilot-scale reactor for gas-phase VOC treatment and E. coli inactivation, based on a photocatalytic textile, optical fiber, illuminated by UV LEDs, achieved a 66 percent degradation of 5 mg.m3 butane-2,3-dione. More traditional geometries such as honeycomb monoliths and tubular flow photoreactors have previously been marketed for photocatalytic air purification.

5.7.2 Energy Applications

(a) *Energy Generation:* Metal oxides have emerged as essential electrode materials in contemporary civilization as a means of avoiding energy and environmental crises through green energy generation. Solar water splitting (WS) in photocatalytic and photoelectrochemical systems is based on common semiconductor materials, particularly metal oxides like TiO_2, WO_3, $BiVO_4$, -Fe_2O_3, and Cu_2O, however efficiency can still be improved. As well as environmental applications, the use of TiO_2 as nanocrystalline electrodes in the field of photovoltaics have been studied widely. For example, dye-sensitized solar cells (DSSC) is one of the developing cheapest photovoltaic technology in this area where dye-sensitized metal oxide photoelectrodes directly convert both artificial and natural (solar) radiation into electric current. In the past decades, research focused on the optimization of the organic dyes in DSSC, while more recently attention has been paid to making use of various types of TiO_2 such as mesoporous, nanotube, anatase-rutile hybrid, core-shell structure and so forth as electrode material. The highest photoconversion efficiency reported for titania-based photoanodes are about (13–15%).

(b) *Energy storage*: Li ion batteries with heteroatom doped metal oxides are found to be a better choice for the purpose of energy storage. Furthermore, graphene-metal oxide nanocomposites, such as capacitors and lithium-ion batteries, offer a lot of potential in energy storage and conversion. Metal oxide nanoparticles were anchored to the metal-decorated surface of graphene 3D structures, making them fully accessible to the electrolyte and providing high reversible lithium storage capacities. To reduce lithium-ion diffusion routes and improve electronic conductivity, nanosized metal oxides and highly conductive graphene are beneficial. Besides, metal oxides can be ionic conductors for lithium-ion batteries. Lithium-ion batteries with $LiCoO_2$, $LiMn_2O_4$, Li [NixCoyMn1-x-y] O_2 (0<x,y<1), and Li4Ti_5O_{12}, have been commercialized so far.

(c) *Water splitting/H_2 evolution:* This is a photocatalytic process that splits water into hydrogen (H_2) and oxygen (O_2) in the presence of a catalyst and natural light. Photocatalytic water splitting (PWS) is considered the best option for H_2 production, due to the fact that it has a good solar hydrogen conversion efficiency, low production costs, easy separation of oxygen and hydrogen. Numerous metal oxide and oxynitride nanosheets are established as suitable photocatalysts for visible-light induced water splitting. Multiwalled carbon nanotube (MWCNT)/graphene/TiO_2, WO_3/g-C_3N_4, $SrTiO_3$:La, MoS_2/ TiO_2, Rh/C/$BiVO_4$:Mo Z-scheme photocatalysts, gC_3N_4/TiO_2, g-C_3N_4/Co_2P, gC_3N_4/Ag/MoS_2, black phosphorus, multishell g-C_3N_4, Ag dimer/$ZnIn_2S_4$ nanosheets/TiO_2 nanofibers, g-C_3N_4 co-modified with MnOx and Au--TiO_2, and boron-doped g-C_3N_4/anatase titania nanocomposites and so forth are a few of them [132–135].

Lab scale analysis of a series of materials shed light on the seeming quantum efficiency data for water splitting, at 420 nm are 27.8 percent on mesoporous N-doped g-C3N4, 28.5 percent on CdS/MoS_2 core-shell nanowires, 57 percent (with NaCl) on polymeric CN with shorter interlayer distance, 8.6 percent on crystalline CN nanosheets, 44.3 percent on α-Fe_2O3/g-C_3N4 Z-scheme photocatalysts, 88 <0.1 percent for NiO/gC_3N_4, 7.1 percent for g-C_3N_4/ZnIn2S_4 nanoleaf, 19 percent on $SrTiO_3$:La, and 2.8 percent on g-C_3N_4 co-modified with MnOx and Au–TiO_2 [136–138]. An exciting contribution of $SrTiO_3$ photocatalyst loaded with Rh, Cr, and Co as cocatalysts is up to 96 percent external quantum efficiency at 360nm, which promoted the commercial importance of this material.

(d) *CO_2 reduction*: Metal oxides, such as TiO_2, are gaining popularity as a potential candidate for photocatalytic CO_2 reduction option for converting anthropogenic CO_2 gas into fuels. TiO_2, on the other hand, has a high band gap (3.2 eV), quick electron and hole recombination, and limited selectivity for CO_2 photoreduction. The photoreduction concept of CO_2/H_2O(g/L) includes the creation of electron holes as a result of light absorption. Following that, the produced electrons and holes are transported separately to the photocatalyst surface. The photogenerated electrons then convert CO2 to value-added compounds, while holes oxidize water to produce oxygen molecules. The location of conduction band (CB) and valence band (VB) of semiconductors are important considerations when choosing a photocatalyst material, as they aid in determining the probability of a redox reaction. Because theoretically its CB is more negative than the reduction potential to create products, but its VB is more positive than the redox potential of the reductant, anatase TiO_2 with CB and VB potentials of 0.5 eV and 2.7 eV, respectively, permits the reduction of CO2 into various fuels (such as CH_4, CH_3OH, and CO) and oxidizes H_2O. Surface modification, heterostructure development, and hybridization with other materials (e.g., graphene, GO, rGO, g-C3N4 and its derivatives, sensitized-materials, and 2D MXene) have all recently made significant progress in improving the CO2 reduction performance of TiO2 photocatalysts.Some of the established visible light inducing photocatalysts for CO_2 reduction are Ti^{3+}–TiO_2, g-C_3N_4/nanocarbon/ZnIn2S_4, α-Fe_2O_3/g-C3N_4, N-doped carbon/NiCo2O_4 double shelled nanoboxes, ZnIn2S_4—In_2O_3, Ag_2CrO_4/g-C3N_4/graphene oxide, Ag dimer/ZnIn2S_4 nanosheets/TiO_2 nanofibers, Au@TiO_2 yolk-shell hollow spheres, carbon-coated In_2O_3 g-C_3N_4 co-modified with MnOx and Au—TiO_2 boron-doped g-C_3N_4/anatase titania nanocomposite, and Au/BiOI/MnOx TiO_2—MnOx—Pt and so forth [139–140]. Due to its improved separation efficiency of photogenerated e/h+ pairs with the recombination of inefficient charge carriers, the construction of an indirect Z-scheme system between TiO_2 and another semiconductor using noble metals such as Pt, Au, and Ag as electron mediators has gained increased attention. Because the stronger interaction between 0D $ZnFe_2O_4$ nanospheres and 1D TiO_2 nanorods is beneficial to the transfer of photogenerated electrons and holes at the interface, a nanocomposite of $ZnFe_2O_4$/Ag/TiO_2 fabricated by physically mixing Ag/TiO_2 nanorods and ZnFe2O_4 nanospheres in methanol solution was found to be a better CO_2 reductant.

5.7.3 Synthetic Applications

Visible-light induced metal oxide based photocatalysis offers exciting opportunities in chemical synthesis. For example, depending on the kind of chemical reaction, some reactions can be used to replace hazardous or expensive reagents or to produce gentler reaction conditions, such as lower synthesis temperature. C-C bond formation, for instance, is a major success in terms of chemical processes that might benefit from visible-light mediated photocatalysis. Carbon-heteroatom bond formation, α-amine functionalization, decarboxylative coupling, cycloadditions, atom-transfer radical addition, and fluorination [141–143]. Attaining a better duplication of Suzuki–Miyaura coupling reaction at room temperature and synthesis of compounds that are relevant to drug development, such as trifluoromethoxy and difluoromethoxy, are some of the interesting findings in addition to the extended transformations and improvement in the reaction conditions. A variety of photocatalyst materials like TiO_2, carbon nitrides and their heterojunctions and so forth can be used for this purpose. In many investigations, simple oxidants or reductants, such as air, oxygen, or amines, are used instead of stoichiometric reagents in photocatalytic processes, which generally take place at room temperature.

The photosensitized, noble metal sensitized and dye sensitized visible light-absorbing catalysts have long tradition in synthetic photochemistry. However, the discovery of catalysts that can efficiently carry out energy transfer processes, along with high-power visible light-producing LEDs, holds the potential of novel transformations. In such circumstances, the method of action is the in situ production of electron donor-acceptor complexes with long wavelength absorption leading to charge separation.

5.8 FUTURE OUTLOOK AND CONCLUSION

In recent years, solar-driven photocatalytic conversion has become very attractive as a means to energy and environmental-related applications. Metal oxides plays a crucial role in simultaneously harvesting and utilizing the visible light in a clean and effective way. Hence, it is critical that researchers understand what has already been established, what the problems are, and what the future prospects are for improving existing technology. Multicomponent systems, such as heterojunction (including Z-scheme) photocatalysts coupled with co-catalysts and/or plasmonic particles, are the most promising techniques for increasing the visible light absorption of photocatalysts for diverse applications. In terms of efficiency no model photocatalyst come closer to the required efficiency level for commercialization of fuel cell has fabricated yet and the estimated solar-to-hydrogen efficiency for practical applications is 10 percent for overall water splitting. In other words, there are significant challenges to overcome for making photocatalysis at its industrial scale and working on increasing the solar light engrossing efficiency is considerably more significant than the developing alternative low-cost materials. The current chapter discussed the recent advances in visible light absorbing photocatalysts, new materials that are experimented, and emerging applications of visible light photocatalysis. In particular, selection of the photocatalyst, advanced TiO_2 nanomaterials and a glimpse on non TiO_2 based materials, problems associated with visible light absorption, photocatalyst design, combination of semiconductors, doping with noble metals, dye sensitization, and so forth are discussed in detail. Furthermore, when compared to standard TiO_2, several additional modified photocatalysts have been discovered to be beneficial for environmental applications under visible light. However, in order to solve various concerns surrounding test procedures, verify real photocatalytic activity, and explore potential commercialization of the material, an effective assessment on the structure-property relationship of nanomaterials is required.

REFERENCES

1. Fujishima, K. Honda H., *Nature* 238, 37, 1972.
2. M. Hoffmann, S. Martin, W. Choi and D. Bahnemann, *Chem. Rev.*, 95, 69–96, 1995.
3. J. Schneider, M. Matsuoka, M. Takeuchi, J. Zhang, Y. Horiuchi, M. Anpo, D. W. Bahnemann, *Chem. Rev.* 114, 9919, 2014.
4. S. Banerjee, S. C. Pillai, P. Falaras, K. E. O'Shea, J. A. Byrne, D. Dionysiou, *J. Phys. Chem. Lett.* 5, 2543, 2014.
5. Lavorato, A., Primo, R., Molinari, H., Garcia, Chem. *A Eur. J.* 20, 187–194, 2014.
6. G. Manna, R. Bose, N. Pradhan, *Angew Chem Int Ed* 53, 6743–6746, 2014.
7. B. O'Regan, and M. Graetzel, *Nature*, vol. 353, pp. 737, 1991.
8. V. Etacheri, D. Sharon, A. Garsuch, M. Afri, A. A. Frimer and D. Aurbach, *J. Mater. Chem.* A, 1, 5021, 2013.
9. V. Etacheri, O. Haik, Y. Gofer, G. A. Roberts, I. C. Stefan, R. Fasching and D. Aurbach, *Langmuir*, 28, 965, 2012.
10. L. Marzo, S. K. Pagire, O. Reiser, and B. König, *Angew. Chem. Int. Ed.* 57, 10034–10072, 2018.
11. M. B. Fisher, D. A. Keane, P. Fernández-Ibáñez, J. Colreavy, S. J Hinder, K. G. McGuigan, S. C. Pillai, *Appl. Catal.*, B 8, 130–131, 2013.
12. M. R. Hoffmann, S. T. Martin, W. Choi and D. W. Bahnemann, *Chem. Rev.*, 95, 69, 1995.
13. R. Nakamura and Y. Nakato, *J. Am. Chem. Soc.,* 126, 1290, 2004.
14. A. J. Hoffmann, E. R. Carraway and M. R. Hoffmann, *Environ. Sci. Technol.*, 1994, 28.

15. V. Etacheri, G. Michlits, M. K. Seery, S. J. Hinder, S. C. Pillai, *ACS Appl. Mater. Interfaces* 5, 1663, 2013.
16. M. Kaneko and I. Okura, *Photocatalysis – Science and Technology*, Springer, 2002.
17. Kormann, D. W. Bahnemann and M. R. Hoffmann, *Environ. Sci. Technol.* 25, 494, 1991.
18. R. Asahi, Y. Taga, W. Mannstadt, A. J. Freeman, *Phys. Rev. B: Condens. Matter* 61, 7459, 2000.
19. N. T. Nolan, M. K. Seery, S. C. Pillai, *J. Phys. Chem. C* 11S3, 16151, 2009.
20. A. L. Linsebigler, G. Lu, Y. T. Yates, *Chem. Rev.* 95, 735, 1995.
21. J. Zhang, Y. Nosaka, *J. Phys. Chem. C* 118, 10824, 2014.
22. Y.H. Sang, H. Liu, A. Umar, *Chemcat chem* 7, 559–573, 2015.
23. G. Reenamole, M. K. Seery, S. C Pillai, *J. Phys. Chem. C*, 112(35), 13563–13570, 2008.
24. Q. J. Xiang, B. Cheng, and J. G. Yu, Angew. *Chem., Int. Ed.* 54, 11350, 2015.
25. A. Dhakshinamoorthy, A. M. Asiri, and H. Garcia, *Angew. Chem., Int. Ed.* 55, 5414, 2016.
26. H. Wang, W. D. Zhang, X. W. Li, J. Y. Li, W. L. Cen, Q. Y. Li, and F. Dong, *Appl. Catal., B* 225, 218, 2018.
27. H. Li, J. Li, Z. H. Ai, F. L. Jia, and L. Z. Zhang, *Angew. Chem., Int. Ed.* 57, 122, 2018.
28. J. D. Xiao, Q. Z. Han, Y. B. Xie, J. Yang, Q. Z. Su, Y. Chen, and H. B. Cao, *Environ. Sci. Technol.* 51, 13380, 2017.
29. A. Behera, D. Kandi, S. Martha, and K. Parida, *Inorg. Chem.* 58, 16592, 2019.
30. B. Aleksandra, He. Yanling, M. C. Alan, Ng, *APL Mater.* 8, 030903, 2020.
31. J. J. Wang, L. Tang, G. M. Zeng, Y. C. Deng, Y. N. Liu, L. L. Wang, Y. Y. Zhou, Z. Guo, J. J. Wang, and C. Zhang, *Appl. Catal., B* 209, 285, 2017.
32. M. Li, F. Y. Liu, Z. Y. Ma, W. Liu, J. L. Liang, and M. P. Tong, *Chem. Eng. J.* 371, 750–758, 2019.
33. Y. N. Chen, G. Q. Zhu, M. Hojamberdiev, J. Z. Gao, R. L. Zhu, C. H. Wang, X. M. Wei, and P. Liu, J. Hazard. *Mater.* 344, 42–54, 2018.
34. S. B. Wang, B. Y. Guan, Y. Lu, and X. W. D. Lou, *J. Am. Chem. Soc.* 139, 17305, 2017.
35. B. Han, S. Q. Liu, N. Zhang, Y. J. Xu, and Z. R. Tang, *Appl. Catal., B* 202, 298, 2017.
36. J. N. Liu, Q. H. Jia, J. L. Long, X. X. Wang, Z. W. Gao, and Q. Gu, *Appl. Catal., B* 222, 35, 2018.
37. J. R. Ran, W. W. Guo, H. L. Wang, B. C. Zhu, J. G. Yu, and S. Z. Qiao, *Adv. Mater.* 30, 1800128, 2018.
38. S. B. Wang, B. Y. Guan, and X. W. D. Lou, *J. Am. Chem. Soc.* 140, 5037–5040, 2018.
39. Zhang, Y. C. Zhang, G. S. Zhang, Z. J. Yang, D. D. Dionysiou, and A. P. Zhu, *Appl. Catal., B* 236, 53–63, 2018.
40. N. Farhadian, R. Akbarzadeh, M. Pirsaheb, T. C. Jen, Y. Fakhri, and A. Asadi, *Int. J. Biol. Macromol.* 132, 360–373, 2019.
41. Zhang, Y. C. Zhang, G. S. Zhang, Z. J. Yang, D. D. Dionysiou, and A. P. Zhu, *Appl. Catal., B* 236, 53–63, 2018.
42. S. Bellamkonda, N. Thangavel, H. Y. Hafeez, B. Neppolian, and G. R. Rao, *Catal. Today* 321, 120–127, 2019.
43. K. K. Das, S. Patnaik, S. Mansingh, A. Behera, A. Mohanty, C. Acharya, and K. Parida, *J. Colloid Interface Sci.* 561, 551–567, 2020.
44. S. Patnaik, K. K. Das, A. Mohanty, and K. Parida, *Catal. Today* 315, 52–66, 2018.
45. A. Chauhan, M. Rastogi, P. Scheier, C. Bowen, R. V. Kumar, and R. Vaish, *Appl. Phys. Rev.* 5, 041111, 2018.
46. D. L. Huang, S. Chen, G. M. Zeng, X. M. Gong, C. Y. Zhou, M. Cheng, W. J. Xue, X. L. Yan, and J. Li, *Coord. Chem. Rev.* 385, 44–80, 2019.
47. L. B. Jiang, X. Z. Yuan, G. M. Zeng, J. Liang, Z. B. Wu, and H. Wang, *Environ. Sci.: Nano* 5, 599–615, 2018.
48. L. B. Jiang, X. Z. Yuan, G. M. Zeng, J. Liang, Z. B. Wu, and H. Wang, *Environ. Sci.: Nano* 5, 599–615, 2018.
49. S. Adhikari, S. Mandal, and D.-H. Kim, *Chem. Eng. J.* 373, 31–43, 2019.
50. T. T. Xiao, Z. Tang, Y. Yang, L. Q. Tang, Y. Zhou, and Z. G. Zou, *Appl. Catal., B* 220, 417–428, 2018.
51. Y. Xu, J. J. Zhang, B. C. Zhu, J. G. Yu, and J. S. Xu, *Appl. Catal., B* 230, 194–202, 2018.
52. A. Y. Meng, B. C. Zhu, B. Zhong, L. Y. Zhang, and B. Cheng, *Appl. Surf. Sci.* 422, 518–527, 2017.
53. J. X. Low, B. Z. Dai, T. Tong, C. J. Jiang, and J. G. Yu, *Adv. Mater.* 31, 1802981, 2019.
54. X. J. She, J. J. Wu, H. Xu, J. Zhong, Y. Wang, Y. H. Song, K. Q. Nie, Y. Liu, Y. C. Yang, M.-T. F. Rodrigues, R. Vajtai, J. Lou, D. L. Du, H. M. Li, and P. M. Ajayan, *Adv. Energy Mater.* 7, 1700025, 2017.

55. Z. F. Jiang, W. M. Wan, H. M. Li, S. Q. Yuan, H. J. Zhao, and P. K. Wong, Adv. Mater. 30, 1706108, 2018.
56. W. L. Yu, J. X. Chen, T. T. Shang, L. F. Chen, L. Gu, and T. Y. Peng, *Appl. Catal., B* 219, 693–704, 2017.
57. Q. Wang, T. Hisatomi, Y. Suzuk, Z. H. Pan, J. Seo, M. Katayama, T. Minegishi, H. Nishiyama, T. Takata, K. Seki, A. Kudo, T. Yamada, and K. Domen, *J. Am. Chem. Soc.* 139, 1675–1683, 2017.
58. W. K. Jo and N. C. S. Selvam, *Chem. Eng. J.* 317, 913–924, 2017.
59. J. Suave, S. M. Amorim, J. Ângelo, L. Andrade, A. Mendes, RFPM Moreira. *J Photochem. Photobiol A Chem* 348, 326–336, 2017.
60. R. Leary, A. Westwood, *Carbon N.Y.* 49, 741–772, 2011.
61. Z. D. Meng, J. G. Choi, J. Park J-Y, *Nano. Res Lett.* 2:29–38, 2011.
62. M. C. Bahome, L .L. Jewell, D. Hildebrandt, D. Glasser, N. J. Coville, *Appl Catal A Gen* 287: 60–67, 2005.
63. H. T. Chung, J. T. Won, P. Zelenay, *Nat Commun.*, 2013.
64. K. K. Paul, P. K. Giri, H. Sugimoto, M. Fujii, and B. Choudhury, *Sol. Energy Mater. Sol. Cells* 201, 110053, 2019.
65. K. M. Choi, D. Kim, B. Rungtaweevoranit, C. A. Trickett, J. T. D. Barmanbek, A. S. Alshammari, P. D. Yang, and O. M. Yaghi, *J. Am. Chem. Soc.* 139, 356–362, 2017.
66. Q. Jiang, C. Ji, D. J. Riley, and F. Xie, *Nanomaterials* 9, 1, 2019.
67. J. Nie, A. O. T. Patrocinio, S. Hamid, F. Sieland, J. Sann, S. Xia, D. W. Bahnemann, and J. Schneider, *Phys. Chem. Chem. Phys.* 20, 5264–5273, 2018.
68. R. Abe, K. Sayama, H. Arakawa, *Chem. Phys. Lett.* 362, 2002, 441, 2002.
69. J. Rehm, G. Mclendon, Y. Nagasawa, K. Yoshihara, J. Moser, M. Gratzel, *J. Phys. Chem.* 100, 9577, 1996.
70. Yan, J. Hupp, *J. Phys. Chem. B* 100, 6867, 1996.
71. V. Etacheri, C. D Valentin, J. Schneider, D. Bahneman, S. C. Pillai, J. photochem. *Photobio C: Photochem Rev.* 25, 1–29, 2015.
72. M. Latorre-Sanchez, C. Lavorato, M. Puche, V.Fornes, R. Molinari, H. Garcia, *Chem. A Eur. J.* 18, 16774–16783, 2012.
73. A. Braga, S. Gimenez, I. Concina, A. Vomiero, I. Mora-Sero, *J. Phys. Chem. Lett.* 2, 454, 2011.
74. W.A. Tisdale, K.J. Williams, B.A. Timp, D.J. Norris, E.S. Aydil, X.Y. Zhu, *Science* 328, 1543, 2010.
75. B. O'Regan, M. Gratzel, *Nature* 353, 737–739, 1991.
76. R. Wang, K. Hashimoto, A. Fujishima, M. Chikuni, E. Kojima, A. Kitamura, *Nature* 388, 431–432, 1997.
77. www.Nature.com, *Scientific Reports*, 7, 15232, 2017.
78. K.T. Ranjit, I. Willner, S.H. Bossmann, A.M. Braun, *Environ. Sci. Technol.* 35, 1544, 2001.
79. Y. Zhu, Y. Wang, Z. Chen, L. Qin, L. Yang, L. Zhu, P. Tang, T. Gao, Y. Huang, Z. Sha, G. Tang, *Appl. Catal. A Gen.* 498, 159–166, 2015.
80. Z. Yue, A. Liu, C. Zhang, J. Huang, M. Zhu, Y. Du, P. Yang, *Appl. Catal. B Environ.* 201, 202–210, 2017.
81. T. Das, X. Rocquefelte, R. Laskowski, L. Lajaunie, S. Jobic, P. Blaha, K. Schwarz, *Chem. Mater.* 29, 3380–3386, 2017
82. J. Rodríguez, F. Paraguay-Delgado, A. López, J. Alarcón, W. Estrada, *Thin Solid Films* 519, 729–735, 2010.
83. Li, X. Nie, J. Chen, Q. Jiang, T. An, P.K. Wong, H. Zhang, H. Zhao, H. Yamashita, *Water Res.* 86, 17–24, 2015.
84. Q. Zhang, X. Quan, H. Wang, S. Chen, Y. Su, Z. Li, *Sci. Rep.* 7, 3128, 2017.
85. Li, B. Liu, Y. Wang, Y. Shi, X. Ma et al., *RSC Advance*, 4, 37992–37997, 2014.
86. R. Asahi, T. Morikawa, T. Ohwaki, K. Aoki and Y. Taga, *Science*, 293, 269, 2001.
87. S. Khan, M. Al-Shahry and W. Ingler, *Science*, 2002, 297, 2243.
88. S. Sakthivel and H. Kisch, *Angew. Chem., Int. Ed.*, 2003, 42, 4908–4911.
89. W. Zhao, W. Ma, C. Chen, J. Zhao and Z. Shuai, *J. Am. Chem. Soc.*, 126, 4782–4783, 2004.
90. X. Hong, Z. Wang, W. Cai, F. Lu and J. Zhang, *Chem. Mater.*, 2005, 17, 1548–1552.
91. H. Irie, Y. Watanabe, K. Hashimoto, *Chemistry Letters* 32, 2003.
92. S. Sakthivel, H. Kisch, *Angewandte Chemie International Edition* 42, 2003.
93. C. Han, M. Pelaez, V. Likodimos, A. G. Kontos, P. Falaras, K. O'Shea, D.D. Dionysiou, *Applied Catalysis B: Environmental* 107, 77–87, 2011.
94. M. Pelaez, A. A. de la Cruz, E. Stathatos, P. Falaras, D.D. Dionysiou, *Catalysis Today* 144, 19–25, 2009.

95. M. Harb, *J. Phys. Chem. C* 117, 25229, 2013.
96. M. K. Seery, R. George, P. Floris, S. C. *Pillai, Journal of Photochemistry and Photobiology A* 189, 258–263, 2007.
97. C. Gunawan, W. Y. Teoh, C. P. Marquis, J. Lifia, R. Amal, *Small* 5, 341–344, 2009.
98. M. J. Yang, C. Hume, S. Lee, Y. H. Son, J. K. Lee, *J. Phys. Chem. C* 114, 15292, 2010.
99. J. W. Liu, R. Han, Y. Zhao, H. T. Wang, W. J. Lu, T. F. Yu, Y. X. Zhang, *J. Phys. Chem. C* 115, 4507, 2011.
100. E. J. Wang, T. He, L. S. Zhao, Y. M. Chen, Y. A. Cao, *J. Mater. Chem.* 21, 144, 2011.
101. H. B. Liu, Y. M. Wu, J. L. Zhang, *ACS Appl. Mater. Interfaces* 3, 1757, 2011.
102. Zhang, Y. Wu, M. Xing, S. A. K. Leghari, S. Sajjad, *Energy Environ. Sci.* 3, 715, 2010.
103. Y. Cong, J. L. Zhang, F. Chen, M. Anpo, *J. Phys. Chem. C* 111, 10618, 2007.
104. S. J. Jum, G. K. Hyun, A. J. Upendra, W. J. Ji, S. L. Jae, Int. *Journal of Hyd. En*, 33, 5975, 2008.
105. J. Savio, A. Monizand, J. Tang, *Chem. Cat. Chem.* 7, 1659–1667, 2015.
106. Li, X.; Lin, H.; Chen, X.; Niu, H.; Liu, J.; Zhang, T.; Qu, F. *Phys. Chem. Chem. Phys.* 18, 9176–9185, 2016.
107. A. Humayun, Z. Zada, M. Li, X. Xie, Y. Zhang, F. Qu, L. Raziq, L Jing L, *Appl Catal B* 180: 219–226, 2016.
108. Ratova, R. Klaysri, P. Praserthdam, P. J. Kelly, *Mater Sci Semicond Process* 71: 188–196, 2017.
109. D. Wu, B. Wang, W. Wang, T. An, G. Li, T. W. Ng, H. Y. Yip, C. Xiong, H. K. Lee, P. K. Wong, *J. Mater. Chem. A* 3, 15148–15155, 2015.
110. D. Wu, L. Ye, S. Yue, B. Wang, W. Wang, H. Y. Yip, P. K. Wong, J. *Phys. Chem. C* 120, 7715–7727, 2016.
111. T. Feng, J. Liang, Z. Ma, M. Li, M. Tong, *Colloids Surf. B Biointerfaces* 167, 275–283, 2018.
112. J. Liang, J. Deng, M. Li, M. Tong, *Colloids Surf. B Biointerfaces*. 138, 102–109, 2016.
113. W. Liu, Y. Feng, H. Tang, H. Yuan, S. He, S. Miao, *Carbon N.Y.* 96, 303–310, 2016.
114. C. Liu, D. Kong, P.-C. Hsu, H. Yuan, H.-W. Lee, Y. Liu, H. Wang, S. Wang, K. Yan, D. Lin, P.A. Maraccini, K. M. Parker, A. B. Boehm, Y. Cui, *Nat. Nanotechnol.* 11, 1098–1104, 2016.
115. P. Cheng, Q. Zhou, X. Hu, S. Su, X. Wang, M. Jin, L. Shui, X. Gao, Y.-Q. Guan, R. Nözel, G. Zhou, Z. Zhang, J.-M. Liu, *ACS Appl. Mater. Interfaces* 10, 23444–23450, 2018.
116. P. Dong, Y. Wang, H. Li, H. Li, X. Ma and L. Han, *J. Mater. Chem. A*, 1, 4651–4656, 2013.
117. Q. Yu, J. Li, Y. Zhang, Y. Wang, L. Liu, M. Li, *Sci. Rep.* 6, 26667, 2016.
118. Y. Takaoka, Y. Hirobe, M. Tomonari, Y. Kinoshita, *Titanium Oxide Photocatalyst and Method of Producing the Same, US5759948A*, 1995.
119. K. B Akshaya, G. Reenamole, K. Sasidharan, T. P. Vinod, V. Anitha and L George *Anal. Methods*. 2019, 11, 537.
120. S. Adhikari, S. Mandal, and D.-H. Kim, Chem. Eng. J. 373, 31–43, 2019.
121. Farhadian, R. Akbarzadeh, M. Pirsaheb, T. C. Jen, Y. Fakhri, and A . Asadi, *Int. J. Biol. Macromol.* 132, 360–373, 2019.
122. S. Hassan, A. A. Jalil, N. F. Khusnun, M. W. Ali, and S. Haron, *J. Alloys Compd.* 789, 221–230, 2019.
123. M. Li, F. Y. Liu, Z. Y. Ma, W. Liu, J. L. Liang, and M. P. Tong, *Chem. Eng. J.* 371, 750–758, 2019.
124. J.L Carra, S. García Sánchez, J. L. Casas López, S. Malato, J. A. Sánchez Pérez, *Sci Total Environ* 478: 123–132, 2014.
125. B. A., Marinho, R. Djellabi, R. O. Cristóvão, J. M. Loureiro, R. A. R. Boaventura, M. M. Dias, J. C. B. Lopes, V. J. P. Vilar. *Chem Eng J* 318: 76–88, 2017.
126. B. M., da Costa Filho, A. L. P. Araujo, G. V. Silva, R. A. R. Boaventura, M. M. Dias, J. C. B. Lopes, V. J. P. Vilar, *Chem Eng J* 310: 331–341, 2017.
127. T. C. Machado, M. A. Lansarin, C. S. Ribeiro, *Photochem Photobiol Sci* 14(3): 501–505, 2015.
128. A. Danion, J. Disdier, C. Guillard, F. Abdelmalek, R. N. Jaffrezic, *Appl Catal B Environ* 52: 213–223, 2004.
129. A. Danion, J. Disdier, C. Guillard, R. N. Jaffrezic, *J Photochem Photobiol A Chem* 190(1): 135–140, 2007.
130. C. L. Bianchi, G. Cerrato, B. M. Bresolin, R. Djellabi, S. Rtimi, *Surfaces*, 3, 11–25, 2020.
131. Horváth, E., Rossi, L., Mercier, C., Lehmann, C., Sienkiewicz, A., Forró, L. *Adv. Funct. Mater.* 30, 2004615, 654, 2020.

132. Li, J. Li, X. Feng, J. Li, Y. Hao, J. Zhang, H. Wang, A. Yin, A.; Zhou, J.; Ma, X, *Nat. Commun.* 10, 1–10, 2019.
133. L. M. Guo, Z. Z. Yang, K. Marcus, Z. Li, B. C. Luo, L. Zhou, X. H. Wang, Y. G. Du, and Y. Yang, Energy Environ. *Sci.* 11, 106–114, 2018.
134. Y. G. Tan, Z. Shu, J. Zhou, T. T. Li, W. B. Wang, and Z. L. Zhao, *Appl. Catal., B* 230, 260–268, 2018.
135. J. R. Ran, G. P. Gao, F. T. Li, T. Y. Ma, A. J. Du, and S. Z. Qiao, *Nat. Commun.* 8, 13907, 2017.
136. Q. Wang, T. Hisatomi, Y. Suzuk, Z. H. Pan, J. Seo, M. Katayama, T. Minegishi, H. Nishiyama, T. Takata, K. Seki, A. Kudo, T. Yamada, and K. Domen, *J. Am. Chem. Soc.* 139, 1675–1683, 2017.
137. X. J. She, J. J. Wu, H. Xu, J. Zhong, Y. Wang, Y. H. Song, K. Q. Nie, Y. Liu, Y. C. Yang, M.-T. F. Rodrigues, R. Vajtai, J. Lou, D. L. Du, H. M. Li, and P. M. Ajayan, *Adv. Energy Mater.* 7, 1700025, 2017.
138. F. Raziq, L. Q. Sun, Y. Y. Wang, X. L. Zhang, M. Humayun, S. Ali, L. L. Bai, Y. Qu, H. T. Yu, and L. Q. Jing, *Adv. Energy Mater.* 8, 1701580. 2018.
139. Y. Bai, L. Q. Ye, L. Wang, X. Shi, P. Q. Wang, and W. Bai, *Environ. Sci.: Nano3*, 902–909. 2016.
140. F. Raziq, Y. Qu, X. L. Zhang, M. Humayun, J. Wu, A. Zada, H. T. Yu, X. J. Sun, and L. Q. Jing, *J. Phys. Chem. C* 120, 98–107, 2016.
141. L. Marzo, S. K. Pagire, O. Reiser, and B. Koenig, *Angew. Chem., Int. Ed.* 57, 10034–10072, 2018.
142. Z. Jiao, Z. Zhai, X. Guo, and X.-Y. Guo, *J. Phys. Chem. C* 119, 3238–3243, 2015.
143. H. Cheng and W. Xu, *Org. Biomol. Chem.* 17, 9977–9989, 2019.

6 Progress in Photocatalysis for Hydrogen Evolution and Environmental Remediation

R. Shwetharani, M.S. Jyothi, M. Dinamani and S. Radoor

CONTENTS

6.1 Introduction ... 131
6.2 Synthesis of Photocatalysts ... 132
6.2.1 Sol-gel Process ... 132
6.2.2 Hydrothermal Process ... 132
6.2.3 Sonochemical Method ... 133
6.2.4 Chemical Vapor Deposition (CVD) ... 133
6.2.5 Physical Vapor Deposition (PVD) ... 134
6.2.6 Electrochemical Deposition ... 134
6.3 Reforms in Photocatalysis ... 134
6.3.1 Doping of Metals or Non-Metals ... 134
6.3.2 Loading of Cocatalysts ... 136
6.3.3 Heterojunction Structures ... 137
6.4 Application ... 138
6.4.1 Hydrogen Evolution ... 138
6.4.2 Waste-Water Treatment ... 145
6.5 Conclusions and Future Prospective ... 147
References ... 148

6.1 INTRODUCTION

A demand for water and energy has been elevating over time, for which, mainly conventional energy sources derived from fossil fuels are deployed extensively. Meanwhile, the classical pathway of its usage results in greenhouse gases, leading to global warming as well as climate changes [1, 2]. The overall outcome of the process is environmental deterioration and reduction in nature that cannot be replenished. The destructive outcome demands a different source of energy that could be harvested from natural resources, such as hydrothermal, geothermal, solar and wind energy. Amongst these, solar energy is most preferred as it offers, not only generation of electricity, but also treatment of wastewater, driving hydrogen evolution and also artificial photosynthesis.

The intrinsic and electronic structures of semiconductors have proven their efficacy for harvesting solar energy. With irradiation of light having an intensity higher than the bandgap energy, the electrons of the semiconductor excite from valence band (VB) to conduction band (CB). The process will generate electron and hole pairs. In case of liquid phase, the holes generated will promote the oxidation of water molecules and generate oxygen molecules in the VB. The electrons will reduce the H^+ ions to hydrogen ions in the CB. The other advantage is that the electron-hole pairs result in formation of reactive oxygen species (ROS) [3]. ROSs are considered to be a strong oxidizing agent, which would oxidize the contaminants of water to result in clean water. This phenomenon lends a

DOI: 10.1201/9781003218708-7

hand in cleaning industrial waste water, and even polluted water from water bodies. In the same way, the photoreduction of CO_2 to form methanol, ethanol, methane and many other smaller units is much appreciated progress in the case of semiconductors.

Owing to these many applications, researchers have developed numerous photocatalysts pertaining to the feasible electronic structures. People have come up with many synthetic strategies, wherein the efforts are being made to tune the electronic features of photoctalytic materials. Anchoring co-catalysts, dopants (intrinsic variations) and adding peripheral electron acceptor (extrinsic variations) are the major methods of tuning the bandgap energy, migration of electrons and holes and even the redox ability of semiconductors. This chapter deals with synthesis and efficiency-enhancing strategies of semiconductors in harnessing solar energy and its applications with respect to hydrogen production and environmental remediation.

6.2 SYNTHESIS OF PHOTOCATALYSTS

Several preparation approaches for synthesizing photocatalysts and doped/composite photocatalysts are reported, such as the sol-gel technique, hydro- and solvothermal methods, direct oxidation reaction, chemical vapor deposition, sonochemical method, electrodeposition process and microwave method, and so on [4–7]. Photocatalytic semiconducting materials can be designed with different morphology such as spherical, fiber, tubes and films [5]. It has been suggested that the photocatalytic activity and physicochemical assets of the semiconducting materials are mainly reliant on the preparation technique used during the synthesis [8]. The details of the synthesis processes are discussed in the following section.

6.2.1 Sol-gel Process

The sol-gel method of synthesis is one of the widely accepted processes for photocatalyst synthesis. The properties of metal oxides can be tuned by regulating reaction conditions such as the nature of precursor, temperature, pH and time of reaction, catalyst concentration and nature, reagent concentration, aging time and temperature, so on. The molecular level homogeneous mixing being one of the major advantages of the sol-gel process, it increases the polycrystalline particles formation and also benefits the preparation of doped materials. The addition of dopants in the sol during gelation process allows dopants to have direct interaction with support material leading to enhanced activity of photocatalyst material [9]. The sol-gel procedure includes alteration of a precursor solution into an inorganic solid material through polymerization reaction influenced by water. Preparation of TiO_2 by titanium (IV) n-butoxide precursor results in formation of anatase or rutile or brookite phase TiO_2 nanostructure depending on the temperature of calcinations and synthesis condition [10]. Tobaldi et al. [11] in 2013 prepared tungsten, silver, and tungsten/silver co-doped TiO_2 nanostructures by the sol-gel method and studied the photocatalytic activity under the illumination of UVA and visible light for methylene blue degradation, which indicates appreciable photocatalytic action under the illumination of visible light for doped and co-doped TiO_2.

6.2.2 Hydrothermal Process

The hydrothermal process refers to the heterogeneous reaction taking place at high temperature and pressure in the attendance of aqueous solvents, which is normally conducted in a steel vessel with or without Teflon coating. A vapor saturation pressure can be reached by increasing the temperature above its boiling point. Attained temperature and solution volume determine the internal pressure generated [12]. WO_3 synthesized by hydrothermal technique was tested for photo-electrochemical water oxidation and rhodamine B (RhB) breakdown under visible light treatment, which presented appreciable photocatalytic activity [13].

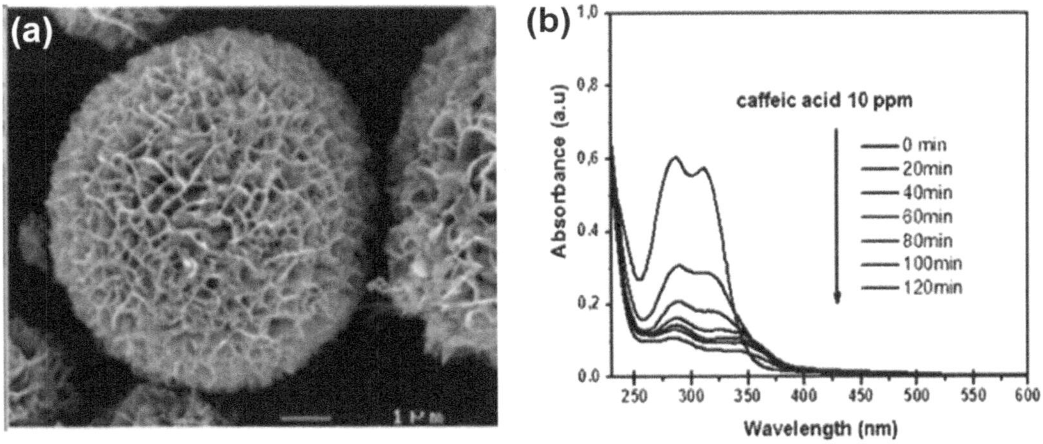

FIGURE 6.1 (a) SEM images of BiOI microspheres synthesized for 24 h; (b) UV-vis spectra during the degradation of caffeic acid in presence of BiOI microspheres synthesized for 12 h [14].

The solvothermal method originated from a hydrothermal procedure, wherein the solvent used is non-aqueous. The temperature can be increased much higher by choosing solvents having high boiling points, and the properties of nanostructured photocatalyst – such as crystallinity, particle size-shape, distribution – can be modified/controlled. Mera et al. [14] in 2014 prepared bismuth oxyiodide through the solvothermal method using ethylene glycol and ionic liquid 1-butyl-3-methylimidazolium iodide at different temperature ranges from 2–48 h resulting in microspheres morphology as revealed in Figure 6.1. Photocatalysts such as TiO_2 are prepared by direct oxidation of titanium metal by means of oxidants or through anodization. Titanium metal plate oxidized with hydrogen peroxide leads to the crystalline nanorods of TiO_2. Anodic oxidation is found to be an electrochemical process in which the oxide film is formed on the metallic substrate.

6.2.3 Sonochemical Method

The power of ultrasound can be used to prepare an extensive choice of photocatalysts comprising transition metals of greater surface area, oxides, alloys, colloids and carbides. In this method, the acoustic cavitation with high temperature of 10,000 K and pressure of 1,000 atm or more breakdowns the chemical bonds. During such intense conditions, diverse physical changes and chemical reactions happen, leading to the development of photocatalysts with specific properties such as particle size and morphology. Anandan et al. [15] in 2014 prepared TiO_2/WO_3 photocatalyst through ultrasound process, where the reaction directed at ultrasonic horn of greater intensity at room temperature in an argon atmosphere for 2.5 h. Prepared photocatalyst exhibited a mixture of square and hexagonal shaped units of dimeter about 8–12 nm and showed higher methylene blue (MB) degradation efficiency in the existence of visible light irradiation. The microwave assisted synthesis of photocatalysts is another type of synthesis process that involves absorption of the microwave radiation by the reaction mixture, which usually depends on the solvent used for the reaction.

6.2.4 Chemical Vapor Deposition (CVD)

CVD is one more technique to prepare semiconductor thin films of high purity and performance. The CVD process involves volatilization followed by reaction of one or more precursors (volatile materials) and deposits on substrate surface at controlled temperature and pressure. Sarantopoulos et al. [16] in 2009 prepared TiO_2 thin films through CVD technique using Si wafers, and borosilicate glass are used as substrates, and titanium isopropoxide is considered as a precursor of titanium. The nature of substrates, temperature of deposition, time and precursors concentration can be altered throughout CVD process in order to get a specific morphology of the films.

6.2.5 Physical Vapor Deposition (PVD)

In distinction to CVD method of preparation, PVD process involves thin film formation from gas phase precursors with no chemical conversion of precursor to final product. There are two well-known PVD processes: (a) the magnetron sputtering method involves fabrication of thin films by sputtering, in which material will be ejected from a target and then deposited on a substrate in the presence of sputtering gas system; (b) pulsed laser deposition is another method in which a laser beam of high power is focused to strike a objective material that has to be deposited followed by deposition on substrate as a thin film. Carneiro et al. [17] in 2005 fabricated Fe doped TiO_2 thin films coated glass and polycarbonate substrates through sputtering of dc reactive magnetron and showed higher photocatalytic deprivation action for rhodamine B.

6.2.6 Electrochemical Deposition

The electrochemical method of photocatalyst preparation involves coating material on conducting or metallic substrates. Typically, two electrodes such as conducting and counter electrodes are arranged in parallel and are immersed in a powder dispersed solvent. Further, an electric potential is given to the arrangement, which makes the particles to transfer towards conducting substrate resulting the formation of thin film on substrate. An et al. [18] in 2005 fabricated TiO_2 films on SnO_2:F conductive glass substrates using $TiCl_4$ and $TiCl_3$ as the precursor materials through cyclic voltammetry. Another physical method of synthesis, such as ball milling, electrospinning and solid state reaction, involves physical techniques – grinding, sputtering and laser techniques – to prepare a photocatalyst. Therefore, various preparation methods for photocatalyst synthesis are available to design a photocatalyst with specific physico-chemical properties [19].

6.3 REFORMS IN PHOTOCATALYSIS

6.3.1 Doping of Metals or Non-Metals

Photocatalytic reactions involve the absorption of light, separation of charge carriers, diffusion to the surface, and the resultant surface redox reactions with adsorbed species at the active sites of the photocatalyst. These steps are greatly affected by bulk, surface and electronic structure of the photo catalyst. Consequently, various methods have been adopted to optimize photocatalytic materials with respect to crystal growth and shape control [20, 21], surface sensitization or modification [22–24], heterostructuring [25], and metal/non-metal doping [4, 26] or plasmonic metallic particle decoration [27–29] for visible light absorption.

In an attempt to enhance the photocatalytic properties of TiO_2, many methods have been explored, such as ion doping [30, 31], metal deposition [32, 33] photosensitization [34, 35], heterojunction construction [36, 37] and the control of morphology [38, 39]. In terms of the ion-doping method, metal-ion doping [40-42], non-metal ion doping [26, 43, 44] along with metal- and non-metal-ion co-doping [45] have been reported as versatile methods for the modification of photocatalysts. Metal-ion doping – such as with Mg [39], Ag [31], Al [46], Cu [47], Mo [48], Ca [41], Pt [49] or Fe [50, 51] – can prevent the carrier recombination through the formation of electron trap centers. Doping of non-metal such as with F [40, 43], N [40, 52], S or H [43] is usually used to broaden the absorption spectrum and a consequent reduction in the band gap energy. In the plethora of all the metal-doping strategies, noble-metal- and transition-metal-doping have been extensively reported; however, only a few publications have reported on alkali-earth-metal doping, along with the mechanism for the enhanced photoactivity (Table 6.1). In order to get the systematic understanding of the doping effect of alkali earth metals on TiO_2 and the relevant mechanism for the enhanced photo activity, AM-TiO_2-x (where AM refers to Mg, Ca, Sr, and Ba) with different doping contents were prepared and tested in RhB photo degradation and H_2 evolution; elaborative characterizations related to the chemical and physical properties as well as photoelectric properties were performed. Based on the investigations performed, a new mechanism for enhanced photo degradation and H_2 evolution by AM-doping is proposed for the first time [30].

The electronic structure of a semiconductor is greatly influenced by doping with metal cations, non-metal anions and non-metal molecules enhancing the absorption of a photocatalyst. The chemical state of the dopant and the location are greatly affecting the effectiveness of the doping procedure. Both metal and non-metal doping, have received much attention recently. Predominantly, non-metal doping with nitrogen, sulphur and carbon has been thoroughly investigated, to lower the E_g of wide band gap semiconductors into the visible light range [29].

Water splitting is a multi-electron process as shown in the following equation:

$$2H_2O\ (l) \rightarrow O_2 + 4H^+\ (aq) + 4e^- \rightarrow 2H_2\ (g) + O_2\ (g)$$

The energy required to generate one molecule of H_2 from water is 2.458 eV, involving a potential of 1.229 V.

Four essential steps are involved in the production of H_2 are,

Stage 1: light irradiation

Stage 2: electron absorption

TABLE 6.1
Some of the Metal Oxides and their Dopants for Various Photocatalytic Degradation

Photo Catalyst	Non-metal Dopant	Doping Procedures	Application	References
TiO_2	N	Hydrothermal	H_2 production	[55]
				[56]
			Dye degradation	[57, 58]
				[59]
				[60]
	B	Hydrothermal	H_2 production	[61]
			Dye degradation	[62]
				[63]
	B-N	Sol–gel	H_2 production from H_2O/ EDTA-2Na (Pt)	[64]
			Dye degradation	[65]
	NH_4TiOF_3 nitridation		H_2 production	[66]
	N-S	Sol–gel $(NH_4)_2TiO(SO_4)_2$ dec.	H_2 production	[67]
		SSR	H_2 production	[68]
$K2La_2Ti_3O_{10}$	N	Annealing in NH_3	H_2 production	[69]
			Dye degradation	[70]
$Sr_2Ta_2O_7$	N	Annealing in NH_3	H_2 production	[71]
$CsCaTa_3O_{10}$- nanosheets	N	Annealing in NH_3	H_2 production	[72]
$Sr_5Ta_4O_{15}$	N	Annealing in NH_3	H_2 production	[73]
$Sr_2Nb_2O_7$	N	Annealing in NH_3	H_2 production	[74]
$Nb_6$0174–	N	Annealing in NH_3	H_2 production	[75]
$CsTaWO_6$	N-S	Annealing S/NH_3	H_2 production	[76]
Co_3O_4F PE-CVD	F	PE-CVD	H_2 production	[77]

Stage 3, 4, and 5: (e^-/h^+) migration and

Stage 6 and 7: reduction and oxidation half reactions [53, 54].

Semiconductors with high Eg are generally quite active under UV light irradiation. But with doping, the electronic structure has been modified to extend the absorption of the photocatalyst into the visible-light region, the activity of the doped photocatalyst decreases. Although the absorption edge of the photocatalyst is shifted to longer wavelength region, meaning that more photons can be absorbed by the semiconductor, the photocatalytic activity decreases under the same irradiation conditions. Various reasons are anticipated for this behavior, the first being that introduced dopants can act as recombination sites. Based on the location of the dopant in the semiconductor crystal, diffusing charge carriers can recombine at dopant sites, and the input energy is lost, thereby, decreasing the photocatalytic activity. Secondly, without considering the dopants as recombination sites, the discrete interstitial states or new conduction band and/or valence band edges can lower the reduction or oxidation potential of the modified semiconductor in comparison with the undoped material, also resulting in decreased activity.

Many examples show increased activities after non-metal doping. It is most important that a direct connection between the homogeneity of the doping, the crystal structure of the doped photocatalyst, and the resulting activity can be correlated. Summarizing the facts about non-metal doping in general, along with the origin of visible-light activity, and the importance of homogeneous nonmetal doping is very important in the literature of photocatalysis. Some of the known non-metal doped transition metal oxide semiconductor materials exhibiting photocatalytic activity are summarized in Table 6.1 along with preparation techniques and the photocatalytic reactions performed under visible light irradiation as reported in the given references.

6.3.2 Loading of Cocatalysts

Solar energy is abundant, and the harvesting process is obviously ecofriendly; the production of H_2 from water by photocatalysis is largely regarded as a remarkable and promising route [78]. In order effectively to utilize solar energy for large-scale application, the construction of a cheap, visible-light-driven photocatalyst with high performance and good durability is of great importance in the processes that involve evolution of H_2. In this context, the performance of a photocatalyst containing the loading metal oxides or metals as cocatalysts dispersed on the surface of the photocatalysts is of vital importance. The electrons generated by the absorption of light migrate through the photocatalyst to the interface that exists between the cocatalyst and the photocatalyst and get captured by the cocatalyst [79], a process which predominantly determines the adsorption as well as activation capabilities of the photocatalytic reactions [80]. Further, the cocatalysts also hinder the adverse electron–hole recombination and speed up the surface chemical reactions by inhibiting the backward reaction [81]. Noble metals, such as Pt [82], Au [83], Pd [84], and Rh [85] with a higher reduction potential, work function and weaker metal–hydrogen bond strength, are the most convenient for hydrogen evolution by photocatalytic activity. These noble metal cocatalysts entrap the photogenerated electrons to prevent electron–hole recombination, thereby reducing the activation energy for the production hydrogen [86]. Similarly, transition metals and their oxides have been used as cocatalysts to increase the rate of oxidation [87]. Liu et al. [88] prepared transition metal oxide clusters, including MnOx, FeOx, CoOx, NiOx, and CuOx, that were loaded in situ into TiO_2 nanosheets by an impregnation method, which promoted the photocatalytic oxidation of water to O_2 significantly, compared to RuO_2/TiO_2 and IrO_2/TiO_2 nanosheets. To achieve higher H_2 generation efficiencies from water splitting, electron donors are usually required to act as sacrificial reagents to consume holes and prevent the recombination of photo induced electrons and holes on the surface of the photocatalyst [89]. The sacrificial agents methanol [90], ethanol [91], triethanolamine [92], disodium ethylenediaminetetraacetic acid [91], and sodium sulfide/

sodium sulfite [80] have been used as hole collectors for photocatalytic evolution of H_2 from water splitting.

$LaFeO_3$ perovskites are successfully studied for the production of H_2 by photocatalysis. Y, Pr, Mg, In, Ca, Eu, and Tb metals were doped into $LaFeO_3$, with RhCrOx as cocatalysts for the evolution of H_2 by photocatalysis through splitting water for the first time. The effects of the doped photocatalysts on different sacrifice by of photocatalysis due to irradiation of visible light. The largest hydrogen productivity obtained 0.5 wt.% RhCrOx loading and 0.1 M Pr-doped $LaFeO_3$ calcined at a temperature of 700 °C in 20 percent TEOA solution [93].

One of the noble metals, Pt is widely used and considered as the best cocatalyst for g-C_3N_4 in H_2 evolution reaction [94], but the disadvantage is high cost and scarcity of Pt. Consequently, the development of cost effective, cheap metal cocatalyst is of current relevance. Transition metal phosphides, such as Ni_2P, CoP, MoP, FeP, and Fe_2P, form the significant class of compounds with metallic properties and good electrical conductivity. They show high electrochemical catalytic activity and good stability for the evolution of H_2 reaction in alkali or acid solutions [95] On the basis of high electron-transfer activity and good stability, Ni_2P and CoP are used as efficient cocatalysts for g-C_3N_4 to speed up the charge transfer and separation, resulting in the enhanced H_2 evolution activity [96].

A new class of dual co-catalysts of black phosphorene (BP) and single Pt atoms on CdS nanospheres. BP/CdS heterostructures is prepared by grinding and sonication. Single layer of Pt is coated onto BP/CdS through a photo-reduction procedure. Further in addition to being anchored on the surface step edges of CdS nanospheres, Pt single-atoms with positive charge are also embedded on Cd vacancies and stabilized by Pt-S covalent bonds. Single Pt atoms are immobilized on the surface of BP also. Greater photoactivity and performance pristine CdS nanospheres by a factor of 96 in terms of hydrogen evolution rate, along with a remarkable apparent quantum efficiency of 46 percent at 420 nm [97].

6.3.3 Heterojunction Structures

For any photocatalysis applications, the electron and hole pairs generated in photocatlysts must be separated efficiently. Heterojunction engineering is found to be the best strategy for the electrons-holes separation, which is due to appropriate, effective and feasible spatial separation of them. Primarily, there are five types of heterojunctions, namely, p-n junction [98, 99], conventional [100, 101], direct Z-scheme [102], graphene-semiconductor [103], and surface heterojunctions [104, 105].

Conventional heterostuctures is an interface between semiconductors of dissimilar band structure leading to varied band arrangements [106, 107]. In conventional structures, there are three types: Type-I, Type-II and Type-III. Let us consider two semiconductors of A and B. In case of Type-1, the CB of A is greater than that of B and VB of A is lower than that of B. When exposed to light, the electrons will gather at CB and holes at VB of B. Since both the energy carriers accumulate on same conductor, their separation is not effective. In case of Type-II, CB and VB of semiconductor A are greater than that of equivalent bands of B. Here, there exists a spatial separation of electrons due to the transfer of electrons to B and hole to A [108, 109]. On semiconductor B, the reduction and oxidation reactions are happening because of low reduction potential and at semiconductor A, because of low oxidation potential. This would effectively reduce the redox ability of Type-II heterojunctions. In Type-II and Type-III, the band alignment is most similar and the exception is that there exists a wide band gap avoiding the overlapping of them [110]. In last case, migration of electrons and holes is quite difficult and makes it unfit for phototacatlytic applications.

Engineering Type-II heterojunctions has gained the interest, wherein, several reports are available for such structures. TiO_2/g-C_3N_4 [111], g-C_3N_4 with $BiPO_4$ [112] and WO_3 [113] and $BiVO_4$/WO_3 [114] are some of them. While, the structuring of heterojunctions has varied from electrophoretic deposition [115] to hydrothermal methods [116]. The study compared the photocatalytic efficiency of Type-I (Ag/AgCl/g-C_3N_4) and Type-II (Ag/AgBr/gC_3N) heterojunction for the production of CH_4 and CO_2 reduction. AgBr/g-C_3N_4 showed a better action than Ag/AgCl/g-C_3N_4. Morphology of the photocatalyst is also important to attain the proper efficiency. Core/shell CdS/ZnO was synthesized by Shen et al., which remarkably increased the charge separation and photocatalytic hydrogen production [117]. The fabricated core/shell structure allowed a greater contact interface and, hence, the enhanced hydrogen production than the individual ZnO or CdS. Heterojunction was also fabricated between two varied phases of TuO_2 (anatase and rutile) [100]. However, despite of many efficient charge separation, the heterojunction photocatalysis will happen in a semiconductor, having lower reduction and oxidation potentials, wherein the reduction and oxidation reactions takes place, respectively. This could remarkably suppress the redox feature. There exists an electrostatic repulsion between similar charge carriers (electron-electron and hole-hole), which further disfavor the migration of electrons from VB of the 1st semiconductor to CB of the 2nd semiconductor (electron rich). And, hence, exploring a material with greater potency of charge separation and redox ability still remains a crucial requirement.

P-n heterojunctions were ideally able to separate the charge carriers in space. It is known to amplify the migration of charge carriers across the heterojunction and enhances the photocatalytic performance [118, 119]. The p-n junction is effectively obtained by the mixture of p-type and n-type semiconductors. The electrons nearby to the p-n interface will migrate to p-type and leave positive holes. In the meantime, p-type semiconductor holes at p-n interface have a tendency to migrate to n-type. This diffusion process of electrons and holes will happen until the attainment of equilibrium of Fermi level. This would result in the charging of the p-n interface and creates the space which is charged and is termed at internal electric field [120, 121]. A p-n heterojunction like MoS_2/CdS [122], NiS/CdS [123], NiO/TiO_2 and many others have been explored.

Surface heterojunctions were first proposed in 2014, wherein, the charge separation was perceived on the single semiconductor's crystal facets [104]. As the semiconductor possesses different crystal facets of varied band structures, there is a possibility of formation of heterojunction between two different facets of the same semiconductor [104, 124]. The formation of heterojunction in [001] and [101] facets of anatase TiO_2 was exposed to photocatalysis and it showed an efficiency 35 times greater than that of the commercial one. Another work deposited Pt nanoparticles, selectively onto the [101] facets (electron rich surfaces) and used for water splitting.

Despite the above-mentioned efficient charge-separation process, the increase in redox ability of the heterojunction remained untouched till Bard et. al., came out with Z-scheme materials [125]. These materials are fabricated by two different semiconductors, separated by electron donor/acceptor medium. There exists no physical contact between the semiconductors, and the migration charge carriers from one semiconductor to other through acceptor/donor pair via redox reaction. Meanwhile, in general Z-scheme heterojunctions are fabricated in liquid phase and there by restricting the wider applications. Tada and group, in 2006, came out with solid state Z-scheme materials, with an aid of solid phase mediator of electron [126]. In later stages, many Z-scheme heterojunctions have appeared, spreading their usage in gas, solid and even in solution media [127, 128]. In 2013, J.G. Yu et al., came across direct Z-scheme heterojunctions, with no electron mediators, which reduced the cost of heterojunction fabrication [102]. g-C_3N_4–TiO_2 [102], WO_3/g-C_3N_4 [129], hierarchical CdS–WO_3 [130] are discovered recently.

6.4 APPLICATION

6.4.1 Hydrogen Evolution

Hydrogen fuel is an ideal solution to tackle the energy crises and global warming due to increased CO_2 gas emission [131]. H_2 is a sustainable, energy dense and eco-friendly clean fuel material produces only H_2O by combustion of H_2 itself with a huge amount of heat release [132].

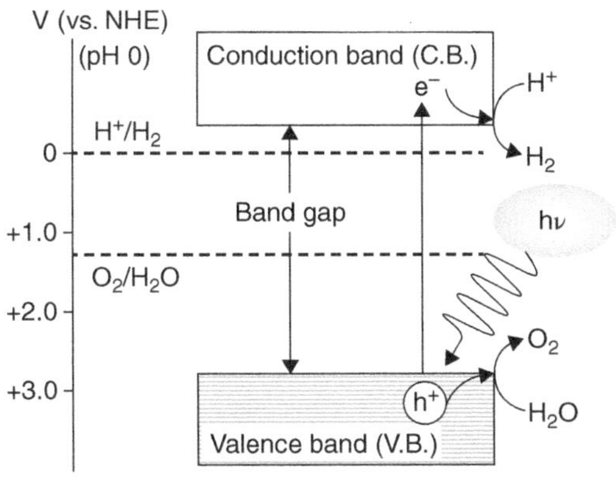

$$2H^+ + 2e^- \rightarrow H_2 \qquad E^0 = 0.00\ eV \qquad (1)$$

$$H_2O + 2H^+ \rightarrow 2H^+ + 1/2O_2 \qquad E^0 = 1.23\ eV \qquad (2)$$

FIGURE 6.2 Band edge position in photocatalyst with respect to redox potential of water and mechanism of H_2 and 0_2 production [133].

Countries like the UK, China and United States are using hydrogen as an alternative fuel in transportation in order to decrease greenhouse gas emissions. Therefore, an easy and efficient process for hydrogen production is necessary. Presently, commercial hydrogen is produced from various techniques, such as cryogenic distillation, gasification of coal, steam reforming of natural gas and water splitting. Among all, the water splitting process is more economical and environmentally friendly. The water splitting process occurs at ambient temperature and pressure, using water and solar light leading to lowering of production costs. The electrochemical water splitting needs a large amount of electric energy, which limits the commercial application of the process. Therefore, hydrogen generation using a photocatalyst is an energy-efficient process that utilizes solar energy.

The photocalytic hydrogen generation in a semiconductor photocatalyst involves irradiation of light on the photocatalyst, which is equivalent to or bigger than the band gap of photocatalyst followed by excitation of electrons from the VB to the CB, leaving holes (h^+) in VB. The e^- and h^+ migrates to the photocatalyst surface and involves in the photocatalytic activity. Electrons on the surface react with protons adsorbed on the photocatalyst resulting the formation of hydrogen (H_2) and holes oxidize water to form oxygen (O_2) followed by desorption of H_2 and O_2 gases from the photocatalyst surface, as depicted in Figure 6.2 [133].

Over four decades of research, various photocatalysts is designed as a potential aspirant for water splitting, such as semiconductors oxides, oxynitrides, oxysulfides [33, 134], graphene based photocatalysts, and so forth. Several reviews have been published of the work on heterogeneous photocatalysts for water splitting.

(a) Oxides-based photocatalysts:
The most widely studied metal oxides for photocatalysis are TiO_2 [135], ZnO, $BiVO_4$, α-Fe_2O_3, and Cu_2O. The ideal characteristics needed for high hydrogen production are, (a) high visible light absorption ability; (b) ideal positioning of the band edges; (c) high stability of the photocatalyst; (d) reduced recombination of charge carriers and reduced defects. TiO_2 was first introduced by Fujishima and Honda [136] for photoelectrochemical water oxidation. It is a potential candidate for converting solar to chemical energy in addition to nontoxicity, stability and abundance. Though the band gap (3.2 eV) confines its photoactivity in visible regions and therefore various modifications to TiO_2 are evaluated to convert the absorption into visible regions. Han et al. [137] prepared N-doped- graphitic-C_3N_4 nanosheets (g-C_3N_4 NSs) hybridized TiO_2 with visible light absorption attributed to increased charge carrier transport and

surface area resulting high photo-conversion and hydrogen evolution. Another work reported a good surface area (80 m^2/g^{-1}) nanostructured TiO_2 with 437.5 μmol h^{-1} hydrogen production. Non-metal like nitrogen, carbon and sulphur doped TiO_2 showed H_2 evolution of 1008 μmol h^{-1} g^{-1}, 2947 μmol h^{-1} g^{-1} and 1269 μmol h^{-1} g^{-1} which is due to the adsorption of visible light instigated by mid band electronic state above TiO_2 VB edge. Raghavan et al. deposited graphene quantum dots (GQDs) on P-25-TiO_2 and studied the photocatalytic hydrogen evolution activity, which depicted enhanced activity attributed to biphasic nature (anatase and rutile) of TiO_2 and dual activity of GQDs as a photosensitizer and cocatalyst in the 0D–0D interface.

Moon et al. [138] have prepared copper phthalocyanine (CuPc) (Figure 6.3a) deposited TiO_2 photocatalyst through a solution reaction method, which showed higher hydrogen evolution activity with a quantum yield of 6.68 percent at 420 nm. CuPc enhances charge transport ability of TiO_2 along with the absorption of visible light. Wang et al. stated that Z-scheme based ZnO–CdS@Cd heterostructure (Figure 6.3b-c) showed higher photocatalytic hydrogen evolution. The superior activity can be ascribed to fast recombination of photoexcited electrons present in CB of ZnO with holes present in VB of CdS leading to the extended lifetime of excited electrons in CB of CdS and holes in VB of ZnO. The

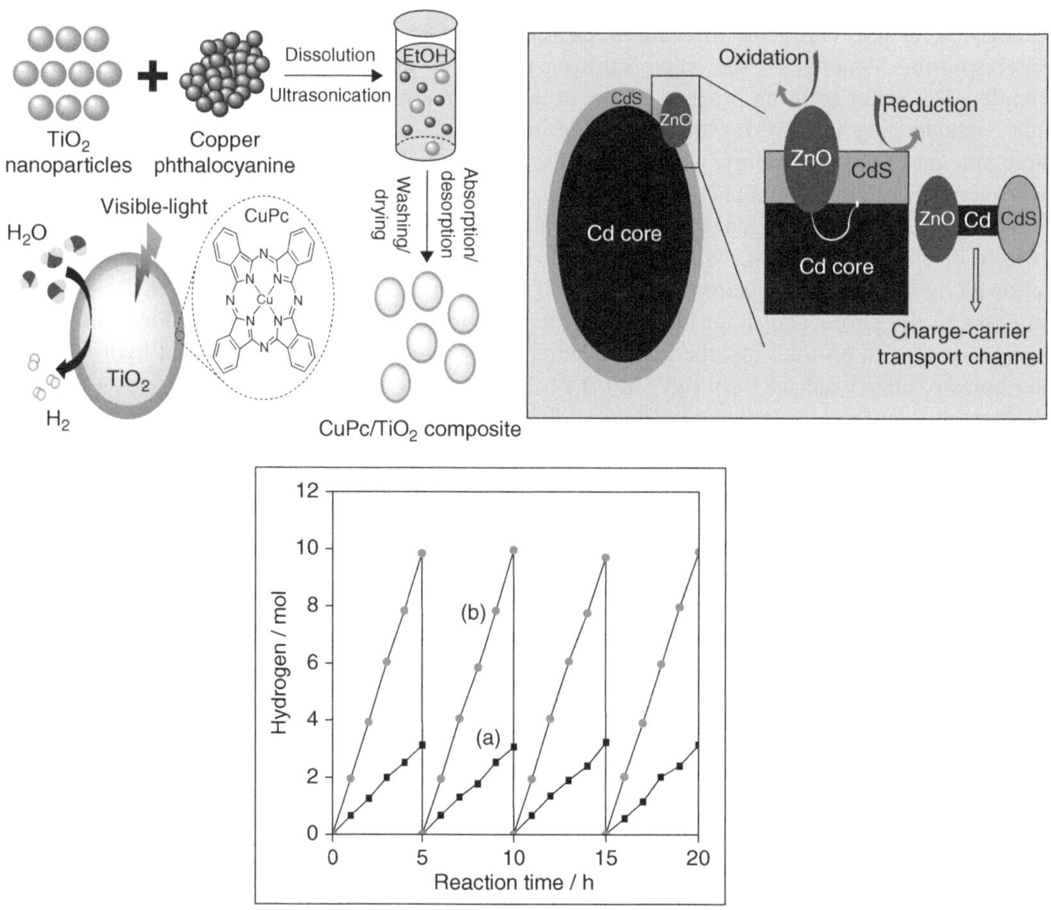

FIGURE 6.3 (a) Schematic of preparation and H_2 generation on CuPc deposited TiO_2; (b–c) schematic of core-shell structure of ZnO-CdS@Cd and H_2 evolution by ZnO-CdS@Cd [138].

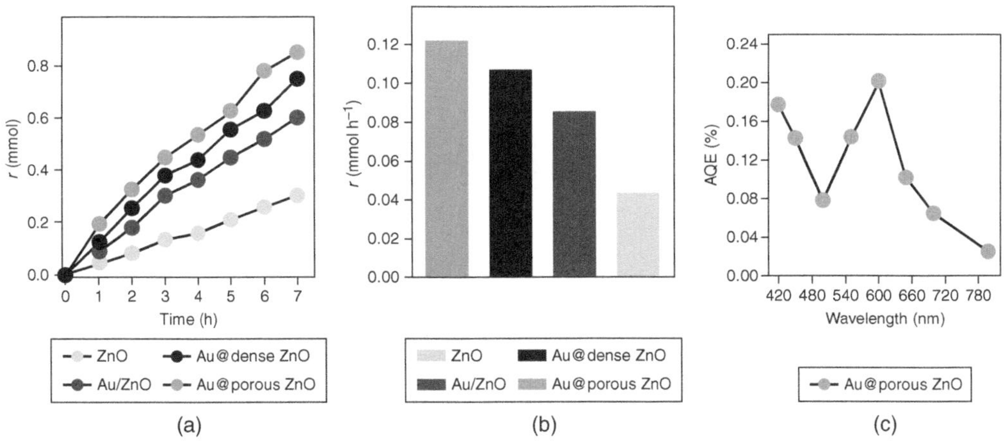

FIGURE 6.4 (a–b) Photocatalytic H_2 production on Au@ porous ZnO; (c) AQE for H_2 evolution on Au@ porous ZnO [140].

presence of Cd metal as a core acts as effectively channels advantageously for such kinds of recombination and the photocatalytic oxidation and reduction process and is similar to TiO_2-Au-CdS [139]. The ZnO–CdS@Cd showed hydrogen evolution of 0.61 mmol h^{-1} (100 mg), that improved to 1.92 mmol h^{-1} with 3 wt% Pt incorporation as a cocatalyst. The reduced charge-carrier transport length in ZnO–CdS heterostructural thin shell also contributes to improved activity. Chen et al. have demonstrated plasmonic Au@ZnO with porous ZnO shells as an active photocatalyst for hydrogen production (Figure 6.4) [140]. Au@ZnO core–shell hybrids had the higher plasmon induced hot electrons and are efficiently utilized by porous ZnO, which might be the reason for improved activity. Hot electrons created from the nonradiative relaxation of localized surface plasmons are more energetic than those from direct photoexcited electrons.

Bi-supported semiconducting photocatalysts such as Bi_2MoO_6, BiOX where, X is Cl, Br and I, Bi_2WO_6, and $Bi_2O_2CO_3$, $BiVO_4$ have drawn considerable attention recently. Hu et al. prepared Z-scheme system of $ZnIn_2S_4/BiVO_4$ (ZIS/BVO) heterojunction through an assembly process and spatially separated redox centers to achieve higher visible light driven hydrogen evolution. In ZIS/BVO hybrid Z-scheme heterojunction interface establishes as soon as the $ZnIn_2S_4$ nanosheets are coated on decahedron $BiVO_4$ surface, which demonstrates unique interface synergistic properties. ZIS/BVO depicts exceptional hydrogen production activity of 5.944 mmol g^{-1} h^{-1} attributed to beneficial reduction and oxidation capability of photoexcited e^- and h^+, respectively (Figure 6.5) [141].

(b) Oxynitrides and Oxysulfides based photocatalysts:

Oxynitrides (Oxy-N) and oxysulfides (Oxy-S) have drawn significant attention in photocatalytic water splitting field as the presence of smaller energy band gap results in visible light absorption. Titanium, barium, tantalum, gallium, zirconium and niobium and their perovskite based structures make important photocatalysts. N 2p of oxy-N and S 3p of oxy-S could form a VB at potentials further negative than that of O 2p orbitals, attributed to the lesser electronegativities of N and S than O. Figure 6.6 depicts variation in band structure with the incorporation of N and S. The three O^{2-} anions are replaced by N^{3-} anions in Ta_2O_5 during nitridation and Ta^{5+} valence state is maintained [142]. In distinction, N 2p orbitals of TaON and Ta_3N_5 are integrated in VB, shifting the top of the VB to +2.0 V and +1.5 V *vs* NHE for TaON and Ta_3N_5 respectively.

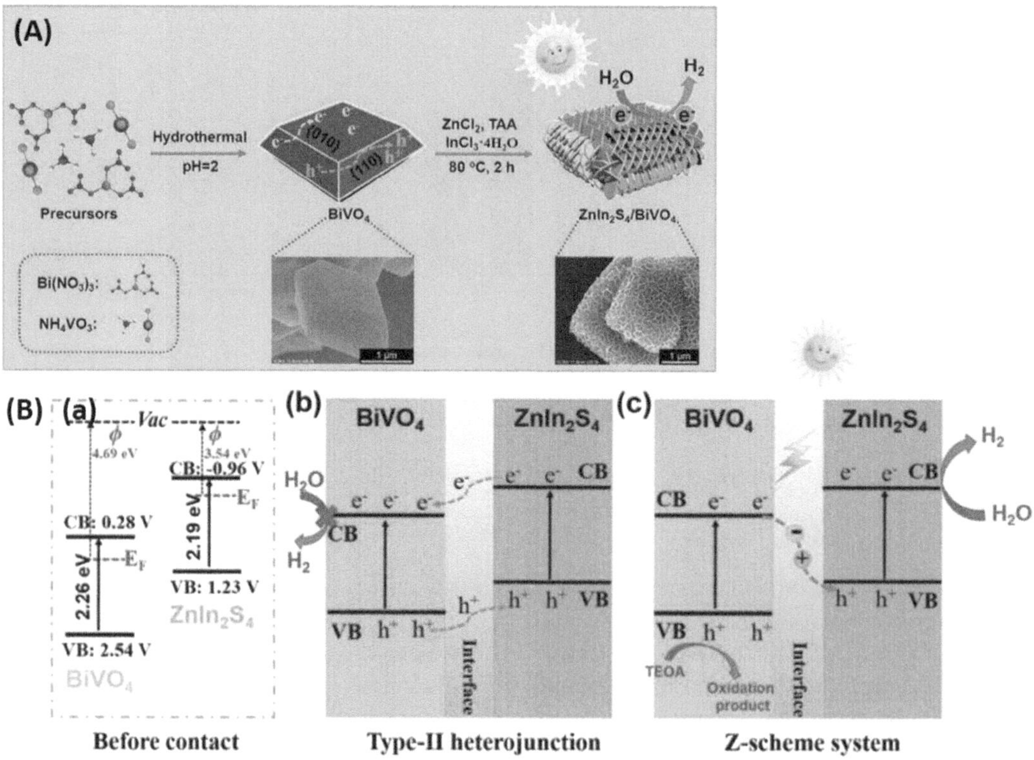

FIGURE 6.5 (A) Schematic of formation of $Znin_2S_4/BiV0_4$; (B) Band structure of $Znin_2S_4/BiV0_4$ (a) before contact; (b) Type II heterojunction; (c) Z-scheme system [141].

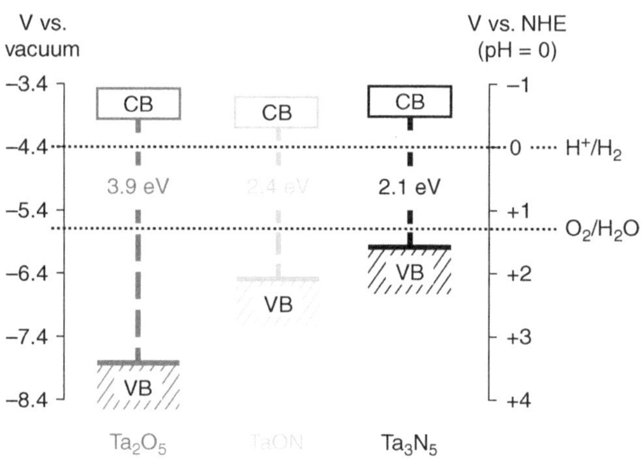

FIGURE 6.6 Energy band structure of $T0_5$, TaON, Ta_3N_5 [142].

Oxy-N and oxy-S can be synthesized by nitriding the appropriate composition of metal oxides in the presence of NH_3 flowing at higher temperatures. Correspondingly, oxy-S can be prepared through sulphurization of metal oxide or metal films in the presence of H_2S gas drift at raised temperatures. For instance, $Sm_2Ti_2S_2O_5$ and Cu_2ZnSnS_4 layers are synthesized through sulphurizing amorphous SmTi-oxide and multilayers of Zn/Sn/Cu electroplated on Mo. Shi et al. have prepared solid solution

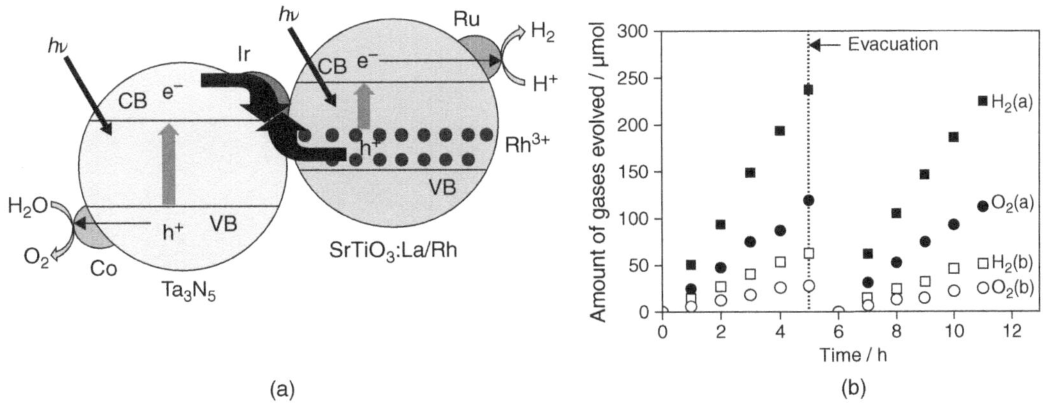

FIGURE 6.7 (A) Schematic of band structure of Ir/Co0_x/Ta_3N_5 and Ru-loaded $SrTi0_3$: La/Rh; (B) Time course of H_2 and 0_2 evolution during Z-scheme water splitting using (a) Ru-loaded $SrTi0_3$:La/Rh; and (b) Ru-loaded $SrTi0_3$:Rh with Ir/Co0 $_x$ /Ta_3N_5 under visible light (420 nm < A < 800 nm) [144].

of perovskite structured $(LaTiO_2N)_{1-x}(LaCrO_3)_x$ (represented as LTON) through a polymerized complex method followed by nitridation process. LTON is studied for sunlight-mediated hydrogen evolution, and the activity was improved with increasing $LaCrO_3$ concentration from 0.0 to 0.3 ascribed to a narrowed band gap resulting in higher charge carrier production and improved lattice distortion promoting the separation and movement of charge carriers [143]. Wang et al. demonstrated core/shell patterned La-Rh codoped $SrTiO_3$ (LR-$SrTiO_3$) for Z-scheme overall water splitting into H_2 and O_2 (Figure 6.7). LR-$SrTiO_3$ showed 3.8 times greater hydrogen evolution and STH improvement by a factor of 3 under visible light irradiation (($\lambda > 420$ nm) in combination with Ir/CoO_x/Ta_3N_5 during overall water splitting compared to $SrTiO_3$:Rh. The increased activity is due to doping of La and Rh, where (a) La doping curbed the development of oxygen vacancies as well as inactive Rh^{4+} (which acts as recombination centers); (b) improved concentration of Rh^{3+} states, which absorbs visible light [144].

Utilization of sacrificial agent/co-catalyst for H_2 and O_2 evolution play a crucial part in water splitting. Sacrificial reagents like methanol or silver nitrate act as a hole or electron scavenger (as depicted in Figure 6.8) that are used to investigate the photocatalytic activity of a material [145]. Maeda et al. prepared $(Ga_{1-x}Zn_x)(N_{1-x}O_x)$ and studied the H_2 evolution in the presence of methanol sacrificial agent, $Rh_{2-y}Cr_yO_3$ as a cocatalyst and O_2 evolution under aqueous silver nitrate solution. Maeda et al. reported that the H_2 evolution and overall water splitting depends on crystallinity and composition of $(Ga_{1-x}Zn_x)(N_{1-x}O_x)$, while the O_2 evolution dependent on specific surface area of the material [146]. Wang et al. [147] have studied the sequential Pt cocatalyst loading on single-crystalline $BaTaO_2N$ towards Z-scheme water splitting and reported an AQY of 6.8 percent with hydrogen evolution of >400 μmol/h. The improved activity is attributed to sequential Pt co-catalyst loading by impregnation-reduction followed by site-selective photodeposition, which contributes to uniform dispersion, and close contact with the photocatalyst is the basis for the enhanced hydrogen evolution.

Further, the oxysulfides show narrow band gaps due to negative shift of valence band edges of oxides by electronegative sulfide ions. Wang et al. prepared $Y_2Ti_2O_5S_2$ with a band gap of 1.9 eV and studied photocatalytic water splitting. $Y_2Ti_2O_5S_2$ directly decomposes water into H_2 and O_2 in a stoichiometric ratio in the presence of IrO_2 and Cr_2O_3/Rh as the oxygen evolution catalyst and hydrogen evolution catalyst, respectively [148]. Metal-free elemental photocatalysts such as B, C, P, S, Si, Se and binary photocatalysts, that is, BC_3, B_4C, C_xN_y, h-BN and so forth also show better activity for hydrogen evolution.

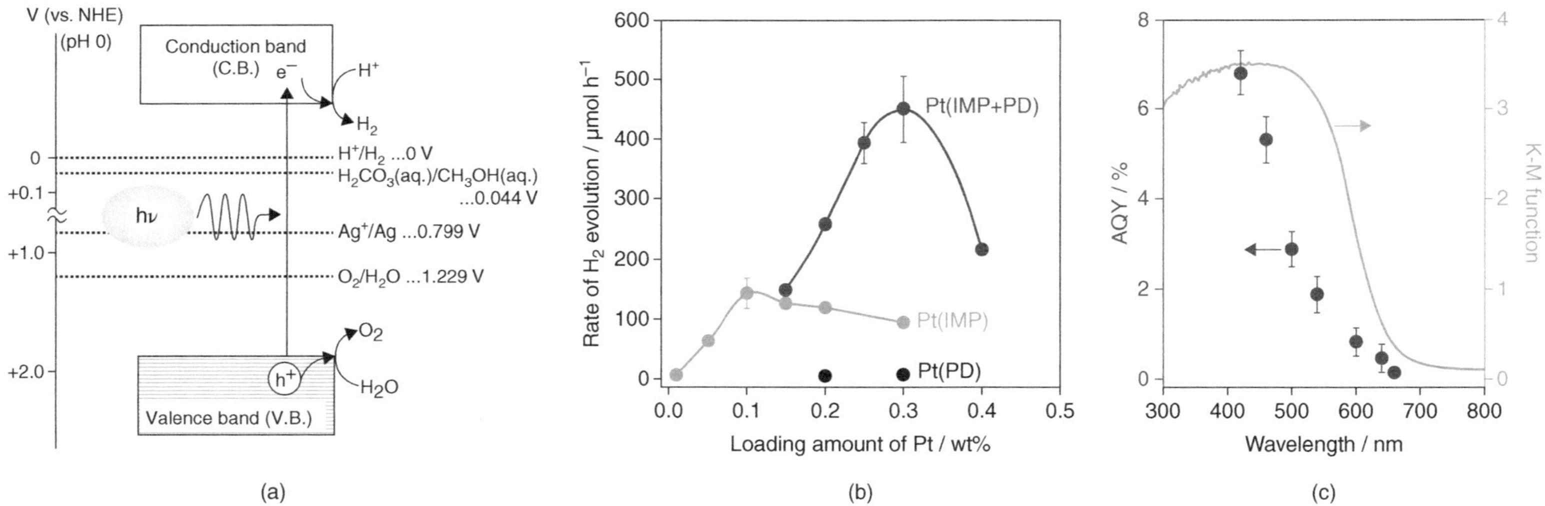

FIGURE 6.8 (A) Band structure and potential for H_2 evolution; (B) (a-b) H_2 evolution on $Y_2Ti_2O_5S_2$ in the presence of Pt as co-catalyst and AQY for the same system [145].

(c) 2-D materials based photocatalysts:
2D nanostructured materials consists of strong in plane chemical bonds along with weak van der waals interaction, making these kind of photocatalysts better for surface-related applications. 2D materials demonstrate distinctive ability of confinement of electrons in the ultrathin layer, contributing excellent opto-electronic properties. The strong in-plane bonding offers sites for the formation of several heterostructures. Furthermore 2-dimensional architecture possess a high specific surface area, which facilitates surface reactions like water splitting. The first 2D material graphene depicts unique electronic, mechanical, and optical properties. Song et al. [149] demonstrated hydrogen evolution of 410.0 μmol g^{-1} h^{-1} from Z-scheme g C_3N_4/Au/$BiVO_4$ photocatalyst ascribed to rapid separation of charge carriers, improved redox driving force by using Au nanoparticles as electron shuttle and surface plasmon effect by Au nanoparticles. The g-C_3N_4/$BiVO_4$ (23.3 μmol g^{-1} h^{-1}), Au/g-C_3N_4 (244.6 μmol g^{-1} h^{-1}) and pristine g-C_3N_4 (2.2 μmol g^{-1} h^{-1}) showed lower H_2 generation than g C_3N_4/Au/$BiVO_4$. Liu et al. [150] prepared core shell structured Cu_2O@g-C_3N_4 octahedra photocatalyst with Cu_2O having (111) surface showed superior activity for H_2 evolution (795 μmol/g) attributed to synergistic effect at the interface between Cu_2O and g-C_3N_4 nanosheets, in addition to large contact area, energy band structure and improved charge carrier separation.

(d) MOFs based
MOFs are constructed by metal clusters and organic linkers, which have attracted extensive interest due to structure diversity, tenability, high surface area and crystallinity. MOFs are considered as microporous semiconductors [151]. Photocatalysts prepared using metal-organic frameworks (MOFs) as template are exhibiting high activity for water splitting. Pullen et al. [152] prepared UiO-66-[FeFe](dcbdt)$(CO)_6$ (dcbdt=1,4-dicarboxylbenzene-2,3-dithiolate) MOF, in which UiO-66 MOF provided a platform to improve [FeFe](dcbdt) $(CO)_6$ moiety structural stability. The as-synthesized MOF exhibited excellent hydrogen production rate in water at pH=5. Liu et al. [153] reported a Z-scheme g-C_3N_4/α-Fe_2O_3 system prepared using MOFs as template for photocatalytic H_2 evolution. Z-scheme system consisting of different semiconductors with a reversible donor-acceptor is attracting great interest. The g-C_3N_4/α-Fe_2O_3 showed high H_2 generation activity of 2066.2 μmol g^{-1} h^{-1} under visible light absorption with high durability (Figure 6.9) [154], which is three times higher than bulk g-C_3N_4 ascribed to efficient separation of photo-excited charge carriers by two narrow band gap semiconductors strongly coupled with the Z-Scheme structural feature.

Several MOFs such as P-MOF with Ti-O cluster, MIL-101(Fe)-NH_2 with Fe-O metal cluster showed exceptionally high photocatalytic activity. Hence, several photocatalysts show high photocatalytic activity for H_2 production from oxides, oxynitrides, oxysulfides to MOFs based photocatalysts.

6.4.2 Waste-Water Treatment

Currently, water contamination and pollution have become serious problems. Water contamination mainly occurs due to rapid industrialization and economic growth. Industries use a wide range of organic synthetic dyes in textiles due to its high stability and their release into water bodies leads to harmful effect on humans and animals. Hence, removal of toxic chemicals from water through a sustainable process is a necessity. A potential technique for water treatment is the photocatalytic degradation due to its capability in direct utilization of solar energy for achieving disinfection as well as chemical detoxification. The mechanism of photocatalytic treatment involves (a) irradiation of light with a wavelength matching the energy band structure of photocatalyst, followed by excitation of

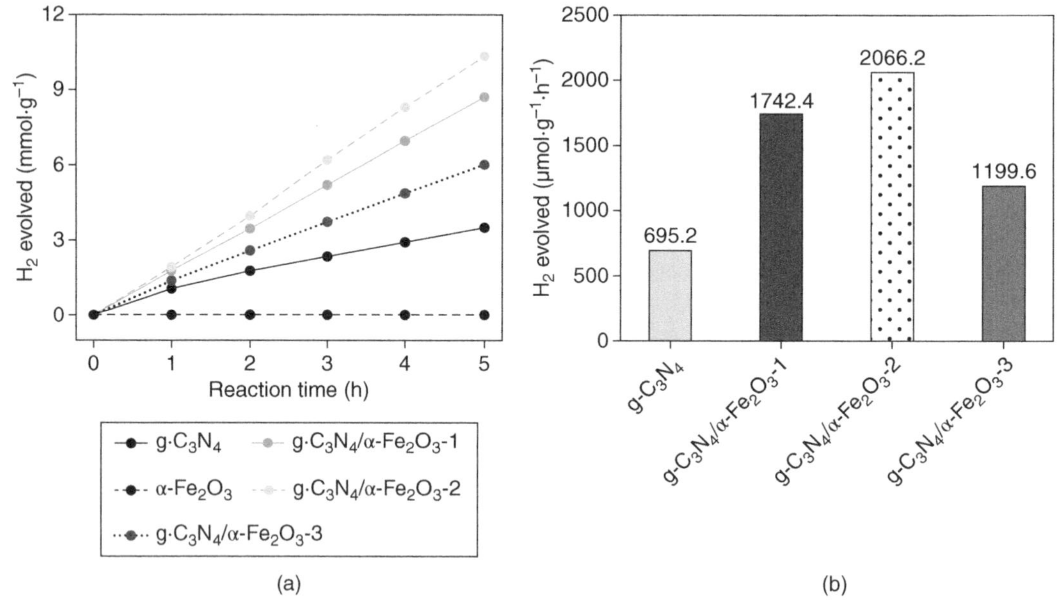

FIGURE 6.9 (a–b) H_2 evolution on Z-scheme g-C_3Nia-Fe_20_3 system and optimization of photocatalyst with different a-Fe203 concentration [154].

TABLE 6.2
Photocatalytic Hydrogen Evolution on Different Photocatalysts

Material	HER	Ref
Pt/TiO_2	2000 μmol/m^2	[155]
Cu-doped TiO_2 film	810 μmol g^{-1} h^{-1}	[156]
g-C_3N_4/α-Fe_2O_3	2066.2 μmol g^{-1} h^{-1}	[153]
MoS_2/CdS	5400 μmol g^{-1} h^{-1}	[157]
ZnS–WS_2/CdS	12 240 μmol g^{-1} h^{-1}	[158]
TiO_2/MoS_2/G	2066.25 μmol g^{-1} h^{-1}	[78]
Bi–Bi_2MoO_6/CdS-DETA	7370 μmol g^{-1} h^{-1}	[159]
Ni_2P/$CaIn_2S_4$	486 μmol g^{-1} h^{-1}	[160]
CdS/MoO_x	573.6 μmol h^{-1}	[161]

electron from the VB to the CB leaving holes in the VB; (b) the holes can oxidize donor molecules and react with H_2O molecules to generate •OH radicals and electrons in the CB react with dissolved O_2 species and forms superoxide ions, which encourages redox reactions. Recently, numerous natural and engineered nanostructured materials have shown strong good degradation and disinfection property, including inorganic nanoparticles like TiO_2, ZnO, GO, rGO, WO_3, g-C_3N_4, Fe_2O_3, $BiVO_4$, Cu, Ag and so forth. The present chapter discusses recent research articles focusing on photocatalytic treatment of contaminated water.

Zhao et al. [162] reported photocatalytic degradation activity of Ag supported z-scheme $WO_{2.9}$/g-C_3N_4 composite. The composite with $WO_{2.9}$ having an oxygen vacancy exhibited better adsorption capability compared to WO_3 and the Ag/$WO_{2.9}$/g-C_3N_4 has low energy of electrons in $WO_{2.9}$, which is neutralized by holes in g-C_3N_4 leaving high energy holes from $WO_{2.9}$ for hydroxyl radical generation and electrons of g-C_3N_4 generate a superoxide radical. Ag/$WO_{2.9}$/g-C_3N_4 showed better photo-degradation performance for methyl orange, Rhodamine B and methylene blue under visible light irradiation ($\lambda > 420$ nm), which is better than Ag/$WO_{2.9}$ and g-C_3N_4.

Tahir et al. prepared rGO/$BiVO_4$ through hydrothermal technique for dyes degradation and hydrogen evolution. The 1.5 percent rGO/$BiVO_4$ depicts transformation of tetragonal phase to monoclinic with high surface area. The hybrid material showed 97 percent degradation of methylene blue in addition to hydrogen energy production attributed to enhanced light absorption [162]. Tahir et al. prepared rGO/$BiVO_4$ through hydrothermal technique for dyes degradation and hydrogen evolution. The 1.5% rGO/$BiVO_4$ depicts transformation of tetragonal phase to monoclinic with high surface area. The hybrid material showed 97 percent degradation of methylene blue in addition to hydrogen energy production attributed to enhanced light absorption [163]. Mushtaq et al. studied the piezoelectrically improved photocatalytic activity in single crystalline $BiFeO_3$ (BFO) nanostructures for water treatment. Piezoelectric materials create electric charges under mechanical stress and change dimensions when electric field applied across the material. BFO nanosheets and nanowires absorb visible light and harness mechanical vibrations contributing to higher reaction rates than photocatalytic degradation. The reaction rates improved further when both solar light and mechanical vibrations are used simultaneously leading to 97 percent of RhB dye degradation in 1h. The higher degradation activity under mechanical vibrations is ascribed to the improved charge carrier separation caused by the internal piezoelectric field of BFO [164].

Mohamed et al. in 2019 reported design of Z-scheme Fe_2O_3/GO/WO_3 for water contaminant degradation. Distinctively, Fe_2O_3/WO_3 showed double absorption edges at 2.0 and 2.3 eV, and Fe_2O_3/GO/WO_3 showed triple absorption at 1.5, 1.8 and 2.1 eV. The Fe_2O_3/GO/WO_3 showed higher photocatalytic activity for degradation of methylene blue, phenol and crystal violet than Fe_2O_3/WO_3, WO_3/GO and Fe_2O_3/GO. The improved activity in ternary system is due to the Z-scheme, where the photoexcited electrons in the conduction band of WO_3 (photosystem I) transferred through GO and recombine with the holes in the VB of Fe_2O_3 (photosystem II), resulting in reduced electron-hole pair recombination [165]. Qiu et al. demonstrated high photocatalytic activity of molybdenum disulfide@ cobaltosic sulfide on nanofibrous aerogel (MoS_2@Co_3S_4/NFA) for degradation of contaminants such as Cr(VI), sulfamethoxazole, and bacteria in addition to high hydrogen evolution [166].

Wu et al. prepared GO/g-C_3N_4/MoS_2 ternary layered nanostructure and studied the photocatalytic degradation ability for methylene blue, rhodamine B and crystal violet. The ternary nanomaterials exhibited significant degradation activity attributed to flower like morphotogy, which effectively enhances the electron and holes collection on MoS2 and g-C3N4 respectively leading to the decreased recombination of charge carriers. Furthermore, GO with higher electrical conductivity transfers the charge carriers faster, thus contributing to improved activity of ternary material [167]. Hence, there are several photocatalysts, which shows better activity for degradation of various toxic chemicals (methylene blue, crystal violet, rhodamine B, phenol along with sustainable fuel hydrogen production.

6.5 CONCLUSIONS AND FUTURE PROSPECTIVE

This chapter has underlined the diverse synthesis strategies of semiconductor nano structures. The processes were aimed to benefit the environmental and energy facets by lending unique features. Though the researchers succeeded in accomplishing the requirements with several semiconductor materials, metal oxides have played a more vital role than any other. The greener methods with less energy consumption and being more economical are in demand. Hydrothermal synthesis is a simple one compared to sonochemical and CVD methods. The photosensitivity and redox ability were enhanced with the aid of several intrinsic and extrinsic modifications, such as co-catalyst loading, metal and non-metal doping and fabrication of heterojunctions. The spatial migration of electrons within the individual semiconductor has resulted in potential photoctalyst offering a pathway to overcome the challenges of lower efficiency. Engineering a heterojunction structure, inducing vacancies, tuning the band gap and exploring the composites are future opportunities.

REFERENCES

1. Sagadevan, S., Recent trends on nanostructures based solar energy applications: A review. *Rev. Adv. Mater. Sci,* 2013. **34**(1): pp. 44–61.
2. Martha, S., P.C. Sahoo, and K. Parida, An overview on visible light responsive metal oxide based photocatalysts for hydrogen energy production. *Rsc Advances*, 2015. **5**(76): pp. 61535–61553.
3. Ohtani, B., Photocatalysis A to Z – What we know and what we do not know in a scientific sense. *Journal of Photochemistry and Photobiology C: Photochemistry Reviews*, 2010. **11**(4): pp. 157–178.
4. Choi, W., A. Termin, and M.R. Hoffmann, Effects of metal-ion dopants on the photocatalytic reactivity of quantum-sized TiO2 particles. *Angewandte Chemie International Edition in English*, 1994. **33**(10): pp. 1091–1092.
5. Shwetharani, R., et al., Recent advances and strategies to tailor the energy levels, active sites and electron mobility in titania and its doped/composite analogues for hydrogen evolution in sunlight. *Catalysis Science and Technology*, 2019. **9**(1): pp. 12–46.
6. Jingfei Luan, K.M., L. Zhang, M. Li, Y. Li and B. Pan, Research on Different Preparation Methods of New Photocatalyst. *Current Organic Chemistry*, 2010. **14**: pp. 683–698.
7. Shwetharani, R., et al., Photocatalytic semiconductor thin films for hydrogen production and environmental applications. *International Journal of Hydrogen Energy*, 2019. **45**(36): pp. 18289–18308.
8. Radhika, N.P., et al., Recent advances in nano-photocatalysts for organic synthesis. *Arabian Journal of Chemistry*, 2019. **12**(8): pp. 4550–4578.
9. Landau, M.V., Sol–Gel Process, in *Handbook of Heterogeneous Catalysis*. pp. 119–160.
10. Shwetharani, R., et al., Photoactive titania float for disinfection of water; evaluation of cell damage by bioanalytical techniques. Photochem Photobiol, 2014. **90**(5): pp. 1099–1107.
11. Tobaldi, D.M., et al., Sol–gel synthesis, characterisation and photocatalytic activity of pure, W-, Ag- and W/Ag co-doped TiO2 nanopowders. *Chemical Engineering Journal*, 2013. **214**: pp. 364–375.
12. Gan, Y.X., et al., Hydrothermal synthesis of nanomaterials. *Journal of Nanomaterials,* 2020. **2020**: p. 8917013.
13. Biswas, S.K. and J.-O. Baeg, A facile one-step synthesis of single crystalline hierarchical WO3 with enhanced activity for photoelectrochemical solar water oxidation. *International Journal of Hydrogen Energy*, 2013. **38**(8): pp. 3177–3188.
14. Mera, A.C., et al., Solvothermal synthesis of BiOI microspheres: Effect of the reaction time on the morphology and photocatalytic activity. *Journal of Photochemistry and Photobiology A: Chemistry*, 2014. **289**: pp. 7–13.
15. Anandan, S., T. Sivasankar, and T. Lana-Villarreal, Synthesis of TiO2/WO3 nanoparticles via sonochemical approach for the photocatalytic degradation of methylene blue under visible light illumination. *Ultrasonics Sonochemistry*, 2014. **21**(6): pp. 1964–1968.
16. Sarantopoulos, C., A.N. Gleizes, and F. Maury, Chemical vapor deposition and characterization of nitrogen doped TiO2 thin films on glass substrates. *Thin Solid Films*, 2009. **518**(4): pp. 1299–1303.
17. Carneiro, J.O., et al., Study of the deposition parameters and Fe-dopant effect in the photocatalytic activity of TiO2 films prepared by dc reactive magnetron sputtering. *Vacuum*, 2005. **78**(1): pp. 37–46.
18. An, H.-J., et al., Cationic surfactant promoted reductive electrodeposition of nanocrystalline anatase TiO2 for application to dye-sensitized solar cells. *Electrochimica Acta*, 2005. **50**(13): pp. 2713–2718.
19. Medina-Ramírez, I., A. Hernández-Ramírez, and M.L. Maya-Treviño, Synthesis Methods for Photocatalytic Materials. *Photocatalytic Semiconductors,* 2015: pp. 69–102.
20. Yang, H.G., et al., Anatase TiO 2 single crystals with a large percentage of reactive facets. *Nature*, 2008. **453**(7195): pp. 638–641.
21. Liu, G., et al., Crystal facet engineering of semiconductor photocatalysts: motivations, advances and unique properties. *Chemical Communications*, 2011. **47**(24): pp. 6763–6783.
22. Tran, P.D., et al., Recent advances in hybrid photocatalysts for solar fuel production. *Energy and Environmental Science*, 2012. **5**(3): pp. 5902–5918.
23. Bledowski, M., et al., Visible-light photocurrent response of TiO 2–polyheptazine hybrids: Evidence for interfacial charge-transfer absorption. *Physical Chemistry Chemical Physics*, 2011. **13**(48): pp. 21511–21519.

24. Ramakrishnan, A., et al., Enhanced performance of surface-modified TiO 2 photocatalysts prepared via a visible-light photosynthetic route. *Chemical Communications*, 2012. **48**(68): pp. 8556–8558.
25. Kronawitter, C.X., et al., A perspective on solar-driven water splitting with all-oxide hetero-nanostructures. *Energy and Environmental Science*, 2011. **4**(10): pp. 3889–3899.
26. Asahi, R., et al., Visible-light photocatalysis in nitrogen-doped titanium oxides. *Science*, 2001. 293(5528): pp. 269–271.
27. Linic, S., P. Christopher, and D.B. Ingram, Plasmonic-metal nanostructures for efficient conversion of solar to chemical energy. *Nature Materials*, 2011. **10**(12): pp. 911–921.
28. Wang, P., et al., Plasmonic photocatalysts: harvesting visible light with noble metal nanoparticles. *Physical Chemistry Chemical Physics*, 2012. **14**(28): pp. 9813–9825.
29. Marschall, R. and L. Wang, Non-metal doping of transition metal oxides for visible-light photocatalysis. *Catalysis Today*, 2014. **225**: pp. 111–135.
30. Lv, C., et al., Alkaline-earth-metal-doped TiO 2 for enhanced photodegradation and H 2 evolution: Insights into the mechanisms. *Catalysis Science and Technology*, 2019. **9**(21): pp. 6124–6135.
31. Md Saad, S.K., et al., Two-dimensional, hierarchical Ag-doped TiO2 nanocatalysts: Effect of the metal oxidation state on the photocatalytic properties. *ACS omega*, 2018. **3**(3): pp. 2579–2587.
32. Zhang, N., et al., Synthesis of M@ TiO2 (M= Au, Pd, Pt) core–shell nanocomposites with tunable photoreactivity. *The Journal of Physical Chemistry C*, 2011. **115**(18): pp. 9136–9145.
33. He, J., et al., In situ synthesis of noble metal nanoparticles in ultrathin TiO2– gel films by a combination of ion-exchange and reduction processes. *Langmuir*, 2002. **18**(25): pp. 10005–10010.
34. Wang, P., et al., A stable quasi-solid-state dye-sensitized solar cell with an amphiphilic ruthenium sensitizer and polymer gel electrolyte. *Nature materials*, 2003. **2**(6): pp. 402–407.
35. Turner, G.M., M.C. Beard, and C.A. Schmuttenmaer, Carrier localization and cooling in dye-sensitized nanocrystalline titanium dioxide. *The Journal of Physical Chemistry B*, 2002. **106**(45): pp. 11716–11719.
36. Wang, X., et al., Photovoltaic properties of titanium dioxide nanowires with different crystal structures. *Chemical Research in Chinese Universities*, 2016. **32**(4): pp. 661–664.
37. Yi, J., et al., AgI/TiO2 nanobelts monolithic catalyst with enhanced visible light photocatalytic activity. *Journal of Hazardous Materials*, 2015. **284** : pp. 207–214.
38. Li, C., et al., Hollowsphere nanoheterojunction of g-C3N4@ TiO2 with high visible light photocatalytic property. *Langmuir*, 2019. **35**(3): pp. 779–786.
39. Zhang, Y., et al., Morphology effect of honeycomb-like inverse opal for efficient photocatalytic water disinfection and photodegradation of organic pollutant. *Molecular Catalysis*, 2018. **444**: pp. 42–52.
40. Wei, F., et al., Preparation of S–N co-doped CoFe 2 O 4@ rGO@ TiO 2 nanoparticles and their superior UV-Vis light photocatalytic activities. *RSC Advances*, 2019. **9**(11): pp. 6152–6162.
41. Castro, Y. and A. Durán, Ca doping of mesoporous TiO 2 films for enhanced photocatalytic efficiency under solar irradiation. *Journal of Sol-Gel Science and Technology*, 2016. **78**(3): pp. 482–491.
42. Sood, S., et al., Efficient photocatalytic degradation of brilliant green using Sr-doped TiO2 nanoparticles. *Ceramics International*, 2015. **41**(3): pp. 3533–3540.
43. Kaviyarasu, K., et al., Synthesis of Mg doped TiO 2 nanocrystals prepared by wet-chemical method: optical and microscopic studies. *International Journal of Nanoscience*, 2013. **12**(05): p. 1350033.
44. Yao, Z., et al., Microporous Ni-doped TiO2 film photocatalyst by plasma electrolytic oxidation. *ACS Applied Materials and Interfaces*, 2010. **2**(9): pp. 2617–2622.
45. Colon, G., et al., Cu-doped TiO2 systems with improved photocatalytic activity. *Applied Catalysis B: Environmental*, 2006. **67**(1–2): pp. 41–51.
46. Lee, H.C., et al., Direct photoelectrochemical characterization of photocatalytic H, N doped TiO2 powder suspensions. *Journal of Electroanalytical Chemistry*, 2018. **819**: pp. 38–45.
47. Liu, R., et al., Visible-light responsive boron and nitrogen codoped anatase TiO2 with exposed {0 0 1} facet: Calculation and experiment. *Applied Surface Science*, 2019. **466**: pp. 568–577.
48. Wang, X.-K., et al., A novel single-step synthesis of N-doped TiO2 via a sonochemical method. *Materials Research Bulletin*, 2011. **46**(11): pp. 2041–2044.
49. Fu, W., et al., F, Ca co-doped TiO 2 nanocrystals with enhanced photocatalytic activity. *Dalton Transactions*, 2014. **43**(43): pp. 16160–16163.

50. Kotzamanidi, S., et al., Solar photocatalytic degradation of propyl paraben in Al-doped TiO2 suspensions. *Catalysis Today*, 2018. **313**: pp. 148–154.
51. Byrne, C., et al., Effect of Cu doping on the anatase-to-rutile phase transition in TiO2 photocatalysts: Theory and experiments. *Applied Catalysis B: Environmental*, 2019. **246**: pp. 266–276.
52. Smirniotis, P.G., et al., Single-step rapid aerosol synthesis of N-doped TiO2 for enhanced visible light photocatalytic activity. *Catalysis Communications*, 2018. **113**: pp. 1–5.
53. Colón, G., Towards the hydrogen production by photocatalysis. *Applied Catalysis A: General*, 2016. **518** : pp. 48–59.
54. Guo, W. and T. Ma, *Nanostructured Nitrogen Doping TiO 2 Nanomaterials for Photoanodes of Dye-Sensitized Solar Cells, in Low-cost Nanomaterials*. 2014, Springer. pp. 55–75.
55. Yuan, J., et al., Preparations and photocatalytic hydrogen evolution of N-doped TiO2 from urea and titanium tetrachloride. *International Journal of Hydrogen Energy*, 2006. **31**(10): pp. 1326–1331.
56. Wu, M.-C., et al., Photo-Kelvin probe force microscopy for photocatalytic performance characterization of single filament of TiO 2 nanofiber photocatalysts. *Journal of Materials Chemistry* A, 2013. **1**(18): pp. 5715–5720.
57. Shieh, D.-L., et al., N-doped, porous TiO2 with rutile phase and visible light sensitive photocatalytic activity. *Chemical Communications*, 2012. **48**(19): pp. 2528–2530.
58. Chen, D., et al., Carbon and nitrogen co-doped TiO2 with enhanced visible-light photocatalytic activity. *Industrial and Engineering Chemistry Research*, 2007. **46**(9): pp. 2741–2746.
59. Bubacz, K., B. Tryba, and A.W. Morawski, The role of adsorption in decomposition of dyes on TiO2 and N-modified TiO2 photocatalysts under UV and visible light irradiations. *Materials Research Bulletin*, 2012. **47**(11): pp. 3697–3703.
60. Cong, Y., et al., Synthesis and characterization of nitrogen-doped TiO2 nanophotocatalyst with high visible light activity. *The Journal of Physical Chemistry C*, 2007. **111**(19): pp. 6976–6982.
61. Liu, G., et al., Heteroatom-Modulated Switching of Photocatalytic Hydrogen and Oxygen Evolution Preferences of Anatase TiO2 Microspheres. *Advanced Functional Materials*, 2012. **22**(15): pp. 3233–3238.
62. Li, F., et al., Investigation on F–B–S tri-doped nano-TiO 2 films for the photocatalytic degradation of organic dyes. *Journal of Nanoparticle Research*, 2011. **13**(10): pp. 4839–4846.
63. Sakthivel, S. and H. Kisch, Daylight photocatalysis by carbon-modified titanium dioxide. *Angewandte Chemie International Edition*, 2003. **42**(40): pp. 4908–4911.
64. Li, Y., et al., Boron and nitrogen co-doped titania with enhanced visible-light photocatalytic activity for hydrogen evolution. *Applied Surface Science*, 2008. **254**(21): pp. 6831–6836.
65. Uddin, M.N., et al., An experimental and first-principles study of the effect of B/N doping in TiO2 thin films for visible light photo-catalysis. *Journal of Photochemistry and Photobiology A: Chemistry*, 2013. **254**: pp. 25–34.
66. Maeda, K., et al., Studies on TiN x O y F z as a visible-light-responsive photocatalyst. *The Journal of Physical Chemistry C*, 2007. **111**(49): pp. 18264–18270.
67. Rengifo-Herrera, J., J. Kiwi, and C. Pulgarin, N, S co-doped and N-doped Degussa P-25 powders with visible light response prepared by mechanical mixing of thiourea and urea. Reactivity towards E. coli inactivation and phenol oxidation. *Journal of Photochemistry and Photobiology A: Chemistry*, 2009. **205**(2–3): pp. 109–115.
68. Martha, S., et al., Facile synthesis of visible light responsive V 2 O 5/N, S–TiO 2 composite photocatalyst: enhanced hydrogen production and phenol degradation. *Journal of Materials Chemistry*, 2012. **22**(21): pp. 10695–10703.
69. Huang, Y., et al., Photocatalytic property of nitrogen-doped layered perovskite K2La2Ti3O10. *Solar Energy Materials and Solar Cells*, 2010. **94**(5): pp. 761–766.
70. Kumar, V. and S. Uma, Investigation of cation (Sn2+) and anion (N3–) substitution in favor of visible light photocatalytic activity in the layered perovskite K2La2Ti3O10. *Journal of Hazardous Materials*, 2011. **189** (1-2): pp. 502–508.
71. Mukherji, A., et al., Nitrogen doped Sr2Ta2O7 coupled with graphene sheets as photocatalysts for increased photocatalytic hydrogen production. *ACS nano*, 2011. **5**(5): pp. 3483–3492.
72. Ida, S., et al., Preparation of Tantalum-Based Oxynitride Nanosheets by Exfoliation of a Layered Oxynitride, CsCa2Ta3O10–x N y, and Their Photocatalytic Activity. *Journal of the American Chemical Society*, 2012. **134**(38): pp. 15773–15782.

73. Chen, S., et al., Nitrogen-doped layered oxide Sr 5 Ta 4 O 15− x N x for water reduction and oxidation under visible light irradiation. *Journal of Materials Chemistry A*, 2013. **1**(18): pp. 5651–5659.
74. Matsumoto, Y., et al., N doping of oxide nanosheets. *Journal of the American Chemical Society*, 2009. **131**(19): pp. 6644–6645.
75. Marschall, R., et al., Preparation of new sulfur-doped and sulfur/nitrogen co-doped CsTaWO 6 photocatalysts for hydrogen production from water under visible light. *Journal of Materials Chemistry*, 2011. **21**(24): pp. 8871–8879.
76. Gasparotto, A., et al., F-doped Co3O4 photocatalysts for sustainable H2 generation from water/ethanol. *Journal of the American Chemical Society*, 2011. **133**(48): pp. 19362–19365.
77. Martha, S. and K. Parida, Fabrication of nano N-doped In2Ga2ZnO7 for photocatalytic hydrogen production under visible light. *International Journal of Hydrogen Energy*, 2012. **37**(23): pp. 17936–17946.
78. Xiang, Q., J. Yu, and M. Jaroniec, Synergetic effect of MoS2 and graphene as cocatalysts for enhanced photocatalytic H2 production activity of TiO2 nanoparticles. *Journal of the American Chemical Society*, 2012. **134**(15): pp. 6575–6578.
79. Ni, M., et al., A review and recent developments in photocatalytic water-splitting using TiO2 for hydrogen production. *Renewable and Sustainable Energy Reviews*, 2007. **11**(3): pp. 401–425.
80. Zhai, Q., et al., Photocatalytic conversion of carbon dioxide with water into methane: platinum and copper (I) oxide co-catalysts with a core–shell structure. *Angewandte Chemie*, 2013. **125**(22): pp. 5888–5891.
81. Chen, S., et al., Interface engineering of a CoOx/Ta3N5 photocatalyst for unprecedented water oxidation performance under visible-light-irradiation. *Angewandte Chemie*, 2015. **127**(10): pp. 3090–3094.
82. Hoffmann, M.R., et al., Environmental applications of semiconductor photocatalysis. *Chemical Reviews*, 1995. **95**(1): pp. 69–96.
83. Daskalaki, V.M., et al., Solar light-responsive Pt/CdS/TiO2 photocatalysts for hydrogen production and simultaneous degradation of inorganic or organic sacrificial agents in wastewater. *Environmental Science and Technology*, 2010. **44**(19): pp. 7200–7205.
84. Jin, Z., et al., Improved quantum yield for photocatalytic hydrogen generation under visible light irradiation over eosin sensitized TiO2: Investigation of different noble metal loading. *Journal of Molecular Catalysis A: Chemical*, 2006. **259**(1–2): pp. 275–280.
85. Sayama, K., et al., Photocatalytic activity and reaction mechanism of Pt-intercalated K4Nb6O17 catalyst on the water splitting in carbonate salt aqueous solution. *Journal of Photochemistry and Photobiology A: Chemistry*, 1998. **114**(2): pp. 125–135.
86. Jang, J.S., et al., Role of platinum-like tungsten carbide as cocatalyst of CdS photocatalyst for hydrogen production under visible light irradiation. *Applied Catalysis A: General*, 2008. **346** (1–2): pp. 149–154.
87. Matsumura, M., Y. Saho, and H. Tsubomura, Photocatalytic hydrogen production from solutions of sulfite using platinized cadmium sulfide powder. *The Journal of Physical Chemistry*, 1983. **87**(20): pp. 3807–3808.
88. Liu, L., et al., In situ loading transition metal oxide clusters on TiO2 nanosheets as co-catalysts for exceptional high photoactivity. *Acs Catalysis*, 2013. **3**(9): pp. 2052–2061.
89. Wang, M., et al., Effects of sacrificial reagents on photocatalytic hydrogen evolution over different photocatalysts. *Journal of Materials Science*, 2017. **52**(9): pp. 5155–5164.
90. Kawai, T. and T. Sakata, Photocatalytic hydrogen production from liquid methanol and water. J*ournal of the Chemical Society, Chemical Communications*, 1980(15): pp. 694–695.
91. Huerta-Flores, A.M., et al., Green synthesis of earth-abundant metal sulfides (FeS 2, CuS, and NiS 2) and their use as visible-light active photocatalysts for H 2 generation and dye removal. *Journal of Materials Science: Materials in Electronics*, 2018. **29**(13): pp. 11613–11626.
92. DeLaive, P.J., et al., Photoinduced electron transfer reactions of transition-metal complexes with amines. Mechanistic studies of alternate pathways to back electron transfer. *Journal of the American Chemical Society*, 1980. **102**(17): pp. 5627–5631.
93. Chiang, T.H., G. Viswanath, and Y.-S. Chen, Effects of RhCrOx Cocatalyst Loaded on Different Metal Doped LaFeO3 Perovskites with Photocatalytic Hydrogen Performance under Visible Light Irradiation. *Catalysts*, 2021. **11**(5): pp. 612.
94. Zhang, J., et al., Nanospherical carbon nitride frameworks with sharp edges accelerating charge collection and separation at a soft photocatalytic interface. *Advanced Materials*, 2014. **26**(24): pp. 4121–4126.

95. Shi, Y. and B. Zhang, Recent advances in transition metal phosphide nanomaterials: synthesis and applications in hydrogen evolution reaction. *Chemical Society Reviews*, 2016. **45**(6): pp. 1529–1541.
96. Zhao, H., et al., Noble-metal-free iron phosphide cocatalyst loaded graphitic carbon nitride as an efficient and robust photocatalyst for hydrogen evolution under visible light irradiation. *ACS Sustainable Chemistry and Engineering*, 2017. **5**(9): pp. 8053–8060.
97. Feng, R., et al., Anchoring single Pt atoms and black phosphorene dual co-catalysts on CdS nanospheres to boost visible-light photocatalytic H2 evolution. *Nano Today*, 2021. **37** : p. 101080.
98. Dai, G., J. Yu, and G. Liu, Synthesis and enhanced visible-light photoelectrocatalytic activity of p– n junction BiOI/TiO2 nanotube arrays. *The Journal of Physical Chemistry C*, 2011. **115**(15): pp. 7339–7346.
99. Wang, M., et al., p–n Heterojunction photoelectrodes composed of Cu 2 O-loaded TiO 2 nanotube arrays with enhanced photoelectrochemical and photoelectrocatalytic activities. *Energy and Environmental Science*, 2013. **6**(4): pp. 1211–1220.
100. Jimmy, C.Y., et al., Preparation of highly photocatalytic active nano-sized TiO2 particles via ultrasonic irradiation. *Chemical Communications*, 2001(19): pp. 1942–1943.
101. Cho, S., et al., Three-dimensional type II ZnO/ZnSe heterostructures and their visible light photocatalytic activities. *Langmuir*, 2011. **27**(16): pp. 10243–10250.
102. Yu, J., et al., Enhanced photocatalytic performance of direct Z-scheme gC 3 N 4–TiO 2 photocatalysts for the decomposition of formaldehyde in air. *Physical Chemistry Chemical Physics*, 2013. **15**(39): pp. 16883–16890.
103. Li, Q., et al., Highly efficient visible-light-driven photocatalytic hydrogen production of CdS-cluster-decorated graphene nanosheets. *Journal of the American Chemical Society*, 2011. **133**(28): pp. 10878–10884.
104. Yu, J., et al., Enhanced photocatalytic CO2-reduction activity of anatase TiO2 by coexposed (001) and (101) facets. *Journal of the American Chemical Society*, 2014. **136**(25): pp. 8839–8842.
105. Lu, D., G. Zhang, and Z. Wan, Visible-light-driven g-C3N4/Ti3+-TiO2 photocatalyst co-exposed {0 0 1} and {1 0 1} facets and its enhanced photocatalytic activities for organic pollutant degradation and Cr (VI) reduction. *Applied Surface Science*, 2015. **358** : pp. 223–230.
106. Vinodgopal, K. and P.V. Kamat, Enhanced rates of photocatalytic degradation of an azo dye using SnO2/TiO2 coupled semiconductor thin films. *Environmental Science and Technology*, 1995. **29**(3): pp. 841–845.
107. Ranjit, K.T. and B. Viswanathan, Synthesis, characterization and photocatalytic properties of iron-doped TiO2 catalysts. *Journal of Photochemistry and Photobiology A: Chemistry*, 1997. **108**(1): pp. 79–84.
108. Marschall, R., Photocatalysis: Semiconductor Composites: Strategies for Enhancing Charge Carrier Separation to Improve Photocatalytic Activity, *Adv. Funct. Mater. 17/2014). Advanced Functional Materials*, 2014. **24**(17): p. 2420.
109. Wang, H., et al., Semiconductor heterojunction photocatalysts: design, construction, and photocatalytic performances. *Chemical Society Reviews*, 2014. **43**(15): pp. 5234–5244.
110. Shi, W. and N. Chopra, Nanoscale heterostructures for photoelectrochemical water splitting and photodegradation of pollutants. *Nanomaterials and Energy*, 2013. **2**(3): pp. 158–178.
111. Li, K., et al., In-situ-reduced synthesis of Ti3+ self-doped TiO2/g-C3N4 heterojunctions with high photocatalytic performance under LED light irradiation. *ACS Applied Materials and Interfaces*, 2015. **7**(17): pp. 9023–9030.
112. Pan, C., et al., Dramatic activity of C3N4/BiPO4 photocatalyst with core/shell structure formed by self-assembly. *Advanced Functional Materials*, 2012. **22**(7): pp. 1518–1524.
113. Huang, L., et al., Visible-light-induced WO 3/gC 3 N 4 composites with enhanced photocatalytic activity. *Dalton Transactions*, 2013. **42**(24): pp. 8606–8616.
114. Hong, S., et al., *Energy Environ. Sci.* 4, 1781 (2011).
115. Zhou, M., et al., Effects of calcination temperatures on photocatalytic activity of SnO2/TiO2 composite films prepared by an EPD method. *Journal of Hazardous Materials*, 2008. **154** (1–3): pp. 1141–1148.
116. Wetchakun, N., et al., BiVO4/CeO2 nanocomposites with high visible-light-induced photocatalytic activity. *ACS Applied Materials and Interfaces*, 2012. **4**(7): pp. 3718–3723.

117. Yang, G., et al., One-dimensional CdS/ZnO core/shell nanofibers via single-spinneret electrospinning: tunable morphology and efficient photocatalytic hydrogen production. *Nanoscale*, 2013. **5**(24): pp. 12432–12439.
118. Yuan, Y., et al., Energy Environ. Sci. 2014, 7, 3934; d) SJ Moniz, SA Shevlin, DJ Martin, Z.-X. Guo, J. Tang. *Energy Environ. Sci*, 2015. **8**: pp. 731.
119. Jiang, D., Y. Zhang, and X. Li, Folded-up thin carbon nanosheets grown on Cu 2 O cubes for improving photocatalytic activity. *Nanoscale*, 2017. **9**(34): pp. 12348–12352.
120. Liu, J., et al., BiOI/TiO 2 nanotube arrays, a unique flake-tube structured p–n junction with remarkable visible-light photoelectrocatalytic performance and stability. *Dalton Transactions*, 2014. **43**(4): pp. 1706–1715.
121. Peng, Y., et al., Novel one-dimensional Bi 2 O 3–Bi 2 WO 6 p–n hierarchical heterojunction with enhanced photocatalytic activity. *Journal of Materials Chemistry A*, 2014. **2**(22): pp. 8517–8524.
122. Zhang, J., Z. Zhu, and X. Feng, Construction of two-dimensional MoS2/CdS p–n nanohybrids for highly efficient photocatalytic hydrogen evolution. *Chemistry–A European Journal*, 2014. **20**(34): pp. 10632–10635.
123. Zhang, J., et al., Fabrication of NiS modified CdS nanorod p–n junction photocatalysts with enhanced visible-light photocatalytic H 2-production activity. *Physical Chemistry Chemical Physics*, 2013. **15**(29): pp. 12088–12094.
124. Sajan, C., et al., TiO2 nanosheets with exposed {001} facets for photocatalytic applications. *Nano Res 9*: 3–27. 2016.
125. Bard, A.J., Photoelectrochemistry and heterogeneous photo-catalysis at semiconductors. *Journal of Photochemistry*, 1979. **10**(1): pp. 59–75.
126. Tada, H., et al., All-solid-state Z-scheme in CdS–Au–TiO 2 three-component nanojunction system. *Nature Materials*, 2006. **5**(10): pp. 782–786.
127. Yin, X.-L., et al., Urchin-like Au@ CdS/WO 3 micro/nano heterostructure as a visible-light driven photocatalyst for efficient hydrogen generation. *Chemical Communications*, 2015. **51**(72): pp. 13842–13845.
128. Huang, Q., et al., One-pot facile synthesis of branched Ag-ZnO heterojunction nanostructure as highly efficient photocatalytic catalyst. *Applied Surface Science*, 2015. **353**: pp. 949–957.
129. Katsumata, H., et al., Z-scheme photocatalytic hydrogen production over WO3/g-C3N4 composite photocatalysts. *RSC advances*, 2014. **4**(41): pp. 21405–21409.
130. Jin, J., et al., A hierarchical Z-scheme CdS–WO3 photocatalyst with enhanced CO2 reduction activity. *Small*, 2015. **11**(39): pp. 5262–5271.
131. Kawai, T. and T. Sakata, Conversion of carbohydrate into hydrogen fuel by a photocatalytic process. *Nature*, 1980. **286**: pp. 474–476.
132. Kazunari Domen, M.H., Junko N. Kondo, Tsuyoshi Takata, Akihiko Kudo, Hisayoshi Kobayashi, Yasunobu Inoue, New Aspects of Heterogeneous Photocatalysts for Water Decomposition. *KoreanJ. Chern. Eng*, 2001. **18(6)**: pp. 862–866.
133. Maeda, K., et al., Photocatalytic Activity of (Ga1-x Zn x)(N1-x O x) for Visible-Light-Driven H2 and O2 Evolution in the Presence of Sacrificial Reagents. *The Journal of Physical Chemistry C*, 2008. **112**(9): pp. 3447–3452.
134. Hisatomi, T., J. Kubota, and K. Domen, Recent advances in semiconductors for photocatalytic and photoelectrochemical water splitting. *Chem Soc Rev*, 2014. **43**(22): pp. 7520–7535.
135. Shwetharani, R., C.A.N. Fernando, and G.R. Balakrishna, Excellent hydrogen evolution by a multi approach via structure–property tailoring of titania. *RSC Advances*, 2015. **5**(49): pp. 39122–39130.
136. Honda, A.F.a.K., Electrochemical Photolysis of Water at a Semiconductor Electrode. *Nature*, 1972. **238**: pp. 37–38.
137. Han, C., R. Luque, and D.D. Dionysiou, Facile preparation of controllable size monodisperse anatase titania nanoparticles. *Chemical Communications*, 2012. **48**(13): pp. 1860–1862.
138. Moon, H.S. and K. Yong, Noble-metal free photocatalytic hydrogen generation of CuPc/TiO2 nanoparticles under visible-light irradiation. *Applied Surface Science*, 2020. **530**: p. 147215.
139. Wang, X., et al., ZnO–CdS@Cd Heterostructure for Effective Photocatalytic Hydrogen Generation. *Advanced Energy Materials*, 2012. **2**(1): pp. 42–46.
140. Chen, Y., L. Ma, and S. Ding, Enhanced Photocatalytic Hydrogen Generation by Optimized Plasmonic Hot Electron Injection in Structure-Adjustable Au-ZnO Hybrids. *Catalysts*, 2020. **10**(4): pp. 376.

141. Hu, J., et al., Spatially Separating Redox Centers on Z-Scheme ZnIn2S4/BiVO4 Hierarchical Heterostructure for Highly Efficient Photocatalytic Hydrogen Evolution. *Small*, 2020. **16**(37): p. 2002988.
142. Hisatomi, T., J. Kubota, and K. Domen, Recent advances in semiconductors for photocatalytic and photoelectrochemical water splitting. *Chemical Society Reviews*, 2014. **43**(22): pp. 7520–7535.
143. Shi, J., et al., LaTiO2N–LaCrO3: continuous solid solutions towards enhanced photocatalytic H2 evolution under visible-light irradiation. *Dalton Transactions*, 2017. **46**(32): pp. 10685–10693.
144. Wang, Q., et al., Core/Shell Structured La- and Rh-Codoped SrTiO3 as a Hydrogen Evolution Photocatalyst in Z-Scheme Overall Water Splitting under Visible Light Irradiation. *Chemistry of Materials*, 2014. **26**(14): pp. 4144–4150.
145. Wang, Z., et al., Sequential cocatalyst decoration on BaTaO 2 N towards highly-active Z-scheme water splitting. *Nature Communications*, 2021. **12**(1): pp. 1–9.
146. Maeda, K., et al., Photocatalytic Activity of (Ga1-xZnx)(N1-xOx) for Visible-Light-Driven H2 and O2 Evolution in the Presence of Sacrificial Reagents. *The Journal of Physical Chemistry C*, 2008. **112**(9): pp. 3447–3452.
147. Wang, Z., et al., Sequential cocatalyst decoration on BaTaO2N towards highly-active Z-scheme water splitting. *Nature Communications*, 2021. **12**(1): p. 1005.
148. Wang, Q., et al., Oxysulfide photocatalyst for visible-light-driven overall water splitting. *Nature Materials*, 2019. **18**(8): pp. 827–832.
149. Song, M., et al., Junction of porous g-C3N4 with BiVO4 using Au as electron shuttle for cocatalyst-free robust photocatalytic hydrogen evolution. *Applied Surface Science*, 2019. **498**: pp. 143808.
150. Liu, L., et al., Efficient visible-light photocatalytic hydrogen evolution and enhanced photostability of core@shell Cu2O@g-C3N4 octahedra. *Applied Surface Science*, 2015. **351**: pp. 1146–1154.
151. Alaasar, M., et al., Controlling liquid and liquid crystalline network formation by core-fluorination of hydrogen bonded supramolecular polycatenars. *Journal of Molecular Liquids*, 2021. **332**: p. 115870.
152. Pullen, S., et al., Enhanced Photochemical Hydrogen Production by a Molecular Diiron Catalyst Incorporated into a Metal–Organic Framework. *Journal of the American Chemical Society*, 2013. **135**(45): pp. 16997–17003.
153. Liu, J., et al., A Metal-Organic-Framework-Derived g-C3N4/α-Fe2O3 Hybrid for Enhanced Visible-Light-Driven Photocatalytic Hydrogen Evolution. *Chemistry–A European Journal*, 2019. **25**(9): pp. 2330–2336.
154. Liu, J., et al., A Metal-Organic-Framework-Derived g-C3N4/α-Fe2O3 Hybrid for Enhanced Visible-Light-Driven Photocatalytic Hydrogen Evolution. *Chemistry–A European Journal*, 2019. **25**(9): pp. 2330–2336.
155. Bashir, S., A.K. Wahab, and H. Idriss, Synergism and photocatalytic water splitting to hydrogen over M/TiO2 catalysts: Effect of Initial Particle size of TiO2. *Catalysis Today*, 2015. **240**: pp. 242–247.
156. Wang, C., et al., Enhanced hydrogen production by water splitting using Cu-doped TiO2 film with preferred (001) orientation. *Applied Surface Science*, 2014. **292**: pp. 161–164.
157. Zong, X., et al., Enhancement of photocatalytic H2 evolution on CdS by loading MoS2 as Cocatalyst under visible light irradiation. *J Am Chem Soc*, 2008. **130**(23): pp. 7176–7177.
158. Chen, G., et al., A novel noble metal-free ZnS–WS2/CdS composite photocatalyst for H2 evolution under visible light irradiation. *Catalysis Communications*, 2013. **40**: pp. 51–54.
159. Lv, J., et al., Bi SPR-Promoted Z-Scheme Bi2MoO6/CdS-Diethylenetriamine Composite with Effectively Enhanced Visible Light Photocatalytic Hydrogen Evolution Activity and Stability. *ACS Sustainable Chemistry and Engineering*, 2018. **6**(1): pp. 696–706.
160. Wang, X.-j., et al., Metalloid Ni2P and its behavior for boosting the photocatalytic hydrogen evolution of CaIn2S4. *International Journal of Hydrogen Energy*, 2018. **43**(1): pp. 219–228.
161. Zhang, H., et al., Ultrasmall MoOx Clusters as a Novel Cocatalyst for Photocatalytic Hydrogen Evolution. *Advanced Materials*, 2019. **31**(6): p. 1804883.
162. Zhao, X., et al., Ag supported Z-scheme WO2.9/g-C3N4 composite photocatalyst for photocatalytic degradation under visible light. *Applied Surface Science*, 2020. **501**: p. 144258.
163. Tahir, M.B., et al., Insighting role of reduced graphene oxide in BiVO4 nanoparticles for improved photocatalytic hydrogen evolution and dyes degradation. *International Journal of Energy Research*, 2019. **43**(6): pp. 2410–2417.

164. Mushtaq, F., et al., Piezoelectrically Enhanced Photocatalysis with BiFeO3 Nanostructures for Efficient Water Remediation. *iScience*, 2018. **4**: pp. 236–246.
165. Mohamed, H.H., Rationally designed Fe2O3/GO/WO3 Z-Scheme photocatalyst for enhanced solar light photocatalytic water remediation. *Journal of Photochemistry and Photobiology A: Chemistry*, 2019. **378**: pp. 74–84.
166. Qiu, J., et al., A novel 3D nanofibrous aerogel-based MoS2@Co3S4 heterojunction photocatalyst for water remediation and hydrogen evolution under simulated solar irradiation. *Applied Catalysis B: Environmental*, 2020. **264**: p. 118514.
167. Wu, M.-h., et al., Fabrication of ternary GO/g-C3N4/MoS2 flower-like heterojunctions with enhanced photocatalytic activity for water remediation. *Applied Catalysis B: Environmental*, 2018. **228**: pp. 103–112.

Section II

Emerging Materials in Nanosensors

This section of the book is dedicated to explaining nanosensors using nanomaterials of different kinds: Detailed information about various kinds of nanosensors, ranging from fluorescent nanosensors, Peptide Luminescent Nanosensors, bio/chemical sensors to electrochemical sensors and so forth. Each nanosensor gives sufficient information about various applications, synthesis, characterization, and end-use.

Finally, we present economic analysis with future prospects and a mechanistic understanding of nanosensors and catalysis.

DOI: 10.1201/9781003218708-8

7 Nanostructured Materials for Sensors Applications

Saravanan Chandrasekaran and Anu Sukhdev

CONTENTS

7.1 Introduction: Background and Driving Forces 159
7.2 Synthesis of Nanostructured Materials 160
7.2.1 2D Layered Inorganic Nanomaterials for Sensor Applications 160
7.2.1.1 Transition Metal Dichalcogenides (TMDs) as Sensor Materials 160
7.2.1.2 Synthesis of 2D Inorganic Materials for Sensor Applications 161
7.2.1.3 Metal Oxides as Sensor Materials 162
7.2.1.4 Synthesis of Metal Oxide Materials for Sensor Applications 162
7.2.2 Carbon and Its Allotropy Materials for Sensor Applications 163
7.2.2.1 Carbon and Graphene Quantum Dots as Sensors 163
7.2.2.2 Synthetic Methods of CQDs and GQDs and Application as Sensor Materials 163
7.2.2.3 Preparation of Carbon Quantum Dots (CQD) for Sensor Applications 164
7.2.2.4 Preparation of Graphene Quantum Dots (GQD) for Sensor Applications 165
7.2.3 Conducting Polymer (CP) Materials for Sensor Applications 166
7.2.3.1 Synthesis of Conducting Polymers (CPs) for Sensor Applications 166
7.2.4 Organic-Inorganic Hybrid Nanomaterials for Sensor Applications 167
7.2.4.1 Perovskites as Sensors 167
7.2.4.2 Synthesis of Perovskites Materials for Sensor Applications 168
7.2.4.3 Metal Organic Frameworks (MOFs) as Sensors 169
7.2.4.4 Synthesis of MOFs for Sensor Applications 170
7.3 Sensor Device Fabrications 171
7.4 Sensing Mechanism 172
7.5 Outlook 173
Acknowledgments 173
References 174

7.1 INTRODUCTION: BACKGROUND AND DRIVING FORCES

Nanostructured materials have attracted intensive scientific attention because, at the nanoscale level, unique electrical, magnetic, optical and mechanical properties emerge. The size-dependent properties of nanostructured materials provide the highest potentials for increasing performance and extending their applications in many diverse fields, such as sensors, display technology, energy storage and energy conversion process, catalysis, pharmaceutical and biomedical. Sensors are devices that are frequently used to detect a variable quantity, commonly electronically, and sensors convert the measurement into specific signals. Utilization of nanostructured materials for sensor applications are technologically advantageous in terms of surface-to-volume ratio and high specific surface area, demonstrating strong potential in improving the performances of the sensors.

DOI: 10.1201/9781003218708-9

Up to now several nanostructured materials – 1D (nanowire, nanotube), 2D (nanosheets), and 3D (nanoparticles) – have been utilized as sensing materials for chemical, gas and biochemical detection applications. Among the various types of nanomaterials, specific types of nanomaterials – such as inorganic nanomaterials, metal oxides, carbon allotropic quantum dots, polymeric materials and organic-inorganic hybrid nanomaterials – are receiving ever-greater attention for sensor applications, because they can remarkably enhance response time, selectivity and sensitivity. As a result, the focus of this chapter is on the synthesis of nanomaterials using recently established techniques for sensor applications, fabrication of sensor devices and the sensing mechanisms involved in sensing various types of sensors, such as gas, chemicals and biosensors.

7.2 SYNTHESIS OF NANOSTRUCTURED MATERIALS

7.2.1 2D Layered Inorganic Nanomaterials for Sensor Applications

Two-dimensional (2D) materials are evolving as competitors for use in sensing devices in the Internet of Things (IoT) platform due to their favorable chemical and physical properties. Their large surface area-to-volume ratio implies that they are extremely sensitive to target substances. Due to their flexibility, high mechanical strength, and optical transparency, 2D material-based sensors can also be miniaturized. They interact with target molecules/ions in two ways: physisorption and chemisorption. Novel 2D-inorganic materials (ZnO, MoS_2, SnS_2, WS_2, etc.) have attracted a lot of attention because of their unique properties (e.g., ultra-thin structure, tunable energy band diagrams and large surface area), which have led to a lot of research into their fundamentals, applications and, more recently, their potential for gas sensing. Several 2D layered inorganic materials, such as MoS_2, WSe_2, WS_2, $GaSe_2$, GaS, FeS_2, $NdSe_2$, CuS, and h-BN, have recently been described in the literature (Figure 7.1), however only a few materials, such as MoS_2, have been employed as gas sensors [1].

7.2.1.1 Transition Metal Dichalcogenides (TMDs) as Sensor Materials

TMDs are an incipient material due to its ease of fabrication (exfoliation process, functionalization), low cost, and great sensitivity. The most typical TMDs is MoS_2, composed of Mo (transition metal) and S (chalcogenide atom), which are covalently bonded. These 2D

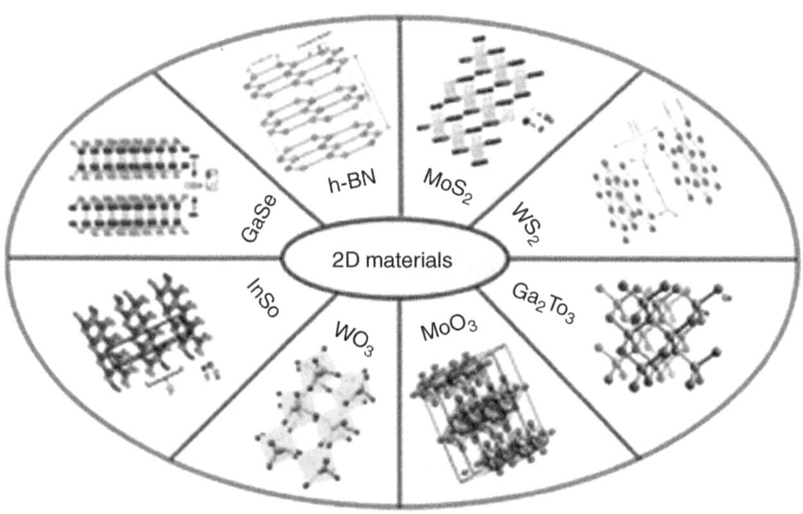

FIGURE 7.1 Examples of inorganic 2D materials (figure is reproduced from Ref. [1]).

stacked layers are piled in a row and held together by a weak van der Waals force. This weak force facilitates exfoliation and allows for single-layered film extraction. TMDs can also be used to detect glucose with decorated nanoparticles (NPs) to enhance the sensing [2]. Parlak et al. used Au NPs to improve glucose selectivity using a sonication method with a mixture of MoS_2 nanosheets and Au nanoparticles [3].

At room temperature, single-layer MoS_2 exhibits greater sensitivity and selectivity toward volatile organic compounds (VOCs) in the concentration range of 50 to 5000 ppm. Various studies that focused on graphene and MoS_2 are reported on sensors, targeting glucose, metal ions, hydrogen ion, organic and bio molecules [4].

Graphene oxide (GO) and TMDs have great potential in metal ion sensing. Their capacity to remove metal ions is owing to their hydrophilic properties and functional groups, which include oxygen atoms that can bind metal ions to the surface. Due to their abundant intrinsic chalcogen atoms, TMDs, on the other hand, have the ability to absorb certain heavy metals. MoS_2-based FETs are used to sense mercury ions. Because of their extremely strong sulphur sites, MoS_2 has a high affinity toward Hg^{2+} [4].

7.2.1.2 Synthesis of 2D Inorganic Materials for Sensor Applications

Various synthetic approaches, such as hydrothermal/solvothermal, chemical /micromechanical exfoliation, solution phase synthesis, chemical vapor deposition, electro deposition, and unzipping nanotubes are used for fabricating 2D layered inorganic nanomaterials (Figure 7.2).

Su et al. used the hydrothermal method to synthesize Au nanoparticles onto MoS_2 and then evaluated them for a variety of analytes, including glucose [5]. Only glucose had sensitivity toward the device. In place of Au, Ni deposited on the nanosheet using the solution method. It was found that the device showed sensitivity toward ascorbic acid, uric acid, dopamine and glucose. The Cu was deposited by the electrodeposition method on the MoS_2 nanosheets and was investigated for various analytes, H_2O_2, ascorbic acid, uric acid, dopamine, lactose and fructose. Only H_2O_2 showed the sensing response.

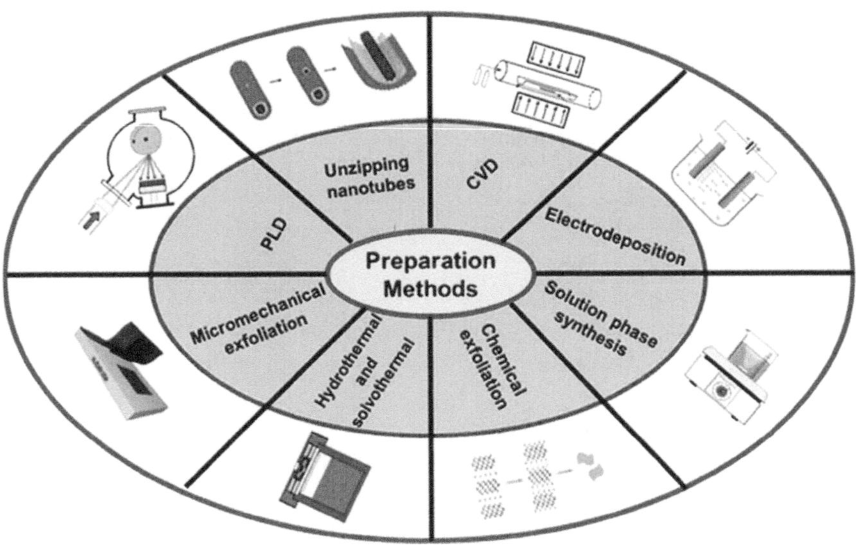

FIGURE 7.2 The range of synthetic approaches used to fabricate 2D layered inorganic Nanomaterials (figure is reproduced from Ref [1]).

7.2.1.3 Metal Oxides as Sensor Materials

Over the last few years there has been a greater emphasis on metal oxides as gas sensors. This is primarily due to metal oxides' property in enhancing the response time, selectivity and sensitivity [6]. Nanomaterials with a wide band gap are good gas-sensing materials with a high response. Metal oxide or doped NPs are utilized as electrode modifiers in sensor devices to detect various analytes in biological and environmental systems.

Transition metals and their oxides with d^0-d^{10} electronic configurations have a wide range of characteristics for sensing applications. These materials' sensing behavior is based on electrochemical changes, catalytic combustion, or resistance modulation. Metal oxide sensors work on the principle of gas adsorption on the surface, which results in a change in electrical resistance or conductivity. The two types of nanomaterials for gas sensing are n-type and p-type. Electrons are the majority charge carriers in n-type nanomaterials, while holes are the majority charge carriers in p-type nanomaterials. Metal oxide nanomaterials gas sensors have a complex gas-sensing mechanism that is based on the adsorption/desorption of test gases on their surfaces [7].

7.2.1.4 Synthesis of Metal Oxide Materials for Sensor Applications

Metal oxide nanoparticles are typically synthesized using a "bottom-up" approach rather than a "top-down" approach (Figure 7.3) [8]. Conventional bottom-up approaches, such as electrochemical, co-precipitation, sol-gel, hydrothermal and solvothermal methods, are used to synthesizing metal oxide NP. Precursor, solvent (water, polar/non-polar organic solvents), capping agents/stabilizing agents, reducing agents (depending on the nature of the precursors), pressure effect, heat treatment, and pH value of solution all play a role in the synthesis of NPs.

NPs of iron oxide are extensively used as electrode modifiers in determination of various analytes such as glucose, urea, hydrogen peroxide, heavy metals, bioactive molecules like dopamine etc. To avoid the pulverization of electrode film, metal ion NP can be impregnated with carbon matrix. Molecularly imprinted electrochemical sensor using Iron oxide NPs functionalized with polyaniline is effectively used for sensing various categories of analytes including bio-active molecules.

TiO_2 NPs are effective electrode modifiers because they are highly conductive and biocompatible. To improve the sensing efficiency, TiO_2 NPs are doped with metal nanoparticles or catalytically active substances. TiO_2 NPs based sensors are evaluated even in samples like water samples, human blood sample serum, fruit juices, milk and tablets samples.

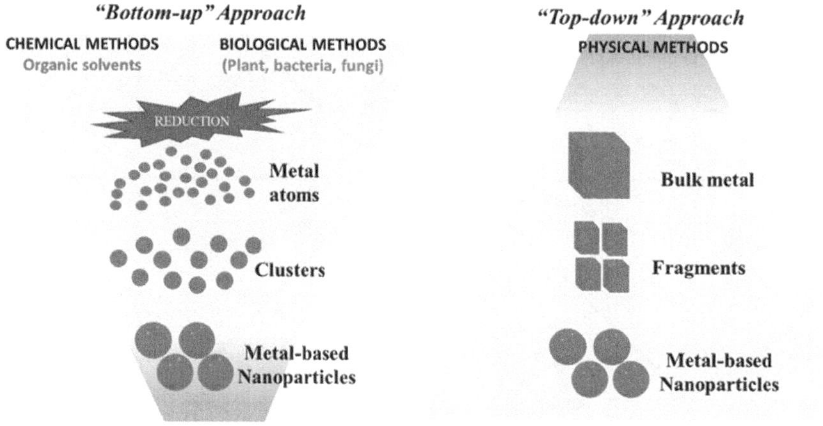

FIGURE 7.3 Different methods used for the synthesis of metal-based nanoparticles (figure is reproduced from Ref. [8]).

One dimensional structured MnO_2 obtained through hydrothermal method with different morphologies are extensively studied as electrochemical sensors. MnO_2 NPs functionalized with reduced graphene oxide (rGO) / nanotubes shows increased sensing application toward glucose quantification in a small detection limits. Morphology of NPs efficiently increases the selectivity and sensitivity of the sensor.

Hydrothermally formed ZnO nanorods were tested as LPG sensors. It was discovered that ZnO nanorods provide a good response to LPG while augmenting the response of zinc stannate micro cubes [9].

For ethanol sensing, photochemically produced Ag NPs embedded ZnO nanorods were investigated. At 10 ppm, Ag-ZnO nanorods showed a better response to ethanol than pure-ZnO nanorods [10]. A special method for depositing Au NPs on the surface of WO_3 nanorods for H_2 gas sensing was developed using an Ion exchange method followed by a hydrothermal method [11].

7.2.2 Carbon and Its Allotropy Materials for Sensor Applications

7.2.2.1 Carbon and Graphene Quantum Dots as Sensors

The carbon-based quantum dots or C-dots or carbon quantum dots (CQD) and graphene quantum dots (GQDs) are categorized as zero dimensional (0D) substances that have fluorescence property with a diameter less than 10 nm. GODs are a subset of CQDs and generally derived from graphene and/or graphene oxide. These QDs have sparked a lot of interest in recent years because of their high chemical stability, low toxicity, higher biocompatibility, good conductivity, luminescence, and broad optical absorption [12].

CQDs-based sensors use mechanisms such as fluorescence quenching, static quenching, dynamic quenching, energy transfer, inner filter effect, photo-induced electron transfer, and fluorescence resonance energy transfer. CQDs have evolved as very accurate sensor systems with limits of detection (LOD) in the nanomolar, picomolar, and even femtomolar levels. Typically, CQDs and GQDs both have a cornucopia of oxygen-containing functional groups on their surfaces. Some of the innovative benefits of using CQDs in sensor applications are ease of synthesis, cost-effective synthesis procedures, simple functionalization routes, high QY, outstanding brightness, chemical stability, and null toxicity.

Using CQDs as the sensing probes, metal ions have become the highest explored analytes. Hence, sensors for a series of metal ions – including Fe^{3+}, Hg^{2+}, Zn^{2+}, Cd^{2+}, Cu^{2+}, Au^{3+}, Co^{2+}, Ni^{2+}, Pd^{2+}, Pb^{2+}, Mn^{2+}, Bi^{3+}, Al^{3+}, K^+, Sn^{2+}, Cr^{6+}, Ag^+ and Be^{2+} – have been developed. CQDs are used as sensing probes in food analysis to identify and analyze functional food components such as protein, phenolic compounds, vitamins and so on [13].

7.2.2.2 Synthetic Methods of CQDs and GQDs and Application as Sensor Materials

CQDs and GQD-based nanomaterials can be synthesized using either bottom-up or top-down methods (Figure 7.4) [14]. Electrochemical carbonization, microwave irradiation synthesis, hydrothermal/solvothermal treatment, and thermal degradation are examples of bottom-up approaches. These procedures typically demand stringent reaction conditions, such as high-grade carbon precursors, high temperatures, concentrated alkali/acid treatments, and hazardous organic solvents. Bottom-up methods have the advantage of offering synthetic control and, as a result, effective size control. The origin of the carbon source has had a significant impact on the properties of CQDs, including sensing properties, where apparently the same CQDs generated with the same procedure but from different precursors will exhibit significantly varied selectivity for different metal ions. Electrochemical oxidation, ultrasonication, chemical ablation or oxidation, laser ablation and arc discharge/plasma treatment are examples of top-down methods [12], [13].

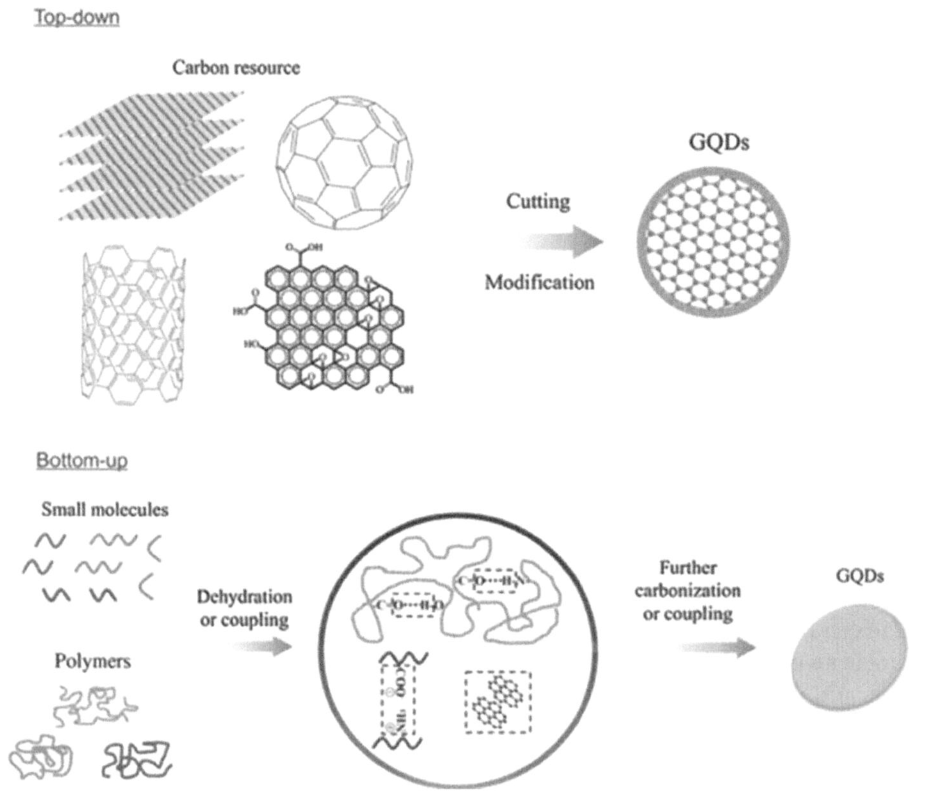

FIGURE 7.4 Preparation of GQDs: the "top-down" and "bottom-up" method (reproduced from Ref. [14]).

7.2.2.3 Preparation of Carbon Quantum Dots (CQD) for Sensor Applications

Long-life photoluminescence in N, S co-doped CQDs made from an ionic liquid has a wide range of applications in pesticide detection. Pesticides used in current agricultural practices can have an adverse effect on the environment and human health. Therefore, it is crucial to develop extremely sensitive detectors for pesticide residues. N and S co-doped CQDs (N,S-CQDs) were produced by a simple pyrolysis method using N-Methylethanolammonium thioglycolate ionic liquid as a single source. The carbon material was subjected to ultrasonication to obtain doped CQDs. The doped CQDs finds application as PL based chemosensors for sensitive detection of pesticides [15]. Wang et al. synthesized photoluminescent carbon dots (CDs). The hydrothermal method was employed to selectively detect Hg^{2+} ions using N-doped CDs. The same has been used to detect Hg^{2+} ions in tap water samples, with a detection limit of 2.8 nM [16].

Chemiluminescence (CL) methods have several advantages over other modes of analysis, including high sensitivity, ease of instrumentation, low detection limits, wide calibration ranges, and short analysis times, making them suitable for environmental analysis. S, N-CQDs can be synthesized using a hydrothermal method using citric acid and L-cysteine to develop CL-based sensors capable of detecting even trace amounts of indomethacin in environmental water and biological samples [17].

The synthesis of copper ion modified CQDs can be done via a microwave-assisted hydrothermal route for drug sensing, which is based on the quenching effect. Wilson's disease (hepatolenticular degeneration), rheumatoid arthritis, primary biliary cirrhosis, cystinuria, scleroderma, progressive systemic sclerosis, and heavy metal poisoning are all treated with D-Penicillamine (DPA). DPA has few side effects at high concentrations, such as myasthenia, ageusia or dysgeusia rash, pemphigus, thrombocytopenia, and so on. Cu^{2+} ions bind to DPA selectively and are released from the surface of CQDs due to their high affinity for DPA [18].

The high energy of an ultrasonic wave is used to crack carbon materials into NPs in the presence of acid, alkali, or oxidants in the synthesis of CQDs using ultrasonic-assisted methods. The synthesized CQDs proved to be an effective fluorescent-sensing probe for the selective detection of Zn^{2+} in aqueous solution [19]. The oxygen-rich groups on the surface of CQDs can be changed using additional materials, which has important applications in sensing and catalysis.

Ming et al. devised a fluorescence sensor for precise detection of heavy metal ions utilizing a simple and low-cost nitrogen doped N-CQDs produced by a one-step hydrothermal technique using thymidine and ethylenediamine as carbon and nitrogen sources, respectively [20]. Lu et al. described a 3D paper-based analytical device for detecting Cr^{6+} in environmental water samples. Carbon dots were prepared through microwave-induced decomposition of precursor materials such as citric acid and N,N′-bis(2-aminoethyl)-1,2-ethanediamine. For ease of use, the C-dots were placed on portable paper strips with novel origami designs [21]. Qiao et al. developed a novel water-soluble CQDS for the detection of $Cr_2O_7^{2-}$ in drinking water using 1,3-phenylenediamine and citric acid as precursors in a one-pot ultrasonic irradiation method [22]. CQDs are also employed for Hg^{2+} detection on numerous occasions. Guo et al. used hydrothermal treatment to prepare CDs from sodium citrate/citric acid for the selective and sensitive detection of Hg^{2+} ions [23]. Though metallic gold is not toxic to humans in and of itself, gold ions may be toxic to the human body. Raji et al. prepared highly luminescent N-doped carbon dots (N-CDs) from jackfruit seeds as a green carbon source using a microwave-assisted method, and the prepared N-CDs were used for the fluorimetric detection of Au^{3+} ions [24].

Fu et al. synthesized novel CQDs co-doped with nitrogen, oxygen, and phosphorous co-doped CDs (NOP-CQDs) using a one-step hydrothermal method and used them to detect ovalbumin (OVA) in egg products [25]. Yang et al. reported the hydrothermal treatment of malic acid and urea to produce water-soluble CQDs that were used for the fluorescent detection of chlorogenic acid in honeysuckle [26].

Curcumin is an acidic polyphenolic compound derived from ginger roots. It is used as a yellow pigment in meat colouring agents and as an acid-base indicator. It also has anti-inflammatory, anti-oxidant, and other medicinal properties. Liu et al. developed N and P dual-doped CQDs (NP-CQDs) with glucose as the carbon source, 1,2-ethylenediamine as the N-dopant, and concentrated phosphoric acid as the P-dopant, which were used as a label-free sensor for Curcumin determination [27]. Guo et al. developed an electrochemical sensor (MIP-Au/CS-CQDs/GCE) for the detection of patulin in fresh apple juice by electrodepositing a molecularly imprinted polymer membrane (MIP) on the glassy carbon electrode (GCE) using a dummy template 2-oxindole [28]. By encapsulating silicon-based CQDs in MIP material, Shao et al. created molecularly imprinted fluorescent quenching particles for sensitive detection of zearalenone in corn (LOD = 0.02 mg L^{-1}) [29].

CQDs were synthesized using a simple one-step microwave irradiation method and were used as electrodes in the development of an electrochemical sensor. In the presence of dopamine and uric acid, the electrochemical sensor demonstrated good selectivity and strong anti-interference ability toward ascorbic acid (AA). As a result, this sensor is a viable tool for detecting AA in real complex systems [30].

7.2.2.4 Preparation of Graphene Quantum Dots (GQD) for Sensor Applications

GQD-sensors based on fluorescence are used to quantify metal ions such as Pb^{2+}, Ag^{+}, Hg^{2+}, Fe^{3+} and Cu^{2+} ions. These sensors have the potential to detect metal ions in water samples as well as to be used in clinical diagnosis. The detection of Fe^{3+} ion levels in human serum can aid in the diagnosis of cancer. Because of its dimensional similarity to antibodies, proteins, and small nucleic acids, it can also be grafted with these molecules. They have the ability to effectively augment the surface of biosensors in order to absorb a large number of receptors. GQDs are used in the development of nucleic acid-based sensors, immunosensors, enzyme-based sensors and MIP-based sensors for biomedical applications such as cardiovascular, autoimmune, and cancer diagnosis, neurodegenerative disease monitoring, pathogen, metal ion, toxin and glucose level detection [31]. Tyramine-functionalized GQDs-based sensors have been reported to screen metabolites (such as glucose, L-lactate, cholesterol, and xanthine) for the detection of various metabolic disorders such as diabetes, lactic acidosis, obesity, gout and hypertension [32].

Li et al. used thermal pyrolysis to create a novel biosensor pentaethylenehexamine and histidine-functionalized graphene quantum dots (PEHA-GQD-His) for detection of biothiols (e.g., GSH, cysteine, or homocysteine) in human serum for cancer diagnosis [33].

Surface modification and element doping with heteroatoms like boron, nitrogen and phosphorous are two methods for improving the bandgap, fluorescence efficiency, and quantum yield of GQDs. Nitrogen (N) functionalization and doping are thought to be extremely useful in modifying the intrinsic properties of GQDs due to valence electrons. In graphene, the integration of electron-donating N atoms causes the neighboring C atoms to have a comparatively high positive-charge density.

The acute administration of angiogenesis pharmaceutical compound such as sunitinib is associated with some toxicities. For estimating sunitinib in real samples, N –GQDs was successfully obtained through one-step pyrolysis of citric acid (CA) as a carbon source and tris(hydroxymethyl) aminomethane (THMA) as a surface passivation agent [34].

Carcinoembryonic antigen (CEA) levels are low in healthy human blood, but when cancer cells form, a concentration of CEA greater than 5 ng mL^{-1} can be detected. The combination of N, S-GQDs and Au-PANI can form a novel nanocomposite with unique properties that individual components lack. Furthermore, the incorporation of N, S-GQDs onto Au-PANI nanocomposites provides the impetus for the development of a label-free immunosensor for ultra-sensitive detection of cancer makers, which can significantly improve the electrochemical and sensing ability toward CEA detection [35].

7.2.3 Conducting Polymer (CP) Materials for Sensor Applications

Since 1977, when polyacetylene, the first CP, was found and reported on in a journal, the field of CPs research has grown and progressed at an unprecedented fast pace. Conducting polymers are one-of-a-kind photonic and electronic functional materials having a unique conducting mechanism as well as reversible oxidation and reduction characteristics. Supercapacitors, solar cells, sensors, actuators, lithium ionic batteries, transistors, electrocatalyst [36] and reusable catalyst [37] are just a few of the applications for CPs.. Because of the unique properties resulting from their nanoscaled size, conducting polymers with nanostructures are expected to perform well in electronic devices. As a result, over the last few decades nanostructured conducting polymers have received a lot of attention in both fundamental research and various application fields.

7.2.3.1 Synthesis of Conducting Polymers (CPs) for Sensor Applications

Conducting polymers, namely, Polyaniline (PANI), Polypyrrole (PPy) and poly(3,4-ethylenedioxythiophene) (PEDOT) and its composites finds potential application as gas sensors, biosensors due to its electrical properties being tunable with ease [38]. The most common methods to synthesize these CPs are in situ polymerization, chemical oxidation and electro chemical polymerization methods [39]. To detect ammonia gas, Chen et al. produced PANI via oxidative polymerization of aniline using ammonium persulfate as an oxidizing agent in an acidic media. The primary mechanism involved is the deprotonation of PANI by ammonia gas, which resulted an increase in resistance [40]. Nano composites of PANI/ZnO synthesized by electrospinning was used to detect HCl and ammonia vapours at room temperature. When exposed to HCl vapors, the resistance of the sensor decreased, whereas it increased when exposed to ammonia vapours [41].

Du et al. developed humid sensors uniting nanostructured polyaniline with carbon nano fibers (CNF). A mechanochemical method was used to functionalize CNF in PANI [42]. Pawar et al. synthesized PANI/g-Fe_2O_3 nanocomposite sensor to detect ammonia gas [43]. Bao et al. functionalized GQDs with PANI synthesized by in situ polymerization. The sensor that was fabricated was utilized to analyze calycosin in complex samples of traditional Chinese medicine [44]. Srivastava et al. developed a PANI/g-Fe_2O_3 nanocomposite sensor to detect LPG at room temperature, which demonstrated a response with traces of 200 ppm LPG [45].

Abdulla S. et al. used carbon nanotubes (CNT) to enhance the flexibility and durability of PANI as well as the number of active sites, which improves intra- and inter-chain charge mobility along the polymer chain in the presence of electron donor or acceptor gases. PANI/ multiwalled CNT (MWCNT) nanocomposites-based sensitive gas sensors for detection of trace-level NH_3 gas [46]. For hydrogen sensing, sensors based on MWCNT and single-walled CNT doped PANI composite thin films perform better than pure PANI [47].

PPy has a high sensitivity to strain and temperature and is commonly used as a gas sensor material [48]. Tiwari et al. was used in situ oxidative polymerization to create a chemiresistive PPy/rGO sensor for NH_3 detection. The sensor's response was three times better than that of a pure PPy thin film sensor [49]. Tang et al. prepared two hybrid sensors, PPy/CVD-G and PPy/rGO, in which an ultrathin PPy layer was electropolymerized on CVD-G and rGO, respectively, to detect ammonia vapors [50]. Electropolymerization was used to create a PPy molecularly imprinted polymeric membrane for an electrochemical sensor to detect herbicide glyphosate in cucumber and tap water [51]. PEDOT and its composites are used as biosensors and chemical sensors. Zhang et al. [52] developed a PEDOT-based biosensor with high sensitivity and promising selectivity for detecting BRCA1 in breast cancer. By incorporating silver nanowire into a PEDOT/PSS composite film, Li et al. reported a flexible electrochemical ammonia sensor. To monitor the freshness of pork, the sensor was combined with a self-designed portable data acquisition system [53].

7.2.4 Organic-Inorganic Hybrid Nanomaterials for Sensor Applications

7.2.4.1 Perovskites as Sensors

Amidst the specified nanomaterials, Perovskites are remarkable hybrid materials with wide applications, such as light-emitting devices, solar cells, sensors, transistors and so forth (Figure 7.5). These are compounds that have an ABX_3 formula with different size cations 'A' and 'B' and anion 'X' [54]. Inorganic Perovskites with the general formula ABO_3 have good thermal stability and a band gap of 3–4 eV and are used for electrochemical sensing of acetone, alcohols, amino acids, gases, H_2O_2 and glucose. The interaction of analyte gases with oxygen at the grain boundaries of perovskite causes a fluctuation in electrical conductivity, excellent reproducibility, sensitivity, unique long-term stability and anti-interference ability. Semiconducting properties allows the detection of gaseous species by means of current–voltage (I–V) responses. The cobaltates, titanates, and ferrites of perovskites were used as gas sensors for sensing CO, NO_2, methanol, ethanol, and hydrocarbons. Perovskite oxides like $LaFeO_3$, $SrTiO_3$, $BaTiO_3$, $CdTiO_3$ are fascinating materials used as gas sensors for their perfect band gap, thermal stability, and size variation in cations of B and A sites in perovskite.

Metal halide perovskite materials have the benefit of providing responses to analyte in solution and solid states. The different crystalline structures of metal halide perovskites play a vital role in sensory studies to detect the following molecules such as CO_2, O_2, NO_2, and so forth. The stability of these sensors depends on several factors, such as solvents, time, temperature and moisture. Precautions are needed while designing these sensors due to disruption in crystallinity, structure and morphology. Due to the stability concern, recycling of these nanomaterials are still challenging in electronic device-based sensors. Many semiconducting perovskites gas sensors are applied in environmental, fire and vehicle monitoring.

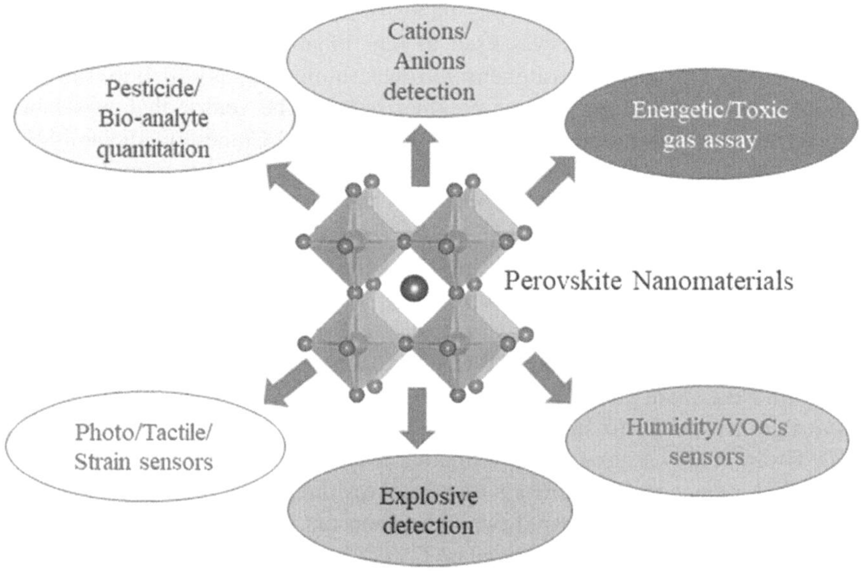

FIGURE 7.5 Schematic illustration of the sensory applications of perovskite nanomaterials (figure is reproduced from Ref. [54]).

Because of their distinctive fluorescence and solvent-tuned aggregation induced emission (AIE) features, organometal halide perovskites are also useful in sensory research. They find wide application as humid sensors and are also used to discriminate VOC gases and toxic gases such as ammonia. One of the pioneer works that have initiated the sensing applications of organometal halide perovskite ($CH_3NH_3PbBr_3$) is found to be in the detection of picric acid.

7.2.4.2 Synthesis of Perovskites Materials for Sensor Applications

Various techniques are used in synthesizing perovskites (nano-) materials. Sol–gel, impregnation, solvothermal, electrospinning and precipitation methods are some of the most promising and efficient methods. Among them the chlorine-doped nanocrystalline $LaFeO_3$ powders synthesized by the sol-gel method and investigated toward ethanol gas-sensing via resistance change [55]. Many metal oxide perovskites are investigated for volatile organic compounds (VOC) sensing applications. Ag-$LaFeO_3$ nanoparticles synthesized from the lotus leaf bio-templated sol–gel process was used to sense Xylene gas. This p-type sensor can work at 125 °C with good humidity and lasting stability [56]. Many metal perovskite nanostructures have also been investigated for flammable/toxic gas detection. Bala and coworkers prepared Ca-doped $BiFeO_3$ using the sol–gel approach to detect flammable H_2 gas sensing. Doping with 15 percent Ca^{2+} in $BiFeO_3$ displayed a good sensor response at 250 °C. due to increase in vacant oxygen sites [57]. Humidity measurement is one of the utmost important issues in many areas of applications such as instrumentation, automated systems, agriculture, climatology, and the Geographic Information System (GIS). Molybdate-, stanate- and titanate-based perovskites constitute an important class of materials used in this area of application. Metal oxide perovskites like Magnesium- (Mg) or Strontium- (Sr) doped nanocrystals of $LaFeO_3$ and $LaMnO_3$ were prepared by the sol-gel method and reported as humidity sensors [58] [59].

Formaldehyde is a biotoxic VOC. Ag-doped $LaFeO_3$ nanofibers. developed by Wei et al. via the electrospinning method was used in formaldehyde (HCHO) gas-sensing studies [60].

Yin et al. synthesized hollow $ZnSnO_3$ nanocubes via solvothermal strategy and functionalized with silver NPs. They have evaluated the sensing performance of Ag functionalized $ZnSnO_3$ nanocubes toward acetone vapor [61].

Organometal halide nanostructured perovskites were also effectually used in the discernment of VOCs and metal ions by means of photoluminescence (PL) and resistance fluctuations. Sheng et al. confirmed the metal ions sensing capability of stable colloidal form of cesium lead halide $CsPbX_3$ (X = Cl, Br or I) perovskites quantum dots via detecting changes in fluorescence. Metal ions play a substantial role in biological and industrial systems, for example, Cu^{2+} and rare earth ions in moderate amounts are important for a healthy life; large amounts or high exposure can be toxic and harmful. It is generally found in many common substances such as petrol, edible oils, skin oils, or lubricating oils [62].

Apart from the above-mentioned synthetic methods, there are various other routes, such as heat-injection processes, solid state mechanical grinding and milling, ligand assisted re-precipitation (LARP) methods, emulsion-LARP, reverse microemulsion – and other indirect methods are available to prepare the inorganic perovskite materials [63–65] and these methods may be utilized to prepare the sensor materials. The heat-injection method is built on the speedy injection of a precursor into a hot solution of the residual precursors, ligands, and a high boiling solvent. The latter method is easy to adopt, sustainable and energy-efficient. The foremost benefit of mechanochemical synthesis is that no solvent is required. This is one of the important aspects of synthesis in green chemistry.

There are several advantages to mechanochemical synthesis, such as exceptional opto-electronic properties and electronic properties of both metal oxide and metal halide/organometallic halide perovskites, allowing them to be used in device-based analyte detection, particularly for humidity and toxic gas quantitation, but there are drawbacks, like reliability and toxicity. Therefore, it still poses challenges in biological and clinical diagnosis. Hence, perovskite-based sensory research still needs a lot of studies and findings with reference to their applicability and future scope.

7.2.4.3 Metal Organic Frameworks (MOFs) as Sensors

Metal organic frame works (MOF) come under the category of crystalline and porous materials. Their distinctive properties have inspired active research for prospective use in gas-sensing, bio-sensing, and chemo-sensing [66]. The judicious self-assembly of organic ligands and inorganic metal ions or clusters enables myriad infinite networks of MOFs of varied physical and chemical properties. The exceptional tunability of MOF structures and properties has a great advantage over other conventional chemo-sensory materials. Any changes in the MOF properties is measured as a sensing signal. Sensing properties are based on signal transduction, which is any change in electrical properties, such as impedance, capacitance, resistance or work function caused by interaction with a target analyte. The nature of the interactions between the target analytes and the active sites of MOFs influences the response time, regeneration, and recyclability of MOF-based sensors [67]. Specific interactions and analyte affinity of a MOF can cause changes in the electrical properties of the MOF layer, which can be precisely determined. As a result, MOF-based sensors have enormous potential for developing strong analytical techniques for quantifying biomolecules in clinical, environmental, and industrial applications.

The functionalization of MOFs is necessary to achieve specific targets. Doping metal ions into the frame of MOF, modifying specific organic ligand, post synthesis modification, entrapping functional groups are the various strategies to functionalize the MOFs to achieve specific targets. Hence, MOFs are considered to be best materials for sensing applications. The advantage of using MOFs as sensors over metal oxides and polymers lies in the preconcentration step that can be totally avoided as MOFs can play the dual role of sensor and adsorbent, thus easing the highly desired portable gas detection systems, whereas, in metal oxides and polymers, in the preconcentration step, target gas is passed through an adsorbent. Following that, the adsorbent is heated to a specific temperature in order to desorb the adsorbate (target gas), allowing the sensor to detect the concentrated target gas. As a result, MOS are regarded as a viable and promising option for electronic gas-sensing applications.

7.2.4.4 Synthesis of MOFs for Sensor Applications

Slow diffusion, hydro or solvothermal, mechanochemical, electrochemical, micro-wave assisted heating, one-pot and ultrasonic methods are various approaches of synthesis [68]. However, selection of suitable synthetic methods is according to the type of sensors. Here, we have discussed a few reported MOFs synthetic methods for sensor applications.

Zeolitic imidazolate frameworks (ZIFs) of MOFs are topologically isomorphic with Zeolites. Very recently, Polyhedral ZIF-8 MOF nanostructures for NO_2 gas-sensing applications were reported by Zhan et al. They have prepared ZIF-8 using solvothermal methods with high selectivity toward NO_2 gas-sensing application [69].

A solvothermal process was used to create In_2O_3/ZIF-8 composite fibers with core-shell structure using In_2O_3/ZnO nanofibers (NFs) as a template and Zn^{2+} as a source to coordinate with 2-methylimidazole in a DMF solution. In_2O_3/ZIF-8 NFs showed a remarkably higher response to relatively low concentration of NO_2 (1 ppm) than the pristine In_2O_3 NFs sensor at 140 °C and displayed an increased sensing response down to 10 ppb [70].

The fabrication and deployment of a selective MOF thin film as an improved chemical capacitive sensor for the sensing/detection of NH_3 at room temperature was reported by Assen and Yassine et al. The solvothermal growth approach was used to build a thin film over a pre-functionalized capacitive interdigitated electrodes (IDE) using a naphthalene-based rare earth-fcu-MOF (NDCY-fcu-MOF) [71].

The first ultrasmall zero-dimensional Zn-MOF-74 [Zn_2(DOBDC), DOBDC=2,5-dihydroxyterephthalic acid] nanodots with the average size within 10 nm was prepared by Wang et al. via size-controlled synthesis by controlling the initial conditions with a diluted material system. This ultra-small MOF nanodots have a highly selective interaction with Fe^{3+} and displayed an explicit blue colorimetric variation in aqueous solution [72].

Zr(IV)-based metal–organic framework (MOF) NH_2-UiO-66 was synthesized through a solvothermal method and their nanocomposite was prepared with reduced graphene oxide (RGO) by ultrasonication followed by copper deposition. This nanocomposite was showing a large surface area with porous structure and high electrical conductivity, which was used as a working electrode to electrochemically detect the ciprofloxacin [73].

Zeinali et al. reported an MOF thin film capacitive humidity sensor was fabricated through electrochemical in situ growth. Electrochemical in situ synthesis and film growth were used to create a nanoporous thin film of MOF, Cu–BTC [1,3,5-benzenetricarboxylate or trimesate (BTC)] on the copper plate electrode as the capacitive sensor's dielectric layer. In this procedure, 1-methyl-3-octylimidazolium chloride ionic liquid (IL) used as conducting salt [74].

The reason for MOFs being used as chemical sensing in aqueous solutions is due to its luminescence property. Luminescence is the phenomenon that occurs when electrons in the excited singlet state return to the ground state via photon emission. Luminescence in MOF is attenuated or quenched upon analyte absorption, which is known as the "turn-off" mechanism. In stark contrast, there are reports that substantiate luminescence enhancement, described as a "turn-on" mechanism. The analysis of analytes can be determined through the aforesaid mechanisms. MOF-based materials are in high demand as multifunctional luminescent materials because both organic and inorganic moieties, as well as foreign moieties, can provide a platform for the generation of a luminescent signal [75].

Lin et al. [76] testified a fluorescent MOF-based probe for copper ion (Cu^{2+}) detection. The encapsulation of branched poly (ethylenimine)-capped CQDs into a zeolitic imidazolate framework (ZIF8) displayed excellent selectivity for sensing Cu^{2+} ions.

Mercury ion (Hg^{2+}) detection is of prime importance as it poses risks to human health and toxicity. A quick and selective sensing approach for detection of Hg^{2+} based on Ru-MOFs was developed by the Chi group [77]. Chi et al. doped the luminescent $Ru(bpy)_3$ in an Ru-MOF framework and encapsulated in the pores of MOF. Under UV light dark yellow powder Ru-MOFs emitting red in color, in the presence of Hg^{2+} and in water the Ru-MOFs were swiftly split, and they released huge volumes of luminescent guest materials ($Ru(bpy)_3^{2+}$) into the water and the water phase turned into yellow. The resulting yellow supernatant showed red light emission under UV light.

7.3 SENSOR DEVICE FABRICATIONS

A sensor fabrication process involves many considerations, and it is an interdisciplinary challenge. The overall sensor design can be simplified, and a broader range of materials can be used in sensor fabrication. Generally, the sensor fabrication process will begin with selection of starting materials and have many integration consequences in later fabrication steps. This chapter gives basic information about the sensor device fabrications techniques for the gas sensors.

Gavgani et al., [78] fabricated a highly efficient room temperature operable volatile organic compounds (VOCs) sensor system from nitrogen-doped graphene quantum dots (N-GQDs) and poly(3,4-ethylenedioxythiophene)–poly(styrenesulfonate) (PEDOT–PSS). Interdigitated Au electrodes (100 nm thickness) were deposited on a SiO_2/Si substrate by a physical vapor deposition method during the fabrication process of an N-GQDs/PEDOT–PSS nanocomposite gas sensor (Figure 7.6a–c). The prepared N-GQDs/PEDOT–PSS nanocomposite solution was drop-casted over an interdigitated electrode with a width of 200 mm and interspacing of 400 mm (Figure 7.6d). Later authors subjected the solution to a baking process for 1 h in a furnace at 80 °C under nitrogen (N_2) atmosphere. The fabricated N-GQDs/PEDOT–PSS nanocomposite gas sensor is displayed in Figure 7.6e and 7.6f and showed improved performance, fast response and recovery behavior indicates a promising application of GQDs for sensor applications.

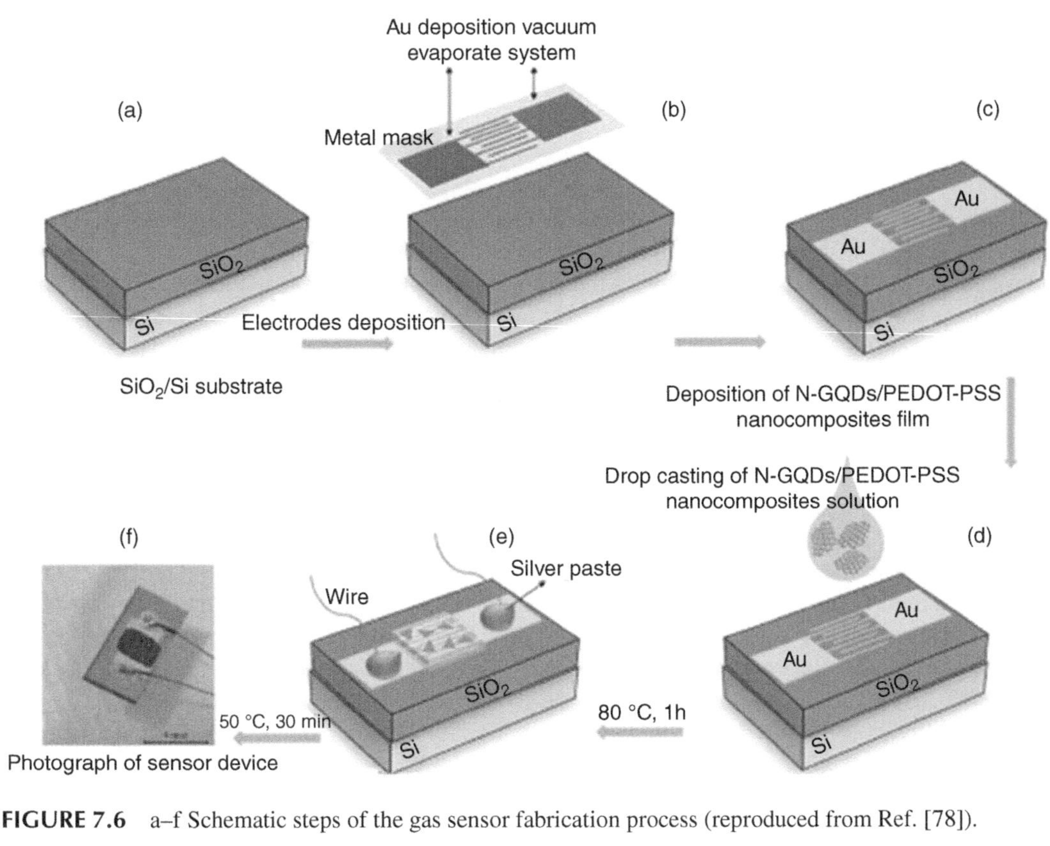

FIGURE 7.6 a–f Schematic steps of the gas sensor fabrication process (reproduced from Ref. [78]).

7.4 SENSING MECHANISM

Gas sensors are of great significance in this day and age due to their widespread use in industrial production and daily life. In this part, ammonia (NH_3) gas-sensing mechanisms of functionalized graphene with conducting polymers are explained in detail, and the scope and relationship of the mechanisms are clarified.

The mechanism that nanomaterials deploy for gas detection is quite convoluted to explain. There are various factors that add to the complexity of the gas-sensing mechanism. Some of these factors include chemical composition, temperature, humidity and surface morphology. The sensing mechanism can be derived from the adsorption ability, surface properties and catalytic effect of the active materials. The adsorption theory explains the gas-sensing mechanism [79]. In line with the nature of the action of the surface property of the active materials on the gas molecules, adsorption can be classified as physical and chemical adsorption. Physical adsorption is attributed to the van der Waals force, which is relatively weak. The surface area of the active materials and the structure of the adsorbent determine the degree of physical adsorption. The sensitivity and response time of the sensors is attributed to the interaction between the analyte and the nanomaterials. In other words, the gas-sensing performance of the sensor is a measure of the gas adsorption property of the gas-sensitive film.

In gas-sensing analysis, the resistance of the rGO-PPy hybrid film improved when exposed to NH_3 gas, which is consistent with the sensing properties of p-type semiconductors (here it is PPy). The detecting mechanism of PPy to NH_3 is proposed to be a deprotonation/protonation process via NH_3 adsorption/desorption on the surface of the rGO-PPy hybrid film (equations 7.1 and 7.2).

Adsorption,

$$PPy^{+} + NH_3 \rightarrow PPy^{0} + NH_3^{+} \tag{7.1}$$

Desorption,

$$PPy^{0} + NH_3^{+} \rightarrow PPy+ + NH_3 \tag{7.2}$$

While the rGO-PPy hybrid film was exposed to NH_3 gas, the following reaction happens between NH_3 gas and PPy. During the interaction, the doublet of nitrogen in NH_3 loses an electron to the N (nitrogen) of the PPy backbone and forms the ammonium ion (NH_3^{+}) due to electron transfers from NH_3 to PPy, which leads to a de-doping or deprotonation process in PPy, which increases PPy resistance [49]. After the NH_3 gas stops flowing and the hybrid film is refreshed with air, the resistance value of the sensor can be totally or partly restored. In addition, the sensing response of PPy-rGO hybrid to NH_3 depends upon several other possible factors. The high electron mobility of graphene at room temperature and greater surface area contribute to the fast carrier transport of hybrid and adsorption process. Thirdly, The electron transfer due to possible π–π stacking and H-bonding exist in the PPy-rGO hybrid film, shown in Figure 7.7, also improves the sensing performance [80]. The exposure of NH_3 gas to PPy-rGO hybrid disrupts the original H-bonding between PPy back bond and rGO results in the increase of resistance [81].

As discussed above, the sensor responses depend on their resistance change caused by electrons transfer between NH_3 and the related sensing layers. High conductance of graphene provides an efficient pathway for electrons transfer, thereby quickening the response and recovery of the sensor. In general, various functional materials perform a crucial role in the sensitivity and selectivity of the sensors. In Table 7.1, the most widely used functional materials, their sensing mechanisms and functionalization methods are listed [82].

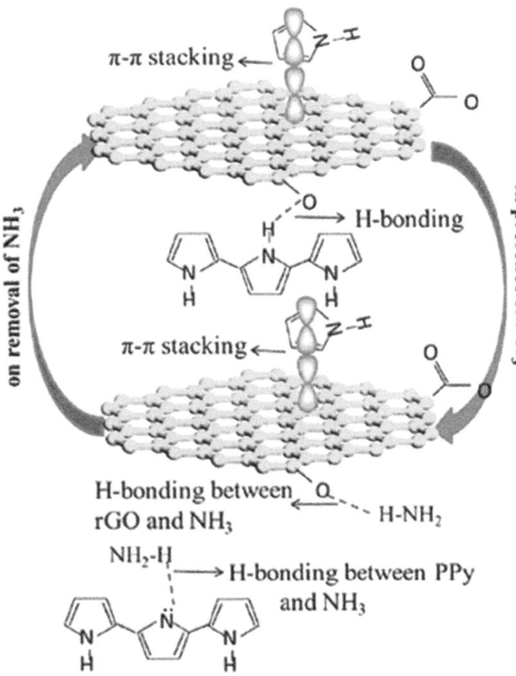

FIGURE 7.7 Schematic presentation depicting interaction of NH_3 with PPy/rGO hybrid (figure is reproduced from [81]).

7.5 OUTLOOK

In this book chapter, we have discussed the synthesis of nanostructured materials, importance of nanomaterials in various types of sensor devices, sensor device fabrication and their sensing mechanism. We focused mainly on 2D materials, TMDs, carbon and their allotropy materials especially CQD and GQD, CPs, Perovskite and MOFs based nanostructured materials for a range of sensors. Among the various nanostructured materials discussed here, carbon-based QDs (CQDs and GQDs) had widespread utilization in constructing the sensors. Because QDs exhibit greater improvements in sensor sensitivity, selectivity, and stability. Many efforts have been done by scientists and technologists in synthesis and fabrication techniques to enhance the sensitivity, selectivity and reversibility of the sensor materials. However, the following points must be addressed: (1) a cost-effective, reproducible, and environmentally friendly procedure is required to produce nanomaterial-based sensors for specific sensor applications (gas, chemical, and biosensors). (2) There are only a few reports on biocompatible and biodegradable sensors, hence more efforts in this research area are required, which may lead to the main research direction of "green electronic" products.

ACKNOWLEDGMENTS

The authors thank Presidency University, Bengaluru management for their motivation and support. Author Saravanan is grateful to the university for providing financial support through a seed grant (File No: (RI&C/Funded Projects/Seed grant/2021/8).

TABLE 7.1
Main Sensing Mechanisms of Functionalized Graphene NH_3 Sensors and Graphene Functionalization Methods Based on Metallic Nanoparticles, Metal Oxides, Organic Molecules, and Conducting Polymers [82]

Functional Materials	Sensing Mechanisms	Functionalization Methods
Metallic nanoparticles: Au Ag Pt Pd	Sensor response depends on electrons transfer from NH_3 to Graphene or rGO. Metallic nanoparticle acts as a catalyst to increase the reaction between Graphene or rGO and NH_3.	Electrochemical method Hydrothermal reduction Physical vapor deposition Layer-by-layer self-assembly Electrostatic interactions Galvanic replacement reaction
Metal oxides: SnO_2 V_2O_2 ZnO TiO_2 Cu_2O WO_3 In_2O_3	Sensor response depends on electrons transfer from NH_3 to C-O-M- bonds. Metal oxide acts as a predominating NH_3 receptor, and G or rGO accelerates sensor response and recovery.	Thermal reduction Hydrothermal reduction Precipitation Electrospinning One-pot polyol Pulse laser deposition
Organic molecules: Rose Bengal (RB) nanocomposite Bromophenol blue (BPB) Co-porphyrin	Sensor response depends on electrons transfer from NH_3 to Graphene or rGO. Functional groups in RB act as extra active sites and facilitate more binding of NH_3. Sensor response depends on electrons transfer from NH_3 to Graphene or rGO. Protonated acidic rings and electrophilic protons in BPB act as NH_3 attractors (electrons transferring between BPB and Graphene or rGO via coupling π bonds).	Drop casting _ coupling Layer-to-layer stacking π coupling Spin coating
Polymers: Polypyrrole Polyaniline Others (PTh, PBuA, and PVDF)	Sensor response depends on electrons transfer from NH_3 to polymers via its conjugated bonds. Graphene or rGO provides an efficient pathway for electron transfer, accelerating sensor response and recovery.	Electropolymerization Electrochemical Chemical Pyrrole reaction

G: graphene, rGO: reduced graphene oxide, M: metal, PTh: polythiophene, PBuA: poly (butyl acrylate), PVDF: poly (vinylidene fluoride).

REFERENCES

1. Kannan, P.K., Late, D.J., Morgan, H., Rout, C.S. Recent developments in 2D layered inorganic nanomaterials for sensing. *Nanoscale* **2015**, *7*, 13293–13312, doi:10.1039/c5nr03633j.
2. Sarkar, D., Liu, W., Xie, X., Anselmo, A.C., Mitragotri, S., Banerjee, K. MoS2 field-effect transistor for next-generation label-free biosensors. *ACS Nano* **2014**, *8*, 3992–4003, doi:10.1021/nn5009148.
3. Parlak, O., İncel, A., Uzun, L., Turner, A.P.F., Tiwari, A. Structuring Au nanoparticles on two-dimensional MoS2 nanosheets for electrochemical glucose biosensors. *Biosens. Bioelectron.* **2017**, *89*, 545–550, doi:10.1016/j.bios.2016.03.024.
4. Lee, C.W., Suh, J.M., Jang, H.W. Chemical Sensors Based on Two-Dimensional (2D) Materials for Selective Detection of Ions and Molecules in Liquid. *Front. Chem.* **2019**, *7*, doi:10.3389/fchem.2019.00708.

5. Su, S., Sun, H., Xu, F., Yuwen, L., Fan, C., Wang, L. Direct electrochemistry of glucose oxidase and a biosensor for glucose based on a glass carbon electrode modified with MoS2 nanosheets decorated with gold nanoparticles. *Microchim. Acta* **2014**, *181*, 1497–1503, doi:10.1007/s00604-014-1178-9.
6. Dey, A. Semiconductor metal oxide gas sensors: A review. *Mater. Sci. Eng. B Solid-State Mater. Adv. Technol.* **2018**, *229*, 206–217, doi:10.1016/j.mseb.2017.12.036.
7. Zappa, D., Galstyan, V., Kaur, N., Munasinghe Arachchige, H.M.M., Sisman, O., Comini, E. Metal oxide -based heterostructures for gas sensors – A review. *Anal. Chim. Acta* **2018**, *1039*, 1–23, doi:10.1016/j.aca.2018.09.020.
8. Sánchez-López, E., Gomes, D., Esteruelas, G., Bonilla, L., Lopez-Machado, A.L., Galindo, R., Cano, A., Espina, M., Ettcheto, M., Camins, A., et al. Metal-based nanoparticles as antimicrobial agents: An overview. *Nanomaterials* **2020**, *10*, 292, doi:10.3390/nano10020292.
9. Sivapunniyam, A., Wiromrat, N., Myint, M.T.Z., Dutta, J. High-performance liquefied petroleum gas sensing based on nanostructures of zinc oxide and zinc stannate. *Sensors Actuators, B Chem.* **2011**, *157*, 232–239, doi:10.1016/j.snb.2011.03.055.
10. Xiang, Q., Meng, G., Zhang, Y., Xu, J., Xu, P., Pan, Q., Yu, W. Ag nanoparticle embedded-ZnO nanorods synthesized via a photochemical method and its gas-sensing properties. *Sensors Actuators, B Chem.* **2010**, *143*, 635–640, doi:10.1016/j.snb.2009.10.007.
11. Pratima Chauhan and Shiva Sharma Nanomaterials for Sensing Applications. *J. Nanomedicine Res.* **2016**, *2*, 00067.
12. Tajik, S., Dourandish, Z., Zhang, K., Beitollahi, H., Le, Q. Van, Jang, H.W., Shokouhimehr, M. Carbon and graphene quantum dots: A review on syntheses, characterization, biological and sensing applications for neurotransmitter determination. *RSC Adv.* **2020**, 10, 15406–15429, doi:10.1039/d0ra00799d.
13. Pan, M., Xie, X., Liu, K., Yang, J., Hong, L., Wang, S. Fluorescent carbon quantum dots-synthesis, functionalization and sensing application in food analysis. *Nanomaterials* **2020**, *10(5)*, 930, doi:10.3390/nano10050930.
14. Zhu, S., Song, Y., Wang, J., Wan, H., Zhang, Y., Ning, Y., Yang, B. Photoluminescence mechanism in graphene quantum dots: Quantum confinement effect and surface/edge state. *Nano Today* **2017**, *13*, 10–14, doi:10.1016/j.nantod.2016.12.006.
15. Li, H., Sun, C., Vijayaraghavan, R., Zhou, F., Zhang, X., MacFarlane, D.R. Long lifetime photoluminescence in N, S co-doped carbon quantum dots from an ionic liquid and their applications in ultrasensitive detection of pesticides. *Carbon N. Y.* **2016**, *104*, 33–39, doi:10.1016/j.carbon.2016.03.040.
16. Wang, B.B., Jin, J.C., Xu, Z.Q., Jiang, Z.W., Li, X., Jiang, F.L., Liu, Y. Single-step synthesis of highly photoluminescent carbon dots for rapid detection of Hg 2+ with excellent sensitivity. *J. Colloid Interface Sci.* **2019**, *551*, 101–110, doi:10.1016/j.jcis.2019.04.088.
17. Hallaj, T., Amjadi, M., Manzoori, J.L., Azizi, N. A novel chemiluminescence sensor for the determination of indomethacin based on sulfur and nitrogen co-doped carbon quantum dot–KMnO4 reaction. *Luminescence* **2017**, *32*, 1174–1179, doi:10.1002/bio.3306.
18. Naghdi, T., Atashi, M., Golmohammadi, H., Saeedi, I., Alanezhad, M. Carbon quantum dots originated from chitin nanofibers as a fluorescent chemoprobe for drug sensing. *J. Ind. Eng. Chem.* **2017**, *52*, 162–167, doi:10.1016/j.jiec.2017.03.039.
19. Qiang, R., Yang, S., Hou, K., Wang, J. Synthesis of carbon quantum dots with green luminescence from potato starch. *New J. Chem.* **2019**, *43*, 10826–10833, doi:10.1039/c9nj02291k.
20. Ming, F., Hou, J., Hou, C., Yang, M., Wang, X., Li, J., Huo, D., He, Q. One-step synthesized fluorescent nitrogen doped carbon dots from thymidine for Cr (VI) detection in water. *Spectrochim. Acta - Part A Mol. Biomol. Spectrosc.* **2019**, *222*, 117165, doi:10.1016/j.saa.2019.117165.
21. Lu, K.H., Lin, J.H., Lin, C.Y., Chen, C.F., Yeh, Y.C. A fluorometric paper test for chromium(VI) based on the use of N-doped carbon dots. *Microchim. Acta* **2019**, *186*, 227, doi:10.1007/s00604-019-3337-5.
22. Qiao, G., Lu, D., Tang, Y., Gao, J., Wang, Q. Smart choice of carbon dots as a dual-mode onsite nanoplatform for the trace level detection of Cr2O72-. *Dye. Pigment.* **2019**, *163*, 102–110, doi:10.1016/j.dyepig.2018.11.049.
23. Guo, Y., Wang, Z., Shao, H., Jiang, X. Hydrothermal synthesis of highly fluorescent carbon nanoparticles from sodium citrate and their use for the detection of mercury ions. *Carbon N. Y.* **2013**, *52*, 583–589, doi:10.1016/j.carbon.2012.10.028.

24. Raji, K., Ramanan, V., Ramamurthy, P. Facile and green synthesis of highly fluorescent nitrogen-doped carbon dots from jackfruit seeds and its applications towards the fluorimetric detection of Au3+ ions in aqueous medium and in: In vitro multicolor cell imaging. *New J. Chem.* **2019**, *43*, 11710–11719, doi:10.1039/c9nj02590a.
25. Fu, X., Sheng, L., Yu, Y., Ma, M., Cai, Z., Huang, X. Rapid and universal detection of ovalbumin based on N,O,P-co-doped carbon dots-fluorescence resonance energy transfer technology. *Sensors Actuators, B Chem.* **2018**, *269*, 278–287, doi:10.1016/j.snb.2018.04.134.
26. Yang, H., Yang, L., Yuan, Y., Pan, S., Yang, J., Yan, J., Zhang, H., Sun, Q., Hu, X. A portable synthesis of water-soluble carbon dots for highly sensitive and selective detection of chlorogenic acid based on inner filter effect. *Spectrochim. Acta - Part A Mol. Biomol. Spectrosc.* **2018**, *189*, 139–146, doi:10.1016/j.saa.2017.07.065.
27. Liu, Y., Gong, X., Dong, W., Zhou, R., Shuang, S., Dong, C. Nitrogen and phosphorus dual-doped carbon dots as a label-free sensor for Curcumin determination in real sample and cellular imaging. *Talanta* **2018**, *183*, 61–69, doi:10.1016/j.talanta.2018.02.060.
28. Guo, W., Pi, F., Zhang, H., Sun, J., Zhang, Y., Sun, X. A novel molecularly imprinted electrochemical sensor modified with carbon dots, chitosan, gold nanoparticles for the determination of patulin. *Biosens. Bioelectron.* **2017**, *98*, 299–304, doi:10.1016/j.bios.2017.06.036.
29. Shao, M., Yao, M., De Saeger, S., Yan, L., Song, S. Carbon quantum dots encapsulated molecularly imprinted fluorescence quenching particles for sensitive detection of zearalenone in corn sample. *Toxins (Basel).* **2018**, *10(11)*, 438, doi:10.3390/toxins10110438.
30. Wei, Y., Zhang, D., Fang, Y., Wang, H., Liu, Y., Xu, Z., Wang, S., Guo, Y. Detection of Ascorbic Acid Using Green Synthesized Carbon Quantum Dots. *J. Sensors* **2019**, doi:10.1155/2019/9869682.
31. Mansuriya, B.D., Altintas, Z. Applications of graphene quantum dots in biomedical sensors. *Sensors (Switzerland)* **2020**, *20*, 1072, doi:10.3390/s20041072.
32. Li, N., Than, A., Wang, X., Xu, S., Sun, L., Duan, H., Xu, C., Chen, P. Ultrasensitive Profiling of Metabolites Using Tyramine-Functionalized Graphene Quantum Dots. *ACS Nano* **2016**, *10*, 3622–3629, doi:10.1021/acsnano.5b08103.
33. Li, N., Li, R., Li, Z., Yang, Y., Wang, G., Gu, Z. Pentaethylenehexamine and histidine-functionalized graphene quantum dots for ultrasensitive fluorescence detection of microRNA with target and molecular beacon double cycle amplification strategy. *Sensors Actuators, B Chem.* **2019**, 283, 666–676, doi:10.1016/j.snb.2018.12.082.
34. Kashani, H.M., Madrakian, T., Afkhami, A. Highly fluorescent nitrogen-doped graphene quantum dots as a green, economical and facile sensor for the determination of sunitinib in real samples. *New J. Chem.* **2017**, *41*, 6875–6882, doi:10.1039/c7nj00262a.
35. Ganganboina, A.B., Doong, R.A. Graphene Quantum Dots Decorated Gold-Polyaniline Nanowire for Impedimetric Detection of Carcinoembryonic Antigen. *Sci. Rep.* **2019**, *9*, doi:10.1038/s41598-019-43740-3.
36. Ibanez, J.G., Rincón, M.E., Gutierrez-Granados, S., Chahma, M., Jaramillo-Quintero, O.A., Frontana-Uribe, B.A. Conducting Polymers in the Fields of Energy, Environmental Remediation, and Chemical-Chiral Sensors. *Chem. Rev.* **2018**, *118*, 4731–4816, doi:10.1021/acs.chemrev.7b00482.
37. Saravanan, C., Palaniappan, S. Synthesis of novel optically active pyrrolidine- Containing polyaniline: a new heterogeneous organo polymeric-base catalyst for direct aldol reaction. *J. Appl. Polym. Sci.* **2010**, *116*, 2536–2540, doi:10.1002/app.31676.
38. Graboski, A.M., Ballen, S.C., Galvagni, E., Lazzari, T., Manzoli, A., Shimizu, F.M., Steffens, J., Steffens, C. Aroma detection using a gas sensor array with different polyaniline films. *Anal. Methods* **2019**, *11*, 654–660, doi:10.1039/c8ay02389a.
39. Wang, Y., Liu, A., Han, Y., Li, T. Sensors based on conductive polymers and their composites: A review. *Polym. Int.* **2020**, *69*, 7–17, doi:10.1002/pi.5907.
40. Chen, D., Lei, S., Chen, Y. A single polyaniline nanofiber field effect transistor and its gas sensing mechanisms. *Sensors* **2011**, *11*, 6509–6516, doi:10.3390/s110706509.
41. Kondawar, S.B., Patil, P.T., Agrawal, S.P. Chemical vapour sensing properties of electrospun nanofibers of polyaniline/ZnO nanocomposites. *Adv. Mater. Lett.* **2014**, *5*, 389–395, doi:10.5185/amlett.2014.amwc.1037.
42. Du, X., Liu, H.Y., Cai, G., Mai, Y.W., Baji, A. Use of facile mechanochemical method to functionalize carbon nanofibers with nanostructured polyaniline and their electrochemical capacitance. *Nanoscale Res. Lett.* **2012**, *7*, 111, doi:10.1186/1556-276X-7-111.

43. Pawar, S.G., Patil, S.L., Chougule, M.A., Godse, P.R., Bandgar, D.K., Patil, V.B. Fabrication of polyaniline/TiO 2 nanocomposite ammonia vapor sensor. *J. Nano- Electron. Phys.* **2011**, *3*, 1056–1063.
44. Bao, Q., Yang, Z., Song, Y., Fan, M., Pan, P., Liu, J., Liao, Z., Wei, J. Printed flexible bifunctional electrochemical urea-pH sensor based on multiwalled carbon nanotube/polyaniline electronic ink. *J. Mater. Sci. Mater. Electron.* **2019**, *30*, 1751–1759, doi:10.1007/s10854-018-0447-5.
45. Srivastava, S., Sharma, S.S., Agrawal, S., Kumar, S., Singh, M., Vijay, Y.K. Study of chemiresistor type CNT doped polyaniline gas sensor. *Synth. Met.* **2010**, *160*, 529–534, doi:10.1016/j.synthmet.2009.11.022.
46. Abdulla, S., Mathew, T.L., Pullithadathil, B. Highly sensitive, room temperature gas sensor based on polyaniline-multiwalled carbon nanotubes (PANI/MWCNTs) nanocomposite for trace-level ammonia detection. *Sensors Actuators, B Chem.* **2015**, *221*, 1523–1534, doi:10.1016/j.snb.2015.08.002.
47. Sen, T., Mishra, S., Shimpi, N.G. Synthesis and sensing applications of polyaniline nanocomposites: A review. *RSC Adv.* **2016**, *6*, 42196–42222, doi:10.1039/c6ra03049a.
48. Potjke-Kamloth, K. Chemical gas sensors based on organic semiconductor polypyrrole. *Crit. Rev. Anal. Chem.* **2002**, *32*, 121–140, doi:10.1080/10408340290765489.
49. Tiwari, D.C., Atri, P., Sharma, R. Sensitive detection of ammonia by reduced graphene oxide/polypyrrole nanocomposites. *Synth. Met.* **2015**, *203*, 228–234, doi:10.1016/j.synthmet.2015.02.026.
50. Tang, X., Lahem, D., Raskin, J.P., Gérard, P., Geng, X., André, N., Debliquy, M. A Fast and Room-Temperature Operation Ammonia Sensor Based on Compound of Graphene with Polypyrrole. *IEEE Sens. J.* **2018**, *18*, 9088–9096, doi:10.1109/JSEN.2018.2869203.
51. Zhang, C., She, Y., Li, T., Zhao, F., Jin, M., Guo, Y., Zheng, L., Wang, S., Jin, F., Shao, H., et al. A highly selective electrochemical sensor based on molecularly imprinted polypyrrole-modified gold electrode for the determination of glyphosate in cucumber and tap water. *Anal. Bioanal. Chem.* **2017**, *409*, 7133–7144, doi:10.1007/s00216-017-0671-5.
52. Wang, G., Han, R., Su, X., Li, Y., Xu, G., Luo, X. Zwitterionic peptide anchored to conducting polymer PEDOT for the development of antifouling and ultrasensitive electrochemical DNA sensor. *Biosens. Bioelectron.* **2017**, *92*, 396–401, doi:10.1016/j.bios.2016.10.088.
53. Li, S., Chen, S., Zhuo, B., Li, Q., Liu, W., Guo, X. Flexible ammonia sensor based on PEDOT:PSS/silver nanowire composite film for meat freshness monitoring. *IEEE Electron Device Lett.* **2017**, *38*, 975–978, doi:10.1109/LED.2017.2701879.
54. Shellaiah, M., Sun, K.W. Review on Sensing Applications of Perovskite Nanomaterials. *Chemosensors* 2020, *8*, 55, doi:10.3390/chemosensors8030055.
55. Cao, E., Wang, H., Wang, X., Yang, Y., Hao, W., Sun, L., Zhang, Y. Enhanced ethanol sensing performance for chlorine doped nanocrystalline LaFeO3-Δ powders by citric sol-gel method. *Sensors Actuators, B Chem.* **2017**, *251*, 885–893, doi:10.1016/j.snb.2017.05.153.
56. Chen, M., Zhang, Y., Zhang, J., Li, K., Lv, T., Shen, K., Zhu, Z., Liu, Q. Facile lotus-leaf-templated synthesis and enhanced xylene gas sensing properties of Ag-LaFeO3 nanoparticles. *J. Mater. Chem. C* **2018**, *6*, 6138–6145, doi:10.1039/c8tc01402g.
57. Bala, A., Majumder, S.B., Dewan, M., Roy Chaudhuri, A. Hydrogen sensing characteristics of perovskite based calcium doped BiFeO3 thin films. *Int. J. Hydrogen Energy* **2019**, *44*, 18648–18656, doi:10.1016/j.ijhydene.2019.05.076.
58. Teresita, V.M., Manikandan, A., Josephine, B.A., Sujatha, S., Antony, S.A. Electromagnetic Properties and Humidity-Sensing Studies of Magnetically Recoverable LaMgxFe1−x O 3−δ Perovskites Nano-photocatalysts by Sol-Gel Route. *J. Supercond. Nov. Magn.* **2016**, *29*, 1691–1701, doi:10.1007/s10948-016-3465-7.
59. Duan, Z., Xu, M., Li, T., Zhang, Y., Zou, H. Super-fast response humidity sensor based on La0.7Sr0.3MnO3 nanocrystals prepared by PVP-assisted sol-gel method. *Sensors Actuators, B Chem.* **2018**, *258*, 527–534, doi:10.1016/j.snb.2017.11.169.
60. Wei, W., Guo, S., Chen, C., Sun, L., Chen, Y., Guo, W., Ruan, S. High sensitive and fast formaldehyde gas sensor based on Ag-doped LaFeO3nanofibers. *J. Alloys Compd.* **2017**, *695*, 1122–1127, doi:10.1016/j.jallcom.2016.10.238.
61. Yin, Y., Li, F., Zhang, N., Ruan, S., Zhang, H., Chen, Y. Improved gas sensing properties of silver-functionalized ZnSnO3 hollow nanocubes. *Inorg. Chem. Front.* **2018**, *5*, 2123–2131, doi:10.1039/c8qi00470f.

62. Sheng, X., Liu, Y., Wang, Y., Li, Y., Wang, X., Wang, X., Dai, Z., Bao, J., Xu, X. Cesium Lead Halide Perovskite Quantum Dots as a Photoluminescence Probe for Metal Ions. *Adv. Mater.* **2017**, 29 (37), doi:10.1002/adma.201700150.
63. Protesescu, L., Yakunin, S., Nazarenko, O., Dirin, D.N., Kovalenko, M. V. Low-Cost Synthesis of Highly Luminescent Colloidal Lead Halide Perovskite Nanocrystals by Wet Ball Milling. *ACS Appl. Nano Mater.* **2018**, *1*, 1300–1308, doi:10.1021/acsanm.8b00038.
64. Hu, Y.L., Wen, Q.L., Pu, Z.F., Liu, A.Y., Wang, J., Ling, J., Xie, X.G., Cao, Q.E. Rapid synthesis of cesium lead halide perovskite nanocrystals by l-lysine assisted solid-phase reaction at room temperature. *RSC Adv.* **2020**, *10*, 34215–34224, doi:10.1039/d0ra07589b.
65. Shamsi, J., Urban, A.S., Imran, M., De Trizio, L., Manna, L. Metal Halide Perovskite Nanocrystals: Synthesis, Post-Synthesis Modifications, and Their Optical Properties. *Chem. Rev.* **2019**, *119*, 3296–3348, doi:10.1021/acs.chemrev.8b00644.
66. Yap, M.H., Fow, K.L., Chen, G.Z. Synthesis and applications of MOF-derived porous nanostructures. *Green Energy Environ.* **2017**, *2*, 218–245, doi:10.1016/j.gee.2017.05.003.
67. Chidambaram, A., Stylianou, K.C. Electronic metal-organic framework sensors. *Inorg. Chem. Front.* 2018, *5*, 979–998, doi:10.1039/c7qi00815e.
68. Silva, P., Vilela, S.M.F., Tomé, J.P.C., Almeida Paz, F.A. Multifunctional metal-organic frameworks: From academia to industrial applications. *Chem. Soc. Rev.* **2015**, *44*, 6774–6803, doi:10.1039/c5cs00307e.
69. Zhan, M., Hussain, S., AlGarni, T.S., Shah, S., Liu, J., Zhang, X., Ahmad, A., Javed, M.S., Qiao, G., Liu, G. Facet controlled polyhedral ZIF-8 MOF nanostructures for excellent NO2 gas-sensing applications. *Mater. Res. Bull.* **2021**, *136*, doi:10.1016/j.materresbull.2020.111133.
70. Liu, Y., Wang, R., Zhang, T., Liu, S., Fei, T. Zeolitic imidazolate framework-8 (ZIF-8)-coated In 2 O 3 nanofibers as an efficient sensing material for ppb-level NO 2 detection. *J. Colloid Interface Sci.* **2019**, *541*, 249–257, doi:10.1016/j.jcis.2019.01.052.
71. Assen, A.H., Yassine, O., Shekhah, O., Eddaoudi, M., Salama, K.N. MOFs for the Sensitive Detection of Ammonia: Deployment of fcu-MOF Thin Films as Effective Chemical Capacitive Sensors. *ACS Sensors* **2017**, *2*, 1294–1301, doi:10.1021/acssensors.7b00304.
72. Wang, J., Fan, Y., Lee, H.W., Yi, C., Cheng, C., Zhao, X., Yang, M. Ultrasmall metal-organic framework zn-mof-74 nanodots: Size-controlled synthesis and application for highly selective colorimetric sensing of iron(III) in Aqueous Solution. *ACS Appl. Nano Mater.* **2018**, *1*, 3747–3753, doi:10.1021/acsanm.8b01083.
73. Fang, X., Chen, X., Liu, Y., Li, Q., Zeng, Z., Maiyalagan, T., Mao, S. Nanocomposites of Zr(IV)-Based Metal-Organic Frameworks and Reduced Graphene Oxide for Electrochemically Sensing Ciprofloxacin in Water. *ACS Appl. Nano Mater.* **2019**, *2*, 2367–2376, doi:10.1021/acsanm.9b00243.
74. Hosseini, M.S., Zeinali, S. Capacitive humidity sensing using a metal–organic framework nanoporous thin film fabricated through electrochemical in situ growth. *J. Mater. Sci. Mater. Electron.* **2019**, *30*, 3701–3710, doi:10.1007/s10854-018-00652-8.
75. Fang, X., Zong, B., Mao, S. Metal–Organic Framework-Based Sensors for Environmental Contaminant Sensing. *Nano-Micro Lett.* **2018**, *10*, 64. doi:10.1007/s40820-018-0218-0.
76. Lin, X.; Gao, G.; Zheng, L.; Chi, Y.; Chen, G. Encapsulation of Strongly Fluorescent Carbon Quantum Dots in Metal–Organic Frameworks for Enhancing Chemical Sensing. *Anal. Chem.* **2014**, *86 (2)*,1223–1228, https://doi.org/10.1021/ac403536a.
77. Lin, X., Luo, F., Zheng, L., Gao, G., Chi, Y. Fast, sensitive, and selective ion-triggered disassembly and release based on tris(bipyridine)ruthenium(II)-functionalized metal-organic frameworks. *Anal. Chem.* **2015**, *87*, 4864–4870, doi:10.1021/acs.analchem.5b00391.
78. Gavgani, J.N., Dehsari, H.S., Hasani, A., Mahyari, M., Shalamzari, E.K., Salehi, A., Taromi, F.A. A room temperature volatile organic compound sensor with enhanced performance, fast response and recovery based on N-doped graphene quantum dots and poly(3,4-ethylenedioxythiophene)-poly(styrenesulfonate) nanocomposite. *RSC Adv.* **2015**, *5*, 57559–57567, doi:10.1039/c5ra08158k.
79. Wang, J., Lin, S., Lin, Y., Wang, X. Preparation of graphene quantum dots and their sensing properties in quartz crystal microbalance acetone sensor. *Sensors Mater.* **2021**, *33*, 499–511, doi:10.18494/SAM.2021.3075.

80. Sun, J., Shu, X., Tian, Y., Tong, Z., Bai, S., Luo, R., Li, D., Liu, C.C. Facile preparation of polypyrrole-reduced graphene oxide hybrid for enhancing NH3 sensing at room temperature. *Sensors Actuators, B Chem.* **2017**, *241*, 658–664, doi:10.1016/j.snb.2016.10.047.
81. Konwer, S., Guha, A.K., Dolui, S.K. Graphene oxide-filled conducting polyaniline composites as methanol-sensing materials. *J. Mater. Sci.* 2013, *48*, 1729–1739, doi:10.1007/s10853-012-6931-z.
82. Tang, X., Debliquy, M., Lahem, D., Yan, Y., Raskin, J.P. A review on functionalized graphene sensors for detection of ammonia. *Sensors* **2021**, *21*, 1–28, doi:10.3390/s21041443.

8 Synthesis of Fluorescent Nanosensor for Biomedical Engineering

Saba Farooq and Zainab Ngaini

CONTENTS

8.1 Introduction 181
8.2 Preparation of Fluorescent Nanosensor 182
8.2.1 Hydrothermal Synthesis 183
8.2.2 Solvothermal Method 184
8.2.3 Ultrasonic Irradiations 185
8.2.4 Microwave Irradiation 185
8.2.5 Precipitation and Coprecipitation 186
8.2.6 Polymerization 186
8.3 Types of Fluorescent Nanosensor 187
8.3.1 pH-Sensitive Nanosensor 187
8.3.2 Temperature-Sensitive Nanosensor 189
8.3.3 Protease-Sensitive Nanosensor 189
8.3.4 Oxygen-Sensitive Nanosensor 190
8.4 Applications of Fluorescent Nanosensor 191
8.4.1 Detection of Microorganisms 191
8.4.2 Detection of Metallic and Nonmetallic Ions 192
8.4.3 Detection of Organic Compounds 192
8.4.3.1 Detection of Amino Acids, Proteins and Vitamins 192
8.4.3.2 Detection of Drugs 194
8.5 Future Perspectives 194
8.6 Conclusion 195
References 195

8.1 INTRODUCTION

Fluorescence sensing materials have gained much attention due to their wide range of applications in nanotechnology biosensors (Tian et al., 2015), aerospace, the food industry (Vyas et al., 2015), agriculture (Guan et al., 2021), military security (Qian et al., 2016) and the biomedical field (Khan et al., 2020). These materials are able to size up to 100 nm (Liu et al., 2016a) with the capability of sensing external changes such as pH, oxygen, temperature, proteins and water (Wu et al., 2021). The fluorescent nanosensor (FNS) is a powerful tool that depicts the advancement in optical-sensor technologies having high selectivity, a wide detection range, low cost and low detection limit (Gong et al., 2015).

The structure of a nanosensor consists of three components, that is: sensor (sensing action, basically analyte, bioreceptor and nanoparticles), transducer (convert I/P to electronic signal) and

DOI: 10.1201/9781003218708-10

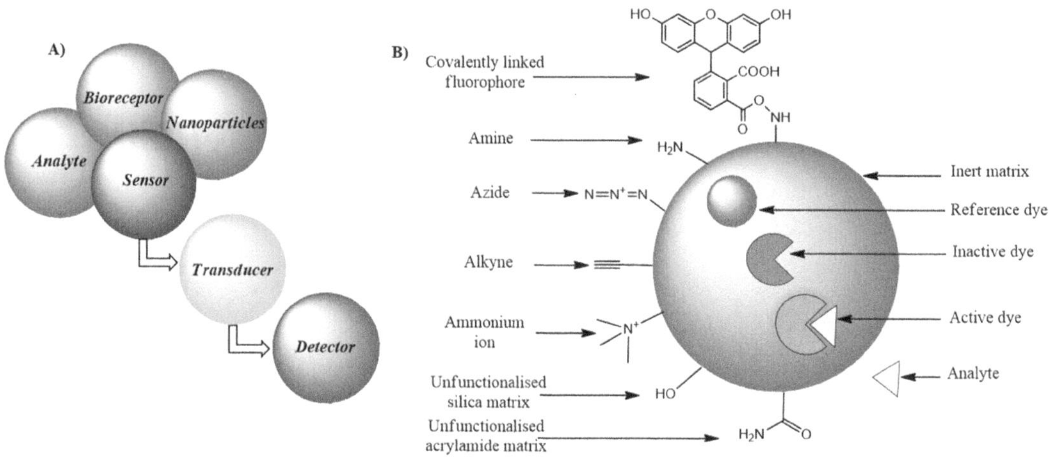

FIGURE 8.1A and 8.1B General components of nanosensor (A) and Functionalized FNSs (B).

detector (Figure 8.1A) (García-Añoveros and Corey, 1997). The fluorescent nanosensor contains fluorescent material that is encapsulated or bound with inserted biocompatible materials with a high level of temporal and spatial resolution (Petrakova et al., 2015). FNS is composed of a reference fluorophore, which contributes toward the stable fluorescence emission and analyte-sensitive fluorophore, a dynamic fluorescence emission depending upon the nature or amount of target materials to achieve both high accuracy and ratiometric output (Figure 8.1B) (Desai et al., 2014). The multiple target materials can be detected by the association of multiple analyte-sensitive fluorophores (Qiu et al., 2015).

Investigation of biological processes is laborious – for example, gene regulation, DNA repair mechanisms, single-cell activity and diagnosis of disease (Francis and Rajith, 2021). Numerous probe sensors are actively used by researchers for disease diagnostic techniques reported in Table 8.1 (Chakraborty et al., 2020). The fluorescent sensor is not, however, straightforward functioning like a naked-eye detected nanosensor. The UV lamp is required for the excitation of fluorophore and the detection of signals. Another drawback encountered by the fluorescent nanosensor is the quenching procedure during the emission of several fluorophores by numerous external factors (i.e., oxygen, light). Quenching of fluorescence can be overcome by doping (Yuan et al., 2016). To date, the development of recyclable, efficient, good stability and sensitivity (Xie et al., 2015), metal-free, eco-friendly and inexpensive high-performance sensors still constitute a challenging process (Wang et al., 2015a). This chapter highlights the fluorescent nanosensor fabrications techniques, response to external stimuli, and applications in the biomedical engineering field. It also attributes the structural attributed significance for sensitive and efficient fluorescence nanomaterials.

8.2 PREPARATION OF FLUORESCENT NANOSENSOR

The fluorescent nanosensor can be naturally present in banana (De and Karak, 2013), sweet potato (Shen et al., 2017), lemon (Das et al., 2019), chitosan (Caprifico et al., 2020), palm kernel shell (Monday et al., 2021)) or created synthetically (i.e., CQDs-RhB (Fu et al., 2017), RbH-CD (Ma et al., 2018)) through various methods, that is, hydrothermal (Shi et al., 2015), microemulsion (Amjadi et al., 2016), solvent displacement (Xie et al., 2016), thermolitic methods (Saenwong et al., 2018), precipitation method (Cun et al., 2016), reverse micelles and arrested precipitates (Zuo et al., 2016). Traditional top-down methods (Farooq et al., 2020) require multisteps that are tedious and

TABLE 8.1
Fluorescent Probes for Intracellular Imaging

Types of Fluorescent Probe (size)	Intracellular Imaging Application	Advantages	Limitations
Fluorescent carbon dots (2–10 nm)	Intracellular organelle, amino acid, RNA, metal ion, redox state	Bright and tunable emission, nontoxic	Larger size, broad emission
Doped semiconductor nanoparticle-based (5–50 nm)	Glucose, drug	Low background fluorescence	Larger size
Quantum dot-based (5–50 nm)	Protein, lipid droplet, nucleus, mitochondria	Bright, narrow and tunable emission, single light for exciting multiple colors	Larger size, cytotoxicity due to Cd
Dye-based (< 2 nm)	Nucleus, mitochondria, lysosome	< 2 nm size for easier cytosolic entry	Photobleaching
Upconversion nanoparticle-based (5–100 nm)	TK1 mRNA and pH detection	Low background noise	Larger size
Aggregation-induced emission active dye-based (< 2 nm)	Mitochondria, lysosome	< 2 nm size for easer cytosolic entry, low photobleaching	Elective chemical structure

Source: Ali et al. 2020.

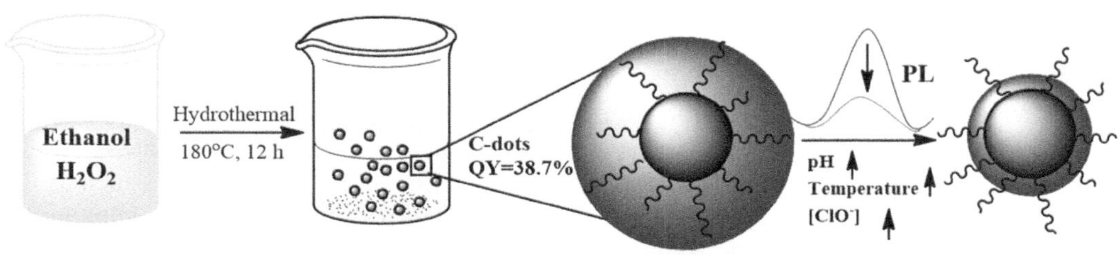

FIGURE 8.2 Hydrothermal synthesis of multifunctional nanosensors.

have high costs. Due to these drawbacks, the latest method (a bottom-up synthesis) is frequently preferred for the designing of the fluorescent nanosensor. The bottom-up method is convenient, cheaper and efficient for designing (Yan et al., 2016b). Some synthetic bottom-up methods are illustrated in detail below.

8.2.1 Hydrothermal Synthesis

Fluorescent nanosensors have been earlier synthesized via a hydrothermal process. The approach is eco-friendly, economical and much greener. It requires affordable and safe reagents, less-expensive equipment, and complicated procedures (Wang et al., 2015b). Hydrothermal synthesis is a simple method to form certain desired fluorescent nanosensors: BZM-G-CDs (Fang et al., 2019), N, S-CQDs (Belal et al., 2021). It involves a one-step approach for the designing of EtOH/H_2O_2-derived fluorescent carbon dots by hydrothermal oxidation. The achieved C-dots have a crystalline nature due to oxygenous groups. The C-dots showed low photoluminescence (PL) quantum yields (QYs = 38.7%), without any heteroatoms doping or surface passivation. These multifunctional fluorescence C-dots nanosensors are able to detect pH, temperature, and the hypochlorite ion (ClO^-) concentration (Figure 8.2) (Hu et al., 2015). However, the surface

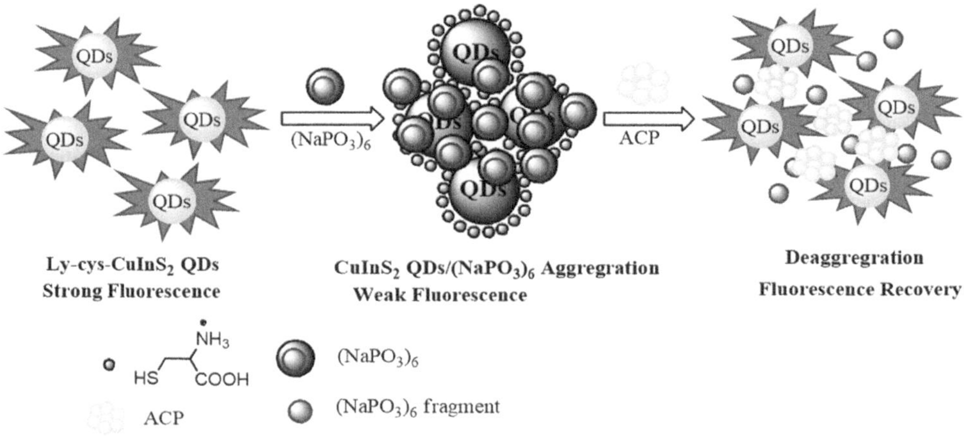

FIGURE 8.3 The fluorescence "turn off-on" nanosensor $CuInS_2$ QDs and $(NaPO_3)_6$ for ACP.

passivation and heteroatoms doping contributing to increasing QY are hazardous, tedious and difficult to obtain (Chen et al., 2018).

Hydrothermal process-based synthesized fluorescent copper-based quantum dots ($CuInS_2$ QDs) (Liu et al., 2012) functioning is described in Figure 8.3. The nanosensor is applied for the detection of acid phosphatase (ACP). ACP detection in serum is significant for diagnosing several diseases, such as kidney disease, hepatitis, prostate cancer, Gaucher disease (Pang and Liu, 2020), thrombophlebitis, Paget's disease, hyperparathyroidism, and multiple myeloma (Li et al., 2020). The detection of ACP through fluorescent nanosensors is more efficient and more convenient than other conventional methods, namely LC-MS/MS (Subhahar et al., 2021) and HPLC, electrophoresis and electrochemistry (Pang and Liu, 2020). The $CuInS_2$ QDs has lower cell toxicity and is eco-friendly in nature compared to other toxic elements such as Cd, Pb, Hg, As, Se and Te. In an aqueous solution, $CuInS_2$ QDs stabilized with L-cysteine with a positive charge at suitable pH, and sodium hexametaphosphate $(NaPO_3)_6$ additionally, with six negative charges favored toward the aggregation for the formation of $CuInS_2$ QDs via electrostatic interactions. Whereas, a strong fluorescence of $CuInS_2$ QDs is decreased due to the quenching property caused by aggregation (Liu et al., 2015b). Aggregation is disturbed due to the ACP between $(NaPO_3)_6$ and $CuInS_2$ QDs, which leads toward disaggregation of assembly and quenched fluorescence restored (Liu et al., 2015b).

8.2.2 Solvothermal Method

The solvothermal method is another non-toxic, cheap, simple and environmentally friendly method for the fabrication of fluorescent nanosensors. This process is based upon the chemical reaction or decomposition between reagents in the presence of a suitable solvent at a temperature higher than the boiling point of a suitable solvent. This methodology has been used for the fabrication of fluorescent nanosensor such as *o*-phenylenediamine derived (N, S)-CDs (Guan et al., 2019), corn bract-based CDs (Zhao et al., 2017), N,O-CDs (Zhang et al., 2021) and CD-tetra (Zhao et al., 2019). Amphiphilic carbon dots-[bis(dimethylamino)methylene]-1H-1,2,3-triazolo[4,5-b]pyridinium 3-oxide hexafluoropho-sphate (ACDs-HATU) obtained from the HATU, benzyltriethylammonium chloride diisopropylamine and 1-benzylimidazole via solvothermal method. The obtained materials, yellow precipitates, were further doped with $AlCl_3·6H_2O$ to enhance the rigidity of the material, to reduce energy loss and to increase luminescent quantum yield for CDs (Figure 8.4) (Zhao et al., 2018).

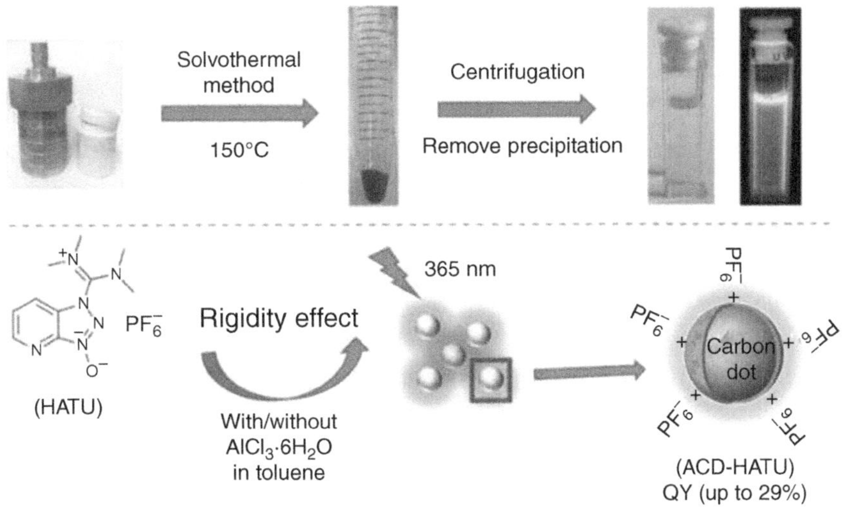

FIGURE 8.4 Synthesis of ACDs –HATU.

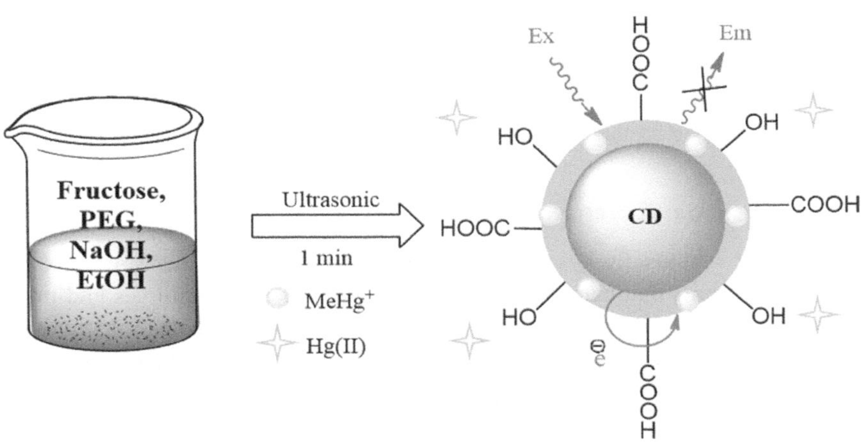

FIGURE 8.5 Synthesis of D-fructose derived CDs *via* ultrasonic irradiations.

8.2.3 Ultrasonic Irradiations

The ultrasonic method is a well-known method that causes various chemical and physical effects through acoustic cavitation to provide the interaction between matter and energy. A variety of fluorescent nanosensors have been designed from this technique, such as lignin-derived CQDs (LCQDs) (Gao et al., 2019), colchicine-derived CDs (El-Malla et al., 2021), and WCDs (Dang et al., 2016). This method has also been used in the synthesis of fluorescent nanosensors such as D-fructose-derived CDs fluorescent probes. The nanosensor has been synthesized under ultrasonic irradiations in the presence of PEG in ethanolic sodium hydroxide solution, which benefits are to indicate the presence of toxic methylmercury (Figure 8.5) (Costas-Mora et al., 2014).

8.2.4 Microwave Irradiation

Microwave irradiation is a fast, efficient and simple method (Yan et al., 2016a) for the preparation of fluorescent carbon dots such as G-BSA-AuNCs (Ryavanaki et al., 2020), nitrogen-doped

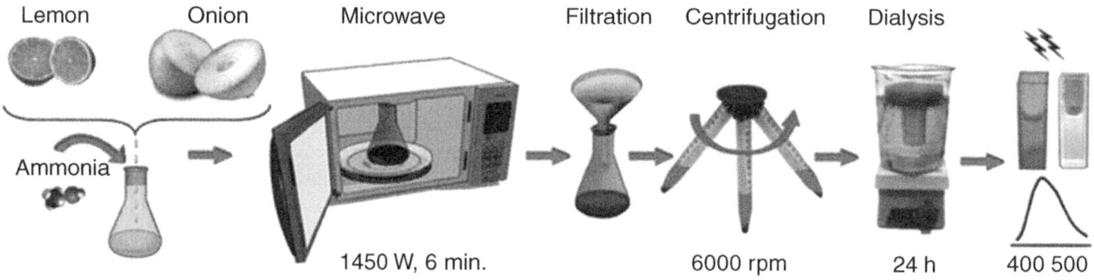

FIGURE 8.6 Lemon and onion extract derived CDs.

LR-CDs (Gu et al., 2016), eggshell-derived CDs (Wang et al., 2012), NCQDs -QPRTase (Singh et al., 2018) and GUCDs (Uriarte et al., 2019). Lemon and onion extract-derived fluorescent CDs are examples of nanosensors prepared through the carbonization microwave-assisted method. The starting material lemon (rich with citric and L-ascorbic acid as carbon precursor) and onion (these have sulfur, ammonium hydroxide useful for N-dopant agents) for CDs synthesis (Figure 8.6). The characterization has been performed by X-ray diffraction (XRD), Transmission electron microscopy (TEM), X-ray fluorescent (XRF), Fourier-transform infrared (FTIR), UV-Vis and fluorescence spectroscopy. The CDs have the capability to determine the presence of vitamin B (riboflavin) in commercial multivitamin/mineral supplements (Monte-Filho et al., 2019). The fluorescent method is fast, responsive, higher sensitivity, lower in cost and simpler than conventional methods used for riboflavin identification (i.e., capillary electrophoresis, spectrophotometry, electrochemistry, immunoassay and high-performance liquid chromatography) (Kundu et al., 2016).

8.2.5 Precipitation and Coprecipitation

For the synthesis of various fluorescent nanosensors, precipitation and/or co-precipitation are straightforward procedures that can reduce the amount of undesired additives and multistep processing. (Wang et al., 2016). Co-precipitation, commonly referred to as the reprecipitation method, is a common technique for creating both organic and inorganic fluorescent nanosensors. For instance, hydrophobic organic molecules are sparingly dissolved in a good solvent before a poor solvent is gradually and excessively added. A smaller number of precipitates has been achieved due to the quick exchange of different solvent environments (Shellaiah and Sun, 2020). Fluorescent materials (i.e., $Ir(CS)_2(acac)$/PViCl-PAN (Borisov et al., 2009), CTNSs (Wang et al., 2011), Fe_3O_4@GQDs@GM (Aladaghlo et al., 2021)) and chemo-dosimeter significant for analyzing chemical hazards (Chen et al., 2019). Coumarin dye-doped PPE P is an example of dots prepared from the mixing of semiconducting polymer, TBDPSCA and PPE via coprecipitation. The obtained TBDPSCA-based PPE Pdots was further coated with layered polydiacetylenes (PDAs) to avoid leakage of imprisoned coumarin dye. Only fluoride ions are infused into the Pdot matrix to give fluorescence by deprotectyl silyl group of TBDPSCA (Figures 8.7 A–8.7C) (Huang et al., 2014).

8.2.6 Polymerization

The coupling process of the low molecular-weight monomers into the macromolecular polymers is known as polymerization (Peng and Chiu, 2015). Polymeric materials obtained from polymeric precursors (e.g., nanoprecipitation, emulsion polymerization) are important in the field of nanomedicine. Whereas, fluorescence imaging is a widespread non-combative visualization technique for the *in vitro*, *ex vivo* and *in vivo* biological diagnosis, tracing, detection studies. Highly sensitive fluorescent polymeric materials such as ZnS:Mn^{2+}@allyl mercaptan (QDs@AM) derived QDs (Bai et al., 2015), red emissive carbon polymer dots (R-CPDs) (Xia et al., 2018) have been

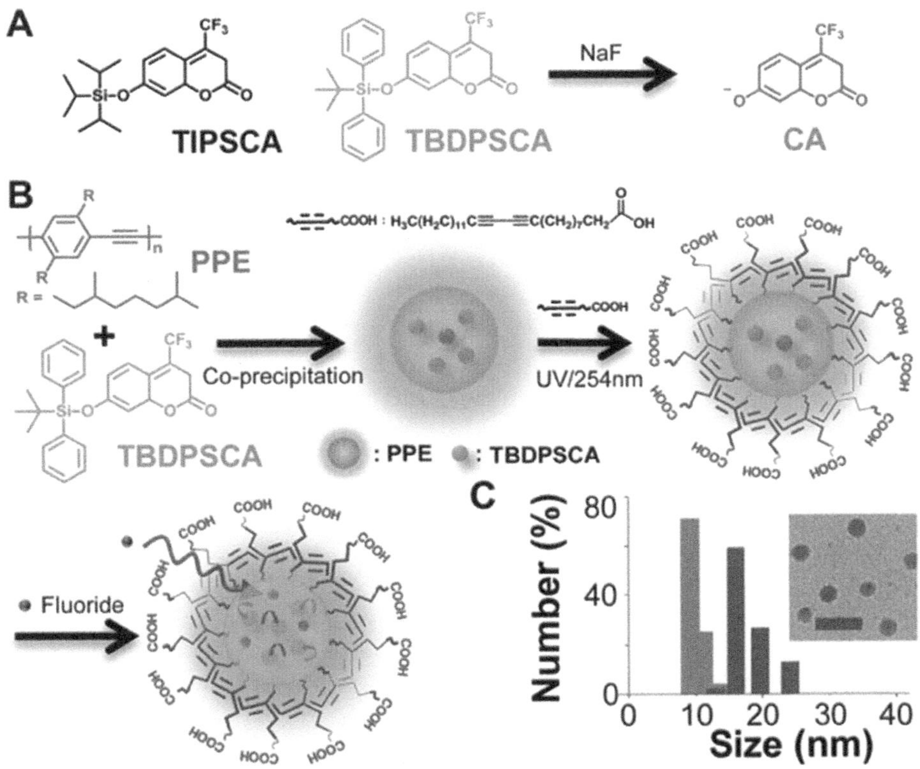

FIGURE 8.7A–8.7C Coumarin dye-doped PPE Pdots for detecting fluoride. Sensing molecules TIPSCA and TBDPSCA for NaF (A), Synthesis of TBDPSCA-based PPE Pdots (B) and Size of CA-doped PPE Pdots without PDAs (red histograms) and with PDAs (blue histograms) and also supported with TEM image (C).

reported for monitoring drug delivery, biodistribution and cellular uptake. The rhodamine-tagged polyacrylamide-polystyrene nanoparticles are an example of nano fluorescent for fluorescent tagging, biodistribution and cellular uptake. Whereas the reversible-addition-fragmentation chain-transfer (RAFT) emulsion polymerization followed for the preparations of nanoparticles (Figure 8.8) (Poon et al., 2017).

8.3 TYPES OF FLUORESCENT NANOSENSOR

There are many types of fluorescent nanosensors prepared and used in various applications (Cheng et al., 2020). The use of fluorescent nanosensors is illustrated as follows:

8.3.1 pH-Sensitive Nanosensor

Intracellular pH homeostasis is vital in numerous biological processes. A slight change in pH can lead to many diseases, such as apoptosis (Chen et al., 2015b), Alzheimer's disease, and others (Chauhan et al., 2013). Investigation of intracellular pH of optic nanosensors with high efficacy gives high spatial resolution due to their small size (Chauhan et al., 2011). The pH sensing materials (e.g., fluorophores-encapsulated nanoliposomes (Zhang et al., 2016), Ada-FITC/ M13-β-CD /Ada-RhB (Chen et al., 2015a), Interestingly, GQDs-GO (Zhu et al., 2011), capt@AgNCs (Zhang et al., 2019b)) are highly sensitive, low-cost and stable for monitoring environmental variations, diagnostics and drug delivery (Barati et al., 2016, Elsutohy et al., 2017). An example of pH probes having

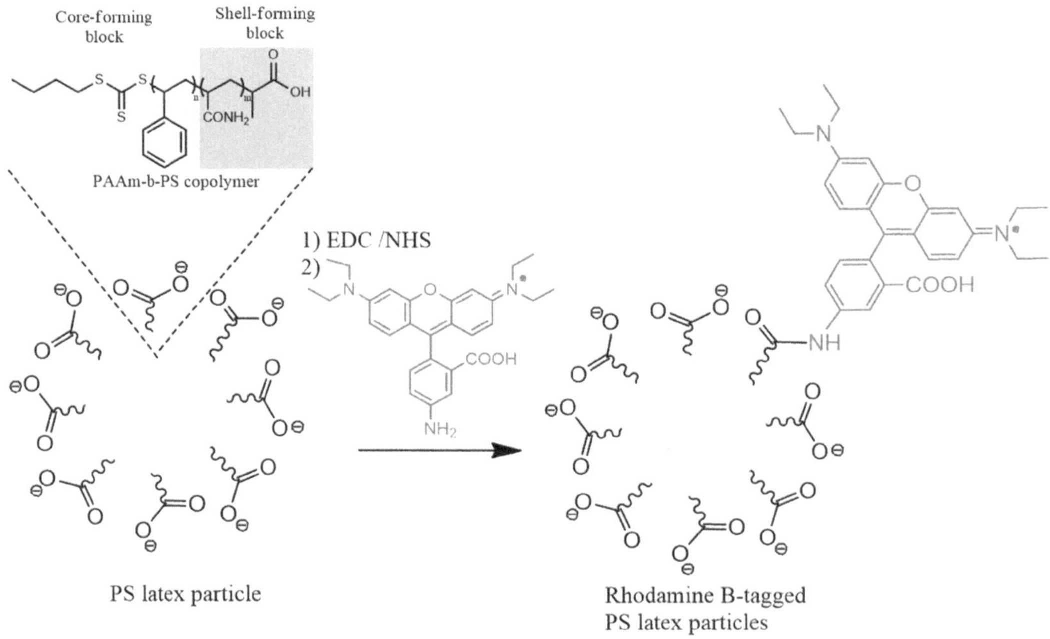

FIGURE 8.8 Rhodamine-labelled polystyrene latex particles.

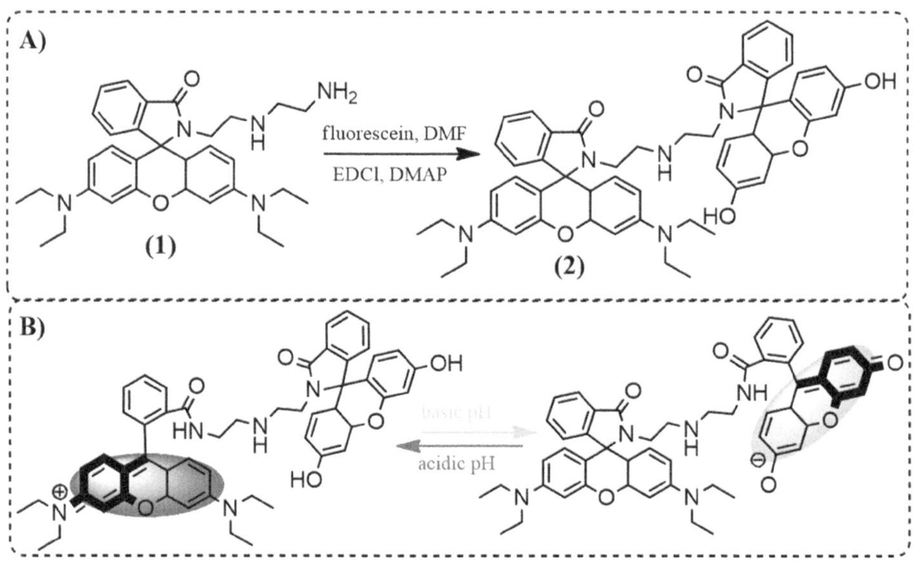

FIGURE 8.9A and 8.9B Rhodamine and fluorescein-based pH probe synthesis (A) and functioning (B).

rhodamine and fluorescein associated with diethylenetriamine spacers is reported in Figures 8.9A and 8.9B. Rhodamine derivative (**1**) was chemically reacted with fluorescein in DMF in the presence of DMAP and EDCI to form (**2**) with 27 percent yield. Spectroscopic analysis of (**2**) in different buffer solutions showed sharp bands at 495 nm and 512 nm attributed to absorption and fluorescence emission, respectively. The acidic pH exhibited 563 nm and 580 nm bands of absorption and emission band respectively. The pH changes have been observed by the ring opening-closing changes in the compound (Lee et al., 2013).

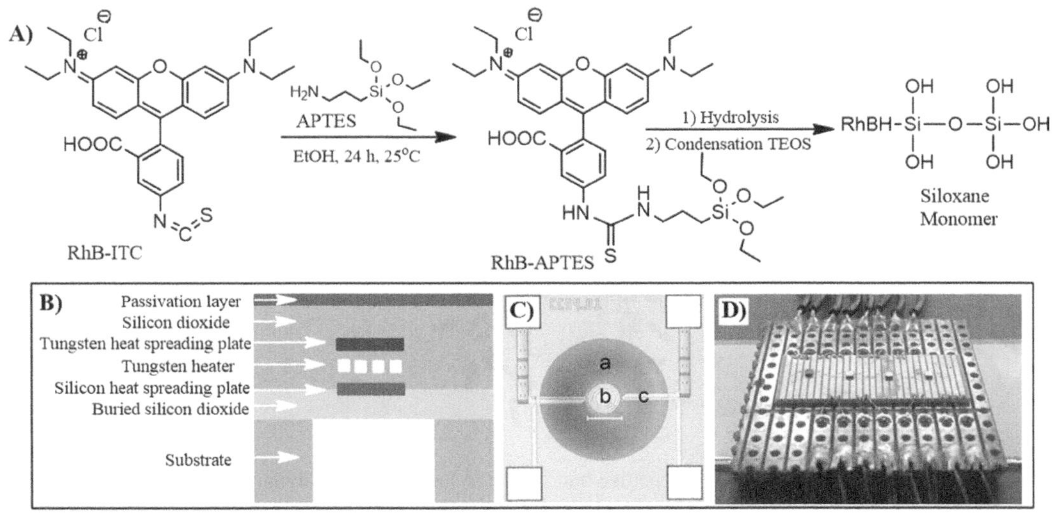

FIGURE 8.10A–8.10D Rhodamine B based temperature-sensitive nanosensor (A), CMOS MEMS micro-hotplate cross section (B), Reflected light image of the silicon wafer of the MEMS device having silicon dioxide membrane (a), tungsten heater (b) and tungsten track (c) (C). MEMS devices image mounted on a circuit board (D).

8.3.2 Temperature-Sensitive Nanosensor

Temperature-sensitive nanosensors have also been reported to monitor the temperature fluctuations in the biological processes because temperature plays key roles in metabolism, pharmacology and cellular physiology processes to diagnosis major diseases (Chen et al., 2017). Abnormal body temperature causes severe hypoxia, tissue acidosis and inflammation of tissues (Zhang et al., 2019a). Advancements in technology have replaced air-based thermometers, mercury-based thermometers, thermocouples, radiation thermometers, with nanothermometers. Non-contact-temperature sensing nanothermometers have been designed from fluorescent nanosensing technology (Mohammed and Omer, 2020). For example RhB@ZIF-71 (Zhang et al., 2020), PP-TzDa (Huang et al., 2020) and Au NCs (Jia et al., 2018). One of the known fluorophore rhodamine B coordinated with silica sol-gel matrix is reported to design temperature-sensitive fluorescent nanosensors from the conjugation of RhB isothiocyanate (RhB-ITC) and (3- aminopropyl)-triethoxysilane (APTES) stirred overnight in absolute ethanol at 4∘C to form the nanoparticles of RhB-APTES. The obtained RhB-APTES hydrolyzed before the condensation with tetraethylorthosilicate (TEOS), which leads toward the RhB-siloxane monomer (Figure 8.10A). These are highly sensitive to identifying the slight variation of temperature used in microelectromechanical systems (MEMS) micro-hotplate (Figures 8.10B–8.10D) (Chauhan et al., 2014).

8.3.3 Protease-Sensitive Nanosensor

Fluorescent nanosensors via bio-imaging play an important role in observing the intracellular changes (i.e., analytes (H^+, Ca^{+2}, O_2)), concentration changes and enzymatic activity changes. The rapid and accurate cellular studies for in vivo proteolytic processes are challenging for researchers. The reason is to strictly control protease activities because any dysregulation can cause disease to cause complications (Watzke et al., 2008). The general depiction of "protease responsive nanoprobe" (PRN) is reported in Figure 8.11A. PRN has bifunctional fluorophore that is attached to the protease peptide as substrate and nanoparticle. Remarkably, the low molecular weight fluorogenic scaffold

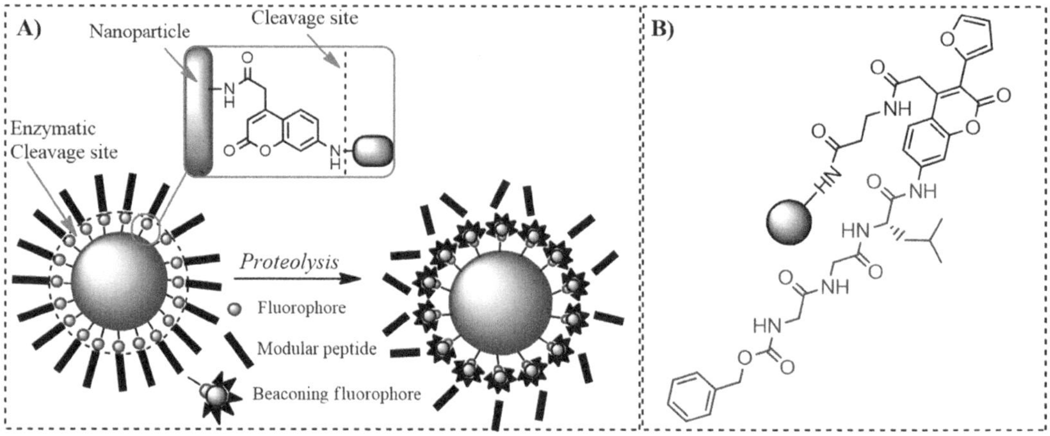

FIGURE 8.11A and 8.11B Protease-sensitive nanoprobe (A) and fluorogenic nanoprobe structure (B).

is covalently attached with a sensor matrix and exhibited the optimized fluorescence properties. The imaging probe via proteolytic activation tends to switch on multiple fluorophores by a single enzyme because of the variety of peptide substrates (Funovics et al., 2003). An example of PRN has been shown in (Figure 8.11B), which is synthesized by the coupling of 7-aminocoumarin-4-acetic acid (ACA) (fluorophore) and Z-Gly-Gly-Leu-AMC (fluorogenic substrate for model protease subtilisin). Furthermore the stability of the nanosensor achieved by spacer β-alanine unit to form an amide bond and the final product as green fluorescent bifunctionalized 3-arylcoumarin dye specifically used as a bio-imaging agent for screening of protease inhibitors (Welser et al., 2009). These fluorogenic moieties are nontoxic and bifunctional reagents with high quantum yield and large Stokes's shift.

8.3.4 Oxygen-Sensitive Nanosensor

The optimum amount of oxygen concentration is required for a living organism. A higher and lower oxygen concentration than the optimum concentration can lead to complications or death (Ashokkumar et al., 2020). Fluorescent nanosensor is one of the suitable techniques to monitor oxygen levels in organisms (Table 8.2) (Wang and Wolfbeis, 2014). Porphyrins is an organic compound (Cammidge et al., 2003) exhibit of intense fluorescence commonly used in fluorescent nanosensors (Negut et al., 2020) that is, Pt-porphyrin-loaded polylactic acid (PLA) nanoparticle (Pandey et al., 2019). Porphyrin-based oxygen nanosensor has been used in tissue engineering for the measurement of oxygen (Giuntini et al., 2014). Metalloporphyrins have also been reported to detect a low concentration of oxygen and to increase the lifetime stability period of phosphorus (Chauhan et al., 2016).

Reactive oxygen species (ROS) is oxygen with reactive species and exists in different interconvertible forms (i.e., hydrogen peroxide (H_2O_2), superoxide anion radical (O_2 •–), electron transfer and hydroxyl radicals (• OH). ROS present as singlet oxygen and responsible for the disease diagnosis and cell signaling for the detection of disorders (i.e., obesity, cancer, diabetes) (Huang et al., 2016a). In humans, specific organelles (i.e., mitochondria) participated to control ROS at the cellular level with the aid of antioxidants to regulate cell processes normally (Yang et al., 2015) FITC@ $mSiO_2$@$hmSiO_2$ (Zhou et al., 2016) and copper and zinc metal-associated cationic porphyrin azide-based nanosensor (Lavado et al., 2015) have high accuracy and selectivity for imaging and biosensing of O_2 •– in mitochondria of living cells has increased their demand (Huang et al., 2016b).

TABLE 8.2
Fluorescent Oxygen Detecting Nanosensors

Nature	Dye /matrix	$\lambda_{exc}/\lambda_{em}$	Sensitivity	Comments
Organic/ Non-metallic	Fluoranthene in polyethylene	330/390	I_0/I_{100} ~ 1.11	First PAH-based optical sensors for oxygen
	Fluorescent Yellow on silica gel	466/519	I_0/I_{21} ~ 3	Used to visualize surface gas flows
	$^{13}C_{70}$ fullerene in polystyrene	470/600–750	$\lambda_0/\lambda_{300ppm}$ ~ 3.5	Suitable for trace oxygen sensing; fully reversible; fluorescence lifetime imaging; temperature influences
Metallic	Ru(8-dpp^i)$_3$(ClO$_4$)$_3$ in PS	460/620	I_0/I_{100} ~ 3.5	Good solubility in polar polymers; QY 0.19; green emitting Ir(III) complexes also reported
	Ir(ppy)$_3$ in poly(styrene-co-TFEM)[a]	376/512	I_0/I_{100} = 15.3	SVPs; almost linear; short response times (t_{95} = 7s); highly photostable
	Os(dpp)$_3$(PF$_6$)$_2$ in PDMS rubber	480(502)/729	I_0/I_{air} = 1.1	PDMS contains (methacroyloxy)propyl side chains; λ_0 = 340 ns
	Tb(III) azaxanthone complex	355/480,540	I_0/I_{100} ~ 1.5	Mixture of Eu(III) and Tb(III) complex used for ratiometric oxygen measurement; efficient generation of singlet oxygen; cell permeable

Source: Wang and Wolfbeis 2014.

8.4 APPLICATIONS OF FLUORESCENT NANOSENSOR

Due to being cheap, easy to prepare and highly sensitive, fluorescent nanosensor has been widely used in wide applications, particularly in the field of biomedical engineering (Figure 8.12).

8.4.1 Detection of Microorganisms

Food safety has been declared a global issue with direct consequences to public health and international trade due to the epidemics of foodborne diseases from microbial contaminants, toxins and chemicals (Zhang et al., 2017). Foodborne infections have caused plenty of hazardous infections due to the pathogenicity and resistivity of microorganisms. The endospores of bacteria have mainly caused pathogenicity, food poisoning, bioterrorism and serious diseases (Donmez et al., 2017), which might lead to human death (Xu et al., 2015).

Due to heavy instrumentation usages, it was previously very hard to identify microorganisms. Nowadays, with the advancement of nanosensors technology (Yin et al., 2020) such as magneto-fluorescent nanosensors (MFnS) of polyacrylic acid (PAA) coated iron-oxide nanoparticles (IONP-COOH) encapsulated with DiI fluorescent dye, the identification of bacterial infections has become easy and precise (Banerjee et al., 2016). UCNPs@NH$_2$ is one of the microorganism detectors, that is easily synthesized via microemulsion from UCNPs, triton X-100 and ammonium ion. The obtained UCNPs@NH$_2$ was further treated with *S*-ethylisothiourea hydrobromide and PBS buffer to produce UCNPs@GDN (Wang et al., 2018). These upconversion-fluorescent nanoparticles (UCNPs@GDN) are highly sensitive for the detection of a variety of foodborne pathogenic bacteria such as *L. monocytogenes*, *E. coli*, *C. sakazakii*, *Salmonella*, *S. flexneri*, *S. aureus* and *V. parahaemolyticus*. The functioning of nanosensors observed by the oxidation of tannic acid by hydrogen peroxide contributed to the detection of bacteria and the fluorescence material, which depicted a high intensity

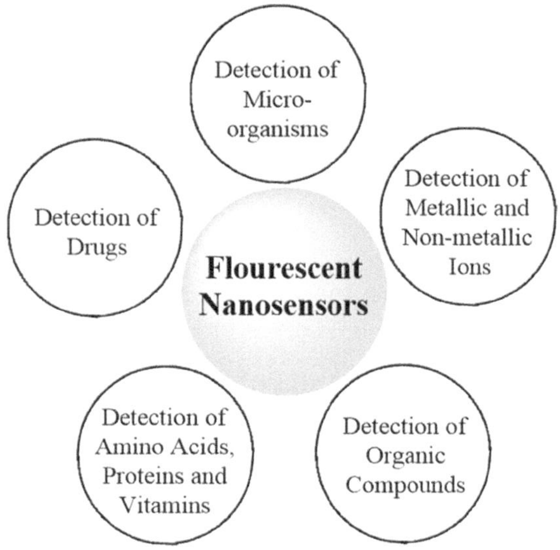

FIGURE 8.12 Applications of fluorescent nanosensors.

of fluorescence (Figures 8.13 A and 8.13B) (Yin et al., 2019). This is a rapid, sensitive and highly effective method to detect bacteria in soil, water and food (Zong et al., 2019).

8.4.2 Detection of Metallic and Nonmetallic Ions

Ion-selective sensors are vital in daily life and useful for environmental analysis or clinical diagnostics (Jokic et al., 2012). The development and exploration of opto-chemical sensors have been focused on the evolution of scientific technologies. Nowadays, a variety of ionophore, fibre-optical sensors, sensor layers, optical chemical probes and micro/nanoparticles exist for the recognition of ions and molecules, for example, MEA-GQDs (Amini et al., 2017), CdTe-L QDs (Elmizadeh et al., 2017). THF cocktail nanoparticles have been synthesized for the detection of alkaline and transition metals (Jarolímová et al., 2016).. A variety of nanosensors have been used for Cu^{+2} identification CdSe@CDs-TPEA, silica-coated CdTe, β-amino alcohol coated nanosensors with high selectivity, and a low detection limit. The thiosemicarbazide incorporated carbon dots have been synthesized via a hydrothermal process from EDTA as a substrate and thiosemicarbazide as identifying moiety (Figure 8.14). Interestingly, the thiosemicarbazide-carbon dots fluorescent probe has excellent detection potency, good biocompatibility and low toxicity (Fu and Cui, 2016).

8.4.3 Detection of Organic Compounds

8.4.3.1 Detection of Amino Acids, Proteins and Vitamins

Amino acids and vitamins are beneficial for the pharmaceutical industry and as food additives to balance diet. An adequate amount is substantial for the human body to maintain good metabolism. The presence of amino acids, that is L-lysin (Ameen et al., 2016), glutathione, homocysteine, cysteine (Guan et al., 2016) can be monitored by a fluorescent nanosensor in eukaryotes and prokaryotes due to their higher-level cause abnormalities. The thiol-based tripeptide glutathione is endogenously produced through L-cysteine, L-glutamic acid and glycine (Yin et al., 2014). Glutathione is participated in maintaining homeostasis and the immune system (Park et al., 2013). In the meantime, many researchers have highlighted the inappropriate concentration of glutathione can cause numerous diseases, such as liver damage, Parkinson's, diabetes, human immunodeficiency virus

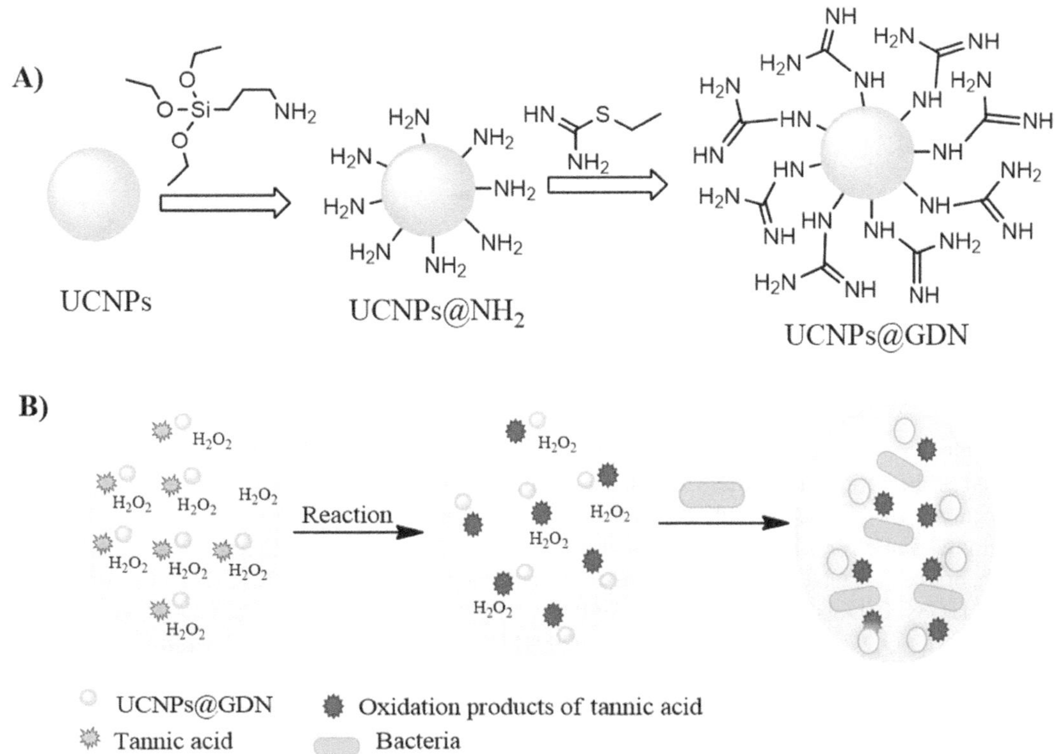

FIGURE 8.13A and 8.13B Identification of bacteria from fluorescent nanosensor.

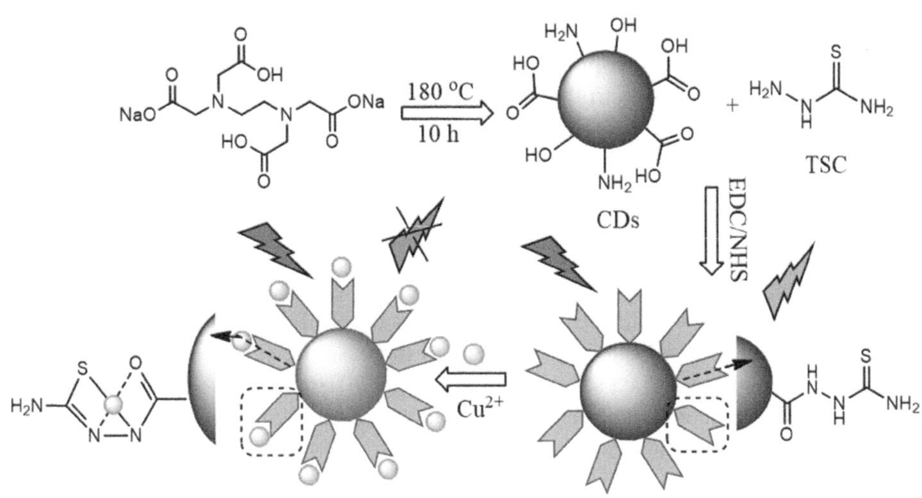

FIGURE 8.14 Thiosemicarbazide-carbon dots for copper ion detection.

(HIV) (Lu, 2009). With the rise of health problems, early detection of amino acids and vitamins is significant for human health. The carbon dots are an updated version of nanoparticles (CD-RhB nonhybrid) and effective for the detection of tripeptide glutathione and mercury(II) ions (Lu et al., 2016).

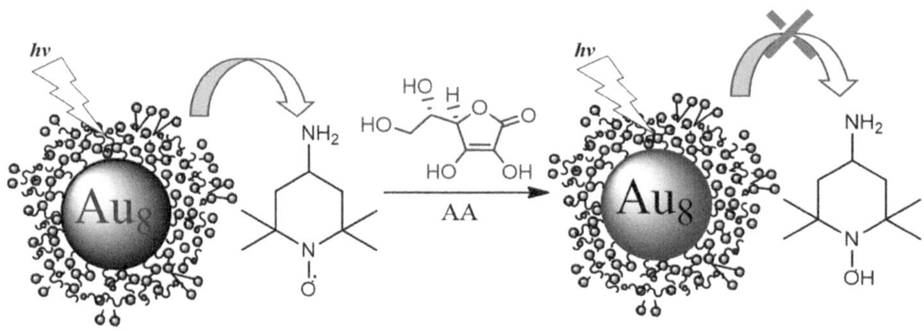

FIGURE 8.15 The schematic depiction of Au_8-cluster functioning for detection of AA.

The telomerase protein is specifically related to DNA-protein used during cell division. The abnormal proliferation of telomerase cells can cause cancer. Telomerase activities can be monitored with accuracy by graphene oxide-derived fluorescent nanosensors (Zhang et al., 2018). Ascorbic acid, another name for vitamin C, can also be detected via intracellular imaging through fluorescent nanosensors (e.g., gCNQDs-TEMPO-ZnPc (Achadu and Nyokong, 2017), CD-MnO_2, PNCQDs/Cr^{6+}, Eu HDSs, C-Dots-$KMnO_4$, CuInZnS QDs/VB1/$KMnO_4$, CQDs/AuNCs, CD-QD@SiO_2, Ag^+/OPD and CD-QD@SiO_2–Fe^{3+} (Zhao et al., 2021)). The Au_8-cluster with the nitroxide radicals will quench and restore fluorescence on the reduction of nitroxide radicals by the ascorbic acid (Figure 8.15) (Liu et al., 2015a). Some other riboflavin (vitamin B) can also be detected from single-walled carbon nanotubes (SWNT) (Iverson et al., 2016).

8.4.3.2 Detection of Drugs

Numerous drugs that are present inside the body can be detected to monitor human fitness tests for agility. Toxic drugs such as methamphetamine and amphetamine cause stimulant psychosis (Pavlova and Petrovska – Jovanović, 2007). Traditionally, the drug presences were detected via conventional methods (i.e., GCMS, HPLC/MS, capillary electrophoresis) that are time-consuming, complicated and expensive as compared to spectrofluorimetric methods (Magdy et al., 2021). The drawbacks of conventional methods can be overcome by rapid, simple and efficient methods via fluorescent nanosensors for the detection of drugs (i.e., florfenicol, chloramphenicol, thiamphenicol (Amjadi et al., 2016), kanamycin (Liu et al., 2016b), cefixime (Akhgari et al., 2017), and nicotine (Xu et al., 2021)). Incorporation of 1,8-naphthaliamide-thiphene derivatives (NTS) with immobilized silica able to form nano-silica particles (NTS@SiO_2), which are highly sensitive nano-sensing material to detect methamphetamine (Figure 8.16) (Rouhani and Haghgoo, 2015). Toxic organic material such as melamine can also be detected via UCNPs-based FRET nanosensors, which are illegally added to pet foods, milk powder and animal feeds to increase the superficial protein content (Wu et al., 2015). Essential nutrients (i.e., selenium (Chen et al., 2016), iron) are required in the adequate amount to maintain the human metabolism. An imbalance in nutrients can be identified via fluorescent nanosensors (Deng et al., 2016).

8.5 FUTURE PERSPECTIVES

Fluorescent nanosensors are promising and emerging materials with a desired and distinctive feature that is significant in biomedicinal applications. The forthcoming chances of a fluorescent nanosensor for biomedical applications are auspicious, particularly for disease diagnosis and treatment. The identification of slight disturbance (i.e., pH, temperature, oxygen) in living organisms was a very crucial issue that is overcome by the development of fluorescent nanosensors. In the future, further modification for highly effective and sensitive nanosensors is required with controlled intelligence

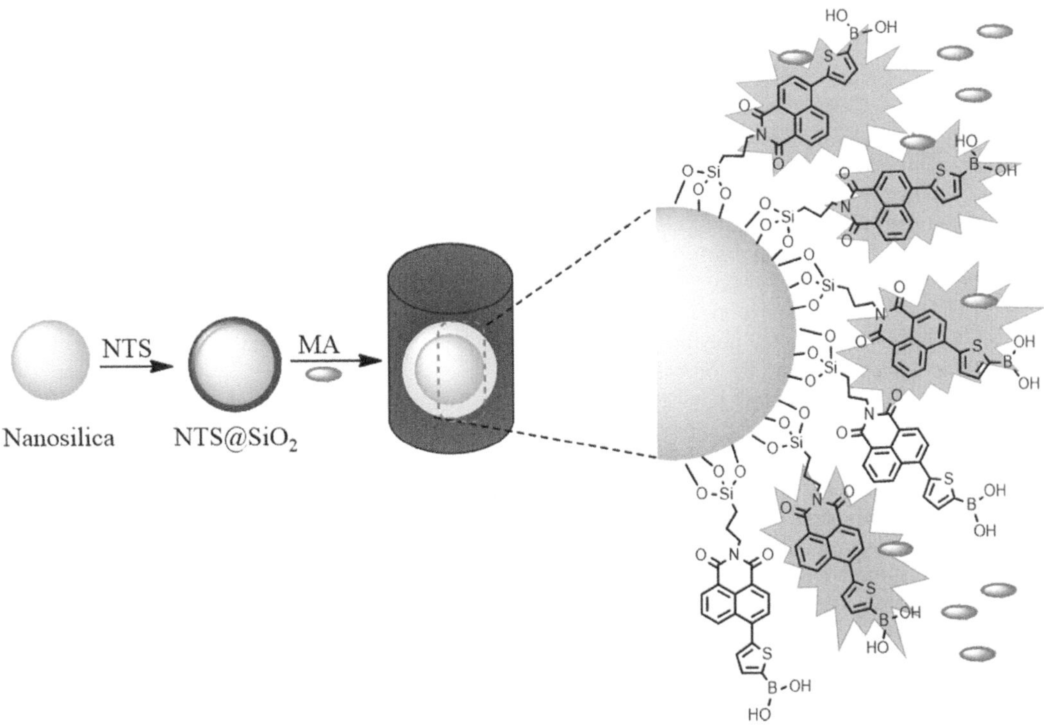

FIGURE 8.16 Methamphetamine detection through nanosensors.

and mental functioning. However, these fields of research are still in their early stages, and untold efforts are still required for the development of biomedical devices with upgraded characteristics. These research areas need to expand and are expected to provide opportunities for further research and development.

8.6 CONCLUSION

In summary, the latest advances in the synthesis and applications of fluorescent nanosensors are comprehensively presented and discussed in this chapter. Several techniques, such as hydrothermal, photoinduced polymerization, precipitation or reprecipitation and others have been applied for the designing of various fluorescent nanosensor materials. Fluorescent nanosensors size or shape or structure can be adjusted by the fabrication methods and easily characterized by FTIR, SEM, AFM, TEM, and others. Moreover, the synthesized fluorescent nanosensor materials exhibit wide applications in the biomedical field to analyze pH, temperature, oxygen, drugs, or disease and are tremendously beneficial for human health.

REFERENCES

Achadu, O.J. and Nyokong, T. 2017. In situ one-pot synthesis of graphitic carbon nitride quantum dots and its 2,2,6,6-tetramethyl(piperidin-1-yl)oxyl derivatives as fluorescent nanosensors for ascorbic acid. *Analytica Chimica Acta*, 991, 113–126.

Akhgari, F., Samadi, N. and Farhadi, K. 2017. Fluorescent carbon dot as nanosensor for sensitive and selective detection of cefixime based on inner filter effect. *Journal of Fluorescence*, 27, 921–927.

Aladaghlo, Z., Javanbakht, S., Fakhari, A.R. and Shaabani, A. 2021. Gelatin microsphere coated Fe_3O_4@ graphene quantum dots nanoparticles as a novel magnetic sorbent for ultrasound-assisted dispersive magnetic solid-phase extraction of tricyclic antidepressants in biological samples. *Microchimica Acta*, 188, 73.

Ali, H., Ghosh, S. and Jana, N.R. 2020. Fluorescent carbon dots as intracellular imaging probes. *WIREs Nanomedicine and Nanobiotechnology*, 12, e1617.

Ameen, S., Ahmad, M., Mohsin, Mohd., Qureshi, M.I., Ibrahim, M.M., Abdin, M.Z. and Ahmad, A. 2016. Designing, construction and characterization of genetically encoded FRET-based nanosensor for real time monitoring of lysine flux in living cells. *Journal of Nanobiotechnology*, 14, 1–11.

Amini, M.H., Faridbod, F., Ganjali, M.R. and Norouzi, P. 2017. Functionalized graphene quantum dots as a fluorescent "off–on" nanosensor for detection of mercury and ethyl xanthate. *Research on Chemical Intermediates*, 43, 7457–7470.

Amjadi, M., Jalili, R. and Manzoori, J. L. 2016. A sensitive fluorescent nanosensor for chloramphenicol based on molecularly imprinted polymer-capped CdTe quantum dots: A fluorescent nanosensor for chloramphenicol. *Luminescence*, 31, 633–639.

Ashokkumar, P., Adarsh, N. and Klymchenko, A.S. 2020. Ratiometric nanoparticle probe based on FRET-amplified phosphorescence for oxygen sensing with minimal phototoxicity. *Small*, 16, 2002494.

Bai, M., Huang, S., Xu, S., Hu, G. and Wang, L. 2015. Fluorescent nanosensors via photoinduced polymerization of hydrophobic inorganic quantum dots for the sensitive and selective detection of nitroaromatics. *Analytical Chemistry*, 87, 2383–2388.

Banerjee, T., Sulthana, S., Shelby, T., Heckert, B., Jewell, J., Woody, K., Karimnia, V., McAfee, J. and Santra, S. 2016. Multiparametric magneto-fluorescent nanosensors for the ultrasensitive detection of Escherichia coli O157:H7. *ACS Infectious Diseases*, 2, 667–673.

Barati, A., Shamsipur, M. and Abdollahi, H. 2016. Carbon dots with strong excitation-dependent fluorescence changes towards pH. Application as nanosensors for a broad range of pH. *Analytica Chimica Acta*, 931, 25–33.

Belal, F., Mabrouk, M., Hammad, S., Barseem, A. and Ahmed, H. 2021. A novel eplerenone ecofriendly fluorescent nanosensor based on nitrogen and sulfur-carbon quantum dots. *Journal of Fluorescence*, 31, 85–90.

Borisov, S.M., Mayr, T., Mistlberger, G., Waich, K., Koren, K., Chojnacki, P. and Klimant, I. 2009. Precipitation as a simple and versatile method for preparation of optical nanochemosensors. *Talanta*, 79, 1322–1330.

Cammidge, A.N., Downing, S. and Ngaini, Z. 2003. Surface-functionalised nano-beads as novel supports for organic synthesis. *Tetrahedron Letters*, 44, 6633–6634.

Caprifico, A.E., Polycarpou, E., Foot, P.J. and Calabrese, G. 2020. Fluorescein isothiocyanate chitosan nanoparticles in oral drug delivery studies. *Trends in Pharmacological Sciences*, 41, 686–689.

Chakraborty, S., Joseph, M.M., Varughese, S., Ghosh, S., Maiti, K.K., Samanta, A. and Ajayaghosh, A. 2020. A new pentacyclic pyrylium fluorescent probe that responds to pH imbalance during apoptosis. *Chemical Science*, 11, 12695–12700.

Chauhan, V.M., Burnett, G.R. and Aylott, J.W. 2011. Dual-fluorophore ratiometric pH nanosensor with tuneable pKa and extended dynamic range. *The Analyst*, 136, 1799.

Chauhan, V.M., Giuntini, F. and Aylott, J.W. 2016. Quadruple labelled dual oxygen and pH-sensitive ratiometric nanosensors. *Sensing and Bio-Sensing Research*, 8, 36–42.

Chauhan, V.M., Hopper, R.H., Ali, S.Z., King, E.M., Udrea, F., Oxley, C.H. and Aylott, J.W. 2014. Thermo-optical characterization of fluorescent rhodamine B based temperature-sensitive nanosensors using a CMOS MEMS micro-hotplate. *Sensors and Actuators B: Chemical*, 192, 126–133.

Chauhan, V.M., Orsi, G., Brown, A., Pritchard, D.I. and Aylott, J.W. 2013. Mapping the pharyngeal and intestinal pH of *Caenorhabditis elegans* and real-time luminal pH oscillations using extended dynamic range pH-sensitive nanosensors. *ACS Nano*, 7, 5577–5587.

Chen, J., Zhang, C., Lv, K., Wang, H., Zhang, P., Yi, P. and Jiang, J. 2017. A silica nanoparticle-based dual-responsive ratiometric probe for visualizing hypochlorite and temperature with distinct fluorescence signals. *Sensors and Actuators B: Chemical*, 251, 533–541.

Chen, L., Tian, X., Zhao, Y., Li, Y., Yang, C., Zhou, Z. and Liu, X. 2016. A ratiometric fluorescence nanosensor for highly selective and sensitive detection of selenite. *The Analyst*, 141, 4685–4693.

Chen, L., Wu, Y., Lin, Y. and Wang, Q. 2015a. Virus-templated FRET platform for the rational design of ratiometric fluorescent nanosensors. *Chemical Communications*, 51, 10190–10193.

Chen, P., Bai, W. and Bao, Y. 2019. Fluorescent chemodosimeters for fluoride ions *via* silicon-fluorine chemistry: 20 years of progress. *Journal of Materials Chemistry C*, 7, 11731–11746.

Chen, T.-T., Tian, X., Liu, C.-L., Ge, J., Chu, X. and Li, Y. 2015b. Fluorescence activation imaging of cytochrome c released from mitochondria using aptameric nanosensor. *Journal of the American Chemical Society*, 137, 982–989.

Chen, Z., Zhao, Z., Wang, Z., Zhang, Y., Sun, X., Hou, L. and Yuan, C. 2018. Foxtail millet-derived highly fluorescent multi-heteroatoms doped carbon quantum dots towards fluorescent ink and smart nanosensor for selective ion detection. *New Journal of Chemistry*, 42, 7326–7331.

Cheng, C., Zhang, R., Wang, J., Zhang, Y., Wen, C., Tan, Y. and Yang, M. 2020. An ultrasensitive and selective fluorescent nanosensor based on porphyrinic metal–organic framework nanoparticles for Cu^{2+} detection. *The Analyst*, 145, 797–804.

Costas-Mora, I., Romero, V., Lavilla, I. and Bendicho, C. 2014. In situ building of a nanoprobe based on fluorescent carbon dots for methylmercury detection. *Analytical Chemistry*, 86, 4536–4543.

Cun, T., Dong, C. and Huang, Q. 2016. Ionothermal precipitation of highly dispersive ZnO nanoparticles with improved photocatalytic performance. *Applied Surface Science*, 384, 73–82.

Dang, H., Huang, L.-K., Zhang, Y., Wang, C.F. and Chen, S. 2016. Large-scale ultrasonic fabrication of white fluorescent carbon dots. *Industrial and Engineering Chemistry Research*, 55, 5335–5341.

Das, P., Ganguly, S., Bose, M., Ray, D., Ghosh, S., Mondal, S., Aswal, V.K., Das, A.K., Banerjee, S. and Das, N.C. 2019. Surface quaternized nanosensor as an one-arrow-two-hawks approach for fluorescence turns "on-off-on" bifunctional sensing and antibacterial activity. *New Journal of Chemistry*, 43, 6205–6219.

De, B. and Karak, N. 2013. A green and facile approach for the synthesis of water soluble fluorescent carbon dots from banana juice. *RSC Advances*, 3, 8286.

Deng, M., Wang, S., Liang, C., Shang, H. and Jiang, S. 2016. A FRET fluorescent nanosensor based on carbon dots for ratiometric detection of Fe^{3+} in aqueous solution. *RSC Advances*, 6, 26936–26940.

Desai, A.S., Chauhan, V.M., Johnston, A.P.R., Esler, T. and Aylott, J.W. 2014. Fluorescent nanosensors for intracellular measurements: synthesis, characterization, calibration, and measurement. *Frontiers in Physiology*, 4, 401.

Donmez, M., Yilmaz, M.D. and Kilbas, B. 2017. Fluorescent detection of dipicolinic acid as a biomarker of bacterial spores using lanthanide-chelated gold nanoparticles. *Journal of Hazardous Materials*, 324, 593–598.

El-Malla, S.F., Elshenawy, E.A., Hammad, S.F. and Mansour, F.R. 2021. N-doped carbon dots as a fluorescent nanosensor for determination of colchicine based on inner filter effect. *Journal of Fluorescence*, 1–10.

Elmizadeh, H., Soleimani, M., Faridbod, F. and Bardajee, G.R. 2017. Ligand-capped CdTe quantum dots as a fluorescent nanosensor for detection of copper ions in environmental water sample. *Journal of Fluorescence*, 27, 2323–2333.

Elsutohy, M.M., Chauhan, V.M., Markus, R., Kyyaly, M.A., Tendler, S.J. and Aylott, J.W. 2017. Real-time measurement of the intracellular pH of yeast cells during glucose metabolism using ratiometric fluorescent nanosensors. *Nanoscale*, 9, 5904–5911.

Fang, B., Lu, X., Hu, J., Zhang, G., Zheng, X., He, L., Cao, J., Gu, J. and Cao, F. 2019. pH controlled green luminescent carbon dots derived from benzoxazine monomers for the fluorescence turn-on and turn-off detection. *Journal of Colloid and Interface Science*, 536, 516–525.

Farooq, S., Ngaini, Z. and Farooq, S. 2020. *Manufacturing and design of smart polymer composites. Smart polymer nanocomposites: biomedical and environmental applications. S.l.*: Woodhead Publishing.

Francis, S. and Rajith, L. 2021. Selective fluorescent sensing of adenine via the emissive enhancement of a simple cobalt porphyrin. *Journal of Fluorescence*, 31, 577–586.

Fu, H., Ji, Z., Chen, X., Cheng, A., Liu, S., Gong, P., Li, G., Chen, G., Sun, Z., Zhao, X. and Cheng, F. 2017. A versatile ratiometric nanosensing approach for sensitive and accurate detection of Hg^{2+} and biological thiols based on new fluorescent carbon quantum dots. *Analytical and Bioanalytical Chemistry*, 409, 2373–2382.

Fu, Z. and Cui, F. 2016. Thiosemicarbazide chemical functionalized carbon dots as a fluorescent nanosensor for sensing Cu^{2+} and intracellular imaging. *RSC Advances*, 6, 63681–63688.

Funovics, M., Weissleder, R. and Tung, C.-H. 2003. Protease sensors for bioimaging. *Analytical and Bioanalytical Chemistry*, 377, 956–963.

Gao, X., Zhou, X., Ma, Y., Qian, T., Wang, C. and Chu, F. 2019. Facile and cost-effective preparation of carbon quantum dots for Fe^{3+} ion and ascorbic acid detection in living cells based on the "on-off-on" fluorescence principle. *Applied Surface Science*, 469, 911–916.

García-Añoveros, J. and Corey, D.P. 1997. The molecules of mechanosensation. *Annual Review of Neuroscience*, 20, 567–594.

Giuntini, F., Chauhan, V.M., Aylott, J.W., Rosser, G.A., Athanasiadis, A., Beeby, A., MacRobert, A.J., Brown, R.A. and Boyle, R.W. 2014. Conjugatable water-soluble Pt(II) and Pd(II) porphyrin complexes: novel nano- and molecular probes for optical oxygen tension measurement in tissue engineering. *Photochem. Photobiol. Sci.*, 13, 1039–1051.

Gong, J., Lu, X. and An, X. 2015. Carbon dots as fluorescent off–on nanosensors for ascorbic acid detection. *RSC Advances*, 5, 8533–8536.

Gu, D., Shang, S., Yu, Q. and Shen, J. 2016. Green synthesis of nitrogen-doped carbon dots from lotus root for Hg(II) ions detection and cell imaging. *Applied Surface Science*, 390, 38–42.

Guan, J., Yang, J., Zhang, Y., Zhang, X., Deng, H., Xu, J., Wang, J. and Yuan, M.S. 2021. Employing a fluorescent and colorimetric picolyl-functionalized rhodamine for the detection of glyphosate pesticide. *Talanta*, 224, 121834.

Guan, Q., Su, R., Zhang, M., Zhang, R., Li, W., Wang, D., Xu, M., Fei, L. and Xu, Q. 2019. Highly fluorescent dual-emission red carbon dots and their applications in optoelectronic devices and water detection. *New Journal of Chemistry*, 43, 3050–3058.

Guan, Y., Qu, S., Li, B., Zhang, L., Ma, H. and Zhang, L. 2016. Ratiometric fluorescent nanosensors for selective detecting cysteine with upconversion luminescence. *Biosensors and Bioelectronics*, 77, 124–130.

Hu, Y., Yang, J., Jia, L. and Yu, J.S. 2015. Ethanol in aqueous hydrogen peroxide solution: Hydrothermal synthesis of highly photoluminescent carbon dots as multifunctional nanosensors. *Carbon*, 93, 999–1007.

Huang, H., Dong, F. and Tian, Y. 2016a. Mitochondria-targeted ratiometric fluorescent nanosensor for simultaneous biosensing and imaging of $O_2^{\bullet-}$ and pH in live cells. *Analytical Chemistry*, 88, 12294–12302.

Huang, M., Chong, J., Hu, C. and Yang, Y. 2020. Ratiometric fluorescent detection of temperature and MnO_4 using a modified covalent organic framework. *Inorganic Chemistry Communications*, 119, 108094.

Huang, S., Wang, L., Huang, C., Su, W. and Xiao, Q. 2016b. Amino-functionalized graphene quantum dots based ratiometric fluorescent nanosensor for ultrasensitive and highly selective recognition of horseradish peroxidase. *Sensors and Actuators B: Chemical*, 234, 255–263.

Huang, Y.-C., Chen, C.-P., Wu, P.-J., Kuo, S.Y. and Chan, Y.H. 2014. Coumarin dye-embedded semiconducting polymer dots for ratiometric sensing of fluoride ions in aqueous solution and bio-imaging in cells. *J. Mater. Chem. B*, 2, 6188–6191.

Iverson, N.M., Bisker, G., Farias, E., Ivanov, V., Ahn, J., Wogan, G.N. and Strano, M.S. 2016. Quantitative tissue spectroscopy of near infrared fluorescent nanosensor implants. *Journal of Biomedical Nanotechnology*, 12, 1035–1047.

Jarolímová, Z., Vishe, M., Lacour, J. and Bakker, E. 2016. Potassium ion-selective fluorescent and pH independent nanosensors based on functionalized polyether macrocycles. *Chemical Science*, 7, 525–533.

Jia, Y., Zhao, T., Jiang, Y., Sun, W., Zhao, Y., Xin, J., Hou, Y. and Yang, W. 2018. Green, fast, and large-scale synthesis of highly fluorescent Au nanoclusters for Cu^{2+} and temperature sensing. *The Analyst*, 143, 5145–5150.

Jokic, T., Borisov, S.M., Saf, R., Nielsen, D.A., Kühl, M. and Klimant, I. 2012. Highly photostable near-infrared fluorescent ph indicators and sensors based on BF_2-chelated tetraarylazadipyrromethene dyes. *Analytical Chemistry*, 84, 6723–6730.

Khan, T., Civas, M., Cetinkaya, O., Abbasi, N.A. and Akan, O.B. 2020. Nanosensor networks for smart health care. *Nanosensors for Smart Cities*, Elsevier, 387–403.

Kundu, A., Nandi, S., Das, P. and Nandi, A.K. 2016. Facile and green approach to prepare fluorescent carbon dots: Emergent nanomaterial for cell imaging and detection of vitamin B2. *Journal of Colloid and Interface Science*, 468, 276–283.

Lavado, A.S., Chauhan, V.M., Zen, A.A., Giuntini, F., Jones, D.R.E., Boyle, R.W., Beeby, A., Chan, W.C. and Aylott, J.W. 2015. Controlled intracellular generation of reactive oxygen species in human mesenchymal stem cells using porphyrin conjugated nanoparticles. *Nanoscale*, 7, 14525–14531.

Li, J., Wei, Y.-Y. and Xu, Z.-R. 2020. Visual detection of acid phosphatase based on hollow mesoporous manganese dioxide nanospheres. *Analytica Chimica Acta*, 1138, 1–8.

Liu, C.P., Wu, T.-H., Liu, C.Y., Cheng, H.J. and Lin, S.Y., 2015a. Interactions of nitroxide radicals with dendrimer-entrapped Au_8-clusters: a fluorescent nanosensor for intracellular imaging of ascorbic acid. *Journal of Materials Chemistry B*, 3, 191–197.

Liu, S., Zhang, H., Qiao, Y. and Su, X. 2012. One-pot synthesis of ternary $CuInS_2$ quantum dots with near-infrared fluorescence in aqueous solution. *RSC Adv.*, 2, 819–825.

Liu, W., Li, C., Ren, Y., Sun, X., Pan, W., Li, Y., Wang, J. and Wang, W. 2016a. Carbon dots: surface engineering and applications. *Journal of Materials Chemistry B*, 4, 5772–5788.

Liu, Z., Lin, Z., Liu, L. and Su, X. 2015b. A convenient and label-free fluorescence "turn off–on" nanosensor with high sensitivity and selectivity for acid phosphatase. *Analytica Chimica Acta*, 876, 83–90.

Liu, Z., Tian, C., Lu, L., and Su, X. 2016b. A novel aptamer-mediated $CuInS_2$ quantum dots@graphene oxide nanocomposites-based fluorescence "turn off–on" nanosensor for highly sensitive and selective detection of kanamycin. *RSC Advances*, 6, 10205–10214.

Lee, M.H., Han, J.H., Lee, J.H., Park, N., Kumar, R., Kang, C. and Kim, J.S. 2013. Two-color probe to monitor a wide range of pH values in cells. *Angewandte Chemie*, 125, 6326-6329.

Lu, S., Wu, D., Li, G., Lv, Z., Chen, Z., Chen, L., Chen, G., Xia, L., You, J. and Wu, Y. 2016. Carbon dots-based ratiometric nanosensor for highly sensitive and selective detection of mercury(II) ions and glutathione. *RSC Advances*, 6, 103169–103177.

Lu, S.C. 2009. Regulation of glutathione synthesis. *Molecular Aspects of Medicine*, 30, 42–59.

Ma, Y., Mei, J., Bai, J., Chen, X. and Ren, L. 2018. Ratiometric fluorescent nanosensor based on carbon dots for the detection of mercury ion. *Materials Research Express*, 5, 055605.

Magdy, G., Abdel Hakiem, A.F., Belal, F. and Abdel-Megied, A.M. 2021. Green one-pot synthesis of nitrogen and sulfur co-doped carbon quantum dots as new fluorescent nanosensors for determination of salinomycin and maduramicin in food samples. *Food Chemistry*, 343, 128539.

Mohammed, L.J. and Omer, K.M. 2020. Carbon dots as new generation materials for nanothermometer: review. *Nanoscale Research Letters*, 15, 182.

Monte-Filho, S.S., Andrade, S.I.E., Lima, M.B. and Araujo, M.C. 2019. Synthesis of highly fluorescent carbon dots from lemon and onion juices for determination of riboflavin in multivitamin/mineral supplements. *Journal of Pharmaceutical Analysis*, 9, 209–216.

Negut, C.C., Stefan-van Staden, R.-I. and van Staden, J.F. 2020. Porphyrins-as active materials in the design of sensors. an overview. *ECS Journal of Solid State Science and Technology*, 9, 051005.

Newman Monday, Y., Abdullah, J., Yusof, N.A., Abdul Rashid S. and Shueb R.H. 2021. Facile hydrothermal and solvothermal synthesis and characterization of nitrogen-doped carbon dots from palm kernel shell precursor. *Applied Sciences*, 11, 1630.

Pandey, G., Chaudhari, R., Joshi, B., Choudhary, S., Kaur, J. and Joshi, A. 2019. Fluorescent biocompatible platinum-porphyrin–doped polymeric hybrid particles for oxygen and glucose biosensing. *Scientific Reports*, 9, 5029.

Pang, S. and Liu, S. 2020. Dual-emission carbon dots for ratiometric detection of Fe^{3+} ions and acid phosphatase. *Analytica Chimica Acta*, 1105, 155–161. doi:10.1016/j.aca.2020.01.033.

Park, K.S., Kim, M.I., Woo, M.A. and Park, H.G. 2013. A label-free method for detecting biological thiols based on blocking of Hg^{2+}-quenching of fluorescent gold nanoclusters. *Biosensors and Bioelectronics*, 45, 65–69.

Pavlova, V. and Petrovska – Jovanović, S. 2007. Simultaneous determination of amphetamine, methamphetamine, and caffeine in seized tablets by high-performance liquid chromatography. *Acta Chromatographica*, 18, 157–167.

Peng, H.-S. and Chiu, D.T. 2015. Soft fluorescent nanomaterials for biological and biomedical imaging. *Chemical Society Reviews*, 44, 4699–4722.

Petrakova, V., Rehor, I., Stursa, J., Ledvina, M., Nesladek, M. and Cigler, P. 2015. Charge-sensitive fluorescent nanosensors created from nanodiamonds. *Nanoscale*, 7, 12307–12311.

Poon, C.K., Tang, O., Chen, X.-M., Kim, B., Hartlieb, M., Pollock, C.A., Hawkett, B.S. and Perrier, S. 2017. Fluorescent labeling and biodistribution of latex nanoparticles formed by surfactant-free RAFT emulsion polymerization. *Macromolecular Bioscience*, 17, 1600366.

Qian, J., Hua, M., Wang, C., Wang, K., Liu, Q., Hao, N. and Wang, K. 2016. Fabrication of L-cysteine-capped CdTe quantum dots based ratiometric fluorescence nanosensor for onsite visual determination of trace TNT explosive. *Analytica Chimica Acta*, 946, 80–87.

Qiu, X., Han, S., Hu, Y. and Sun, B. 2015. Ratiometric fluorescent nanosensors for copper(II) based on bis(rhodamine)-derived pmos with J-type aggregates. *Chemistry - A European Journal*, 21, 4126–4132.

Rouhani, S. and Haghgoo, S. 2015. A novel fluorescence nanosensor based on 1,8-naphthalimide-thiophene doped silica nanoparticles, and its application to the determination of methamphetamine. *Sensors and Actuators B: Chemical*, 209, 957–965.

Ryavanaki, L., Tsai, H. and Fuh, C.B. 2020. Microwave synthesis of gold nanoclusters with garlic extract modifications for the simple and sensitive detection of lead ions. *Nanomaterials*, 10, 94.

Saenwong, K., Putthasehn, C., Tunsawat, A., Nuengmatcha, P. and Chanthai, S. 2018. Using thermolytic solution of anionic – decorated Gqds as fluorescence turn on-off sensor for selective screening test of metal ions. *Oriental Journal of Chemistry*, 34, 55–63.

Shellaiah, M. and Sun, K.W. 2020. Review on sensing applications of perovskite nanomaterials. *Chemosensors*, 8, 55.

Shen, J., Shang, S., Chen, X., Wang, D. and Cai, Y. 2017. Facile synthesis of fluorescence carbon dots from sweet potato for Fe^{3+} sensing and cell imaging. *Materials Science and Engineering: C*, 76, 856–864.

Shi, D., Yan, F., Zheng, T., Wang, Y., Zhou, X. and Chen, L. 2015. P-doped carbon dots act as a nanosensor for trace 2,4,6-trinitrophenol detection and a fluorescent reagent for biological imaging. *RSC Advances*, 5, 98492–98499.

Singh, R., Kashayap, S., Singh, V., Kayastha, A.M., Mishra, H., Saxena, P.S., Srivastava, A. and Singh, R.K. 2018. QPRTase modified N-doped carbon quantum dots: A fluorescent bioprobe for selective detection of neurotoxin quinolinic acid in human serum. *Biosensors and Bioelectronics*, 101, 103–109.

Subhahar, M.B., Karakka Kal, A.K., Philip, M., K. Karatt, T., Vazhat, R.A. and MP, M.A. 2021. Detection and identification of ACP-105 and its metabolites in equine urine using LC/MS/MS after oral administration. *Drug Testing and Analysis*, 13, 299–317.

Tian, J., Cheng, N., Liu, Q., Xing, W. and Sun, X. 2015. Cobalt phosphide nanowires: efficient nanostructures for fluorescence sensing of biomolecules and photocatalytic evolution of dihydrogen from water under visible light. *Angewandte Chemie*, 127, 5583–5587.

Uriarte, D., Domini, C. and Garrido, M. 2019. New carbon dots based on glycerol and urea and its application in the determination of tetracycline in urine samples. *Talanta*, 201, 143–148.

Vyas, S.S., Jadhav, S.V., Majee, S.B., Shastri, J.S. and Patravale, V.B. 2015. Development of immunochromatographic strip test using fluorescent, micellar silica nanosensors for rapid detection of B. abortus antibodies in milk samples. *Biosensors and Bioelectronics*, 70, 254–260.

Wang, C., Xu, Z., Cheng, H., Lin, H., Humphrey, M.G. and Zhang, C. 2015a. A hydrothermal route to water-stable luminescent carbon dots as nanosensors for pH and temperature. *Carbon*, 82, 87–95.

Wang, J., Su, S., Wei, J., Bahgi, R., Hope-Weeks, L., Qiu, J. and Wang, S. 2015b. Ratio-metric sensor to detect riboflavin via fluorescence resonance energy transfer with ultrahigh sensitivity. *Physica E: Low-dimensional Systems and Nanostructures*, 72, 17–24.

Wang, Q., Liu, X., Zhang, L., and Lv, Y. 2012. Microwave-assisted synthesis of carbon nanodots through an eggshell membrane and their fluorescent application. *The Analyst*, 137, 5392.

Wang, W., Li, H., Yin, M., Wang, K., Deng, Q., Wang, S. and Zhang, Y. 2018. Highly selective and sensitive sensing of 2,4,6-trinitrophenol in beverages based on guanidine functionalized upconversion fluorescent nanoparticles. *Sensors and Actuators B: Chemical*, 255, 1422–1429.

Wang, X., Meier, R.J., Schäferling, M., Bange, S., Lupton, J.M., Sperber, M., Wegener, J., Ondrus, V., Beifuss, U., Henne, U. and Klein, C. 2016. Two-photon excitation temperature nanosensors based on a conjugated fluorescent polymer doped with a europium probe. *Advanced Optical Materials*, 4, 1854–1859.

Wang, X.D., Song, X.H., He, C.Y., Yang, C.J., Chen, G. and Chen, X. 2011. Preparation of reversible colorimetric temperature nanosensors and their application in quantitative two-dimensional thermo-imaging. *Analytical Chemistry*, 83, 2434–2437.

Wang, X. and Wolfbeis, O.S. 2014. Optical methods for sensing and imaging oxygen: materials, spectroscopies and applications. *Chem. Soc. Rev.*, 43, 3666–3761.

Watzke, A., Kosec, G., Kindermann, M., Jeske, V., Nestler, H.P., Turk, V., Turk, B. and Wendt, K.U. 2008. Selective activity-based probes for cysteine cathepsins. *Angewandte Chemie International Edition*, 47, 406–409.

Welser, K., Grilj, J., Vauthey, E., Aylott, J.W. and Chan, W.C. 2009. Protease responsive nanoprobes with tethered fluorogenic peptidyl 3-arylcoumarin substrates. *Chemical Communications*, 6, 671–673.

Wu, L., Zou, H., Wang, H., Zhang, S., Liu, S., Jiang, Y., Chen, J., Li, Y., Shao, M., Zhang, R. and Li, X. 2021. Update on the development of molecular imaging and nanomedicine in China: optical imaging. *WIREs Nanomedicine and Nanobiotechnology*, 13, e1660.

Wu, Q., Long, Q., Li, H., Zhang, Y. and Yao, S. 2015. An upconversion fluorescence resonance energy transfer nanosensor for one step detection of melamine in raw milk. *Talanta*, 136, 47–53.

Xia, J., Chen, S., Zou, G.Y., Yu, Y.L. and Wang, J.H. 2018. Synthesis of highly stable red-emissive carbon polymer dots by modulated polymerization: from the mechanism to application in intracellular pH imaging. *Nanoscale*, 10, 22484–22492.

Xie, X., Gutiérrez, A., Trofimov, V., Szilagyi, I., Soldati, T. and Bakker, E. 2015. Charged solvatochromic dyes as signal transducers in pH independent fluorescent and colorimetric ion selective nanosensors. *Analytical Chemistry*, 87, 9954–9959.

Xie, X., Szilagyi, I., Zhai, J., Wang, L. and Bakker, E. 2016. Ion-selective optical nanosensors based on solvatochromic dyes of different lipophilicity: from bulk partitioning to interfacial accumulation. *ACS Sensors*, 1, 516–520.

Xu, L., Callaway, Z.T., Wang, R., Wang, H., Slavik, M.F., Wang, A. and Li, Y. 2015. A fluorescent aptasensor coupled with nanobead-based immunomagnetic separation for simultaneous detection of four foodborne pathogenic bacteria. *Transactions of the ASABE*, 58, 891–906.

Xu, X., Chen, Z., Li, Q., Meng, D., Jiang, H., Zhou, Y., Feng, S. and Yang, Y. 2021. Copper and nitrogen-doped carbon dots as an anti-interference fluorescent probe combined with magnetic material purification for nicotine detection. *Microchemical Journal*, 160, 105708.

Yan, F., Shi, D., Zheng, T., Yun, K., Zhou, X. and Chen, L. 2016a. Carbon dots as nanosensor for sensitive and selective detection of Hg^{2+} and L-cysteine by means of fluorescence "Off–On" switching. *Sensors and Actuators B: Chemical*, 224, 926–935.

Yan, X., Song, Y., Zhu, C., Song, J., Du, D., Su, X. and Lin, Y. 2016b. Graphene quantum dot–MnO_2 nanosheet based optical sensing platform: a sensitive fluorescence "turn off–on" nanosensor for glutathione detection and intracellular imaging. *ACS Applied Materials and Interfaces*, 8, 21990–21996.

Yang, L., Li, N., Pan, W., Yu, Z. and Tang, B. 2015. Real-time imaging of mitochondrial hydrogen peroxide and ph fluctuations in living cells using a fluorescent nanosensor. *Analytical Chemistry*, 87, 3678–3684.

Yin, J., Kwon, Y., Kim, D., Lee, D., Kim, G., Hu, Y., Ryu, J.H. and Yoon, J. 2014. Cyanine-based fluorescent probe for highly selective detection of glutathione in cell cultures and live mouse tissues. *Journal of the American Chemical Society*, 136, 5351–5358.

Yin, M., Jing, C., Li, H., Deng, Q. and Wang, S. 2020. Surface chemistry modified upconversion nanoparticles as fluorescent sensor array for discrimination of foodborne pathogenic bacteria. *Journal of Nanobiotechnology*, 18, 1-14.

Yin, M., Wu, C., Li, H., Jia, Z., Deng, Q., Wang, S. and Zhang, Y. 2019. Simultaneous sensing of seven pathogenic bacteria by guanidine-functionalized upconversion fluorescent nanoparticles. *ACS Omega*, 4, 8953–8959.

Yuan, H., Yu, J., Feng, S. and Gong, Y. 2016. Highly photoluminescent pH-independent nitrogen-doped carbon dots for sensitive and selective sensing of p-nitrophenol. *RSC Advances*, 6, 15192–15200.

Zhang, B., Li, H., Pan, W., Chen, Q., Ouyang, Q. and Zhao, J. 2017. Dual-color upconversion nanoparticles (UCNPs)-based fluorescent immunoassay probes for sensitive sensing foodborne pathogens. *Food Analytical Methods*, 10, 2036–2045.

Zhang, L., Peng, J., Hong, M.F., Chen, J.Q., Liang, R.P. and Qiu, J.D. 2018. A facile graphene oxide-based fluorescent nanosensor for in situ "turn-on" detection of telomerase activity. *Analyst*, 143, 2334–2341.

Zhang, Q., Song, H., Yu, M., Zhang, H. and Li, Z. 2021. Preparation of yellow fluorescent N,O-CDs and its application in detection of ClO^-. *Journal of Fluorescence*, 1–8.

Zhang, X., Zhang, W., Wang, Q., Wang, J., Ren, G. and Wang, X.D. 2019a. Quadruply-labeled serum albumin as a biodegradable nanosensor for simultaneous fluorescence imaging of intracellular pH values, oxygen and temperature. *Microchimica Acta*, 186, 584.

Zhang, Y., Guo, X., Li, G. and Zhang, G. 2019b. Photoluminescent Ag nanoclusters for reversible temperature and pH nanosenors in aqueous solution. *Analytical and Bioanalytical Chemistry*, 411, 1117–1125.

Zhang, Y., Gutiérrez, M., Chaudhari, A.K. and Tan, J.C. 2020. Dye-encapsulated zeolitic imidazolate framework (ZIF-71) for fluorochromic sensing of pressure, temperature, and volatile solvents. *ACS Applied Materials and Interfaces*, 12, 37477–37488.

Zhao, D., Liu, X., Wei, C., Qu, Y., Xiao, X. and Cheng, H. 2019. One-step synthesis of red-emitting carbon dots via a solvothermal method and its application in the detection of methylene blue. *RSC Advances*, 9, 29533–29540. doi:10.1039/C9RA05570C.

Zhao, J., Huang, M., Zhang, L., Zou, M., Chen, D., Huang, Y. and Zhao, S. 2017. Unique approach to develop carbon dot-based nanohybrid near-infrared ratiometric fluorescent sensor for the detection of mercury ions. *Analytical Chemistry*, 89, 8044–8049.

Zhao, P., Li, X., Baryshnikov, G., Wu, B., Ågren, H., Zhang, J. and Zhu, L. 2018. One-step solvothermal synthesis of high-emissive amphiphilic carbon dots via rigidity derivation. *Chemical Science*, 9, 1323–1329.

Zhao, T., Zhu, C., Xu, S., Wu, X., Zhang, X., Zheng, Y., Wu, M., Tong, Z., Fang, W. and Zhang, K. 2021. Fluorescent color analysis of ascorbic acid by ratiometric fluorescent paper utilizing hybrid carbon dots-silica coated quantum dots. *Dyes and Pigments*, 186, 108995.

Zhou, Y., Ding, J., Liang, T., Abdel-Halim, E.S., Jiang, L. and Zhu, J.J. 2016. FITC doped rattle-type silica colloidal particle-based ratiometric fluorescent sensor for biosensing and imaging of superoxide anion. *ACS Applied Materials and Interfaces*, 8, 6423–6430.

Zhu, S., Zhang, J., Qiao, C., Tang, S., Li, Y., Yuan, W., Li, B., Tian, L., Liu, F., Hu, R., Gao, H., Wei, H., Zhang, H., Sun, H. and Yang, B. (2011). Strongly green-photoluminescent graphene quantum dots for bioimaging applications. *Chemical Communications*, 47, 6858–6860.

Zong, C., Fang, L., Song, F., Wang, A. and Wan, Y. 2019. Fluorescent fingerprint bacteria by multi-channel magnetic fluorescent nanosensor. *Sensors and Actuators B: Chemical*, 289, 234–241.

Zuo, P., Lu, X., Sun, Z., Guo, Y. and He, H. 2016. A review on syntheses, properties, characterization and bioanalytical applications of fluorescent carbon dots. *Microchimica Acta*, 183, 519–542.

9 Applications of Peptide Luminescent Nanosensors

Tatiana Duque Martins Ertner de Almeida,
Diéricon Sousa Cordeiro, Ramon Miranda Silva, and
Igor Lafaete de Freitas Pereira Morais

CONTENTS

9.1 The Advent of Sensors and Nanosensors: Impact on Society 203
9.2 Classification of Nanosensors 207
9.3 Optical Nanosensors; Luminescent Nanosensors 207
9.4 Mechanisms of Operation of Luminescent Peptide Nanosensors 211
9.5 Future Perspectives of Peptide Nanosensors 212
9.6 Acknowledgments 213
References 213

9.1 THE ADVENT OF SENSORS AND NANOSENSORS: IMPACT ON SOCIETY

Since the Industrial Revolution, the need for optimized production lines and for efficiently provided quality control has boosted the development of a wide variety of sensors, based in several functioning concepts. This need led to a new technological field with great perspectives to enter a market, absent at that moment. Nowadays, illustrating the increasing sensing market, there are more than 7 million patent applications that mention the word "sensor" in their titles [1] since the first patent application was filed in 1856 by L. Tillotson, in the United States (Figure 9.1). In the document, Tillotson describes his invention as being "a new and useful apparatus for employing electricity in connection with cupping for the purpose of operating on diseased parts of the human body;[...] The nature of my invention consists in employing the electric, electro-magnetic, voltaic, magneto, and thermo-electric forces, in connection with cupping for the purpose of operating on diseased parts of the human body,[...] the cup of a cupping instrument the apparatus presently described, consisting of a metal -wire placed -within it, to the end of which is fastened a metallic plate or disk (to be used as the positive or negative pole as the case requires), the other end passing through a hermetically sealed aperture and terminating in a loop on the outside [...] It will be seen from the above description in what manner the cup may be made to Id the electrical forces, a both positive and negative, also the way in which the positive and negative forces may aid the action of the cup, in the combined action of both in loo their application to diseased parts of the human body, producing such results as neither one nor the other could produce when used separately" [2].

Due to the increasing ability sensors show of changing their design, structure, functioning mechanisms, supporting and functional materials, since their first conception, their applications always were unlimited, until the need for reliable, biocompatible and miniaturized sensors became a barrier. From this increasing need for miniaturization with applications mainly in medicine, the field of nanotechnology was required to participate in new creations and designs to meet new needs for sensors: nanosensors.

DOI: 10.1201/9781003218708-11

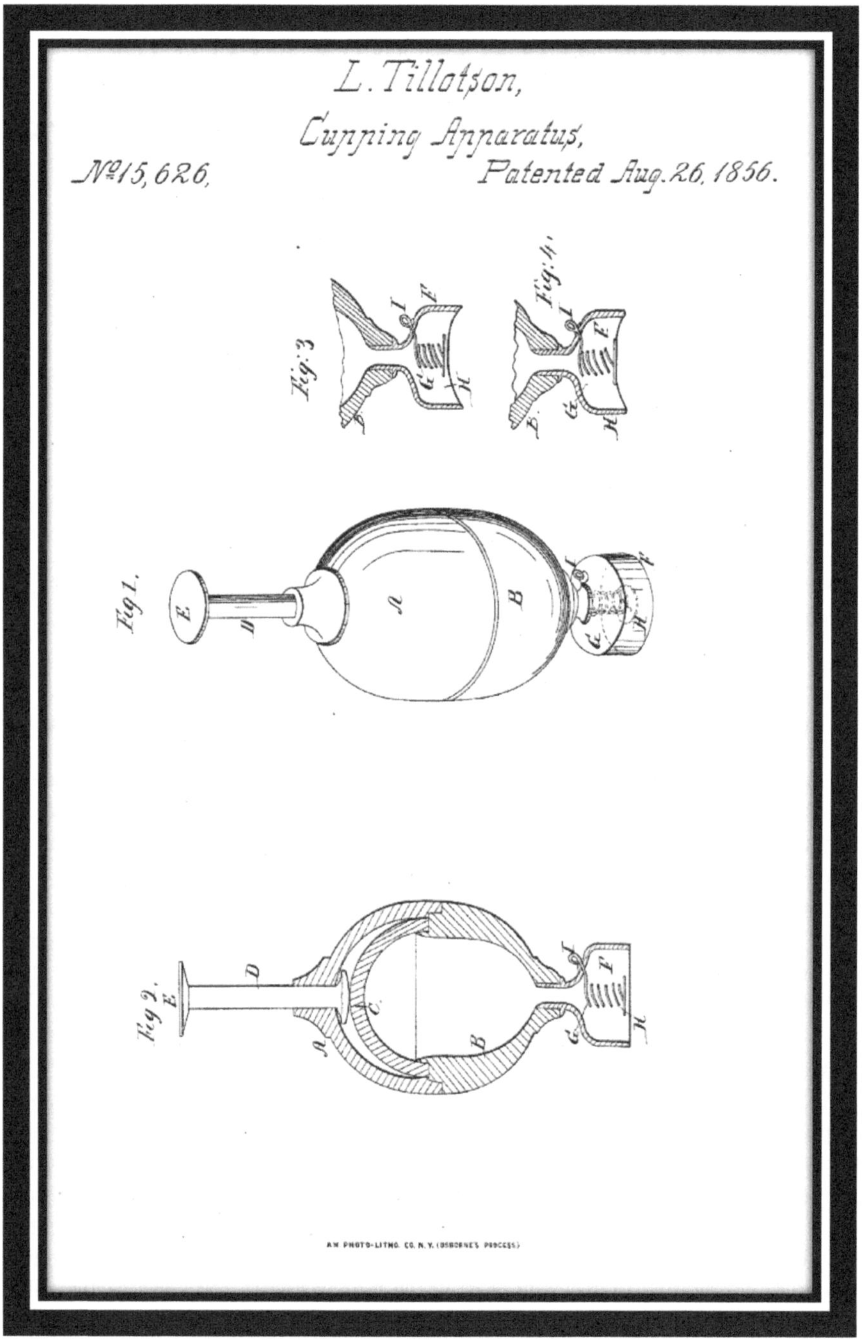

FIGURE 9.1 First patent application on "sensor," by L. Tillotson (1856) [2].

Simply, nanosensors are sensing devices built to present dimensions inferior to 100 nm and still presenting the ability to vary a property in the presence of an analyte, in a measurable manner. They are developed to mainly attend the needs of technological fields for miniaturization, aiming at new searching environments and enhanced sensibility and selectivity. They can monitor physico-chemical phenomena in a variety of environments, since nanoscopic inorganic materials can be applied to living tissues and organs and they are able to adapt to any needs. Therefore, many of the technological advances that are available to society have only come to be due to the construction of nanosensors.

Since the first patent application for nanosensors, in 1973 (patent document SU453598A1), over 1,800 applications claimed nanosensors for several purposes. For almost thirty years, this technological field was insipient and since 2006, when nanomaterials started to be used in sensors, in medical, biological and materials fields, patent applications exponentially grew (see Figure 9.2), evidencing the efficient technology transfer from the academy to industry. The technological fields most frequently cited in nanosensors patents and scientific publications are shown in Figure 9.3.

From Figure 9.3, it can be seen that the technological field of most nanosensors is related to *measuring or testing processes involving enzymes, nucleic acids or microorganisms* – C12N and

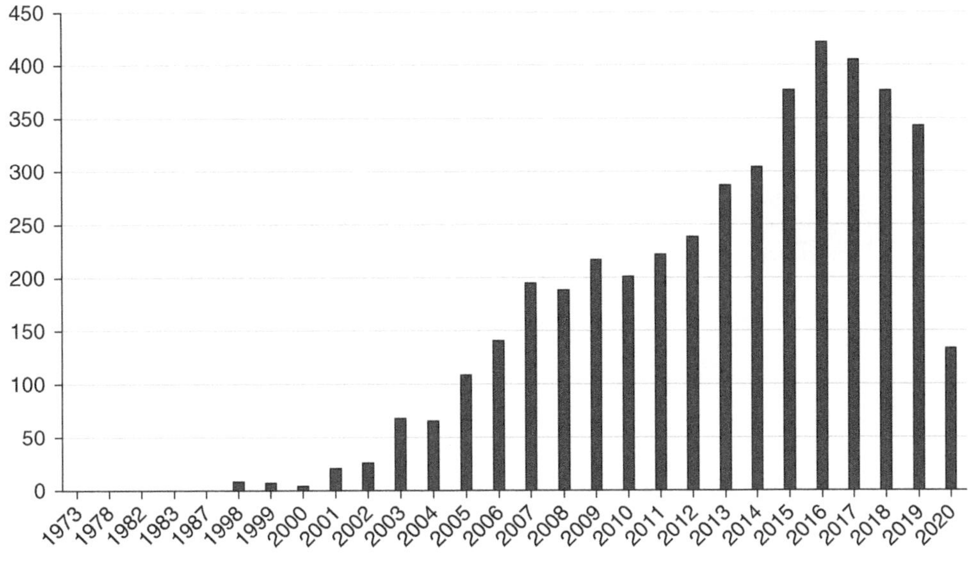

FIGURE 9.2 Nanosenor patent application evolution.

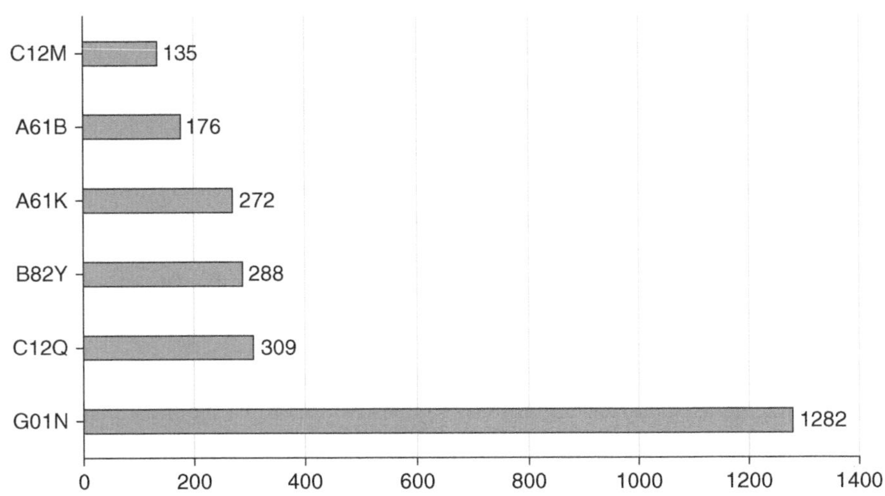

FIGURE 9.3 Main technological fields of patent applications on nanosensors.

TABLE 9.1
International Patent Classification (IPC) of Most Frequently Technological Fields of Patent Applications on Nanosensors

IPC	Meaning
G01N	*Investigating or analysing materials by determining their chemical or physical properties*
C12Q	*Measuring or testing processes involving enzymes, nucleic acids or microorganisms*
A61B	*Diagnosis; surgery; identification*
B82Y	*Specific uses or applications of nanostructures; measurement or analysis of nanostructures; manufacture or treatment of nanostructures*
C12M	*Apparatus for enzymology or microbiology*
A61K	*Preparations for medical, dental, or toilet purposes*

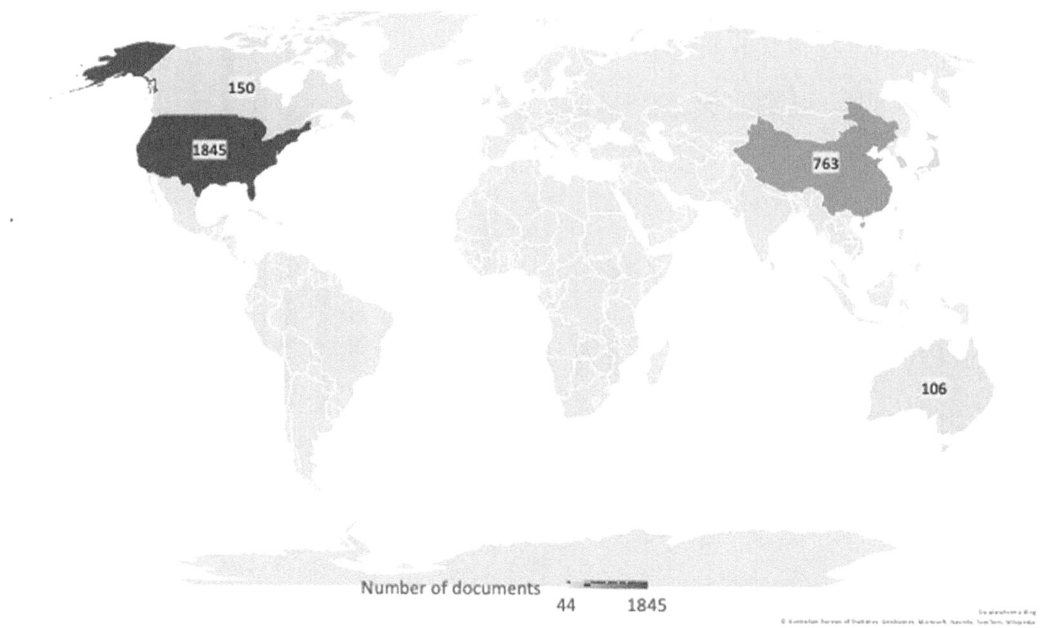

FIGURE 9.4 Countries that file most patent applications on nanosensors.

most of the IPC that are given to these technologies refer to a medical need. However, 13 percent of the patent applications are related to "specific uses or applications of nanostructures; measurement or analysis of nanostructures; manufacture or treatment of nanostructures" – B82Y, indicating that a significant quantity of nanosensors have been built on nanostructures and nanomaterials, which evidences the perspective of even more important developments in this technological field. The meaning of the IPC numbers is shown in Table 9.1.

In addition, the countries that have filed the most patent applications are the United States, China, South Korea and Japan, as shown in Figure 9.4, which also identifies the most promising markets for nanosensors nowadays.

Due to the variety of applications and materials used to build nanosensors, some desired properties have been pointed out as essential to a reliable nanosensor and need to be considered when preparing a nanomaterial for this purpose. They are: (1) size and shape of nanomaterials must be controlled to guarantee the optimal performance of a nanosensor; (2) combinations of nanomaterials must be considered and well planned to achieve the desired specificity, sensibility and stability of the future nanosensor; (3) in this regard, many multifunctional structures have been developed, specifically to function in nanosensors, considering the close relationship between composition, structure and reactivity; (4) offer a variety of responses that can be obtained from a nanosensor and qualitatively and quantitatively related to an analyte; and also (5) improve the sensing property to achieve even more accurate responses by correlating the characteristics of the developed nanomaterial – always aiming a new application for the nanosensor [3, 4].

Based on the characteristics and strategies of operation designed to a nanosensor, it is classified to better attend the requirements of an application.

9.2 CLASSIFICATION OF NANOSENSORS

Depending on the signal type delivered by the nanosensor, its energy source, structure and applications, it finds a classification that is able to direct the following developments it can be subject to. When classification is based on energy source, nanosensors are either *active nanosensors*, when they need an energy source, or *passive nanosensors*, when no energy source is needed. Examples of active nanosensors are the thermistors, and an example of a passive nanosensor is a piezoelectric.

When classification is based on structure, there can be three types of sensors: (i) *organic*, when based on purely organic nanostrucutres, including polymers, enzymes, peptides, DNA, RNA, and proteins; (ii) *inorganic nanosensors*, which are based on metallic and/or inorganic nanoparticles, nanocrystals; (iii) *hybrid,* which are based on nanostructures containing organic groups combined into inorganic compounds. There is a great variety of nanomaterials in this class, such as quantum dots, functionalized or doped carbon nanostructures and also polymeric structures.

Based on the signal they can process, they can be: (i) *optical*, (ii) *electrical*, (iii) *magnetic*, (iv) *mechanical*, (v) *chemical* or (vi) *biological*. However, there are nanosenors that are able to process more than a single type of signal.

Also, they can be classified, based on application ,as (i) *chemical nanosensors*, (ii) *deployable nanosensors*, (iii) *electrometers* and (iv) *biosensors*, although biosensors have particular requirements of composition.

Moreover, a biosensor can be classified under distinct classifications, which is essential to provide information on uses, building materials, availability, reliability and possible next developments.

As an example of classification, Quantum Dot–based luminescent sensors are classified as optical nanosensors. Depending on the nature of the nanomaterial used to build it, it can be either inorganic or a hybrid sensor and if it is applied to monitor cancer cells, it is considered a biosensor.

9.3 OPTICAL NANOSENSORS; LUMINESCENT NANOSENSORS

Optical sensors depend on optical properties of nanomaterials to be capable of monitoring chemical or physical changes, in several technological fields, such as medicine, environmental sciences, foods, chemical processes, biotechnology, health and hygiene. The first reported optical nanosensor was based on polyacrylamide nanoparticle doped with fluorescein, for pH measurement [5]. However, as fluorescein is strongly fluorescent compound, this advance gave rise to novel nanosensors that explored the fluorescence of their constituents: *fluorescent nanosensors*. These fluorescent nanosensors are formed by nanoparticles that include at least one binding component and photoactive moieties, and they correlate the qualitative and quantitative responses to a signal in the light-emitted form, exploring luminescent phenomena of the analyte and the transducer of the nanosensor [6, 7]. In the luminescent phenomenon a fluorescence chemical compound, named fluorophore, absorbs light of a certain wavelength to be promoted to an electronic excited state and, after some non-radiative deactivation processes, the compound emits a quantum of light to return to the electronic ground state from which it was promoted to the excited state. This quantum of light emitted by the fluorophore presents precisely the energy corresponding to the energetic difference between the ground and the singlet electronic excited state with less energy [8, 9].

Therefore, the wavelength of the emitted light is characteristic of this energy difference and, hence, can inform about the identity of a compound or a specific electronic excited state, in an atom, molecule of group of molecules. Organic, inorganic, hybrid and biological compounds can be excited by light absorption, and this characteristic makes this phenomenon promising for sensing purposes due to the wide range of applications it can find.

From that, the first nanosensors based on fluorescent responses were applied to biological environments, and many compounds acting as markers were developed to be applied in organs and tissues. The most significant advance this type of nanosensor presents is that it is less invasive than other sensing methods used previously. However, many markers can be toxic to cells or can be interferential in several biological processes because of interactions with proteins in cells. Sasaki et al. [10] reported a three-dimensional nanoprobe that is able to determine local pH changes. The probe is based on a fluorescein-based polymer nanoparticle that presents its fluorescence influenced by pH changes, which is a characteristic of the fluorescence compound chosen to dope the particle. This approach is based in a free fluorophore, acting as doping agent. However, a very popular approach is to label polymers or particles with a fluorescent marker, which is mostly used to analyze biological environments. Among the advantages of developing labelled inert or biocompatible particles to build nanosensors is the fact that it reduces toxicity, amplifies the possibilities of applications, and enhance the solubility of fluorescent compounds that, in general, are poorly soluble in water and body fluids. It can also improve the fluorophore photostability, prolonging the lifetime of the sensor and making t more reliable.

Sensing also can be improved by co-incorporating reference fluorophores, which extend response ranges and enhance sensing accuracy. Examples are the noninvasive nanosensor produced by Lee et al. [11] and based on a polyacrylamide hydrogel containing fluorescent dyes and surface-conjugated tumor-specific peptides, which were efficient in oxygen-level determinations in living cells to diagnose cancer, for instance. Another example is the nanosensor produced by Zheng et al. [12], in which a tripeptide (glycine (G)-histidine (H)-lysine) is co-assembled with dopamine to yield a fluorescent nanostructure, sensible to pH changes.

In fact, nanosensors based on peptides present several advantages, among them: they are chemically and thermally stable, they are biocompatible, present low toxicity, many are soluble in water and offer a variety of structures design, they can be easily combined to other materials to generate new materials with the desired properties and all these characteristics make them excellent optional material to be used in nanosensors. Also, a variety of structures with distinct properties and abilities can be easily produced by supramolecular approaches. Some examples of self-assembled nanostructures based on peptides are illustrated in Figure 9.5.

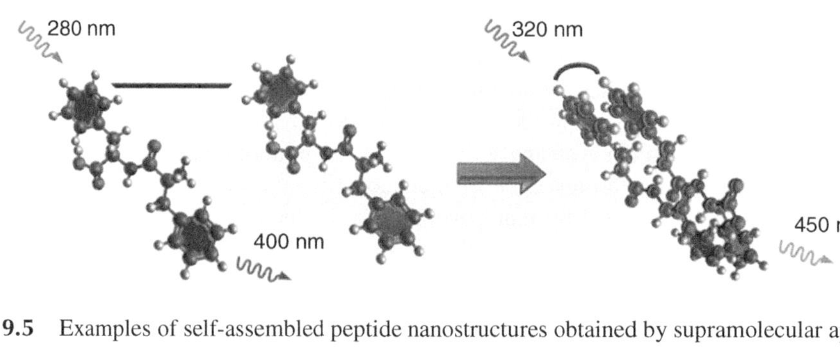

FIGURE 9.5 Examples of self-assembled peptide nanostructures obtained by supramolecular approaches.

Peptide-based nanosensors have been developed to detect several chemicals, especially in the human body, such as several metal ions, whose detection can inform about different biological processes. Also, they have been developed to detect oxygen [13] to inform about pH changes in biological environments [12, 14]. Also, biomimetic nanosensors based on peptides have been proposed as alternatives for the usual detection techniques benefiting from the selectivity and specificity of the lock-and-key interactions of proteins and DNA. In this alternative, as proposed by Cui et al. [15], the nanosensors use peptide aptamers to mimic the proteins binding domains, as observed by Wu et al. [16]. In the sensors, these peptides are designed to bind specific targets, and it is possible that due to the diversity of physical–chemical characteristics of peptides, such as solubility, hydrophilicity, acidity, self-assembling capability and diverse composition allied to their compact size relative to proteins, antibodies and other structures. They also are much more tough than proteins or antibodies, enabling their use in a variety of environments in which proteins would be denatured [15].

The applications of peptide-based nanosensors go further, as demonstrated by the alternative proposed by Wang et al., in which a multipurpose nanosensor was presented [17]. Another interesting application of peptide-based nanosensors is for imaging of living cells, organs and tissues, as the nanosensor developed by Ge et al. [18], which used a peptide-fluorophore combined with a gold nanoparticle to process the imaging of living cells through a two-photon excitation followed by fluorescence emission and the sensor developed by Li et al. [19] that imaged glutathione in living cells and zebrafish. They can also be used in combination with a variety of materials to achieve perspectives of applications. In their approach, Wu et al. [20] developed a quantum dot nanosensor in which the quantum dot is capped with a denatured human albumin peptide combined with polyethylene oxide to detect pH variations via fluorescence emission. In fact, the variety of nanosensors that can be built using peptides is due to the distinct recognition mechanisms that can be exploited in materials containing peptides. Most frequent mechanisms exploited in fluorescent nanosensors based on peptides are photoinduced electron transfer (PET), fluorescence resonance energy transfer (FRET) and chelation enhanced fluorescence (CHEF) [21].

To understand these mechanisms, especially CHEF, it is crucial to have in mind that fluorescence is often quenched by lone pairs of electrons, among other quenchers. However, if a fluorescent compound is coordinated to some metal ions this effect is inhibited and fluorescence from the metal complex that is formed is observed and even enhanced. This is what happens, for instance, when N,N′-Bis(2-dimethylaminoethyl)-N,N′-dimethyl-9,10-anthracenedimethanamine is complexed with cation Zn^{2+}. It is non-fluorescent when it is in aqueous solution alone, but when complexed with Zn $^{2+}$ through a chelate bond with the nitrogen atoms, it becomes highly fluorescent due to occurrence of the CHEF effect, and this phenomenon is observed when fluorophores containing heteroatoms interacts with metal ions, changing their electronic states energies. When the ligand is a peptide, sensors are developed through a design strategy that considers peptide sequences that are conjugated to fluorophores or presenting fluorescence themselves. These fluorophores moieties must coordinate to metal ions.

It is well known that coordination with metal cations severely impact secondary and tertiary structures of peptides, to render highly stable conformations, which are explored for sensing by turning the combination peptide-fluorophore sensitive to local changes in the microenvironment (Figure 9.6). It is noteworthy that in compounds presenting heavy atoms in their composition, another quenching effect that is often occurring is the inner filter effect or trivial energy transfer, in which fluorescence emission of the donor is quenched because it is absorbed by the acceptor. To occur, the absorption spectrum of the acceptor must overlap in some degree to the fluorescence spectrum of the donor. Then the donor absorbs light from an excitation source, which does not excite the acceptor but is promoted to an electronic excited state and emits its fluorescence, which is promptly absorbed by the acceptor. In a sensor based on this phenomenon, the acceptor is the analyte-sensitive fluorophore, the donor is the analyte-insensitive fluorophore, and the spectral changes resulting from this effect are analyzed.

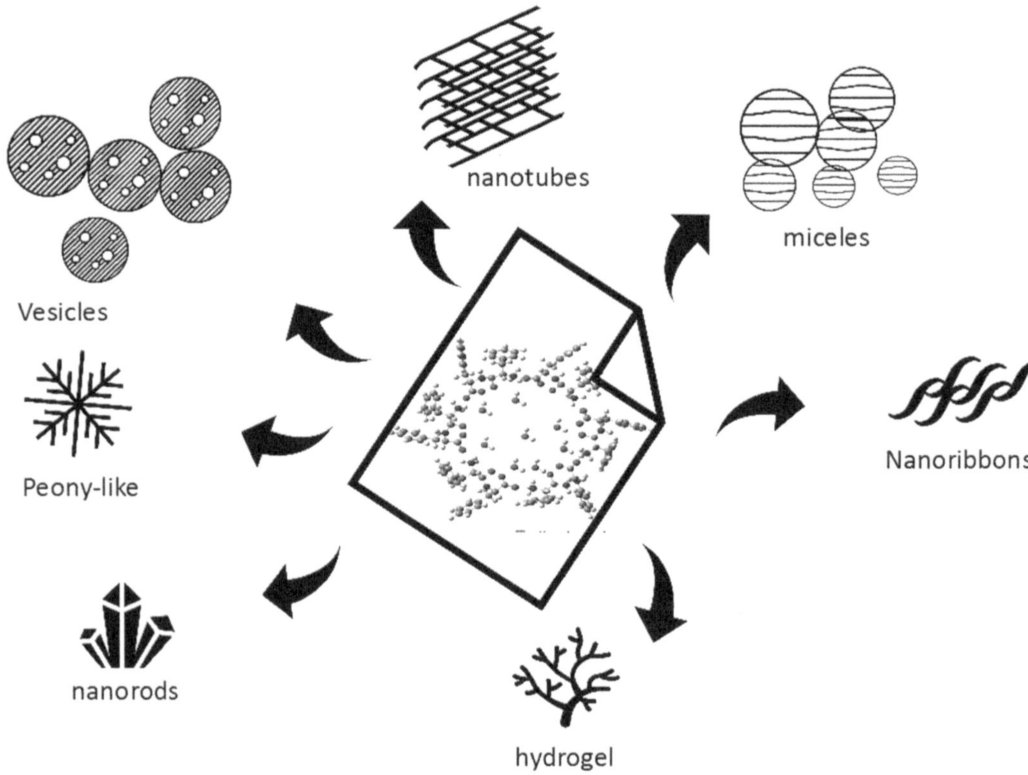

FIGURE 9.6 Illustration of fluorescence modulation effect of peptide fluorophores by conformational changes.

The trivial energy transfer has been also exploited as a sensing method because it does not require that the analyte and the analyzer are in the same vessel, or container, which is interesting for non-invasive analysis. Considering peptide self-assembled structures used in nanosensors, the trivial energy transfer is promoted by significant changes in structures due to aggregation, distinct interactions with the analyte, environment and so forth, all able to modulate the intensity of the energy transfer, which is interpreted as a fluorescence signal [8, 22–31].

Another common mechanism of activity of fluorescent peptide-based nanosensors is through photoinduced electron transfer (PET), in which energy transfer occurs due to a redox reaction involving the electronic excited state. In this process, the electron transfer may occur from the conduction band of fluorophores to the acceptor moieties or by the donation of an electron from the donor moieties to the valence band holes of the fluorophore. It is important that the donor and acceptor be at a close distance, therefore, they have to be attached to the peptide at sites where they are able to interact to promote the excited state redox reaction. This process has been proposed for decades to be used in peptide sensors and, in the work of Torrado et al. [32], it is shown that peptide fluorescence is modulated by its binding to a metal, leading to a conformational change that favor the fluorescence quenching by PET.

Recently, Wang et al. developed a highly sensitive peptide-based nanosensor for detection of zinc and copper ions in living cells [33]. The multifunctional sensor they developed is based on a tetrapeptide combined to dansyl groups, selectively responding to zinc ions and this selectivity is possible because the sensor explores the photoinduced electron transfer effect (PET). Also, An et al. [34] developed a tetrapeptide (Dansyl-Gly-Pro-Trp-Gly-NH2) – based nanosensor to diagnose prostate cancer that is based on the complexation with zinc ions, also based on the PET mechanism. Their sensor is highly selective and sensitive, able to inform about slight Zn^{2+} concentration differences in cancerous and healthy tissues, which makes it promising for early prostate cancer diagnosis.

In a very promising study, Wang et al. [35] developed another tetrapeptide-based sensor very sensitive to Cd^{2+} cations and cysteine in water at neutral pH, also with the ability to detect over twenty amino acids, with very low detection limits, being a candidate to multifunctional peptide-based sensor.

9.4 MECHANISMS OF OPERATION OF LUMINESCENT PEPTIDE NANOSENSORS

The most prominent mechanism of electronic excited states deactivation that has been widely exploited by peptide-based nanosensors is Forster's resonance energy transfer (FRET) [8]. It consists of a non-radiative energy transfer process involving the long-range dipole–dipole interaction between a donor moiety, which is in its electronic excited state and an acceptor moiety in its ground state. In fact, this is a virtual process in which the donor is excited by light absorption and, due to the overlap between its emission spectrum and the acceptor absorption spectrum, energy is transferred from the former to the latter. However, distinctly from the trivial energy transfer, there is no fluorescent emission from the donor: This energy transfer occurs during the permanence of the donor in its electronic excited state, occurring at the same instant in which excitation takes place and the energy transfer results in a decrease of the fluorescence lifetime of the donor. The acceptor is promoted to its electronic excited state by a resonant effect that arose from the dipole-dipole interaction between these two excited states, and fluorescence emission can only be detected from the acceptor moiety, presenting its emission characteristics: wavelength and quantum yield [8]. In Forster's theory, FRET rate depends on the extent of spectral overlap between the donor emission and the acceptor absorption spectra, the relative orientation of the donor and acceptor transition dipoles, the distance between the donor and acceptor, which have to be at a long-range distance (known as Forster Distance), typically of tens of nanometers, and the effects of the dipoles of solvents, when the fluorophores are in solution, that is, wrapped by molecules in the environment. It has been widely exploited in peptide-based nanosensors mainly due to the possibility of modulating the energy transfer by FRET in these sensors, conferred by the conformational changes in peptide self-assembled structures, which are able to interfere with the overlap of donor emission – acceptor absorption spectra. As explained by Shamsipur et al. [36], the effects that the conformational changes have on fluorescence properties of peptides, due to the FRET modulation possibility, the self-assembling mechanisms have been exploited to convert the structural changes of, for instance, pH-responsive materials into the fluorescence signal that can be related to distinct conformers and be used as multiple pH indicators.

Also, over the past decades, several peptide-based nanosensors have been proposed, especially because they enable a variety of applications due to their photostability, selectivity, sensitivity, and also because they are easy to functionalize, opening the opportunity to create a variety of materials to be applied as nanosensors.

As taught by Gazit et al. [37], peptide self-assembling nanostructures are naturally considered for multifunctional materials preparation due to their exceptional characteristics of chemical and thermal stability, biocompatibility and the wide range of physical–chemical properties that can be tailored by supramolecular approaches of preparation, which enable means for bottom-up fabrication of several tailored multifunctional materials. Since the peptide self-assembling is a is dynamic and reversible process, it enables multiple applications and designs, through a perspective of creating materials with tunable properties. Unfortunately, because of the variety of peptide-based nanosensors that rely on FRET, fluorescence intensity measurements are not always informative enough or sensitive enough to enable a careful analysis. Due to this limitation, several techniques of detection even more sensitive and selective have been thought to expand peptide nanosensors applications, including imaging strategies. Among them, fluorescence lifetime imaging microscopy (FLIM) have been widely exploited, mainly due to the biological applications this technique has helped discover. It is an imaging technique that relates pixels with distinct fluorescence exponential decay rates of fluorophores in a sample. Its efficiency is based on the fact that fluorescence lifetime of the fluorophore depends on the local microenvironment of the fluorophore and, therefore, is highly sensitive to the local changes that self-assembling can cause on the photophysical properties of the peptide or of the fluorophore attached to it. Also, it is not affected by the common error sources that fluorescence intensity-based detecting methods are, such as background light intensity or changes in brightness of the light source, and it is not limited by quenching or photobleaching, nor to concentration effects and scattering of light in tissue, turning it into a robust method for imaging of local environments [38].

In their work, Buschhaus et al. [39] used FLIM to quantify apoptosis of cancer cells at single-cell resolution; Okkelman et al. [40] proposed a low affinity Ca^{2+} biosensor based on a twitch-2B fluorescent protein combined to cellulose and collagen-binding peptides to be applied in multiplexed FLIM analysis in 3D tissue environments.

Jena et al. [41] developed fluorophore-labelled peptide substrates for FLIM imaging of kinases activities. In their approach, the selected probes do not rely on FRET pairs or probe multiplexing, and pixel-by-pixel quantification are used to estimate the relative proportion of modified probe to achieve single-cell mapping of multiple kinases in live cells.

In an interesting approach of a novel material for intramitochondrial pH nanosensor, Ripoll et al. [42] developed a highly specific nanosensor using fluorescence lifetime imaging microscopy. The quantum dot-based nanosensor presents the advantage of longer fluorescence lifetimes, being hundreds of nanoseconds against the common fluorescence lifetimes of fluorophores for biological environments, which are of 2 to 5 ns. With this longer lifetime, their nanosensor is more sensitive and accurate than the normally used fluorophores. Their sensor is based on a double modification of a commercial quantum dot nanoparticle surface. First, mercaptopropionic acid is attached to the quantum dot as a pH-sensitive group, then Szeto-Schiller peptides are added to the surface for specific mitochondrial delivery. They claim that their nanosensor is able to establish differences between metabolic pathways in breast cancer cells.

Damayanti et al. [43] proposed an approach to monitor acetyltransferase acetylation in live single cells, using FLIM with peptide biosensors with specific response to acetyltransferase enzyme activity in a fluorescence lifetime-dependent manner. They were able to perform real-time longitudinal measurement of acetylation activity with high spatial and temporal resolution in live single cells. With that, they claim that they can monitor cell function and even evaluate drug effects in cancer treatments.

9.5 FUTURE PERSPECTIVES OF PEPTIDE NANOSENSORS

Arguably, the use of peptides in the development of nanosesnors for the most varied purposes and based on the combination of peptides with the most varied materials encouraged the emergence of a new technological field that combines the advantages of nanotechnology with the simplicity, robustness and biocompatibility of inspired devices in biological processes and compositions [44]. Undoubtedly, this is the path for future developments in the field of sensing. The search is now for multifunctional sensors of simple build procedures, low cost, durable, with high perspective of new developments and that can be applied to solve several questions in the medical field but also in the materials and technological fields, at once. With this regard, peptide-based nanosensors had been excellent candidates for the technological market that is arising.

The perspectives are multiple, as demonstrated by the diversity of multifunctional applications that can be found for peptide nanosensors. For instance, Shi et al. [45] developed a multifunctional humidity nanosensor – based on a gold electrode drop-coated by self-assembled peptide nanofibers – that acts as the humidity-sensing material. Because they present good optical, mechanical and semiconductive properties and are intrinsically biocompatible, they are thought, in this work, to monitor human health and activities through humidity measurements. Self-assembled peptide nanofiber networks greatly absorb water from any environment, which is the reason for their excellent humidity sensitivity and ultrafast response. They also demonstrate multifunctional possibilities by applying their peptide humidity nanosensor to respiration monitoring and baby diaper wetting monitoring, providing good responses on detecting humidity changes in physiological activities.

Pang et al. [46] developed a multifunctional sensor based on a fluorescent pentapeptide (dansyl-Gly-His-Gly-Gly-Trp) that is selective and sensitive to Hg^{2+} and Cu^{2+} among a variety of metal ions and 6 anions. It also detected S^{2-} anion, making it multifunctional.

Zhang et al. [47] showed that sensors based on self-assembling of gold nanoparticles in the assistance of designed peptides, which can drive the gold nanoparticles assembling into fibrils, as peptides self-assemble. This hybrid nanosensor exhibits excellent electrochemical and colorimetric sensing performances and is able to detect H_2O_2 by electrochemical means and Hg2+ by colorimetric means with good limits of detection.

Wang et al. [48] proposed an electrochemical sensing platform based on a multifunctional peptide design, considering several desired characteristics of peptides conferred by distinct sequences, such as anchoring, doping, linking, and antifouling sequences. Their achievements showed that it is possible to strategically design biosensors to be used or in real biological systems, and it is the focus also of fluorescent nanosensors.

As it is demonstrated recently, the interest in developing new materials and sensing strategies based on the singular properties of peptides has boosted several scientific and technological advances worldwide [49].

However, as the use of peptides as part of luminescent nanosensor is a relatively new concept, it is expected that the remarkable results of such researches may quickly lead to an increasing market interest, which will improve the actual applications possibilities. As has been demonstrated, applications in health, diagnosis, environmental control, food quality control, drugs developments and so many others are already being considered for devices used in such fields of interest. However, as research is performed, new features of these nanosensors and new applications are still to be proposed, establishing a new generation of peptide-based sensing.

9.6 ACKNOWLEDGMENTS

Authors thank CNPq and CAPES for scholarships.

REFERENCES

1. ESPACENET PATENT DATABASE, available at: https://worldwide.espacenet.com/.
2. Tillotson, L. Cupping. (Patent Number US15626A). Espacenet Patent search. 1856.
3. Abdel-Karim, R., Reda,Y., Abdel-Fattah, A. Review – nanostructured materials-based nanosensors. *J Electrocheml Soc* 167, 037554, 2020. DOI: 10.1149/1945-7111/ab67aa.
4. National Research Council. expanding the vision of sensor materials. Washington, DC: The National Academies Press. 1995. https://doi.org/10.17226/4782.
5. de Silva, A.P., Gunaratne, H. Q., Gunnlaugsson, T., Huxley, A. J., McCoy, C. P., Rademacher, J. T., Rice, T. E. Signaling recognition events with fluorescent sensors and switches. *Chem Rev* 97(5), 1515–1566, 1997. DOI: 10.1021/cr960386p.
6. Czarnik, A. W. Fluorescent chemosensors for ion and molecule recognition. Ed. American Chemical Society. Division of Organic Chemistry. *American Chemical Society. Meeting* (204th: 1992: Washington, DC) Series. R857.B54F58 1993 681'.761—dc20 93-21067 CIP.
7. Kulmala, S., Suomi, J. Current status of modern analytical luminescence methods. *Anal Chim Acta* 500 (1–2), c21-69, 2003. https://doi.org/10.1016/j.aca.2003.09.004.

8. Lakowicz, J. R. *Principles of Fluorescence Spectroscopy*. 3rd ed., Berlin, New York, Springer USA, XXVI, 954, 1999.
9. Wood. E. J. Molecular probes: handbook of fluorescent probes and research chemicals: by R. P. Haugland. p. 390. Interchim (Molecular Probes, Eugene, OR, USA). 1992–1994. *Biochem Education*, 22 (2), 83, 1994. https://doi.org/10.1016/0307-4412(94)90083-3.
10. Sasaki, K., Shi, Z. Y., Kopelman, R., Masuhara, H. Three-dimensional ph microprobing with an optically-manipulated fluorescent particle. *Chem. Lett.* 25, 141–142, 1996. https://doi.org/10.1246/cl.1996.141.
11. Lee, Y. E. K., Ulbrich, E. E., Kim, G., Hah, H., Strollo, C., Fan, W., Gurjar, R., Koo, S. M., Kopelman, R. Near infrared luminescent oxygen nanosensors with nanoparticle matrix tailored sensitivity. *Anal. Chem.* 82, 8446–8455, 2010. https://doi-org.ez49.periodicos.capes.gov.br/10.1021/ac1015358.
12. Zheng, F., Guo, J., Khan, A. J., Miao, P., Zhang, F. Coassemble dopamine and GHK tripeptide into fluorescent nanoparticles for pH sensing. *Luminescence* 36, 28–34, 2021 https://doi-org.ez49.periodicos.capes.gov.br/10.1002/bio.3907.
13. Dmitriev, R. I., Ropiak, H. M., Yashunsky, D. V., Ponomarev, G. V., Zhdanov, A. V., Papkovsky, D. B. Bactenecin 7 peptide fragment as a tool for intracellular delivery of a phosphorescent oxygen sensor. *FEBS J* 277(22), 4651–4661, 2010, DOI:10.1111/j.1742-4658.2010.07872.x.
14. Oliveira, E., Genovese, D., Juris, R., Zaccheroni, N., Capelo, J. L., Raposo, M. M. M., Costa, S. P. G., Prodi, L., Lodeiro, C. Bioinspired systems for metal-ion sensing: new emissive peptide probes based on benzo[d]oxazole derivatives and their gold and silica nanoparticles. *Inorganic Chem* 50(18), 8834-8849, 2011, https://doi.org/10.1021/ic200792t.
15. Cui, Y., Kim, S. N., Naik, R. R., Mcalpine, M. C. Biomimetic peptide nanosensors, *Acc Chem Res* 45(5), 696–704, 2012, DOI: 10.1021/ar2002057.
16. Wu, T. Z., Lo, Y. R., Chan, E. C .Exploring the recognized bio-mimicry materials for gas sensing. *Biosens. Bioelectron.* 16, 945–953, 2001. DOI: 10.1016/s0956-5663(01)00215-9.
17. Wang, P., Liu, L., Zhou, P., Wu, W., Wu, J., Liu, W., Tang, Y. A peptide-based fluorescent chemosensor for multianalyte detection. *Biosensors and Bioelectronics* 72, 80-86, 2015. https://doi.org/10.1016/j.bios.2015.04.094.
18. Ge, Q., Wang, N., Li, J., Yang, R. Peptide-fluorophore/AuNP conjugate-based two-photon excited fluorescent nanosensor for caspase-3 activity imaging assay in living cells and tissue. *MedChemComm* 8(7), 1435–1439, 2017. DOI: 10.1039/c7md00177k.
19. Li, Y., Di, C., Wu, J., Si, J., Chen, Y., Zhang, H., Ge, Y., Liu, D., Liu, D. A peptide-based fluorescent sensor for selective imaging of glutathione in living cells and zebrafish. *Anal Bioanal Chem* 412, 481–488,2020. https://doi.org/10.1007/s00216-019-02257-4.
20. Wu, Y., Chakrabortty, S., Gropeanu, R. A., Wilhelmi, J., Xu, Y., Er, K. S., Kuan, S. L., Koynov, K., Chan, Y., Weil, T. pH-responsive quantum dots via an albumin polymer surface coating. *J. Am. Chem. Soc.* 132(14), 5012–5014, 2010. https://doi.org/10.1021/ja909570v.
21. Zhang, L., Cao, J., Chen, K., Liu, Y., Ge, Y., Wu, J., Liu, D. A selective and sensitive peptide-based fluorescent chemical DSH sensor for detection of zinc ions and application in vitro and in vivo. *New J. Chem.*, 43, 3071–3077, 2019. DOI: 10.1039/C8NJ06552G.
22. Zheng, M., Xie, Z., Qu, D., Li, D., Du, P., Jing,X., Su, Zn. On-off-on fluorescent carbon dot nanosensor for recognition of chromium (VI) and ascorbic acid based on the inner filter effect. *ACS Appl. Mater. Interfaces* 5, 24, 13242–13247, 2013. https://doi.org/10.1021/am4042355.
23. Akhgari, F., Samadi, N., Farhadi, K. Fluorescent carbon dot as nanosensor for sensitive and selective detection of cefixime based on inner filter effect. *J Fluoresc* 27, 921–927, 2017. https://doi.org/10.1007/s10895-017-2027-0.
24. Bao, Q., Lin, D., Gao, Y., Wu, L., Fu,J., Galaa, K., Lin,X., Lin, L. Ultrasensitive off-on-off fluorescent nanosensor for protamine and trypsin detection based on inner-filter effect between N,S-CDs and gold nanoparticles *Microchem J* 168, 106409 (9p), 2021. https://doi.org/10.1016/j.microc.2021.106409.
25. Yuan, P., Walt, D. R. Calculation for fluorescence modulation by absorbing species and its application to measurements using optical fibers, *Anal. Chem.* 59(19), 2391–2394, 1987. https://doi.org/10.1021/ac00146a015.
26. Ghosh, M., Ta, S., Banerjee, M., Mahiuddin, Md., Das, D. exploring the scope of photo-induced electron transfer – chelation-enhanced fluorescence–fluorescence resonance energy transfer processes for

recognition and discrimination of Zn2+, Cd2+, Hg2+, and Al3+ in a ratiometric manner: Application to sea fish analysis. *ACS Omega* 3, 4262–4275, 2018. https://doi.org/10.1021/acsomega.8b00266.
27. Schneide, H. J., Dürr, H. Frontiers in supramolecular organic chemistry and photochemistry. Michigan, VCH, 485 p. 1991. ISBN 3527280162,9783527280162.
28. Czarnik, A. W. Chelation-enhanced fluorescence detection of metal and nonmetal ions in aqueous solution. Proc. SPIE 1648, Fiber Optic Medical and Fluorescent Sensors and Applications, (16 April 1992); https://doi.org/10.1117/12.58297.
29. Gan, W. *Synthesis and Design of Fluorescence Ligands to Act as Sensor for Zinc*. Thesis. University of North Carolina. Wilmington. 2004.
30. Pazos, E., Vazquez, O.P., Mascarenas J. L., Vazquez, M. E. Peptide-based fluorescent biosensors. *Chem. Soc. Rev*., 38, 3348–3359, 2009. DOI: 10.1039/b908546.
31. Deems, J. C., Reibenspies, J. H., Lee, H. S., Hancock, R. D. Strategies for a fluorescent sensor with receptor and fluorophore designed for the recognition of heavy metal ions. *Inorganica Chim Acta*. 499, 119181, 2020. https://doi.org/10.1016/j.ica.2019.119181.
32. Torrado, A., Imperiali, B. New synthetic amino acids for the design and synthesis of peptide-based metal ion sensors. *J Org Chem* 61(25), 8940–8948, 1996. DOI: 10.1021/jo961466w.
33. Wang, P., Wu, J. Highly selective and sensitive detection of Zn(II) and Cu(II) ions using a novel peptide fluorescent probe by two different mechanisms and its application in live cell imaging. *Spectrochim Acta Part A-Mol Biomol Spectroscopy*. 208, 140–149, 2019. DOI: 10.1016/j.saa.2018.09.054.
34. An, Y., Chang, W., Wang, W., Wu, H., Pu, K., Wu, A., Qin, Z., Tao, Y., Yue,Z., Wang, P., Wang, Z. A novel tetrapeptide fluorescence sensor for early diagnosis of prostate cancer based on imaging Zn2+ in healthy versus cancerous cells. *J Adv Res* 24, 363-370, 2020. https://doi.org/10.1016/j.jare.2020.04.008.
35. Wang, P., Duan, L. P., Liao, Y. W.. A retrievable and highly selective peptide-based fluorescent probe for detection of Cd2+ and Cys in aqueous solutions and live cells. *Microchem J* 146, 818–827, 2019. https://doi.org/10.1016/j.microc.2019.02.004.
36. Shamsipur, M., Barati, A., Nematifar, Z. Fluorescent pH nanosensors: Design strategies and applications. *J Photochem Photobiol C: Photochem Rev* 39, 76–141, 2019. https://doi.org/10.1016/j.jphotochemrev.2019.03.001.
37. Aizen, R., Tao, K., Rencus-Lazar, S., Gazit, E. Functional metabolite assemblies – A review. *J Nanopart Res* 20, 125 (9p), 2018. https://doi.org/10.1007/s11051-018-4217-3.
38. Datta, R., Heaster, T. M., Sharick, J. T., Gillette, A. A., Skala, M. C. Fluorescence lifetime imaging microscopy: Fundamentals and advances in instrumentation, analysis, and applications *J Biomed Optics*. 25(7), 071203(44p), 2020 https://doi.org/10.1117/1.JBO.25.7.071203.
39. Buschhaus, J. M., Gibbons, A. E., Luker, K. E., Luker, G. D. Fluorescence lifetime imaging of a caspase-3 apoptosisreporter. *Curr Protocols Cell Biol* 77, 21.12.1–21.12.12, 2017. DOI: 10.1002/cpcb.36.
40. Okkelman, I. A., McGarrigle, R., O'Carroll, S., Berrio, D. C., Schenke-Layland, K., Hynes, J., Dmitriev, R. I. Extracellular Ca2+-sensing fluorescent protein biosensor based on a collagen-binding domain. *ACS Appl. Bio Mater*. 3, 5310–5321, 2020. https://doi.org/10.1021/acsabm.0c00649.
41. Jena, S., Damayanti, N. P., Tan, J., Pratt, E. D., Irudayaraj, J. M. K., Parker, L. L. Multiplexable fluorescence lifetime imaging (FLIM) probes for Abl and Src-family kinases. *Chem Communications* 56(87), 13409–13412, 2020. DOI:10.1039/d0cc05030j.
42. Ripoll, C., Roldan, M., Contreras-Montoya, R., Diaz-Mochon, J., Martin, M., Ruedas-Rama, M. J., Orte, A. Mitochondrial ph nanosensors for metabolic profiling of breast cancer cell lines C. *Int J Mol Sci* 21(10), 3731 (14p), 2020. DOI: 10.3390/ijms21103731.
43. Damayanti, N. P., Buno, K., Harbin, S. L. V., Irudayaraj, J. M. K. Epigenetic process monitoring in live cultures with peptide biosensors. *ACS Sensors* 4(3), 562–565, 2019. DOI: 10.1021/acssensors.8b01134.
44. Šmidlehner, T., Rožman, A., Piantanida, I. Advances in cyanine – amino acid conjugates and peptides for sensing of DNA, RNA and protein structures. *Curr Protein & Peptide Sci* 20 (11), 1040–1045, 2019. DOI: 10.2174/1389203720666190513084102.
45. Shi, H., Wang, R., Yu, P., Shi, J., Liu, L. Facile environment-friendly peptide-based humidity sensor for multifunctional applications. Appl Nanosci 11, 961–969, 2021. DOI: 10.1007/s13204-021-01683-0.
46. Pang, X., Wang, L., Gao, L., Feng, H. Y., Kong, J. M., Li, L. Z. Multifunctional peptide-based fluorescent chemosensor for detection of Hg2+, Cu2+ and S2- ions. *Luminescence* 34(6), 585–594, 2019. https://doi.org/10.1002/bio.3641.

47. Zhang, W. S., Xi, J. D., Zhang, Y. C., Su, Z. Q., Wei, G. Green synthesis and fabrication of an electrochemical and colorimetric sensor based on self-assembled peptide-Au nanofibril architecture. *Arab J Chem* 13, 1406–1414, 2020. https://doi.org/10.1016/j.arabjc.2017.11.012.
48. Wang, G. X., Han, R., Li, Q., Han, Y. F., Luo, X. L. Electrochemical biosensors capable of detecting biomarkers in human serum with unique long-term antifouling abilities based on designed multifunctional peptides. *Anal. Chem*. 92, 10, 7186–7193, 2020. https://doi.org/10.1021/acs.analchem.0c00738.
49. Weaver, E., Uddin, S., Cole, D. K., Hooker, A., Lamprou, D. A. The present and future role of microfluidics for protein- and peptide-based therapeutics and diagnostics. *Appl. Sci.* 11, 4109–4125, 2021. https://doi.org/10.3390/app1109410.

10 Graphene-Based Hybrid Nano Composites for Bio/Chemical Sensors

C. Kavitha and Rapela R. Maphanga

CONTENTS

10.1 Introduction ... 217
10.1.1 History of Graphene and its Derivatives ... 218
10.1.2 Graphene Oxide (GO)/Reduced Graphene Oxide (rGO) Hybrid with Metal (M)/Metal Oxide (MO) Nano Structures ... 219
10.1.3 Sensors ... 220
10.1.4 Surface Enhanced Raman Spectroscopy (SERS) Based Sensors ... 220
10.1.5 Electrochemical Sensors ... 221
10.2 Types of GO/rGO Hybrid M/MO Nanocomposites ... 222
10.2.1 GO/rGO-Au Hybrid Nanocomposites ... 223
10.2.2 GO/rGO-Ag Hybrid Nanocomposites ... 225
10.2.2.1 GO/rGO–Os Hybrid Nanocomposites ... 228
10.2.3 GO/rGO-Cu Hybrid Nanocomposites ... 230
10.2.4 GO/rGO-ZnO Hybrid Nanocomposites ... 233
10.2.5 GO/rGO-SnO_2 Hybrid Nano Composites ... 237
10.2.6 GO/rGO-Fe_2O_3 Hybrid Nanocomposites ... 239
10.3 Summary ... 240
References ... 241

10.1 INTRODUCTION

Graphene (G) and its derivatives, Graphene Oxide (GO) and reduced Graphene Oxide (rGO), are playing a key role in vast applications in emerging fields such as Optoelectronics, sensors, energy storage, water purification and so on. This is due to their unique properties: homogeneous surface, mobility of electrons, mechanical strength, electrical and thermal conductivity [1–7]. Because of these wonderful unique properties, there is a great interest in using these materials in analytical chemistry [8–14]. Specifically, the superior optical and electronic properties of graphene materials make them apt for sensor application, which use electromagnetic waves called surface plasmons for conducting material, which reacts with light exposure [15–17]. Further, molecule absorption on the surface of the one-atom thick materials shows significant changes in conductivity and electrical capacity of the material. This mechanism is used for sensor fabrication to detect the chemicals/gases in lower concentration. In this case Surface Enhanced Raman Spectroscopy (SERS) based optical sensors play a vital role in detecting chemicals, biochemical, insecticides, water pollutants and so forth in very low concentrations. Electrochemistry is an emerging technology for detecting materials of interest with various modified electrodes. Here, GO and rGO act as excellent mediators where they facilitate the excellent electron transfer between electrochemically active material and

DOI: 10.1201/9781003218708-12

the surface of the electrode. SERS and electrochemical detection are the low-cost, environment-friendly, non-destructive methods to detect chemicals, biochemicals and gases with excellent sensitivity and selectivity.

10.1.1 History of Graphene and its Derivatives

In 2004, the graphene was discovered by Andre Geim and Konstantin Novoselov. This motivated the scientific community to explore extensively the potential applications of graphene and its derivatives as shown in Figure 10.1. Graphene is the mother of all other carbon 2-D materials. It is a nano material consisting of atom-thick planar sheets of sp^2 bonded carbon atom arranged in a honeycomb lattice. It is made up of pure carbon whereby each carbon atom is covalently bonded together in the same planar, and the monolayer graphene sheets are linked by van der Waals forces. Graphene oxide (GO) is considered as the oxidized form of graphene. In 1859, GO was first synthesized by oxidation and exfoliation of graphite. The discovery of graphene oxide was much before the discovery of graphene. The synthesis of GO mainly achieved by the top-down approach, which includes the treatment of graphite with strong oxidants or sulfuric acid and potassium permanganate, and the subsequent exfoliation step achieved by mechanical peeling methods such as sonication and shearing stress.

Through this treatment process, the sp2 structure of graphite layers is disrupted and acquires several different oxygen-containing functional groups such as carboxyl, hydroxyl, or epoxy groups. The oxidation of graphite layers increases the interplanar spacing of the graphite structure. The subsequent exfoliation step separates the layers of graphite oxide to obtain a solution of homogenous graphene oxide layers. The sp2 bonding network disruption leads to prominent properties. Depending on degree of oxidation, GO shows low electrical conductivity, as a result behaving like insulating or semiconducting materials. The existence of a number of other functional groups results in hydrophilic behavior of GO. So they show good dispersibility and stability toward aqueous solutions in all concentrations. They do disperse in organic solvents like ethylene glycol, dimethyl formamide (DMF), n-methyl-2-pyrrolidone (NMP), and tetrahydrofuran (THF) due to hydrogen bonding between surface and solvent interface. GO also shows good optical transparency because of the presence of oxidized components.

The transformation of Graphene Oxide (GO) to reduced Graphene Oxide (rGO) can be achieved through chemical, thermal, photo thermal reduction and so on. rGO can be obtained from chemical-reducing agents like NaB, phenyl-hydrazine hydrate, and so forth. Thermal reduction can occur at temperatures of 300°C–2000°C with an inert atmosphere. Further, photo thermal reduction can be achieved by irradiation of 390 nm laser wave length. As a result of the reduction process, the oxygenous functional groups in the GO are eliminated to form rGO with a carbon to oxygen (C/O) ratio of 8:1–246:1 [19]. The restoration of sp2 structure after the reduction process increases the electric conductivity, high mobility and surface area. Due to the increased C/O ratio, rGO acquires a hydrophobic behavior. Though the reduction process is not completely restoring the graphene structure

FIGURE 10.1 Schematic representation of Graphene and its derivatives.

Source: G. Reina 2017 [18].

in rGO, it still retains significant properties like controllable functionality, high electric and thermal conductivity, low cost and a large-scale preparation process.

10.1.2 Graphene Oxide (GO)/Reduced Graphene Oxide (rGO) Hybrid with Metal (M)/Metal Oxide (MO) Nano Structures

Presently emerging research is devoted to graphene and its derivative hybrids with noble metal (M) and metal oxide (MO) nanocomposites. These hybrid nano composites have wider applications than individual components of their own. These GO/rGO hybrids with M/MO have improved mechanical, thermal, chemical, electrical and optical properties, at the most extent. Particularly, these graphene-based hybrid materials have high sensitivity and selectivity in all type of sensors due to the synergic effect of the compound configuration [20].

The solution-phase method is a comfortable technique for synthesizing graphene-based hybrid metal nanocomposites. Figure 10.2 shows the schematic representation of graphene-based materials hybrid with gold metal nano composites. This can be applicable to all noble metals hybrids with graphene and its derivatives. A schematic illustration of the binding mechanisms of nanoparticles (NPs) onto highly reduced graphene (HRG) sheets is shown in Figure 10.3. In situ binding and ex-Situ hybridization are the two ways to load or bind metal/metal oxide NPs on graphene derivatives to synthesis graphene-based hybrid nanocomposites. In in situ chemical reduction methods, the preparation of graphene-based metal and metal oxide nanocomposites using metal precursors such as $HAuCl_4$, $AgNO_3$, K_2PtCl_4, and H_2PdCl_6 with reductants like hydrazine hydrate, amines, and $NaBH_4$ are more common and widely applied [22, 23]. This method has an advantage of having high density and uniform coverage of nanostructures by graphene sheets.

But, ex situ hybridization involves mixing of separate solution of surface-functionalized graphene nanosheets and pre-synthesized NPs. The functionalization of graphene and NPs greatly enhances their solubility [25, 26]. This broadens the opportunities for the preparation of graphene-based composites. However, this method has the disadvantage of having low density and non-uniform coverage of nanostructures by graphene sheets.

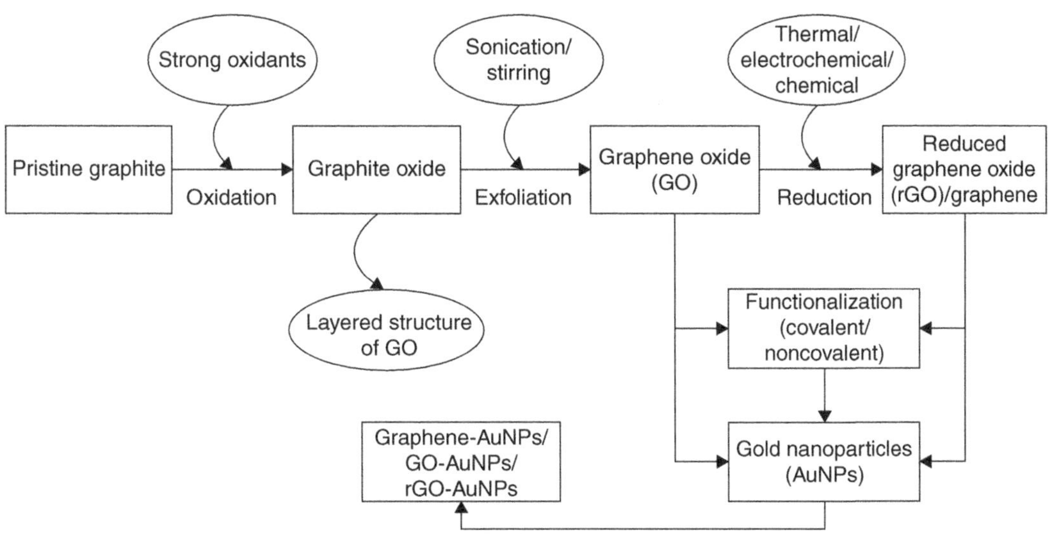

FIGURE 10.2 Schematic flow chart of Graphene derivatives-Au nano composite preparation.

Source: Khalil 2016 [21].

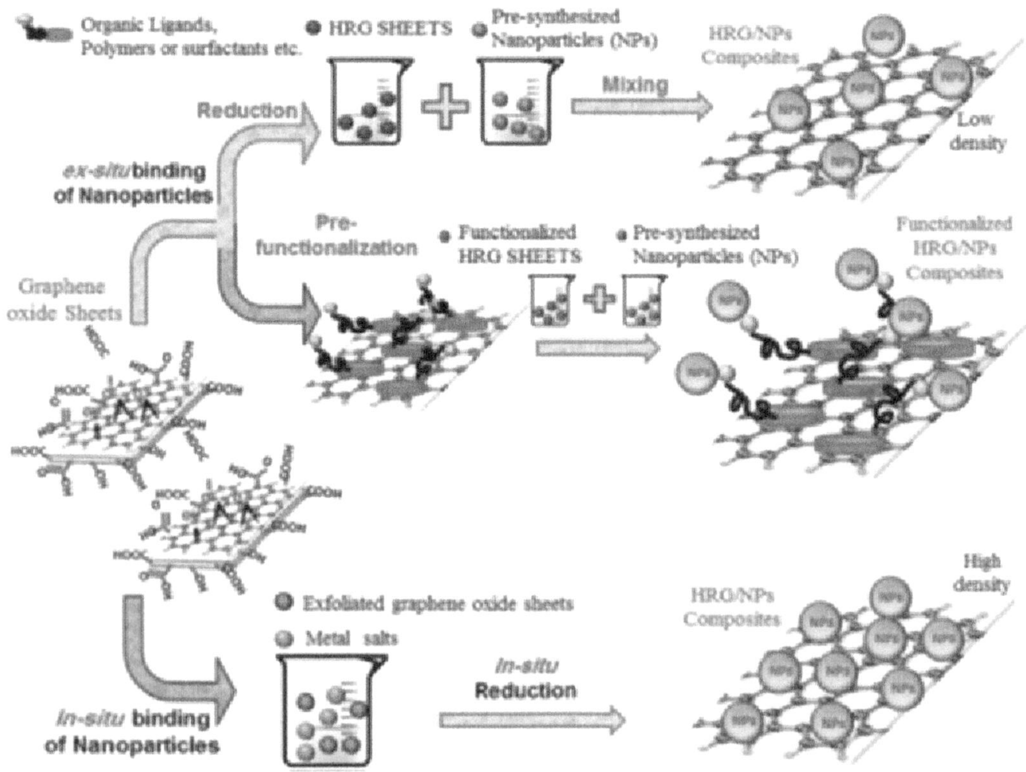

FIGURE 10.3 Schematic illustration of the binding mechanisms of nanoparticles (NPs) onto HRG sheets. The NPs can be loaded onto HRG sheets in two different ways by in situ or ex-situ binding.

Source: Mujeeb Khan 2015 [24].

10.1.3 Sensors

Sensor technology is a vast area and has been widely used as effective analytical tools to achieve the accurate detection of various pollutants. Environment pollution leads to greater risk to human health. So environmental safety is one of the major concerns for our well-being. The beauty of nanomaterial in nanotechnology can resolve the pollution problem and improve the quality of human life. SERS-based optical sensors, gas sensors, biosensors, and electrochemical sensors constitute the recent emerging sensor technology to detect various pollutants like dyes, pesticides, toxins, pathogens, pharma contaminants, heavy metals and so on. Spectroscopy-based optical sensors, electrochemical sensors, are low cost, environment friendly and an easy technique. The sensitivity and selectivity are very accurate even in very low concentration of pollutants.

10.1.4 Surface Enhanced Raman Spectroscopy (SERS) Based Sensors

Surface-enhanced Raman spectroscopy (SERS) is a surface-sensitive technique that enhances Raman scattering by molecules adsorbed on rough metal surfaces. The enhancement factor can be as much as 10^{10} to 10^{11}, which means the technique may detect single molecules. Recently, researchers are giving more attention to preparing coinage metals, metal oxides, and their hybrids with carbon materials as a SERS substrate to detect low concentration pollutants.

Surface enhancement is influenced by two broad mechanisms (see Figure 10.4),

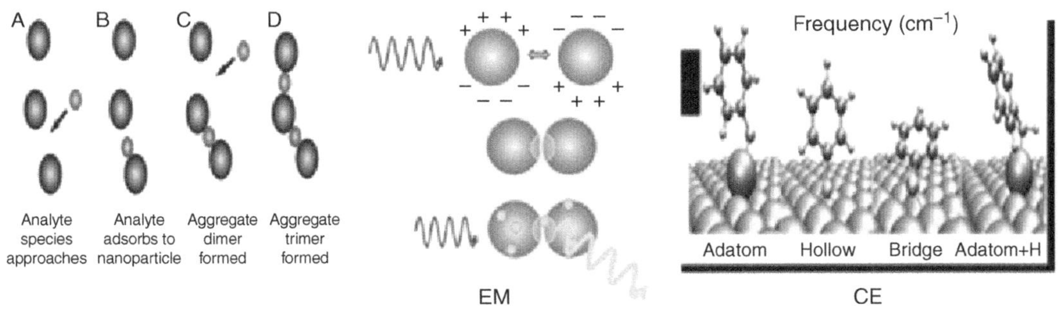

FIGURE 10.4 Schematic formation of nanoparticle aggregates by the stepwise addition of adsorbate molecules (red spheres), Electromagnetic enhancement (EM), Chemical Enhancement (CE) Mechanism of surface enhancement.

Source: Brian D. 2007 [27a].

(1). Electromagnetic enhancement (EM)
On a noble metal nanoaggregate, incident light is resonant with plasmon due to dipolar oscillation of conduction band electrons – localized surface plasmon resonance (LSPR). Then, an electromagnetic (EM) field is enormously enhanced at a junction of the nanoaggregate. The enhanced EM field causes the enormous enhancement of effective Raman scattering cross-section of molecules adsorbed on the junction of the Ag nanoaggregate.

(2). Charge transfer (chemical enhancement -CE)
This involves changes to the adsorbate electronic states because of chemisorption of the analyte.

From Figure 10.5, it is clear that at the resonant wavelength excitation of R6G (Figure 10.5b), 514.5 nm, the clear spectra with strong Raman signals and a high signal-to-noise ratio were observed on graphene, while only the fluorescence background was observed for 10μM R6G aqueous solution when collecting Raman spectra from the SiO_2/Si substrate [27]. Due to the graphene-induced fluorescence quenching [28], the baselines for graphene and the SiO_2/Si substrate are at different levels.

10.1.5 Electrochemical Sensors

Basically, the devices that can sense the physical, chemical, or biological changes of analytes and convert them into measurable amperometry, voltammetry, impedance, or other electrochemical signals are electrochemical sensors and biosensors (see Figure 10.6). This electrochemical sensing system usually contains three electrodes: working electrode, counter electrode, and reference electrode. The redox reaction that occurs on the surface of working electrodes is the especially the basis of electrochemical detection. So, environmental contaminants with electrochemical activities can be directly sensed by electrochemical sensors. Voltammetry is the most well-known technology for electrochemical sensors. This can measure the current response to achieve a precise quantitative analysis of target analytes under an applied potential. To improve the sensitivity and selectivity of particular environmental pollutants, one has to modify the working electrode with recognition element such as antibodies, enzymes and nucleic acids. These recognition elements specifically recognize one or several particular pollutants from complex real sample mixture without any interference.

So, the efficiency of electrochemical biosensors depends on the search of novel materials to improve the sensitivity and selectivity in detection of particular analyte. Recently many emerging materials such as conducting polymers, M/MO NPs, CNTs, graphene and Metal Organic Framework

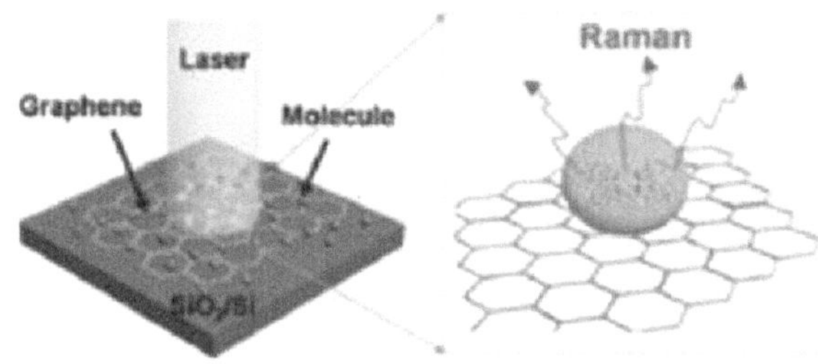

(a)

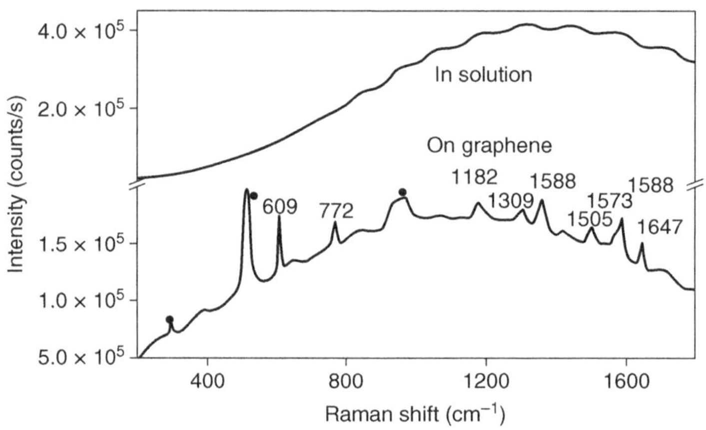

(b)

FIGURE 10.5 Graphene as a substrate for enhanced Raman spectroscopy. (a) Schematic illustration of the molecules on a graphene/SiO_2 /Si substrate. (b) Raman-fluorescence spectra of rhodamine 6G (R6G) in water (10μM) (blue line) and R6G on monolayer graphene (red line) at 514 nm excitation.

Source: (a) Xi Ling 2010 [27] and (b) Xie L. M. 2009 [28].

are explored to modify the surface of the working electrode. This modified electrode leads to good stability, a huge specific area, improved redox performance and recyclability.

In electroanalysis, the electrical and electronic properties of graphene play a great role (see Figure 10.6). It has a wide electrochemical potential window, excellent charge-transfer resistance and highly efficient electron transfer behavior, which are very much needed for analytical sensing systems. Graphene has excellent biosensor applications due to excellent electrochemical performance toward biomolecules and high electrocatalytic activity. So graphene and graphene-based materials are emerging trend materials for the fabrication of electrochemical sensors/biosensors.

10.2 TYPES OF GO/RGO HYBRID M/MO NANOCOMPOSITES

The advent of methods for the preparation of stable and homogeneous dispersions of graphene on a large scale has made the researcher in making graphene-based nanocomposites and exploring them for different applications. In the last few decades, researchers made huge efforts to synthesize composites of graphene with inorganic NPs, mostly based on transition metal and metal oxide NPs [30, 31]. Most of graphene based Nano composites are mainly hybrids of two components.

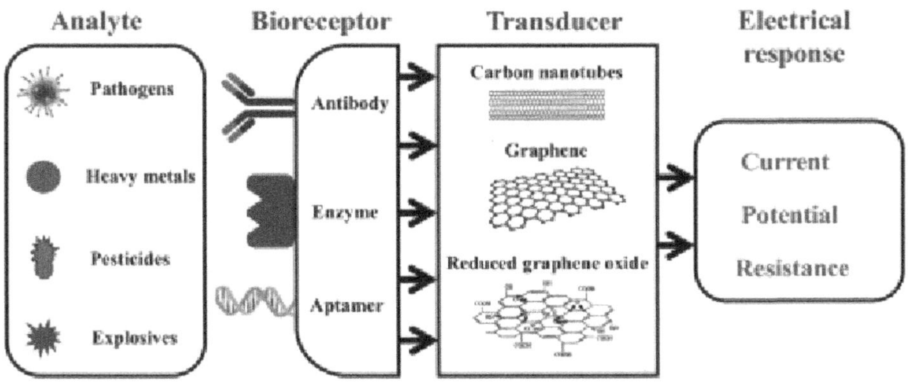

FIGURE 10.6 Schematic representation of carbon materials based electrochemical sensors/biosensors.

Source: Pankaj R 2016 [29].

Multi- component composites are also under investigation depending on the special applications. There are three types of graphene-based nanocomposites [30]. Type 1: Graphene sheet acts as a substrate for supporting a second component that attaches to the graphene sheets. Type 2: Typically, the second components are inorganic (M/MO) NPs, polymeric nanostructures, CNTs and so forth. Type 3: Graphene sheets act as nano-fillers, where they behave as the continuous matrix of the second component.

10.2.1 GO/rGO-Au Hybrid Nanocomposites

Graphene and its derivatives are attracted as a superior substrate for SERS. This was investigated with single-layer graphene as well as with multilayer graphene. Xie and co-workers reported that the single-layered graphene showed the highest SERS intensity over the multilayer graphene or graphites [32]. It was explained by Yu et al. that a mildly reduced GO (RGO) is superior to the pristine graphene due to a strong local electric field induced from the electronegative oxygen species [33]. Chemical vapor deposition and UV/ozone-based oxidation methods are used to derive a highly oxidized large-scale graphene and showed high chemical enhancement in SERS due to the structural disorder and defects on the surface [34].

Graphene/Metal NPs hybrid composites also play a major role in enhancing the Raman signal due to their synergic property.

Han et al. explained unique synthetic scheme of preparing the 2D graphene substrates for SERS measurements. (See Figure 10.7a). Initially, 2D GO sheet were generated by spin-coating a water-soluble GO solution at 4000 rpm. This gave a single-layered (~0.8 nm thickness) and uniform GO sheet. The formation of Au NPs on the GO surfaces mainly relies on the presence of negatively charged oxygen functional groups. Through the reduction reaction by a sodium citrate solution (a reducing agent), the gold(III) ions turned into the Au NPs that were grown on the GO surface. To obtain RGO, the GO at 200°C for 20 h was incubated with hydrazine hydrate. Since the reduction process is not harsh enough to remove the attached Au NPs, the position and the number of Au NPs on the GO would be identical with those on the RGO. As a result, the 2D structural graphenes (2D GO, RGO, GO/Au NP and RGO/Au NP) were generated and then utilized as SERS substrates to detect R6G molecules. From Figure 10.7b,c, it is clear that synergic effect of 2D rGO/Au hybrid substrate was lesser than 2D GO/Au hybrid substrate. The 2D GO/Au, 2D rGO/Au NPs composite film shows 8×10^2 fold, 1.4×10^2 fold improved SERS enhancement respectively. This is due to synergistic effect of graphene's chemical enhancement and electromagnetic effect by Au NPs [35].

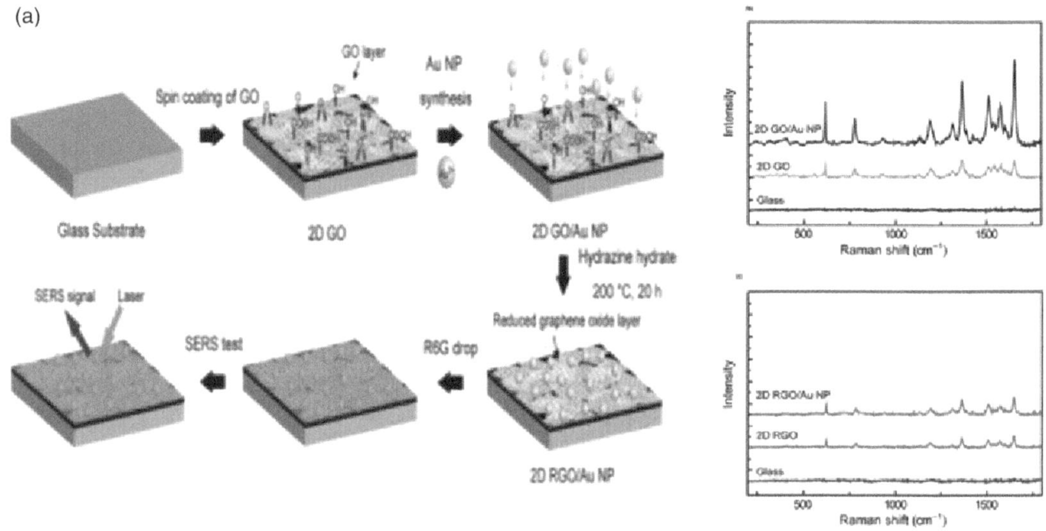

FIGURE 10.7 Fabrication scheme of (a) the 2D GO, 2D GO/Au NP, 2D RGO/Au NP composite as SERS substrate. The single-layered 2D GO sheets were generated by using a spin-coating method on the glass wafer, Raman spectra of Rhodamine 6G (R6G) by using (b) a bare glass, the 2D GO and the 2D GO/Au NP, and (c) a bare glass, the 2D RGO and the 2D RGO/Au NP as a SERS substrate.

Source: Dong Ju Han 2013 [35].

Wang et al. explained the effect of a Au film deposited on a single-layer graphene as a SERS substrate. They showed that a 7-nm thick gold film produced strongest Raman signals for detecting R6G molecules [36]. Hao et al. synthesized high-quality, single-layer graphene sheets by chemical vapor deposition (CVD) and transfer them from copper foils to gold nanostructures, that is, nanoparticle or nanohole arrays. The combined graphene nanostructure substrates show about threefold or ninefold enhancement in the Raman signal of MB, compared with the bare nanohole or nanoparticle substrates, respectively [37]. Huang et al. developed controlled size, size distribution, and morphology of the metal nanoparticles in the metal-GO nanohybrids. They described the formation of Au nanoparticle-graphene oxide (Au-GO) and -reduced GO (Au-rGO) composites by noncovalent attachment of Au nanoparticles pre-modified with 2-mercaptopyridine to GO and rGO sheets, respectively, via π–π stacking and other molecular interactions. They showed excellent improvement toward SERS, Catalysis measurement [38]. Deng et al. explained about rGO can be decorated by bovine serum albumin (BSA) at suitable pH and temperature. This results in bio conjugates of BSA and GO or rGO and act as an ideal template for different sizes and shapes of nanoparticle. This method helps to investigate structure-performance relationship of hybrid nanomaterials for special applications [39]. Piao et al. represented, the electrochemical behavior of dopamine (DA) on the surface of the graphene/gold nanocomposite modified Glassy carbon electrode was investigated by cyclic voltammetry (CV) and differential pulse voltammetry (DPV). In comparison to the reduced graphene oxide (rGO) modified electrode, the graphene/gold nanocomposite modified electrode shows good performance for the selective determination of DA [40]. The nanocomposite modified electrode showed a low detection limit (0.095 μM). Berry et al. demonstrated controlled diffusion and catalytic reduction of gold ion on GO. This results in "snowflake shaped" gold nano structures (SFGN) was template on graphene. The Growth mechanism of SFGNs on GO sheets and their structural dependence on the synthesis temperature is shown in Figure 10.8a. SFGN interaction with GO was probed by SERS on bare GO, GO-SFGN sheets. The prominent D band (~1340 cm^{-1})

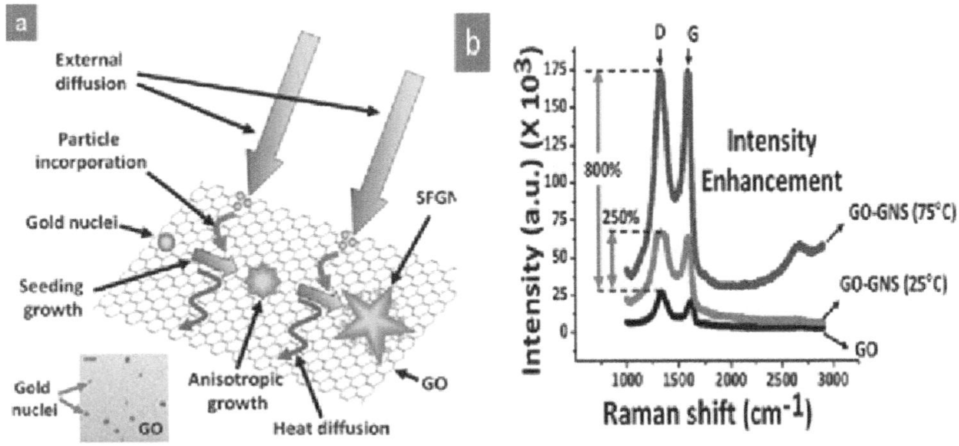

FIGURE 10.8 (a) Schematic showing the elementary steps involved in the seeding growth of SFGNs on GO template. Au ions diffuse from the bulk to the GO sheet where they are catalytically reduced and incorporated in the growing Au nuclei. Bottom inset shows seed particles on GO that were prepared by sodium borohydride-assisted reduction of gold salt in the presence of sodium citrate and GO. Scale bar =10 nm. (b) Raman Spectra for GO_SFGN and GO showing the presence of SFGNs on GO enhances the intensity of D and G bands by ~250% and ~800% suggesting chemical enhancement, and hence a chemical-bond formation between SFGNs and GO.

Source: Kabeer Jasuja 2009 [41].

and G band (~1590 cm^{-1}) of GO was observed with different substrates. SNGO hybrids with GO prepared by 25°C synthesis temperature enhanced the intensity of these bands by > 250 percent. Later, enhancement was increased up to ~800 percent when the synthesis temperature was increased up to 75°C (see Figure 10.8b) (Table 10.1).

10.2.2 GO/rGO-Ag Hybrid Nanocomposites

Among the coinage metals, Silver (Ag) is the most conductive and reactive material, due to its germicidal nature recently it has much attention toward biomedical applications like anti-bacterial activity. Further Ag hybrid with GO nano composites enhances the properties of GO with good dispersion, low resistance and enhanced mechanical strength and so on. Further Ag intercalated with GO can improve the electrochemical properties and photocatalytic activities.

Majumder et al. demonstrated the sonochemical method to prepare GO-Ag nano composites for enhanced Raman and electrochemical activity. Further it proved as a good substrate for enhanced thermoluminescence and dye degradation properties as shown in Figure 10.9 [56].

P.K. Dwivedi et al. explained the wet-chemical polyol method and C. Kavitha et al. demonstrated low-cost, simple liquid/liquid interface method for the preparation of rGO-Ag nano hybrids for effective R6G at lower concentrations till 1 nm [57–58]. These substrates show enhanced sensitivity and selectivity with various analytes like 2, 4 dinitro toluene (2, 4-DNT), 4, Mercapto Benzoic Acid (4-MBA). The enhancement factors for rGO-Ag nano composites shows 2.3×10^{8}, 1.29×10^{5} by liquid/liquid interface method, Polyol method respectively. Chao Xu et al. reported silver mirror reaction for flexible silver films assembled on GO sheets for generic SERS applications [59]. Wei Fan et al. prepared GO hybrids with different shape and size of the Ag NPs for ultra-sensitive, single particle SERS sensors [60]. Metal NPs, normally shows fluorescence background when you record Raman spectra. As a result, valuable Raman information will be suppressed by that fluorescence

TABLE 10.1
GO/rGO Hybrid with Au Nanostructures for Electrochemical, SERS Based Sensor Applications

Materials	Synthetic routes	Applications	References
Au/HRG	In situ electrochemical synthesis, Ex-situ layer-by-layer elfassembly	Bio Sensors	42
Au/HRG	In situ Chemical Synthesis	Electrochemical sensor for dopamine	43
Au/HRG	In situ Chemical synthesis	Electrochemical sensor for epinephrine	44
Au/HRG	In situ Chemical synthesis	Electrochemical sensor for levofloxacin	45
Au/HRG	In situ Chemical synthesis	Electrocatalytic application	46
Au/GO or rGO	Ex situ: p–p stacking via 2-mercaptopyridine	Catalysis, SERS	38
Au/GO or rGO	Ex situ: p–p stacking via bovine serum albumin	Catalysis, SERS	39
Au/GO or rGO	In situ: photochemical reduction	Catalysis, SERS	47
Au/GO or rGO	In situ: reduction by hydroxyl-amine	Raman enhancement	41
Au/GO or rGO	In situ: sonolytic reduction in poly(ethylene glycol) at 211 kHz	Raman enhancement	48
Au/rGO	In situ: reduction by $NaBH_4$	Plasmonics	49
Au/rGO	In situ: reduction by amino terminated ionic liquid; perylene-modified rGO	electroanalysis	50
Au/rGO	In situ: reduction by sodium citrate	SERS	51
Au/rGO	In situ: reduction by ascorbic acid in the presence of CTAB	SERS	52
Au/rGO	In situ: microwave assisted reduction	SERS	53, 54
Au/pristine graphene	In situ: thermal evaporation	Identification of layer number of pristine graphene	55

Table 10.1 Displays the previous reports on electrochemical, SERS based sensor applications of GO/rGO-Au hybrid nano structures prepared by different synthesis methods.

background. But J. Huang et al. explained that Graphene and its derivatives are excellent fluorescence quencher so that they become active SERS substrate for fluorescence molecules [61]. D.H. Chen et al. demonstrated a facile and rapid microwave-assisted green route for the formation of Ag/rGO hybrid nano structures. They have used L-Arginine as a reducing agent. They found that the substrate has high sensitivity and homogeneous with very high enhancement factor as a SERS substrate [62]. P.K. Dwivedi demonstrated an ultrafast laser-ablated hierarchically patterned Ag/GO nano hybrid SERS substrate for highly sensitive and reproducible detection of an explosive marker 2,4-dinitrotoluene (2,4-DNT) [63].

Since synergic effect of GO/M nano composites exhibit improved performance than NPs and graphene separately, more efforts have been taken by researchers recently to decorate metal NPs on Graphene and its derivatives with various synthesis routes [64]. They would like to explore these hybrid materials applications toward photo thermal therapy, drug delivery, catalysis and biosensors (see Figure 10.11).

Tuan Le et al. has shown an easy and effective photo chemical method of preparing Ag-GO aqueous solution and investigated their optical properties of photoluminescence and UV visible spectra. The excellent optical properties of Ag-GO aqueous solution may open up an effective path to optoelectronic devices [65]. Rahman et al. performed rapid and green microwave irradiation

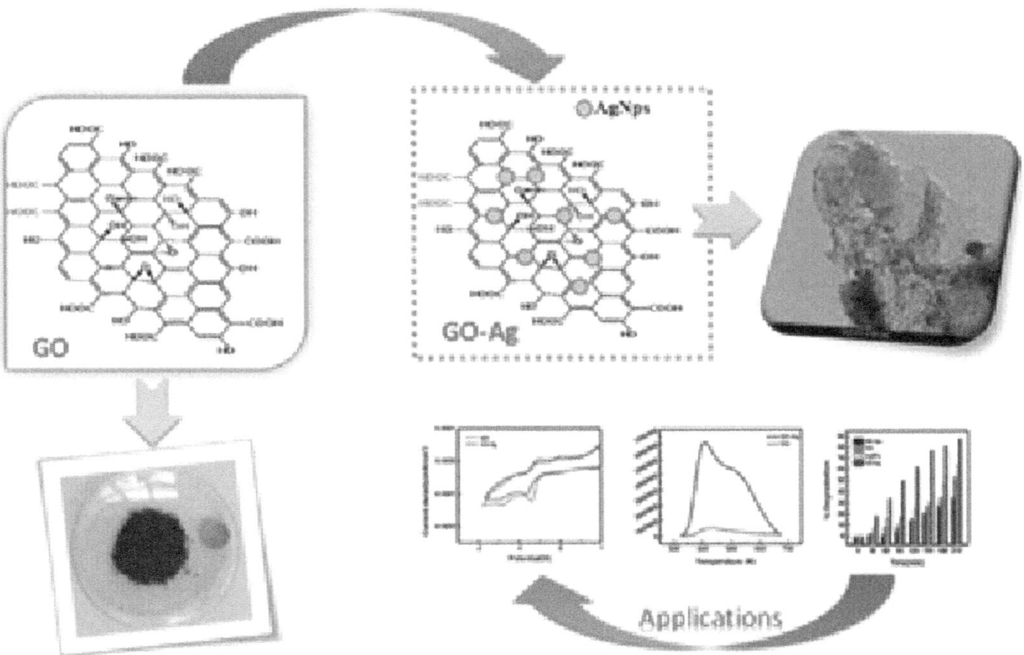

FIGURE 10.9 Graphical representation of GO-Ag hybrid nanocomposites for unique physiochemical applications.

Source: Sujata Kumari 2020 [56].

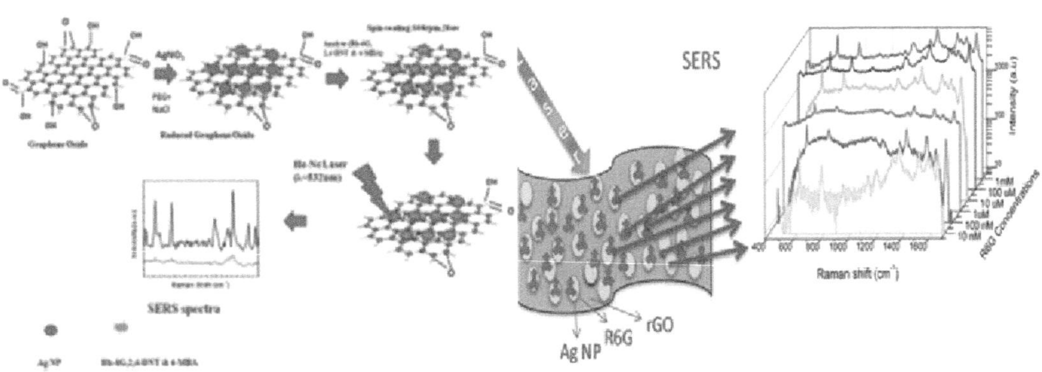

FIGURE 10.10 Pictorial representation of rGO-Ag hybrid nano composites for R6G detection prepared by (a) wet chemical Polyol method, (b) Liquid/liquid interface method.

Source: (a) Tania K. N. 2019 [57] and (b) C. Kavitha 2015 [58].

method to synthesize Ag-GO hybrid nanostructures. The synergic effect of Ag-GO shows excellent antibacterial properties [66].

Nagaraja. H.S. et al. demonstrated a galvanic replacement method for the preparation of Ag dendrites and its hybrid with GO. This AgD-GO substrate shows excellent electrochemical paracetamol sensor with high sensitivity of 2.511×106μA /mM/g with a lower detection limit of 0.025 μM [67]. A.M. Noor et al. fabricated a glassy carbon electrode (GCE) was modified with the

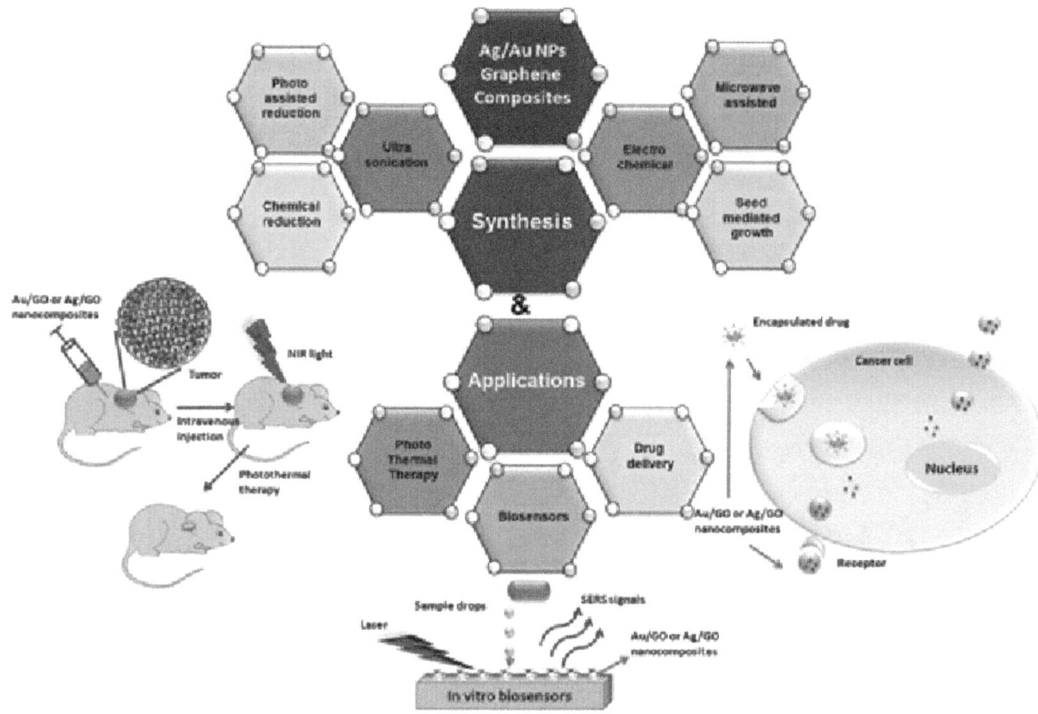

FIGURE 10.11 Schematic representation of Ag/Au NPs/rGO composite materials: Possible synthesis and applications.

Source: D. Gitashree 2019 [64].

rGO-Ag nanocomposite and it was used for the effective detection of 4-Nitrophenol (4-NP). The sensor shows very good selectivity for 4-NP even in the presence of 50-fold higher concentrations of analogues such as bromophenol blue, 2-amino-4-nitrophenol, 2-chlorophenol and 2,4-dinitrophenol [68] (Table 10.2).

10.2.2.1 GO/rGO–Os Hybrid Nanocomposites

Hybrid nanoparticles of various sizes and shapes have received tremendous interest because of their distinguished properties like electrical, optical and electro chemical properties. These properties make them as unique materials for photonics, electronics, sensors and catalysis areas. Among the coinage metals, Os recently attracted attention due to its electro and hetero catalysts properties. Surface structure and morphology are the important factors for the selectivity and activity of the catalyst material. Though the Os cost is high, it is a good catalyst for alcohol to aldehyde oxidation, CO oxidation, borohydride oxidation and hydrogen oxidation and so on [78–81]. There is a very little literature on optical and electrochemical sensor properties of Osmium. Recently Sampath et al. explored an interconnected, ultrafine Os cluster prepared by simple hydrothermal method with ascorbic acid as reducing agent. In Figure 10.12, these Os nano chains have been demonstrated as excellent SERS based optical sensor to detect 4-Mercapto Pyridine (non-fluorescent) and R6G (fluorescent) analytes in low concentrations [82].

C. Kavitha et al. demonstrated rGO/Os hybrid nano dendrites prepared by liquid/liquid interface method for the first time. As shown in Figure 10.13, they proved that the rGO-Os hybrid shows improved catalytic activity and 14-fold enhancement in SERS activity to detect R6G compared with Os alone [83].

TABLE 10.2
GO/rGO Hybrid with Ag Nanostructures for Electrochemical, SERS Based Sensor Applications

Materials	Synthetic Routes	Applications	References
Ag-Dentrite /rGO	Galvanic replacement synthesis	SERS-Rhodamine B	69
Ag-rGO/GCE	Green synthesis	EC sensor-dopamine, uric acid	70
Ag/GO or rGO	Ex situ: π–π stacking via bovine serum albumin	SERS	71
Ag/GO or rGO	In situ: reduction by $NaBH_4$	SERS	72
Ag/GO or rGO	In situ: electroless depositionSERS	SERS	73
Ag/GO/Au electrode	Method for effective immobilization of Ag NPs on GO with single strand DNA decorated Au electrode	EC-sensor-H_2O_2 detection	74
Ag/rGO/GCE	one-pot synthesis of (AgNPs/rGO) nanocomposites, diethylenetriamine as a reducing agent	EC-sensor-H_2O_2 detection	75
Ag/rGO/GCE	fabricated through a one-step strategy, carried out by hydrothermal treatment	EC-sensor-H_2O_2 detection	76
Ag/GN/GCE	(AgNPs/GN) composites have been rapidly prepared by a one-pot microwave-assisted reduction method	EC-sensor-H_2O_2 detection	77

Table 10.2 clearly shows different synthesis methods of Ag hybrids with GO/rGO and their applications in SERS optical and electrochemical sensors.

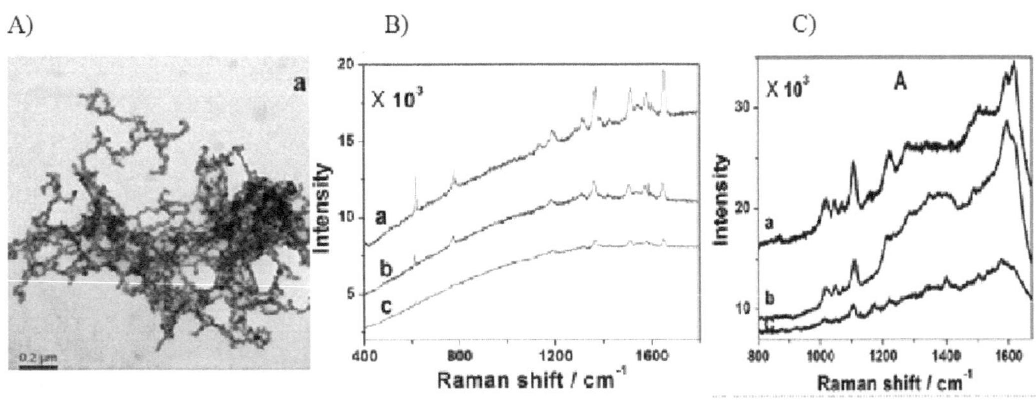

FIGURE 10.12 (A) Bright field TEM picture of Os nanochain, (B) SERS spectra of B)R6G, (C) 4-MPy adsorbed on Os nanochains from (a) 1mM, (b) 100 µM and (c) 10 µM solutions. The excitation wavelength used is 514 nm.

Source: K. Chakrapani 2013 [82].

Neena et al. demonstrated rGO-Os hybrid prepared by simple liquid/liquid interface method (LLI) as a quantitative electrochemical sensor for Rhodamine B (RhB) (See Figure 10.14a). The free standing rGO-Os films can be lifted directed on the desired electrode without the help of any binders.

From Figure 10.14B, it is clear that RhB oxidation peak observed for bare and modified electrodes. This is reversible in all cases. For Os NPs modified electrode the peak observed at 0.325 V is attributed to self-oxidation of Os NPs. On the other hand when Os Nps hybrids with rGO sheet,

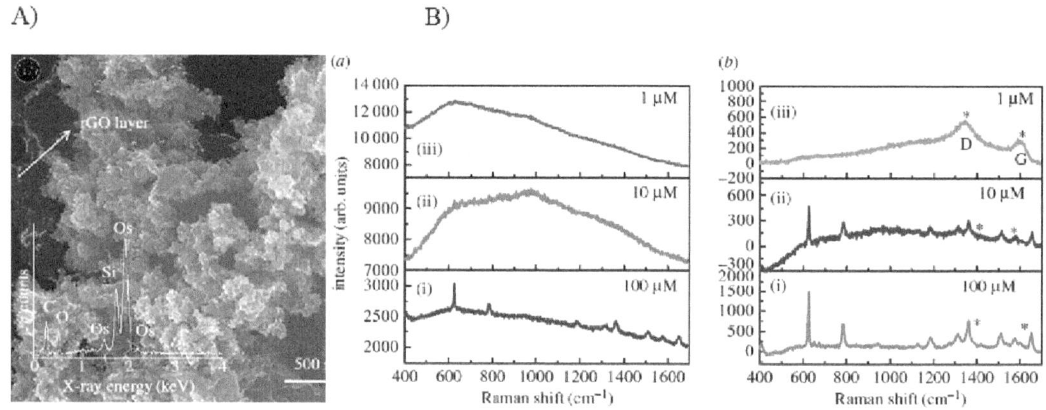

FIGURE 10.13 A) FESEM Image and EDS analyses of Os-rGO hybrid thin film, B) SERS spectra of R6G with various concentrations on (a) Os nano dendrite, (b) rGO-Os hybrid nano thin film.

Source: C. Kavitha 2017 [83].

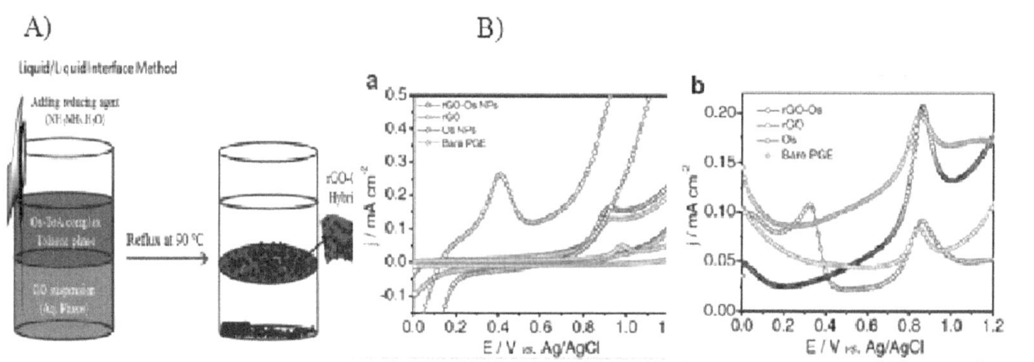

FIGURE 10.14 (A) rGO-Os hybrid thin film prepared by LLI method, (B) electrochemical studies, (a) cyclic voltammetry curves, (b) Differential Pulse Voltametry curves of RhB on bare PGE and modified electrodes with films of Os NPs, rGO, rGO-Os hybrid in Phophate buffere solution (5.8 pH) at scan rate of 100 mVs^{-1}.

Source: K. Bramhaiah 2018 [84].

it protects the Os from self-oxidation and made them available for RhB oxidation [84]. So, the well-defined RhB dye oxidation peak observed at 0.87. The electrochemical current responses of DPV are also consistent with CV, as shown in Figure 10.14 B.

10.2.3 GO/rGO-Cu Hybrid Nanocomposites

Among precious metals like Au, Ag, Os, Copper (Cu) is cheap, with a broad range of potential applications like catalysis, photovoltaics, electronics, optics and sensors. The industrial applications of Copper is as follows: (1) Cu catalyzed methanol synthesis [85]. (2) Hydrogen production by water-gas shift [86]. Copper exhibits localized surface plasmon resonance (LSPR) in the visible range. This optical property may be used in diagnostic and potential sensor applications. Apart from advantages, copper nanoparticles have disadvantages such as the stabilization of particles against agglomeration, the achievement of monodisperse size distributions and oxidative stability have

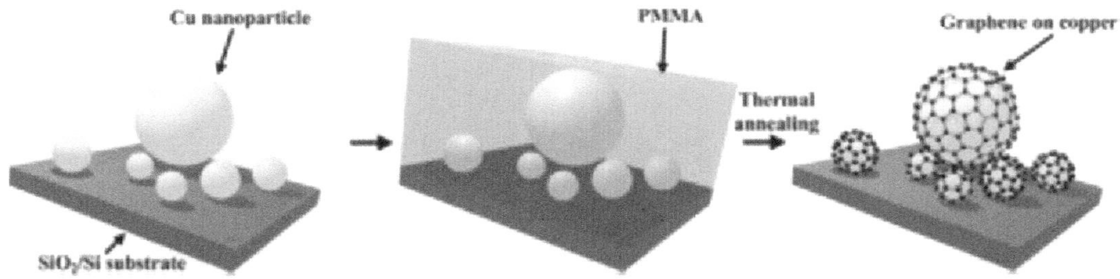

FIGURE 10.15 Schematic synthesis method of Cu encapsulated Graphene Nano Balls (GNBs) using a solid carbon source poly (methyl methacrylate) (PMMA).

Source: S. Lee 2013 [90].

presented considerable obstacles for their practical application. In order to utilize the properties of Cu NPs, it is necessary to control well the chemical stability of Cu NPs. In order to overcome this drawback, scientist have taken an effort to hybrid Cu with Carbon materials because the carbon shells can serve as shields to protect the Cu NPs from oxidation. It is well known that graphene and its derivatives can act as a template to protect NPs from agglomeration and oxidation.

Yuehui et al. reported metal organic chemical vapor deposition to produce thermally stable Cu NPs encapsulated by multi-layer graphene [87]. Stark et al. explained about the synthesis of high air stable metallic Cu NPs through bottom-up approach and in situ coated with graphene. These highly stable hybrid Cu/G NPs are economically more attractive for industrial applications [88]. Again, Stark et al. demonstrated large-scale production of Cu NPs coated with carbon about 1 nm by modified flame spray synthesis unit under highly reducing conditions. The pressed hybrid Cu/C NPs shows highly pressure, temperature dependent electrical conductivity with sensitivity comparable with commercial materials. These properties made Cu/C hybrid as novel, low-cost sensor materials [89].

Lee et al. demonstrated a facile synthesis of fully encapsulate Cu NPs with few layers of graphene using a solid phase carbon source of poly (methyl methacrylate) (PMMA) resulting in the fabrication of graphene nanoballs (GNBs) with Cu cores. The solid source of PMMA was first converted to amorphous carbon layers through a pyrolysis process at the temperature regime of 400°C. This prevented the Cu NPs from agglomeration. Further they were converted to a few-layered graphene (FLG) at elevated temperatures. Raman and transmission electron microscope analyses confirmed the synthesis of FLG with thickness of approximately 3 nm directly on the surface of the Cu NPs. X-ray diffraction and X-ray photoelectron spectroscopy analyses, along with electrical resistance measurement according to temperature changes showed that the FLG-encapsulated Cu NPs were highly resistant to oxidation even after exposure to severe oxidation conditions [90].

Song et al. demonstrated the facile preparation of well-defined Cu/reduced graphene oxide (rGO) hybrid materials by controlling the amount of ascorbic acid and maintaining an apt pH value [91]. They explained that rGO helps to hamper the aggregation of copper nanoparticles, resulting in a small size of the copper nanoparticles. The schematic synthesis method and the TEM image of as prepared Cu/rGO Nano composites with uniform distribution is shown in Figures 10.16a,b. Further the as prepared rGO/Cu hybrid Nano composites were tested for well-known catalytic activity. The reduction of 4-nitrophenol into 4-aminophonol by $NaBH_4$ is employed as a model reaction. In Figure 10.17a,b,c, it is clear that $NaBH_4$ is unable to reduce 4-NP into 4-AP without catalyst even after 1 hour also. Further it is clear from Figures 10.17b,c, after adding Cu NPs, Cu/rGO nano composites to the solution, the complete reduction occurs within 30 minutes and 15 minutes, respectively. Especially rGO/Cu act as an excellent SERS substrate for crystal violet (CV). It is clear from the Figures 10.17d that 0.01 mM CV on Cu/rGO composites show enhanced Raman spectra than pure Cu NPs alone.

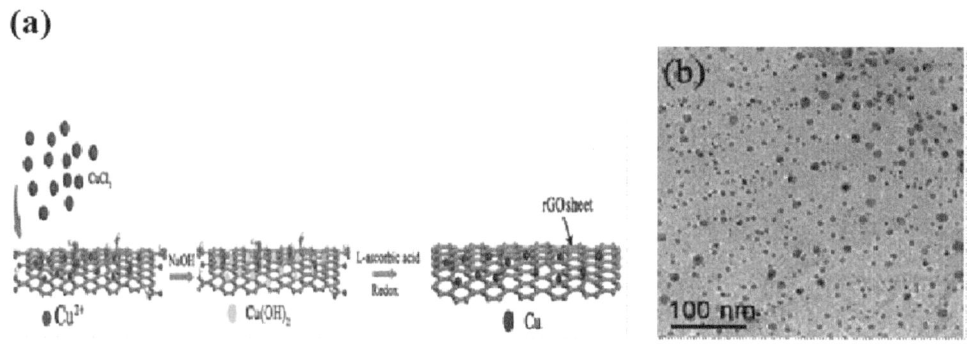

FIGURE 10.16 (a) The schematic illustration of the Cu/rGO hybrids formation, (b) TEM image of Cu/rGO with pH-10.

Source: Meizhen Guo 2016 [91].

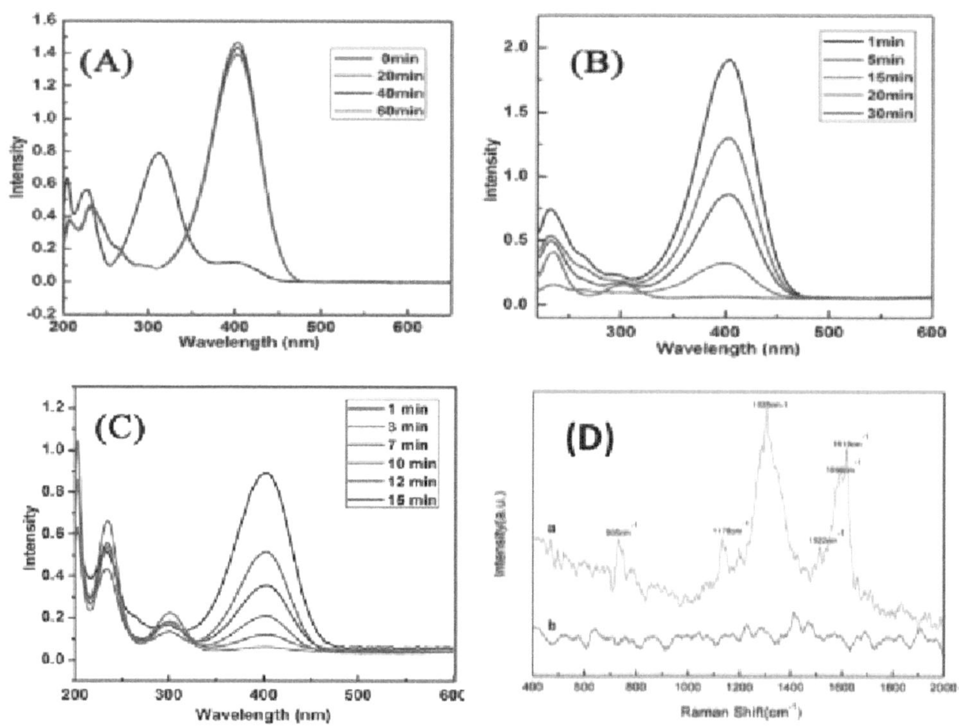

FIGURE 10.17 UV-Vis spectra of the 4-Nitro phenol reduction reaction (A) In the absence of any catalyst, (B) pure copper nanoparticles, (C) Cu/rGO nanocomposite, (D) Raman spectrum of 10^{-5} M crystal violet (CV) solution adsorbed on the a. rGO/Cu Nano composites, b. pure copper nanoparticles.

Source: Meizhen Guo 2016 [91].

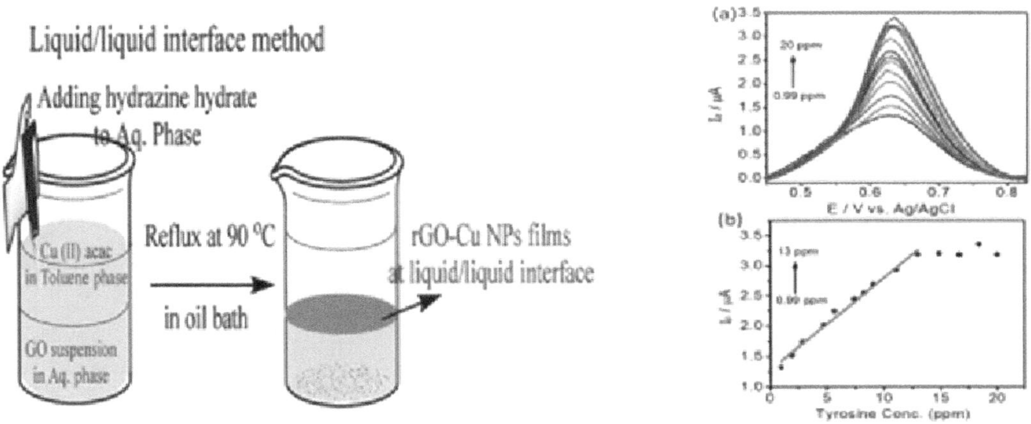

FIGURE 10.18 (A) Schematic representation of the liquid/liquid interface method, (B) Detection of L-Tyrosine with different concentrations on rGO–Cu NP film-coated PGE (a) Differential Pulse Voltammetry (DPV) (b) peak currents (Ip) versus concentration of L-Tyrosine.

Source: C. Kavitha 2020 [92].

C. Kavitha et al. showed facile synthesis of rGO/Cu ultra-thin-film prepared by liquid/liquid interface method. As prepared rGO/Cu modified Pencil graphite electrode was used for the effective detection of L-Tyrosine amino acid.

The schematic representation of a liquid/liquid interface method to fabricate rGO/Cu hybrid nano thin film is shown in Figure 10.18a. It is clear from Figure 10.18b that one can observe that the peak current increases linearly with an increase in L-tyrosine concentration range from 0.99–13 ppm, which can be a calibration curve for the linear range detection of L-tyrosine as an analyte [92].

Bagheri et al. demonstrated a novel electrochemical sensor based on Cu/MWCNT/rGO modified GCE for the effective detection of nitrite and nitrate ion simultaneously [93]. Bian et al. synthesized Cu/rGO, Au/rGO and Cu/Au/rGO using chemical and fractional reduction process. The Cu-Au/rGO exhibited the best catalytic properties. Under optimized conditions (pH 6.5, scan rate 100 mV/s), a sensor based on a glassy carbon electrode modified with this material exhibited a low 4-ChloroPharm detection limit of 0.17 μmol/L. These nano composites were demonstrated for good reproducibility and long-term stability [93].

10.2.4 GO/rGO-ZnO Hybrid Nanocomposites

In general, the semiconductor nanoparticles such as ZnO, TiO_2, have a definite energy band gap suitable for adsorbing/emitting light in the UV visible range and act as a photocatalyst. The plasmonic nanoparticles absorb/scatter light in UV-visible region and used in surface enhanced Raman spectroscopy, electrochemical and other detection applications. Graphene and graphene oxide nano composites have been used to improve signal reproducibility and detection sensitivity. Graphene's honeycomb structure avoids nanoparticle aggregation effectively and allows attaching a large number of nanoparticles on their surface. As a result more electromagnetic hot spots will be formed, which are responsible for SERS. The detection of dye molecule, toxic metal ions and biomolecules are possible using this approach. The detection sensitivity can reach in the nM to pM range.

Wang et al. fabricated noble metal free substrate using annealed Graphene oxide/ZnO nano composites for efficient chemical SERS substrate [95].

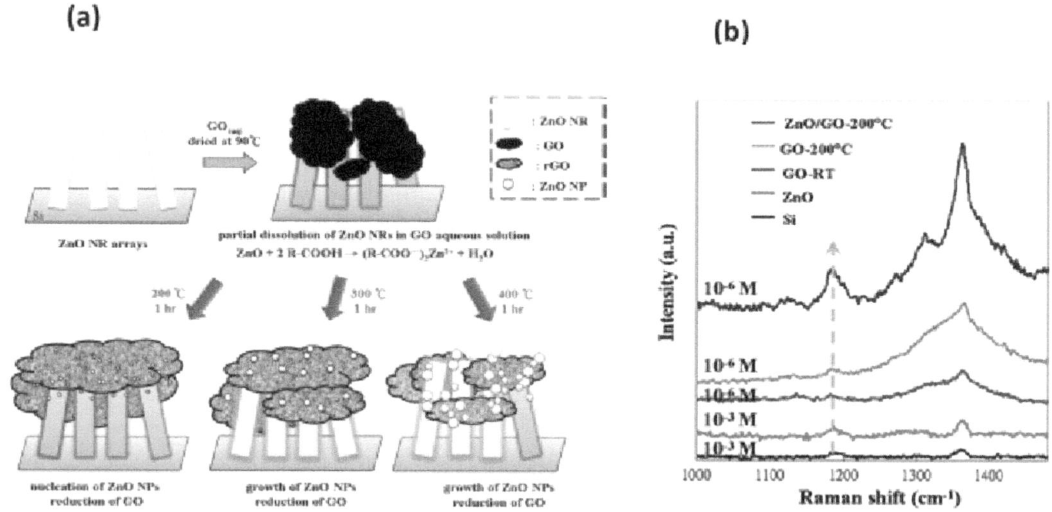

FIGURE 10.19 (a) Growth mechanisms of the various rGO–ZnO composites by annealing. (b) Raman spectra of R6G molecules on various substrates.

Source: Ruey-Chi Wang 2014 [95].

The significant changes in morphologies, structures and functional groups of the rGO/ZnO composites after different heat treatments indicate strong reactions between rGO and ZnO. The growth mechanism with different annealing temperature are shown in Figure 10.19a. Figure 10.19b shows the Raman spectra of R6G molecules absorbed on the Si, ZnO NRs, GO, and GO/ZnO composites with annealing treatments. It is clear from the figure that at lower concentration of R6G (10^{-3} M), the Si, ZnO NRs shows very low SERS effect. But the rGO/ZnO sample treated at 200°C has the highest intensity [95]. Bhat, explained the Tang Lau synthesis technique for the preparation of rGO-ZnO nano composites. It shows excellent photocatalytic activity in the photodegradation of methyl orange (MO) dye from water under UV light [96]. N.S. John et al. prepared free-standing ZnO-rGO hexagonal cylinders using a simple one-step water/toluene liquid/liquid interface (LLI) method. It shows excellent SERS activity on R6G due to the charge transfer synergic effect of ZnO and rGO [97].

Hexagonal ZnO nano structures are interspersed with rGO layers. These length and width of hexagonal nanostructures are polydisperse as shown in Figure 10.20a. It is clear from Figure 10.20b, the enhancement is very less with fluorescence back ground on ZnO substrate. But in the case of rGO-ZnO, the fluorescence background have been quenched by rGO and a clear enhancement of R6G has been shown. These ZnO/rGO metal oxides were explored for reusable SERS substrate by using their photodegradation catalysis property [97].

ZnO is one of the transition metals semiconducting material widely used as chemical sensor due to its significant advantages, such as good chemical stability, easy production and low cost. The sensing mechanism of ZnO- electro chemical gas sensor is based on the variation of its electrical conductivity in the presence of reducing and oxidizing gases. Further, the addition GO and its derivatives will greatly enhance the response of ZnO for the detection of H_2, N_2 and CH_4 gases.

Giorgio et al. reported the fabrication of RGO/ZnO composites via thermal reduction of GO at 250°C under argon gas atmosphere by incorporating the GO sheets to the ZnO nanostructures. The schematic sample preparation and device fabrication are as shown in Figure 10.21. The NO2, H2 and CH4 gas sensing properties of the obtained nanocomposites were demonstrated by exposing them to gases. Interestingly the RGO/ZnO composites showed better gas-sensing performance

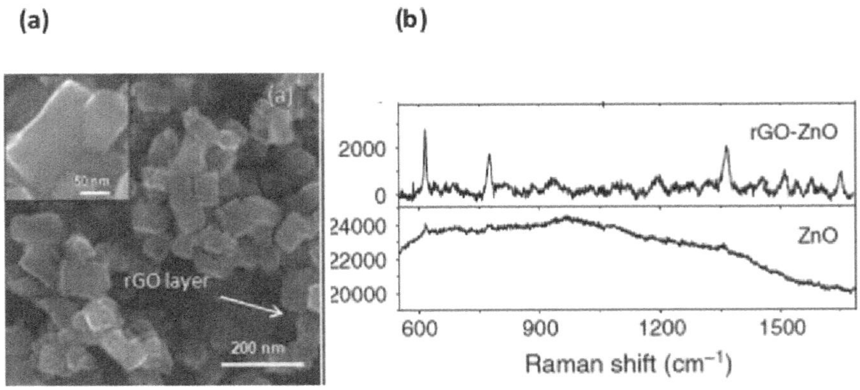

FIGURE 10.20 (a) FESEM Images of rGO-ZnO nano hybrid thin film prepared by LLI method, Inset:- high magnification image of a rGO-ZnO hexagonal cylinder, (b) SERS spectra of 1mM R6G dye on ZnO and rGO-ZnO respectively.

Source: K. Bramhaiah 2017 [97].

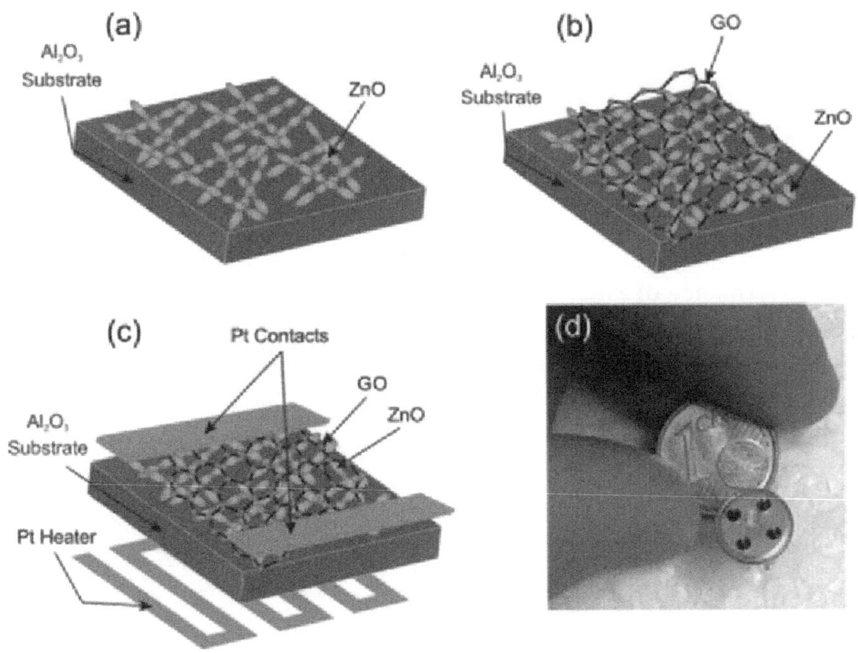

FIGURE 10.21 The schematic of the sample and device fabrication steps for gas sensing measurements. (a) Synthesis of ZnO nanostructures on an alumina substrate, (b) preparation of RGO/ZnO nanocomposite and (c) deposition of Pt electrodes and the heater for gas sensing measurements. (d) Digital photograph of the gas sensing device.

Source: Vardan Galstyana 2016 [98].

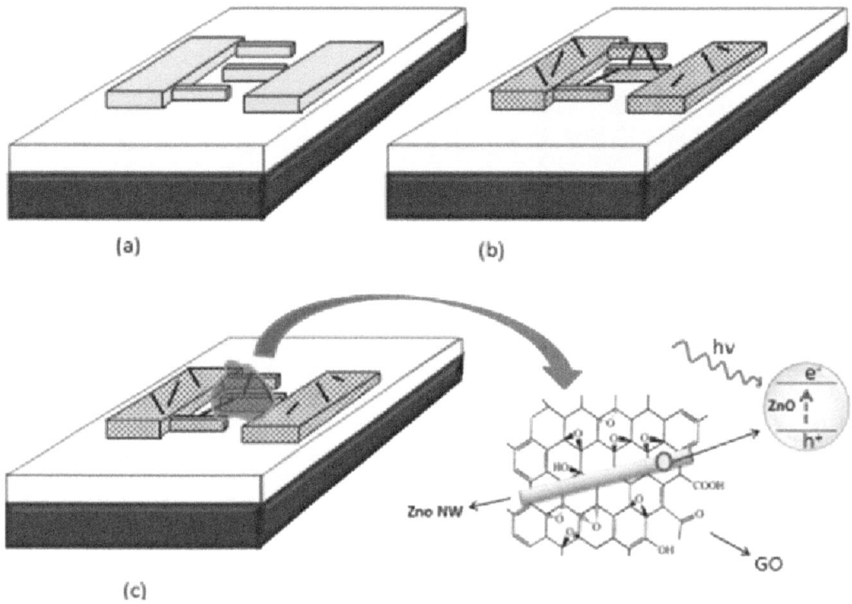

FIGURE 10.22 Schematics showing the process flow to achieve G/ZnO based sensor. (a) Patterning the comb-like Zn electrodes, (b) oxygen plasma bombardment to form ZnO nanowires, (c) coating the graphene oxide sheets on the obtained structure, and (d) photo-catalytic reduction of grapheme oxide using ZnO nanowires.

Source: ParvanehAfzali 2014 [99].

TABLE 10.3
Gas Sensing Performance of ZnO/Graphene Hybrids with Different Morphology

Structure	Shape of ZnO	Operating temperature	Test gas	Ref
Graphene/ZnO	Nanorods	300	Ethanol	100
GO/ZnO	Nanorods/nanoplates	340	Ethanol	101
GO/ZnO	Nanorods/nanoparticle	150	H_2	102
p-RGO/ZnO	Nanofibers	300	H_2	103
RGO/ZnO	Nanorods	190	CH_4	104
Graphene/ZnO	Thin film	200	NO_2	105
RGO/ZnO	Nanoparticles	25	NO_2	106
RGO/ZnO	Nanoparticles	Room temperature	NO_2	107

compared to the pure ZnO nanostructures for NO2 at relatively low working temperatures and for H2 at 250 °C. The architecture of rGO/ZnO hybrid nano composites can be an essential candidate for high-performance gas sensors [98].

Improved sensitivity of graphene-based oxygen sensor using photocatalytic activity of ZnO nanowires is reported by Arzi et al.

Here, they have used a new approach to create laterally grown ZnO nanowires using oxygen plasma treatment [99].

The schematic sample patterning, oxygen plasma bombardment and phtoto catalytic reduction of graphene oxides were shown in Figure 10.22.

The Table 10.3 compares the gas-sensing performance of the ZnO, graphene derivatives nano composite structure proposed in recently reported results.

10.2.5 GO/rGO-SnO2 Hybrid Nano Composites

Being a wide-band gap semiconductor, SnO_2 is proved to be an excellent sensor for gases with high sensitivity. Especially, different shapes and sizes of SnO_2 which includes nanowires, nanorods, nanofibers, nanobelts, nanotubes, and so forth shows considerable interest in the field of sensor research. Further, the unique carbon materials especially, graphene and its derivatives supports the sensing component fabrication due its unique properties like large specific surface area, high electron mobility, high density electronic states and low electric noise. Especially GO and rGO show excellent synergic properties in sensing field due to its remarkable dispersity in aqueous solution, availability to be chemically modified with metal and metal oxide or semiconductors.

Yang et al. prepared SnO_2 nano fibers by electrospinning method followed by calcination treatment. GO hybridization with SnO_2 has been done to improve the sensing properties of formaldehyde (HCHO) [108].

The sensing mechanism of SnO_2 was shown in Figure 10.23. The mechanism includes change in electrical conductivity caused by chemical interaction of gas molecules with the surface of metal oxides. The 100 nm porous structure of SnO_2 nano fiber is helping the target gases to pass through. Introducing GO helps in reducing the energy barrier by keeping its Fermi level between LUMO level of formaldehyde and conduction band level of SnO_2. This improves the response of the material with favorable electron transfer [108].

As per Figure 10.24a, the sensing performance of HCHO has the following order: GO/SnO_2> CNTs/SnO_2 > G/SnO_2> SnO_2 with the equal amount of carbon material hybridization. The GO had abundant oxide functional groups and unique 2-D structure which are the key properties responsible for the significant enhancement of HCHO sensing. The selectivity of other organic vapors, including ethanol, ammonia, acetone, methanol and toluene in optimum temperatures is shown in Figure 10.24b. GO/SnO2 sensors showed selective sensitivity to HCHO, which is proved by their relative low response to other interference gases at the same temperature [108].

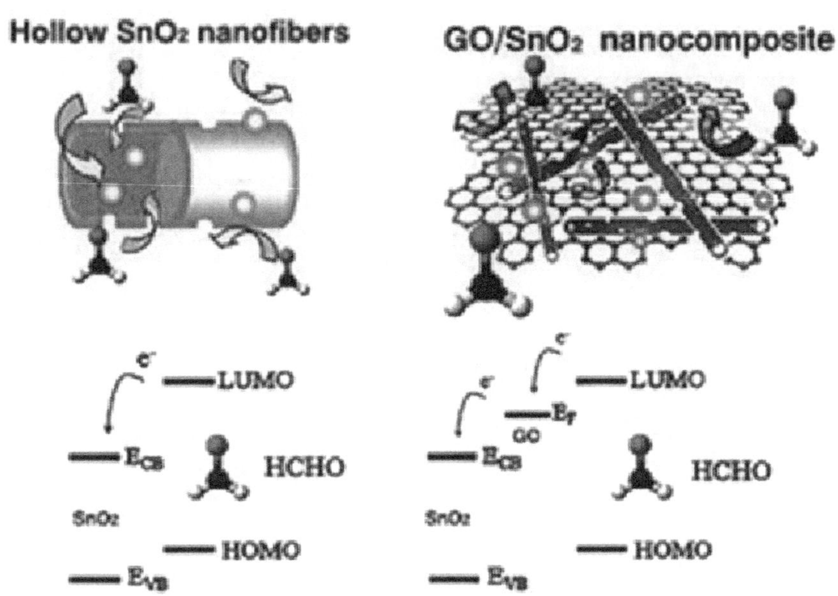

FIGURE 10.23 Schematic picture of GO/SnO_2 sensing mechanism.

Source: DingWang 2017 [108].

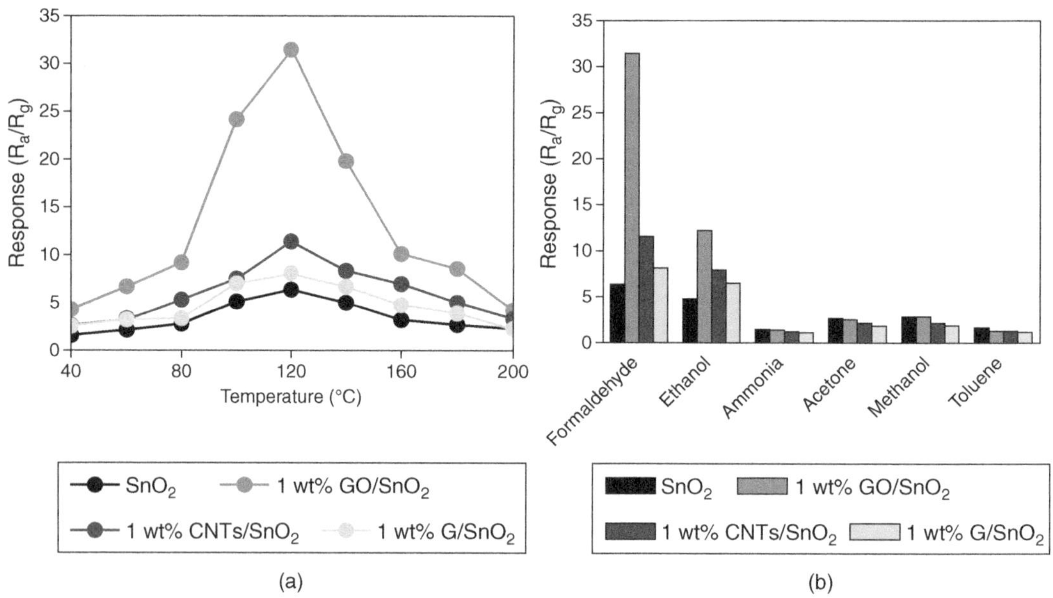

FIGURE 10.24 (a) Response of the sensors versus operating temperature to 100 ppm HCHO, (b) The response of sensors to 100 ppm different gases at 120 °C.

Source: DingWang 2017 [108].

FIGURE 10.25 FESEM images of rGO-metal oxide hybrid films (a) rGO-SnO_2; inset gives high magnification image of core–shell SnO_2@rGO spheres (b) high magnification image of the thinner area of rGO-SnO_2 marked in (a).

Source: K. Bramhaiah 2017 [97].

C. Kavitha et al. explained the simple liquid/liquid interface method to synthesis SnO_2 and rGO/SnO_2 for SERS based sensor applications. This rGO/SnO_2 substrate was demonstrated as reusable SERS substrate for studying different dye molecules.

As is shown by Figure 10.25a, the SEM image of rGO-SnO_2 film reveals the presence of 300–500 nm size SnO_2 spheres distributed over a thin film of rGO. These SnO_2 spheres are seen covered by a thin layer of rGO and hence can be considered as core–shell nanoparticles (Figure 10.25a inset).

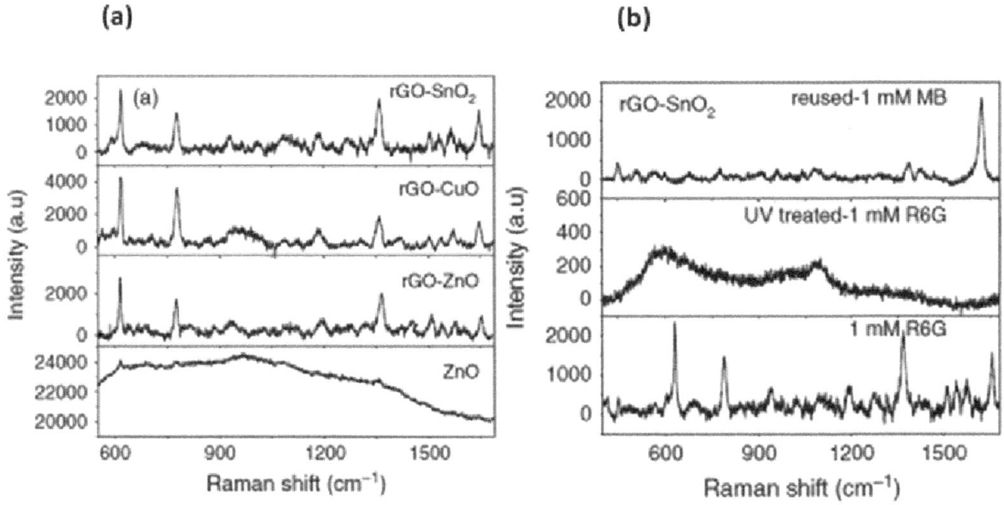

FIGURE 10.26 (a) SERS for 1mM R6G dye adsorbed on rGO/metal oxide NPs thin films, (b) SERS for 1mM R6G before and after UV treatment and that of 1mM MB dye on the UV treated rGO/SnO_2 hybrid thin films.

Source: K. Bramhaiah 2017 [97].

Metal oxides are biocompatible. So researchers are interested to study SERS on metal oxide substrate. Further it is interesting to study dye adsorbate interaction with metal oxides to understand the functioning of dye based solar cell. Figure 10.26a shows the SERS enhancement of different metal oxide substrate of 1mM R6G. Just a metal oxide shows huge fluorescence from dye, which masks the Raman details of the dye. When it combined with rGO, fluorescence quenching occur due to the contribution of local electric field dipole moment from oxygen functional group of rGO. The SERS spectra of before and after UV treated rGO/SnO_2 for two different analyte are shown in Figure 10.26.b. Photo degradation catalysis mechanism is used for this purpose. Metal oxide coupling with rGO facilitate the efficient absorption of dyes and improved charge separation boosting degradation kinetics [97].

The characteristic sensor materials depens on following properties, which include, morphology of structure as surface reaction, material as the medium of sensing and process preparation. For SnO_2 based humidity sensor, to improve the sensor efficiency, recently researchers involved in doping different metals with SnO_2. Further the doped SnO_2 will be hybrid with graphene derivatives for excellent low humidity detection. Popa et al. demonstrated reduced graphene oxide decorated with Fe doped SnO_2 nanoparticles for humidity sensor [109]. Zhang et al. explained One-Step Hydrothermally Synthesized Tin Dioxide-Decorated Graphene Nanocomposite on Polyimide Substrate for better humidity sensor applications [110]. Lu et al. demonstrated the SnO_2 double decorated with G–GO Nano composites by classical electro spinning and solution evaporation. It shows very high sensitivity, fast response and good stability [111].

10.2.6 GO/rGO-Fe_2O_3 Hybrid Nanocomposites

Iron (III) oxide or ferric oxide is the inorganic compound with the formula Fe_2O_3. It is one of the three main oxides of iron, the other two being iron (II) oxide (FeO), which is rare; and iron (II, III) oxide (Fe_3O_4), which also occurs naturally as the mineral magnetite. Fe_2O_3 can be obtained in various polymorphs, α-Fe_2O_3 has the rhombohedral, corundum (α-Al_2O_3) structure and is the most common form. It occurs naturally as the mineral hematite, which is mined as the main ore of iron. γ-Fe_2O_3 has a cubic structure. It is metastable and converted from the alpha phase at high temperatures. It occurs naturally as the mineral maghemite.

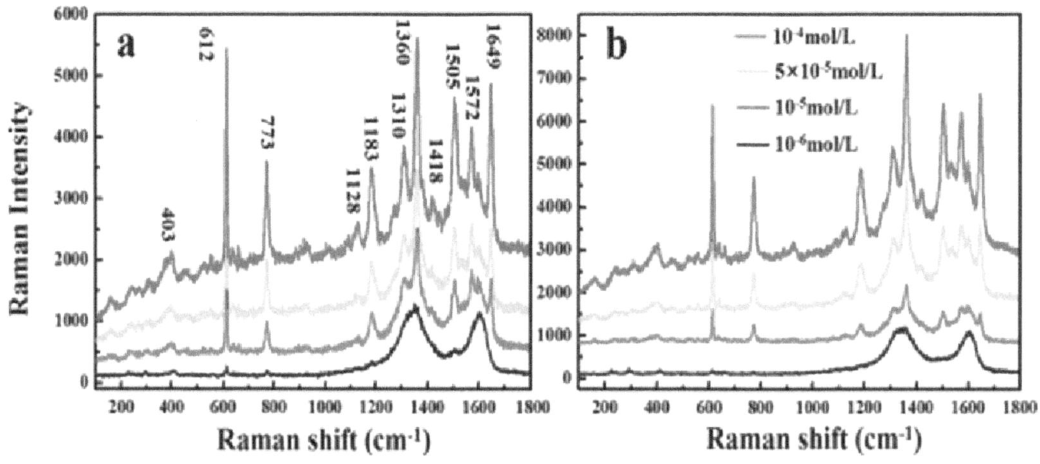

FIGURE 10.27 (a) SERS spectra of R6G with different concentrations using α-Fe_2O_3/rGO as substrates; (b) SERS spectra of R6G with different concentrations using α-Fe_2O_3/rGO as substrates after five cycling photodegradation of R6G measurements using α-Fe_2O_3/rGO as photocatalysts.

Source: Lili Zhang 2015 [112].

α-Fe_2O_3 is a cheap and widely used photocatalyst with band gap of ~2.2 eV. It has an application in the field of water treatment because of its good chemical stability, low toxicity, and ease of preparation. However, the fast electron-hole pair recombination significantly reduces its photocatalytic activity. This difficulty is overcome by graphene and its derivatives hybridization with metal oxide semiconductors. Though α-Fe_2O_3/rGO composites have been investigated in the field of lithium ion batteries, super capacitors and photo catalytic water oxidation, Yu , demonstrated α-Fe_2O_3/rGO as a SERS substrate for detection of R6G at lower concentration and exhibit much higher photo catalytic activity [112].

It is clear from Figure 10.27a,b, As the R6G concentration decreases, the Raman signal intensities decrease and the detection limit of α Fe_2O_3 /rGO can be as low as ~10^{-6} mol/L. Even after five cycling photo degradation of R6G measurements using α Fe_2O_3 /rGO as photo catalysts. No noticeable change was observed in the SERS signal, and the detection limit can still reach ~10^{-6} mol/L. This ascertains the stability and recyclability of sample [112]. Wang et al. explained the preparation of sub-monolayers of *α*-Fe_2O_3 nanocrystals (sphere, spindle, and cube) and their usage of detecting low concentration 4-mercaptopyridine (4-MPy) probed by SERS [113]. Bei et al. synthesized 2-D - α-Fe_2O_3 monolayers by assembling of single crystalline α-Fe_2O_3 nanospheres onto the surfaces of quartz slides. Hexamethylene diisocyanate (HDI) has been used as "bridge" to link between quartz surface and single-crystalline α-Fe_2O_3 nanospheres. This highly stable, uniform, rough substrate shows highly enhanced Raman signal for 4-MPy [114].

10.3 SUMMARY

Electrochemical SERS-based optical sensors are the non-destructive, low-cost, efficient and eco-friendly methods to detect low concentrations of chemical and bio chemicals in recent years. Contamination of food stuffs, drinking water and air with hazardous pollutants and other foreign substances are real and a direct threat to human health. Early detection of biomolecules like life-threatening cancer cells will help mankind to cure the disease completely. The prime concern should be for their rapid and reliable detection by a convenient method. Graphene and its derivative hybrids with metal and metal oxides are opt effective nano composites substrate for SERS and electrochemical detection. This is due to their specific characteristic like high surface-to-volume ratio, good electrical conductivity, catalytic action, beneficial biocompatibility and they can be simply modified with functional groups. The progress in fabricating low-cost reusable SERS substrate, portable Raman spectrometer and electrochemical workstation are not limited to not limiting the instant detection of chemicals and biomolecules. This chapter gives an overview of sensitivity and selectivity of SERS and electrochemical detection methods by employing graphene oxide-based hybrid nano composite substrate utilizing synergic electromagnetic and charge transfer mechanism for excellent detection sensitivity. The analytes, like dyes, amino acids, hazardous chemicals and gases have been detected with excellent sensitivity and selectivity. This proves the potential commercial applications of these lo- cost detection methods.

REFERENCES

1. Geim, A. K.; Novoselov, K. S. The Rise of Graphene. *Nat. Mater.* 2007, 6, 183–191. DOI: 10.1038/nmat1849.
2. Brumfiel, G. Graphene Gets Ready for the Big Time. *Nature* 2009, 458, 390–391. DOI: 10.1038/458390a.
3. Novoselov, K. S.; Geim, A. K.; Morozov, S. V.; Jiang, D.; Zhang, Y.; Dubonos, S. V.; Grigorieva, I. V.; Firsov, A. A. Electric Field Effect in Atomically Thin Carbon Films. *Science* 2004, 306, 666–669. DOI: 10.1126/science.1102896.
4. Novoselov, K. S.; Geim, A. K.; Morozov, S. V.; Jiang, D.; Katsnelson, M. I.; Grigorieva, I. V.; Dubonos, S. V.; Firsov, A. A. Two-Dimensional Gas of Massless Dirac Fermions in Graphene. Nature 2005 438, 197–200. DOI: 10.1038/nature04233.
5. Lee, C.; Wei, X.; Kysar, J. W.; Hone, J. Measurement of the Elastic Properties and Intrinsic Strength of Monolayer Graphene. *Science* 2008, 321, 385–388. DOI: 10.1126/science.1157996.
6. Bolotin, K. I.; Sikes, K. J.; Jiang, Z.; Klima, M.; Fudenberg, G.; Hone, J.; Kim, P.; Stormer, H. L. Ultrahigh Electron Mobility in Suspended Graphene. *Solid State Commun.* 2008, 146, 351–355. DOI: 10.1016/j.ssc.2008.02.024.
7. Balandin, A. A.; Ghosh, S.; Bao, W.; Calizo, I.; Teweldebrhan, D.; Miao, F.; Lau, C. N. Superior Thermal Conductivity of Single-Layer Graphene. *Nano Lett.* 2008, 8, 902–907. DOI: 10. 1021/nl0731872.
8. Loh, K. P.; Bao, Q.; Eda, G.; Chhowalla, M. Graphene Oxide as a Chemically Tunable Platform for Optical Applications. *Nature Chem.* 2010, 2, 1015–1024. DOI: 10.1038/nchem.907.
9. Shao, Y.; Wang, J.; Wu, H.; Liu, J.; Aksay, I. A.; Lin, Y. Graphene Based Electrochemical Sensors and Biosensors: A Review. *Electroanalysis* 2010, 22, 1027–1036. DOI: 10.1002/ elan.200900571.
10. Wang, X.; Zhang, W. Application of Graphene Derivatives in Cancer Therapy: A Review. *Carbon* 2014, 67, 795–797. DOI: 10.1016/j.carbon.2013.10.043.
11. Toda, K .; Furue, R .; Hayami, S. Recent Progress in Applications of Graphene Oxide for Gas Sensing: A Review. *Anal. Chim. Acta* 2015, 878, 43–53. DOI: 10.1016/j.aca.2015. 02.002.
12. Suvarnaphaet, P.; Pechprasarn, S. Graphene-Based Materials for Biosensors: A Review. *Sensors* 2017, 17, 2161. DOI: 10. 3390/s17102161.
13. Rowley-Neale, S. J.; Randviir, E. P.; Dena, A. S. A.; Banks, C. E. An Overview of Recent Applications of Reduced Graphene Oxide as a Basis of Electroanalytical Sensing Platforms. *Appl. Mat. Today* 2018, 10, 218–226. DOI: 10.1016/ j.apmt.2017.11.010.
14. Guo, H.; Jiao, T.; Zhang, Q.; Guo, W.; Peng, Q.; Yan, X. Preparation of Graphene Oxide-Based Hydrogels as Efficient Dye Adsorbents for Wastewater Treatment. *Nanoscale Res. Lett.* 2015, 10, 272. DOI: 10.1186/s11671-015-0931-2.
15. Liu, X.; Kim, H.; Guo, L. J. Optimization of Thermally Reduced Graphene Oxide for an Efficient Hole Transport Layer in Polymer Sola r Cells. *Org. Electron.* 2013, 14, 591–598. DOI: 10.1016/j.orgel.2012.11.020.
16. Sa, K.; Mahakul, P. C.; Das, B.; Subramanyam, B. V. R. S.; Mukherjee, J.; Saha, S.; Raiguru, J.; Patra, K. C.; Nanda, K. K.; Mahanandia, P. Large Scale Synthesis of Reduced Graphene Oxide Using Ferrocene and HNO3. *Mater. Lett.* 2018, 211, 335–338. DOI: 10.1016/j.matlet.2017.10.031.

17. Sun, X.; Liu, Z.; Welsher, K.; Robinson, J. T.; Goodwin, A.; Zaric, S.; Dai, H. Nano-Graphene Oxide for Cellular Imaging and Drug Delivery. *Nano Res*. 2008, 1, 203–212. DOI: 10. 1007/s12274-008-8021-8.
18. G. Reina, J. M. González-Domínguez, A. Criado, E. Vázquez, Promises, A. Bianco, M. Prato. Facts and challenges for graphene in biomedical applications, *Chem. Soc. Rev*. 2017, 46, 4400. https://doi.org/10.1039/C7CS00363C.
19. Shang Y. U., Zhang D, Liu Y, Guo C. Preliminary comparison of different reduc- tion methods of graphene oxide. *Bull Mater Sci* 2015;38:7–12. https://doi.org/10.1007/s12034-014-0794-7.
20. Meng, F.-L., Guo, Z., Huan, X.-J. Graphene-based hybrids for chemiresistive gas sensors. *TrAC Trends Anal. Chem*. 2015, 41, 37–47. https://doi.org/10.1016/j.trac.2015.02.008.
21. Khalil, I.; Julkapli, N.M.; Yehye, W.A.; Basirun, W.J.; Bhargava, S.K. Graphene – Gold Nanoparticles Hybrid-Synthesis, Functionalization, and Application in a Electrochemical and Surface-Enhanced Raman Scattering Biosensor. *Materials* 2016, 9, 406. doi:10.3390/ma9060406.
22. E. Nossol, A. B. S. Nossol, S.-X. Guo, J. Zhang, X.-Y. Fang, A. G. J. Zarbin and A. M. Bond, Synthesis, characterization and morphology of reduced graphene oxide–metal–TCNQ nanocomposites, *J. Mater. Chem. C*, 2014, 2, 870–878. https://doi.org/10.1039/C3TC32178A.
23. C. Kavitha, K. Bramhaiah, Neena S. John, Reduced Graphene oxide / Nanoparticle hybrid structures: A new generation smart materials for optical sensors, *Materials Today: Proceedings* 5 (2018) 2609–2618. https://doi.org/10.1016/j.matpr.2018.01.040.
24. Khan, M., Tahir, M. N., Adil, S. F., Khan, H. U., Siddiqui, M. R. H., Al-warthan, A. A., & Tremel, W.. Graphene based metal and metal oxide nanocomposites: synthesis, properties and their applications, *J. Mater. Chem. A*, 2015, 3, 18753–18808. DOI: 10.1039/c5ta02240a..
25. K. P. Loh, Q. Bao, K. P. Ang and J. Yang, The chemistry of graphene. *J. Mater. Chem*., 2012, 20, 2277–2289. DOI: 10.1039/b920539j.
26. H. Bai, C. Li and G. Shi, Functional Composite Materials Based on Chemically Converted Graphene Adv. *Mater*., 2011, 23, 1089–1115. DOI: 10.1002/adma.201003753.
27. (a) B. D. Piorek, S. Joon Le, J. G. Santiago, M. Moskovits, S. Banerjee, and C. D. Meinhart, Free-surface microfluidic control of surface-enhanced Raman spectroscopy for the optimized detection of airborne molecules, 2007, *PNAS*, 104, 18898**,** DOI: 10.1073/pnas.0708596104. (b) X. Ling, L. Xie, Y. Fang, H. Xu, H. Zhang, J. Kong, M. S. Dresselhaus, J. Zhang, Z. Liu, Can Graphene be used as a Substrate for Raman Enhancement? 2010 *Nano Lett*. 10, 553. DOI: 10.1021/nl903414x.
28. Xie L. M, Ling X., Fang Y., Zhang J. and Liu Z. F. Graphene as a Substrate to Suppress Fluorescence in Resonance Raman Spectroscopy 2009. *J. Am. Chem. Soc*. 131, 9890. DOI: 10.1021/ja9037593.
29. P. Ramnani, N. M. Saucedo, A. Mulchandani, Carbon nanomaterial-based electrochemical biosensors for label-free sensing of environmental pollutants, *Chemosphere*, 143, 2016, 85–98. http://dx.doi.org/10.1016/j.chemosphere.2015.04.063.
30.. S. Bai and X. Shen, Graphene–inorganic nanocomposites, *RSC Adv*., 2012, 2, 64–98, DOI: 10.1039/c1ra00260k.
31.. X. Huang, X. Qi, F. Boey and H. Zhang, Graphene-based composites, *Chem. Soc. Rev*. 2012, 41, 666–686. https://doi.org/10.1039/C1CS15078B.
32.. L. Xie, L. Xi, Y. Fang, J. Zhang, and Z. Liu, Graphene as a Substrate To Suppress Fluorescence in Resonance Raman Spectroscopy, *J. Am. Chem. Soc*. 131, 9890. 2009, https://doi.org/10.1021/ja9037593.
33. X. Yu, H. Cai, W. Zhang, X. Li, N. Pan, Y. Luo, X. Wang, and J. G. Hou, Tuning Chemical Enhancement of SERS by Controlling the Chemical Reduction of Graphene Oxide Nanosheets, *ACS Nano* 5, 952. 2011, https://doi.org/10.1021/nn102291j.
34. 26. S. Huh, J. Park, Y. S. Kim, K. S. Kim, B. H. Hong, and J.-M. Nam, UV/Ozone-Oxidized Large-Scale Graphene Platform with Large Chemical Enhancement in *Surface-Enhanced Raman Scattering, ACS Nano* 5, 9799. 2011, https://doi.org/10.1021/nn204156n.
35. D. Ju Han, K. Seok Choi, F. Liu, and T. Seok Seo, Effect of Chemical and Structural Feature of Graphene on Surface Enhanced Raman Scattering, *Journal of Nanoscience and Nanotechnology*, Vol. 13, 8154–8161, 2013, doi:10.1166/jnn.2013.7927.
36. Y. Wang, Z. Ni, H. Hu, Y. Hao, C. P. Wong, T. Yu, J. T. L. Thong, and Z. X. Shen, Gold on graphene as a substrate for surface enhanced Raman scattering study, *Appl. Phys. Lett*. 97, 163111. 2010, https://doi.org/10.1063/1.3505335.

37. Q. Hao, B. Wang, J. A. Bossard, B. Kiraly, Y. Zeng, I-K. Chiang1, L. Jensen, D. H. Werner, and T. Jun Huang, *J Phys Chem C Nanomater Interfaces*. 2012 April 5; 116(13): 7249–7254. doi:10.1021/jp209821g.
38. J. Huang, L. Zhang, B. Chen, N. Ji, F. Chen, Y. Zhang and Z. Zhang, Nanocomposites of size-controlled gold nanoparticles and graphene oxide: Formation and applications in SERS and catalysis, *Nanoscale*, 2010,2, 2733-2738, https://doi.org/10.1039/C0NR00473A.;l
39. J. Liu, S. Fu, B. Yuan, Y. Li, Z. Deng, Toward a Universal “Adhesive Nanosheet” for the Assembly of Multiple Nanoparticles Based on a Protein-Induced Reduction/Decoration of Graphene Oxide, *J. Am. Chem. Soc*. 2010, 132, 21, 7279–7281, https://doi.org/10.1021/ja100938r.
40. M. Kwak, S. Lee, D. Kim, S.-K. Park, Y. Piao, Facile synthesis of Au-graphene nanocomposite for the selective determination of dopamine, *Journal of Electroanalytical Chemistry*, Vol. 776, 1 September 2016, Pages 66–73, https://doi.org/10.1016/j.jelechem.2016.06.047.
41. K. Jasuja, V. Berry, Implantation and Growth of Dendritic Gold Nanostructures on Graphene Derivatives: Electrical Property Tailoring and Raman Enhancement, *ACS Nano*, 3, 2358–2366, 2009, https://doi.org/10.1021/nn900504v.
42. B. Zhang, Y. Cui, H. Chen, B. Liu, G. Chen and D. Tang, A New Electrochemical Biosensor for Determination of Hydrogen Peroxide in Food Based on Well-Dispersive Gold Nanoparticles on Graphene Oxide, Electroanalysis, 2011, 23, 1821–1829, https://doi.org/10.1002/elan.201100171.
43. S. Liu, G. Yan, G. He, D. Zhong, J. Chen, L. Shi, X. Zhou, H. Jiang, Layer-by-layer assembled multilayer films of reduced graphene oxide/gold nanoparticles for the electrochemical detection of dopamine, *J. Electroanal. Chem*., 2012, 672, 40–44, https://doi.org/10.1016/j.jelechem.2012.03.007.
44. F. Cui and X. Zhang, Electrochemical sensor for epinephrine based on a glassy carbon electrode modified with graphene/gold nanocomposites, J. *Electroanal. Chem*., 2012, 669, 35–41, https://doi.org/10.1016/j.jelechem.2012.01.021.
45. F. Wang, L. Zhu and J. Zhang, Electrochemical sensor for levofloxacin based on molecularly imprinted polypyrrole–graphene–gold nanoparticles modified electrode, Sens. *Actuators, B*, 2014, 192, 642–647, https://doi.org/10.1016/j.snb.2013.11.037.
46. J. Hu, F. Li, K. Wang, D. Han, Q. Zhang, J. Yuan and L. Niu, One-step synthesis of graphene–AuNPs by HMTA and the electrocatalytical application for O_2 and H_2O_2, *Talanta*, 2012, 93, 345–349, https://doi.org/10.1016/j.talanta.2012.02.050.
47. X. Huang, X. Z. Zhou, S. X. Wu, Y. Y. Wei, X. Y. Qi, J. Zhang, F. Boey and H. Zhang, Reduced Graphene Oxide-Templated Photochemical Synthesis and in situ Assembly of Au Nanodots to Orderly Patterned Au Nanodot Chains, *Small*, 2010, 6, 513–516, https://doi.org/10.1002/smll.200902001.
48. K. Vinodgopal, B. Neppolian, I. V. Lightcap, F. Grieser, M. Ashokkumar and P. V. Kamat, Sonolytic Design of Graphene–Au Nanocomposites. Simultaneous and Sequential Reduction of Graphene Oxide and Au (III), *J. Phys. Chem. Lett*., 2010, 1, 1987–1993, https://doi.org/10.1021/jz1006093.
49. R. Muszynski, B. Seger and P. V. Kamat, Decorating Graphene Sheets with Gold Nanoparticles, *J. Phys. Chem. C*, 2008, 112, 5263–5266. https://doi.org/10.1021/jp800977b.
50. F. H. Li, H. F. Yang, C. S. Shan, Q. X. Zhang, D. X. Han, A. Ivaska and L. Niu, The synthesis of perylene-coated graphene sheets decorated with Au nanoparticles and its electrocatalysis toward oxygen reduction, *J. Mater. Chem*., 2009, 19, 4022–4025, https://doi.org/10.1039/B902791B.
51. G. Goncalves, P. Marques, C. M. Granadeiro, H. I. S. Nogueira, M. K. Singh and J. Gracio, Surface Modification of Graphene Nanosheets with Gold Nanoparticles: The Role of Oxygen Moieties at Graphene Surface on Gold Nucleation and Growth Chem. *Mater*., 2009, 21, 4796–4802, https://doi.org/10.1021/cm901052s.
52. Y.-K. Kim, H.-K. Na, Y. W. Lee, H. Jang, S. W. Han and D.-H. Min, The direct growth of gold rods on graphene thin films, *Chem. Commun*., 2010, 46, 3185–3187, https://doi.org/10.1039/C002002H.
53. H. M. A. Hassan, V. Abdelsayed, A. Khder, K. M. AbouZeid, J. Terner, M. S. El-Shall, S. I. Al-Resayes and A. A. El-Azhary, Microwave Synthesis of Graphene Sheets Supporting Metal Nanocrystals in Aqueous and Organic Media, *Mater. Chem*., 2009, 19, 3832–3837, https://doi.org/10.1039/B906253J.
54. K. Jasuja, J. Linn, S. Melton and V. Berry, Microwave-Reduced Uncapped Metal Nanoparticles on Graphene: Tuning Catalytic, Electrical, and Raman Properties, *J. Phys. Chem. Lett*., 2010, 1, 1853–1860, https://doi.org/10.1021/jz100580x.

55. H. Q. Zhou, C. Y. Qiu, Z. Liu, H. C. Yang, L. J. Hu, J. Liu, H. F. Yang, C. Z. Gu and L. F. Sun, Thickness-Dependent Morphologies of Gold on N-Layer Graphenes, *J. Am. Chem. Soc.*, 2010, 132, 944–946, https://doi.org/10.1021/ja909228n.
56. S. Kumari, P. Sharma, S. Yadav, J. Kumar, A. Vij, P. Rawat, S. Kumar, C. Sinha, J. Bhattacharya, C. Mohan Srivastava, S. Majumder, A Novel Synthesis of the Graphene Oxide-Silver (GO-Ag) Nanocomposite for Unique Physiochemical Applications, *ACS Omega*. 2020, 5, 10, 5041, https://doi.org/10.1021/acsomega.9b03976.
57. T. K. Naqvia, A. K. Srivastavac, M. M. Kulkarnia, A. M. Siddiquib, P. K. Dwivedia, Silver nanoparticles decorated reduced graphene oxide (rGO) SERS sensor for multiple analytes, *Applied Surface Science*, 478, 2019, pp. 887–895, https://doi.org/10.1016/j.apsusc.2019.02.026.
58. C. Kavitha, K. Bramhaiah, N. S. John, B. E. Ramachandran, Low cost, ultra-thin films of reduced graphene oxide–Ag nanoparticle hybrids as SERS based excellent dye sensors, *Chemical Physics Letters*, 629, 2015, 81–86, https://doi.org/10.1016/j.cplett.2015.04.026.
59. C. Xu, X. Wang, Fabrication of flexible metal-nanoparticle films using graphene oxide sheets as substrates, *Small* 5, 2009, 2212–2217, https://doi.org/10.1002/smll.200900548.
60. W. Fan, Y. H. Lee, S. Pedireddy, Q. Zhang, T. Liu, X. Y. Ling, Graphene oxide and shape-controlled silver nanoparticle hybrids for ultrasensitive single-particle surface-enhanced Raman scattering (SERS) sensing, *Nanoscale* 6, 2014, 4843–4851, https://doi.org/10.1039/c3nr06316j.
61. J. Kim, L. J. Cote, F. Kim, J. Huang, Visualizing graphene based sheets by fluorescence quenching microscopy, *J. Am. Chem. Soc.* 132, 2009 260–267, https://doi.org/10.1021/ja906730d.
62. K.-C. Hsu and D.-H. Chen, Microwave-assisted green synthesis of Ag/reduced graphene oxide nanocomposite as a surface-enhanced Raman scattering substrate with high uniformity, *Nanoscale Research Letters*, 2014, 9:193, doi: 10.1186/1556-276X-9-193.
63. T. K. Naqvi, M. S. S. Bharati, A. K. Srivastava, M. M. Kulkarni, A. M. Siddiqui, S. Venugopal Rao, P. K. Dwivedi, Hierarchical Laser-Patterned Silver/Graphene Oxide Hybrid SERS Sensor for Explosive Detection, *ACS Omega* 2019, 4, 17691–17701, DOI: 10.1021/acsomega.9b01975.
64. G. Darabdhara, M. R. Das, S. P. Singh, A. K. Rengan, S. Szunerits, R. Boukherroub, Ag and Au nanoparticles/reduced graphene oxide composite materials: Synthesis and application in diagnostics and therapeutics, *Advances in Colloid and Interface Science* 271, 2019, 101991, https://doi.org/10.1016/j.cis.2019.101991.
65. N. T. Lan, D. T. Chi, N. X. Dinh, N. D. Hung, H. Lan, P. A. Tuan, L. H. Thang, N. N. Trung, N. Q. Hoa, T. Q. Huy, N. V. Quy, T.-T. Duong, V. N. Phan, A.-T. Le, Photochemical decoration of silver nanoparticles on graphene oxide nanosheets and their optical characterization, *Journal of Alloys and Compounds* 615 (2014) 843–848, http://dx.doi.org/10.1016/j.jallcom.2014.07.042.
66. S. W. Chook, C. H. Chia1, S. Zakaria, M. K. Ayob, K. L. Chee, N. M. Huang, H. M. Neoh, H. G. Lim, R. Jamal, R. M. F. R. A. Rahman, Antibacterial performance of Ag nanoparticles and AgGO nanocomposites prepared via rapid microwave-assisted synthesis method, *Nanoscale Research Letters* 2012, 7:541, doi:10.1186/1556-276X-7-541.
67. Dhanush S., Sreejesh M., Bindu K., Chowdhury P., Nagaraja H. S., Synthesis and electrochemical properties of Silver Dendrites and Silver Dendrites/rGO composite for applications in paracetamol sensing, *Materials Research Bulletin*, 100, 2018, 295–301, https://doi.org/10.1016/j.materresbull.2017.12.044.
68. A. M. Noor, P. Rameshkumar, N. Yusoff, H. N. Ming, M. S. Sajab, Microwave synthesis of reduced graphene oxide decorated with silver nanoparticles for electrochemical determination of 4-nitrophenol, *Ceramics International*, 42, 2016, 18813–18820, https://doi.org/10.1016/j.ceramint.2016.09.026.
69. L. Fu, DeM. Zhu, A. Yu, Galvanic replacement synthesis of silver dendrites-reduced graphene oxide composites and their surface-enhanced Raman scattering characteristics, *Spectrochimica Acta Part A: Molecular and Biomolecular Spectroscopy*, 149, 2015, http://dx.doi.org/10.1016/j.saa.2015.04.049.
70. S. P. Nayak, S. S. Ramamurthy, J. K. K. Kumar, Green synthesis of silver nanoparticles decorated reduced graphene oxide nanocomposite as an electrocatalytic platform for the simultaneous detection of dopamine and uric acid, Materials Chemistry and Physics, Vol. 252, 2020, 123302, https://doi.org/10.1016/j.matchemphys.2020.123302.
71. J. Liu, S. Fu, B. Yuan, Y. Li and Z. Deng, Toward a Universal "Adhesive Nanosheet" for the Assembly of Multiple Nanoparticles Based on a Protein-Induced Reduction/Decoration of Graphene Oxide J. *Am. Chem. Soc.,* 2010, 132, 7279–7281, https://doi.org/10.1021/ja100938r.

72. T. Cassagneau and J. H. Fendler, Preparation and Layer-by-Layer Self-Assembly of Silver Nanoparticles Capped by Graphite Oxide Nanosheets, *J. Phys. Chem. B*, 1999, 103, 1789–1793, https://doi.org/10.1021/jp984690t.
73. X. Zhou, X. Huang, X. Qi, S. Wu, C. Xue, F. Y. C. Boey, Q. Yan, P. Chen and H. Zhang, In Situ Synthesis of Metal Nanoparticles on Single-Layer Graphene Oxide and Reduced Graphene Oxide Surfaces *J. Phys. Chem. C*, 2009, 113, 10842–10846, https://doi.org/10.1021/jp903821n.
74. Lu, W., Chang, G., Luo, Y., Liao, F., Sun, X. Method for effective immobilization of Ag nanoparticles/graphene oxide composites on single-stranded DNA modified gold electrode for enzymeless H_2O_2 detection. *J Mater Sci*, 2011, 46, 5260–6, DOI 10.1007/s10853-011-5464-1.
75. Li, Q., Qin, X., Luo, Y., Lu, W., Chang, G., Asiri, A. M., One-pot synthesis of Ag nanoparticles/reduced graphene oxide nanocomposites and their application for nonenzymatic H_2O_2 detection. *Electrochim Acta* 2012; 83: 283–287, https://doi.org/10.1016/j.electacta.2012.08.007.
76. Qin, X., Luo, Y., Lu, W., Chang, G., Asiri, A. M., Al-Youbi, A. O., One-step synthesis of Ag nanoparticles-decorated reduced graphene oxide and their application for H_2O_2 detection. *Electrochim Acta* 2012; 79: 46–51, https://doi.org/10.1016/j.electacta.2012.06.062.
77. Liu, S., Tian, J., Wang L, Sun, X. Microwave-assisted rapid synthesis of Ag nanoparticles/graphene nanosheet composites and their application for hydrogen peroxide detection. *J Nanopart Res* 2011;13: 4539–48, https://doi.org/10.1007/s11051-011-0410-3.
78. M. Zahmakran, S. Akbayrak, T. Kodaira, and S. Ozkar, Osmium (0) nanoclusters stabilized by zeolite framework; highly active catalyst in the aerobic oxidation of alcohols under mild conditions, *Dalton. Trans*., 2010, 39, 7521, https://doi.org/10.1039/C003200J.
79. C. Li, W. K. Leong and Z. Zhong, Metallic osmium and ruthenium nanoparticles for CO oxidation, J. *Organomet. Chem*., 2009, 694, 2315, https://doi.org/10.1016/j.jorganchem.2009.03.038.
80. V. W. S. Lam and E. L. Gyenge, High-Performance Osmium Nanoparticle Electrocatalyst for Direct Borohydride PEM Fuel Cell Anodes, *J. Electrochem. Soc*., 2008, 155, B1155, https://doi.org/10.1149/1.2975191.
81. J. U. Godinez, R. H. Catellanos, E. B. Arco, A. A. Gutierrez and O. J. Sandoval, Novel osmium-based electrocatalysts for oxygen reduction and hydrogen oxidation in acid conditions, *J. Power Sources*, 2008, 177, 286, https://doi.org/10.1016/j.jpowsour.2007.11.063.
82. K. Chakrapani, S. Sampath, Interconnected, ultrafine osmium nanoclusters: preparation and surface enhanced Raman scattering activity, *Chem. Commun*., 2013, 49, 6173. DOI: 10.1039/c3cc41940a.
83. C. Kavitha, K. Bramhaiah, Neena S. John, Shantanu Aggarwal, Improved surface-enhanced Raman and catalytic activities of reduced graphene oxide–osmium hybrid nano thin films, *R. Soc. Open Sci.* 4, 170353. http://dx.doi.org/10.1098/rsos.170353.
84. K. Bramhaiah, I. Pandey, V. N. Singh, C. Kavitha, N. S. John, Enhanced electrocatalytic activity of reduced graphene oxide-Os nanoparticle hybrid films obtained at a liquid/liquid interface, *J Nanopart Res* (2018) 20: 56, https://doi.org/10.1007/s11051-018-4150-5.
85. S. Schimpf, A. Rittermeier, X. Zhang, Z. Li, M. Spasova, M. van den Berg, M. Farle, Y. Wang, R. Fischer and M. Muhler, Stearate-Based Cu Colloids in Methanol Synthesis: Structural Changes Driven by Strong Metal–Support Interactions, *ChemCatChem*, 2010, 2, 214, https://doi.org/10.1002/cctc.200900252.
86. J. B. Park, J. Graciani, J. Evans, D. Stacchiola, S. D. Senanayake, L. Barrio, P. Liu, J. F. Sanz, J. Hrbek and J. A. Rodriguez, Gold, Copper, and Platinum Nanoparticles Dispersed on $CeO_x/TiO_2(110)$ Surfaces: High Water-Gas Shift Activity and the Nature of the Mixed-Metal Oxide at the Nanometer Level, *J. Am. Chem. Soc*., 2010, 132, 356, https://doi.org/10.1021/ja9087677.
87. S. Wang, X. Huang, Y. He, H. Huang, Y. Wu, L. Hou, X. Liu, T. Yang, J. Zou, B. Huang, Synthesis, growth mechanism and thermal stability of copper nanoparticles encapsulated by multi-layer graphene, *Carbon*, 2012, 50, 2119, doi:10.1016/j.carbon.2011.12.063.
88. N. A. Luechinger, E. K. Athanassiou, W. J. Stark, Graphene-stabilized copper nanoparticles as an air-stable substitute for silver and gold in low-cost ink-jet printable electronics, *Nanotechnology*, 2008, 19, 445201, DOI: 10.1088/0957-4484/19/44/445201.
89. E. K. Athanassiou, R. N. Grass, W. J. Stark, Large-scale production of carbon-coated copper nanoparticles for sensor applications, *Nanotechnology*, 2006, 17, 1668, DOI: 10.1088/0957-4484/17/6/022.

90. S. Lee, J. Hong, J. H. Koo, H. Lee, S. Lee, T.-j. Choi, H. Jung, B. Koo, H. Kim, J. Park, Synthesis of Few-Layered Graphene Nanoballs with Copper Cores Using Solid Carbon Source ACS Appl. *Mater. Interfaces*, 2013, 5, 2432. dx.doi.org/10.1021/am3024965.
91. M. Guo, Y. Zhao, F. Zhang, L. Xu, H. Yang, X. Song, Y. Bu, Reduced graphene oxide-stabilized copper nanocrystals with enhanced catalytic activity and SERS properties, *RSC Adv.*, 2016, 6, 50587, DOI: 10.1039/c6ra05186c.
92. C. Kavitha,, K. Bramhaiah, N. S. John, Low-cost electrochemical detection of L-tyrosine using an rGO–Cu modified pencil graphite electrode and its surface orientation on a Ag electrode using an ex situ spectroelectrochemical method, *RSC Adv.*, 2020, 10, 22871, DOI: 10.1039/d0ra04015k.
93. H. Bagheri, A. Hajian, M. Rezaei, A. Shirzadmehr, Composite of Cu metal nanoparticles-multiwall carbon nanotubes-reduced graphene oxide as a novel and high performance platform of the electrochemical sensor for simultaneous determination of nitrite and nitrate, *Journal of Hazardous Materials*, 324, 2017, 762–772, https://doi.org/10.1016/j.jhazmat.2016.11.055.
94. Y. Yang, N. Ma, Z. Bian, Cu-Au/rGO Nanoparticle Based Electrochemical Sensor for 4- Chlorophenol Detection, *Int. J. Electrochem. Sci.*, 14, 2019, 4095–4113, doi: 10.20964/2019.05.04.
95. R.-C. Wang, Y.-C. Chen, S.-J. Chen, Y.-M. Chang, Unusual functionalization of reduced graphene oxide for excellent chemical surface-enhanced Raman scattering by coupling with ZnO, *Carbon* Vol. 70, April 2014, pp. 215–223, http://dx.doi.org/10.1016/j.carbon.2013.12.110.
96. S. Meti, M. R. Rahman, Md. I. Ahmad, K. U. Bhat, Chemical free synthesis of graphene oxide in the preparation of reduced graphene oxide-zinc oxide nanocomposite with improved photo catalytic properties, *Applied Surface Science* 451, 2018, 67–75, https://doi.org/10.1016/j.apsusc.2018.04.138.
97. K. Bramhaiah, V. N. Singh, C. Kavitha, N. S. John, Films of Reduced Graphene Oxide with Metal Oxide Nanoparticles Formed at a Liquid/Liquid Interface as Reusable Surface Enhanced Raman Scattering Substrates for Dyes, *Journal of Nanoscience and Nanotechnology* 17(4): 2711–2719, 2017, DOI: 10.1166/jnn.2017.13431.
98. V. Galstyana, E. Cominia, I. Kholmanova, G. Fagliaa, G. Sberveglieri, Reduced graphene oxide/ZnO nanocomposite for application in chemical gas sensors, *RSC Adv.*, 6, 34225–34232, 2016. https://doi.org/10.1039/C6RA01913G.
99. P. A., Y. Abd, E Arzi, Directional reduction of graphene oxide sheets using photocatalytic activity of ZnO nanowires for the fabrication of a high sensitive oxygen sensor, *Sensors and Actuators B: Chemical*, 195, 92–97, 2014, https://doi.org/10.1016/j.snb.2013.12.105.
100. J. Yi, J. M. Lee and W. Il Park, Vertically aligned ZnO nanorods and graphene hybrid architectures for high-sensitive flexible gas sensors, *Sensors and Actuators B-Chemical,* 155, 264–269, 2011, https://doi.org/10.1016/j.snb.2010.12.033.
101. N. Song, H. Fan and H. Tian, PVP assisted in situ synthesis of functionalized graphene/ZnO (FGZnO) nanohybrids with enhanced gas-sensing property, *Journal of Materials Science*, 50, 2229–2238, 2015, https://doi.org/10.1007/s10853-014-8785-z.
102. K. Anand, O. Singh, M. P. Singh, J. Kaur, and R. C. Singh, Hydrogen sensor based on graphene/ZnO nanocomposite, Sensors and Actuators *B Chemical*, 195, 409–415, 2014, https://doi.org/10.1016/j.snb.2014.01.029.
103. Z. U. Abideen, H. W. Kim, and S. S. Kim, An ultra-sensitive hydrogen gas sensor using reduced graphene oxide-loaded ZnO nanofibers, *Chemical communications* (Cambridge, UK), 51, 15418–15421, 2015, DOI: 10.1039/c5cc05370f.
104. D. Zhang, N. Yin and B. Xia, Facile fabrication of ZnO nanocrystalline-modified graphene hybrid nanocomposite toward methane gas sensing application, *Journal of Materials Science-Materials in Electronics*, 26, 5937–5945, 2015, https://doi.org/10.1007/s10854-015-3165-2.
105. H. Xie, K. Wang, Z. Zhang, X. Zhao, F. Liu and H. Mu, Temperature and thickness dependence of the sensitivity of nitrogen dioxide graphene gas sensors modified by atomic layer deposited zinc oxide films, *Rsc Advances*, 5, 28030–28037, 2015, https://doi.org/10.1039/C5RA03752B.
106. N. Kumar, A. K. Srivastava, H. S. Patel, B. K. Gupta, and G. D. Varma, Facile synthesis of ZnO–Reduced Graphene Oxide Nanocomposites for NO_2 Gas Sensing Applications, *European Journal of Inorganic Chemistry*, 1912–1923, 2015, DOI: 10.1002/ejic.201403172.
107. S. Liu, B. Yu, H. Zhang, T. Fei and T. Zhang, Enhancing NO_2 gas sensing performances at room temperature based on reduced graphene oxide-ZnO nanoparticles hybrids, *Sensors and Actuators B: Chemical*, 202, 272–278, 2014, https://doi.org/10.1016/j.snb.2014.05.086.

108. D. Wang, M. Zhang, Z. Chen, H. Li, A. Chen, X. Wang, J. Yang, Enhanced formaldehyde sensing properties of hollow SnO_2 nanofibers by graphene oxide, *Sensors and Actuators B: Chemical*, 250, 533–542, 2017. https://doi.org/10.1016/j.snb.2017.04.164.
109. D. Toloman, A. Popa, M. Stan, C. Socaci, A. R. Biris, G. Katona, F. Tudorache, I. Petrila, F. Iacomi, Reduced graphene oxide decorated with Fe doped SnO_2 nanoparticles for humidity sensor, *Applied Surface Science*, 402, 410–417, 2017, https://doi.org/10.1016/j.apsusc.2017.01.064.
110. D. Zhang, H. Chang, and R. Liu, Humidity-Sensing Properties of One-Step Hydrothermally Synthesized Tin Dioxide-Decorated Graphene Nanocomposite on Polyimide Substrate, *J. Electron. Mater.*, 45, (2016) 4275–4281, https://doi.org/10.1007/s11664-016-4630-2.
111. Jing Xu, Shaozhen Gu, Bingan Lu, Graphene and Graphene Oxide Double Decorated SnO_2 Nanofibers with Enhanced Humidity Sensing Performance RSC *Adv.*, 5, 72046–72050, 2015, https://doi.org/10.1039/C5RA10571D.
112. L. Zhang, Z. Bao, X. Yu, P. Dai, J. Zhu, M. Wu, G. Li, X. Liu, Z. Sun, Changle Chen Rational Design of α-Fe_2O_3/Reduced Graphene Oxide Composites: Rapid Detection and Effective Removal of Organic Pollutants, *ACS Appl. Mater. Interfaces* 2016, 8, 6431–6438, DOI: 10.1021/acsami.5b11292.
113. Fu, X., Bei, F., Wang, X., Yang, X., Lu, L. Surface-Enhanced Raman Scattering of 4-Mercaptopyridine on Sub-Monolayers of α-Fe2O3 Nanocrystals (Sphere, Spindle, Cube). *J. Raman Spectrosc.* 2009, 40, 1290–1295, https://doi.org/10.1002/jrs.2281.
114. Fu, X.; Bei, F.; Wang, X.; Yang, X.; Lu, L. Two-Dimensional Monolayers of Single-Crystalline α-Fe2O3 Nanospheres: Preparation, Characterization and SERS Effect. *Mater. Lett.* 2009, 63, 185–187, https://doi.org/10.1016/j.matlet.2008.09.027.

11 Nanomaterial-Based Electrochemical Sensors for Vitamins and Hormones

Anila Rose Cherian, Rijo Rajeev, M. Nidhin, Anitha Varghese, and Gurumurthy Hegde

CONTENTS

11.1 Introduction 249
11.2 Nanomaterials – Principle and Conduction Properties for Electrochemical Sensing 250
11.3 Electrochemical Sensors for Water-Soluble Vitamins 250
11.4 Electrochemical Sensors for Hormone 253
11.5 Conclusions and Future Perspectives 262
References 262

11.1 INTRODUCTION

Vitamins are a collection of chemical substances that people obtain from food consumption and are essential for humanity [1–4]. Even though small levels of vitamins are present, they fulfill an important function in maintaining a lively metabolism in the human body [5, 6]. For instance, vitamins assists in activating enzymes thereby supporting various biochemical activities in our bodies that are mediated by enzymes [7]. Vitamin deficiency in our body is known to cause a variety of disorders. Because most vitamins are not synthesized by the human body, they must be obtained through food sources [8, 9]. As a result, vitamin detection has gained a lot of recognition in the recent years. High-performance liquid chromatography techniques [10–14], fluorescence [15–18], and capillary electrophoresis [19–21] are all extensively used methods for detecting vitamins. Although these approaches have numerous capabilities, including superior sensitivity along with selectivity, a few of the associated drawbacks are that they are time-consuming techniques, with laborious and complicated sample preparation and a need for large sample sizes, Instrumentation is costly, and staff must be well-equipped and educated [22–30]. Electrochemical sensors are a viable alternative to the technologies outlined above, as they are economical, facile, and showcase high selectivity and sensitivity – they are reproducible, can be miniaturized, and offer superior dependability as well [31–41]. Electrochemical sensors have thus been broadly used in a variety of industries, such as food and pharmaceuticals, and in labs and clinical workspaces [42, 43]. Electrochemical sensors work on the principle whereby the analyte undergoes electrochemical oxidation or reduction reaction and may gauge the analyte concentration by monitoring working electrode current. In addition, the current response generated is generally proportional to the analyte concentration [44, 45]

Amperometry and voltammetry are two of the most widely utilized and popular electrochemical techniques right now. Similarly, two straightforward voltammetry procedures are cyclic voltammetry (CV) and linear scan voltammetry (LSV). In addition, two approaches that can be utilized for enhancing the overall sensitivity by tuning different parameters are square wave voltammetry (SWV) and differential pulse voltammetry (DPV), both of which assist by improving sensitivity [46]. Hormones released by endocrine cells are delivered to target tissues via the circulatory system.

DOI: 10.1201/9781003218708-13

These compounds are found in very low concentrations (1 nM or less), and their identification requires a high level of sensitivity. Endocrine cells, which are found in the endocrine glands, generate and secrete hormones. Electrochemical sensing is a considerably easier and faster method for hormone measurement than other methods [47–50], such as high-performance liquid chromatography/tandem mass spectrometry (LC-MS/MS). In terms of cost-effectiveness, time performance, enhanced accessibility, high selectivity, and sensitivity, the electrochemical method for hormone sensing is highly promising and significantly overcomes the limitations associated with traditional methods, including enzyme-linked immunosorbent assay (ELISA). In most cases, enzymes or antibodies are used as biorecognition elements in electrochemical biosensors to detect hormones. A chemical reaction between immobilized molecules and target substances produces electrons that modify the characteristics of the solution in electrochemical biosensors [51–53].

11.2 NANOMATERIALS – PRINCIPLE AND CONDUCTION PROPERTIES FOR ELECTROCHEMICAL SENSING

Over the last several decades, there have been significant attempts to build an extensive range of sensing platforms for the detection of various metabolites which are being currently employed as disease biomarkers toward the assessment of overall health. Electrochemical sensors have earned a significant amount of curiosity as they are economical, offer rapid response time, and possess high selectivity and sensitivity. Nanomaterials, including carbon and polymeric nanomaterials, along with noble metallic and metal oxide nanoparticles, could improve the performance of electrochemical sensors. Carbon nanomaterials and metallic nano particles [NPs] in particular, have been generally utilized toward biomolecule electrochemical sensing, owing to their large surface-active area, exceptional electrical conductivity, and catalytic activity, and they even encompass high biocompatibility, which has resulted in increased Limit of Detection [LOD] and sensitivity. This chapter mainly discusses electrochemical sensors that utilize voltammetric approaches to target tiny metabolites such as vitamins and hormones – approaches that have been created during recent years employing carbon nanomaterials and metallic NPs. Moreover, the limitations and ways of mastering them for use in electrochemical sensing of tiny metabolites have been discussed.

11.3 ELECTROCHEMICAL SENSORS FOR WATER-SOLUBLE VITAMINS

Vitamins found to be partially, or highly, soluble in water compensate for nutrient deficiencies in foods including milk powder, and drinks. These vitamins cannot be kept in the human body for lengthy periods because they are quickly disposed of as part of the urine. A lack of these vitamins can have various undesirable effects on human health and is frequently the cause of sickness. As a result, maintaining normal human body function necessitates a continuous and adequate dietary intake and food allowance of water-soluble vitamins.

In the present scientific community, sensors based on electrochemical detection of water-soluble vitamins have achieved significant success.

(A) Electrochemical Sensors for Thiamine (VB_1)
Thiamine, also identified as VB_1, is an example of a water-soluble vitamin that is commonly present in rice, eggs, and milk [54]. It is vital to the human body's metabolism, notably in the neurological system. A variety of illnesses are associated with VB_1 deficiency, such as beriberi and other neurological ailments [55–57]. Muppariqoh and co-workers developed a carbon paste electrode (CPE) designed for VB_1 detection as the paste may be quickly transformed to an electrochemical activity displaying species media [58]. In another study, polypyrrole-modified molecularly imprinted polymers (MIP) were utilized for altering CPE for better performance. As part of the study, CV was run for detecting thiamine present in

potassium chloride (KCl) tris buffer (pH 10). The developed sensor showcased a low-level detection limit and linear range of 0.069 mmol L^{-1} and 0.256–10 mmol L^{-1} respectively. Following that, another work reported the creation of an MIP-based electrochemical sensor that senses VB_1 by making use of different monomers, as illustrated in Figure 11.1 [59]. The research work conducted involved the combination of Ni nanomer with acryloylated Ni NPs and functionalized multi-walled carbon nanotubes (f-MWCNTs) for enriching the molecular recognition capabilities along with the N-methacryloyl glutamic acid (NMGA) monomer. In addition, the f-MWCNTs were used to improve the adhesion of MIP film coating on the pencil graphite electrode's surface (PGE). The developed sensor showcased linear ranges of -0.63–19.63, 0.6–19.43, 0.7–19.36, and 0.75–21.05 ng mL^{-1} with a low-limit detection limits of 0.18, 0.17, 0.2, 0.19 ng mL^{-1} in pharmaceutical, aqueous, urine, and blood serum samples.

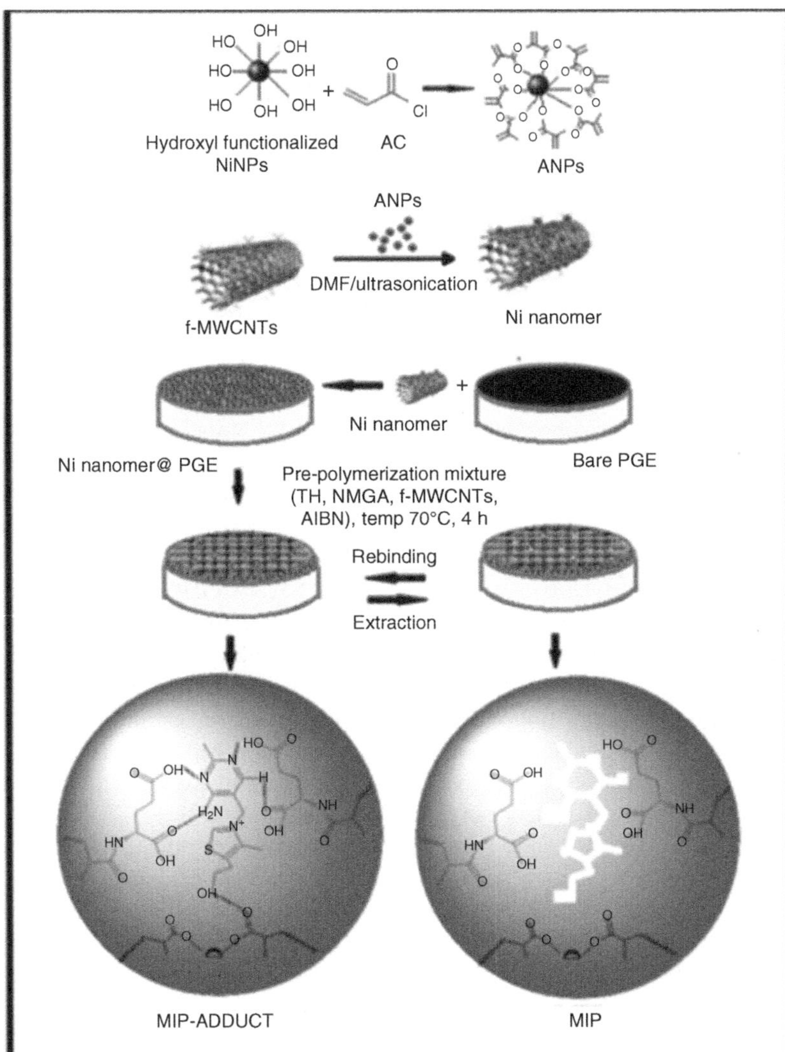

FIGURE 11.1 Scheme for the preparation of MIP/PGE. Reproduced with permission from Reference 59.

In another work, researchers developed a screen-printed electrochemical sensor toward VB_1 detection not long after [60]. The developed screen-printed carbon-based electrode's mediator was cobalt phthalocyanine (CoPC) (SPCE). In a phosphate buffer solution (PBS) (pH 12), Amperometry [AMP] was used to test VB_1. A LOD of 6.3 ng m L^{-1} and a linear range of 0.1–20 g m L^{-1} for VB_1 was detected in a multivitamin tablet and energy drinks using the sensor. In another work, Wahyuni and co-workers developed a glass carbon electrode (GCE) toward the detection of VB_1 and modulation of pH controlling the solution pH is critical for VB_1 detection [61]. Later, CV was employed for detecting thiamine in KCl solution having a pH regulated by the addition of different amounts of KOH. As a result, the VB_1 amount was found to be 35 μmol L^{-1} in brown rice samples.

(B) Electrochemical sensors for Riboflavin (VB_2)

Riboflavin (VB_2) can be found in abundance in most dairy farm products including eggs, milk, and green leafy vegetables. It is mainly associated with the synthesis of flavin mononucleotide and flavin adenine dinucleotide (co-enzymes) that assists our bodies in converting macromolecules to energy. Skin illnesses and eye difficulties, including stomatitis, and sensitivity toward sunlight, can occur if it is deficient [62, 63]. By applying SWV, Gribat et al. created colloidal hematite (-Fe_2O_3) film for modifying the GCE (-Fe_2O_3/GCE) toward VB_2 detection in sodium chloride (NaCl) electrolyte. The LOD and the linear range were found to be 8.4 mol L^{-1} and 1.3–100 mol L^{-1} respectively [64].

Selvarajan and co-workers then created a ZnO-manganese hexacyanoferrate (ZnO/MnHCNF) nanocomposite for altering the GCE (ZnO/MnHCNF/GCE) for VB_2 detection in pH 6.4 PBS using DPV [65]. The developed sensor was used for assessing riboflavin in milk powder and pharmaceutical samples with great results. Another group developed flower-shaped Fe_3O_4 attached reduced graphene oxide (Fe_3O_4/rGO) nanocomposites, that were subsequently placed onto the GCE (Fe_3O_4/rGO/GCE) for VB_2 quantification in in pharmaceutical samples using DPV in PBS at pH 5 [66]. With a LOD of 89 nmol L^{-1}, two linear ranges of 0.3 to 1 mol L^{-1} and 1–100 mol L^{-1} were found. The developed sensor was used to test VB_2.

Soon after, an economic electrochemical sensor based on MIP-modified GCE was designed by Zhang and co-workers toward the quantification of VB_2 in pharmaceutical samples [67]. In Britton-Robison buffer (BRB) at pH 4, LSV was used to identify VB_2. The linear range and LOD, respectively, were 0.002–0.9 mol L^{-1} and 0.7 nmol L^{-1}. To build the SPCE (N-CQD/SnO_2/SPCE) for VB_2 quantification, Muthusankar and his team used nitrogen-doped carbon quantum dots/SiO_2 (N-CQD/SnO_2) nanocomposite [68]. The developed sensor showcased a low-level detection limit of 8 nmol L^{-1} and the linear range was 0.05–306 mol L^{-1}. VB_2 was measured in milk powder samples and pharmaceutical by utilizing the fabricated sensor. As a result, Sangili and colleagues created a new nanocomposite by depositing palladium-copper nanoparticles atop nutshell carbon (Pd–Cu@NSC), which was then drop cast to create the modified SPCE (Pd–Cu@NSC/SPCE) toward the quantification of riboflavin in PBS (pH 7) using DPV [69]. With a low-detection limit of 7.6 pmol L^{-1}, the linear range was 0.004–0.1 mol L^{-1}

In addition to the GCE and SPCE, Yu and colleagues developed a carbon cloth electrode (CCE) for VB_2 quantification, using bacteria Shewanella oneidensis MR1 as a biocatalyst [70]. After then, AMP was employed for detecting riboflavin in PBS having a pH of 7. The electrochemical sensor showcased a linear range of 2–100 nmol L^{-1} and a low-level limit of detection of 0.85 nmol L^{-1} with a linear range of 2–100 nmol L^{-1}. Tigari has developed a sodium lauryl sulphate modified carbon nanotube paste electrode (SLS/CNTPE) for VB_2 detection in pH 7 PBS using CV [71]. The linear range was 2–200 μmol L^{-1} along with a low-level detection limit of 0.0925 μmol L^{-1}.

(C) **Electrochemical Sensors for Pyridoxine (VB_6)**

VB_6, also known as pyridoxal or pyridoxine, is a coenzyme involved in more than a hundred metabolic processes involving amino acids, glucose, and lipids. Its presence helps with red blood corpuscle creation, and neurological and immune system support. Anaemia, skin conditions such as dermatitis, and melancholy and mental instability can all be caused by a lack of VB_6. For the detection of VB_6, Yao and colleagues developed highly oxidized carbon nanosheets (hCNSs) to change the GCE (hCNSs/GCE) [72].

To detect VB_6, DPV was performed in acetate buffer (ACB) (pH 5) as an electrolyte, which led to a low-level detection limit of 5.4 nmol L^{-1}, with the linear range 0.02–1 mol L^{-1}. The designed sensor was utilized for determination of the presence of pyridoxine in plasma and urine samples. Manoj and co-workers created heterostructures of mesoporous TiO_2 and SnO_2 nanoparticles to change the GCE (SnO_2–TiO_2/GCE) toward VB_6 quantification [73]. PBS having a pH of 7 was utilized as a supporting electrolyte in a DPV experiment to evaluate VB_6. The LOD and linear range were 35 nmol L^{-1} and 0.1–31.4 mol L^{-1} respectively. VB_6 was detected in pharmaceutical tablets using the sensor. The presence of pyridoxine in plasma samples and pharmaceutical was detected. To modify the PGE (PGE/AuNPs/POAP), Rejithamol and co-workers produced gold nanoparticles (AuNPs) along with a non-conducting polymeric film of ortho-aminophenol (POAP) toward VB_6 detection by making use of DPV in PBS having a pH of 7 [74]. The LOD and linear range were found to be 0.3 mol L^{-1} and 5–200 mol L^{-1} respectively. The electrochemical sensor was used to assess the concentration of VB_6 in human blood samples. For achieving the quantification of VB_6 employing DPV in PBS (pH 7), Nejad et al. produced $La3^+/Co_3O_4$ nanoflowers for modifying graphite screen-printed electrode ($La3^+/Co_3O_4$NFs/GSPE) [75]. The LOD was 0.4 mol L^{-1} and the linear range was 1–600 mol L^{-1}. The sensor was used to measure VB_6 levels in ampoule and urine samples.

Porto et al. modified a pyrolytic graphite electrode (PYGECoPC/MWCNT) for VB_6 determination using DPV in PBS having a pH of 5.5 by synthesizing a multi-walled carbon nanotube (MWCNT) and CoPC composite [76]. A linear range of 10–400 mol L^{-1} was obtained, with a LOD of 0.5 mol L^{-1}. The designed electrochemical sensor was used to measure VB_6 levels in pharmaceutical samples.

11.4 ELECTROCHEMICAL SENSORS FOR HORMONE

(A) Electrochemical Sensors for Insulin

Insulin, a polypeptide hormone produced in the pancreas, acts as a predictive marker for type 1 or type 2 diabetes [77, 78] The standard fasting insulin level in healthy human blood is below 25 mIU/L (0.86 ng/mL) [79] Therefore, insulin detection and other related diseases such as obesity, kidney failure, and insulinoma are clinically significant. Even though several methods – radioimmunoassay [80, 81], fluorescence resonance energy transfer (FRET) [82], (ELISA) [83], (MALDI-TOF mass spectrometry [84] – have been documented, certain drawbacks such as complexity, time-consuming procedures, and requirement of well-trained lab technicians have led to their underutilization. Electrochemical sensing has, in contrast, a plethora of advantages such as cost-effectiveness, facile operation, high selectivity, and sensitivity, and designing of miniaturized sensors [85-87].

Insulin detection can be classified into two major groups of analytical methods – immune and non-immune [88]. Immune methods of insulin determination techniques that have been employed since the 1950s include RIA, [89] ELISA, [90] and luminescent immunoassay (LIA) [91]. The utilization of immune methods in clinical laboratories is hindered by their limitations, such as low sensitivity, the requirement of an enormous amount of time, and multistage steps [92] The non-immune methods that are generally being utilized in clinical laboratories are revealed to have certain limitations, such as low sensitivity, are laborious and has very slow operational rate. The electrochemical insulin determination when compared with various non-immune methods is much more significant and facile, and ensures better sensitivity [93]. Owing to these advantages, electrochemical detection of insulin using chemically modified electrodes (different metal nanomaterials) is gaining attention among the scientific community [94]. Various advantageous properties expected to be possessed by an insulin sensor include high sensitivity and selectivity (>100 pA/pM), long-term durability and stability, and ability to work in a particular concentration range and biocompatibility. Also, the sensor must ensure facile and rapid preparation and fabrication.

Jaafariasl et al. developed an electrochemical sensor consisting of silica gel-modified carbon paste electrode (Si-CPE) toward selective and responsive insulin detection. The developed electrode showcased excellent reproducibility, stability, and even a unique property of renewability by means of a simple polish during electrode fouling. The effect of different types of experimental studies toward voltammetric response showcased insulin was studied and analyzed. The LOD and sensitivity of the developed amperometric technique were revealed to be 36 pM and 107.3 pA/pM (1511.3 pA/pM cm^2), respectively. Also, at optimum operational conditions, employing hydrodynamic amperometry resulted in the calibration plot of insulin to linear varying between 90–1400 pM [87]. In a similar study, Noorbakhsh and Alnajar reported the construction of a novel, facile, economic rGO-modified GC electrode for insulin electrochemical determination. The analytical parameters of the developed rGO-modified GC (GC/rGO) electrode toward insulin oxidation were analyzed using CV and hydrodynamic amperometry. Both these methods were also employed for gaining a better understanding of antifouling and signal amplification properties by comparing the developed electrode with bare glassy carbon electrode and glassy carbon electrodes modified with GO and G films. As expected, the electrooxidation of insulin on the GC electrode coated with rGO showcased a steady insulin oxidation current, whereas, in the case of the counterpart electrodes the insulin electroxidation led to the surface deactivation after a short period. The inclusion of rGO was found to be significantly improving insulin electroxidation. The high electroactive surface area of rGO nanosheets ensured limited resistance during electron transfer and improved the ease of reach of insulin to the rGO nanosheets. The amperometric response of the developed electrode toward insulin at optimum operational conditions was revealed to be linear over the concentration range of 4-640 nM with a detection limit of 350 pM and sensitivity of 7.1254 nA/nM, all of which signifies the usability of the fabricated sensor for insulin determination in human serum matrix in clinical laboratories [95]. In another study, Lin and co-workers reported a simple one-pot method for synthesizing $Ni(OH)_2$–GN nanocomposite toward insulin electrooxidation. The synthesized composite materials consist of $Ni(OH)_2$ nanoparticles that are consistently distributed and directly produced on graphene surface possessing heightened conductivity. Overlapping of the graphene sheets onto each other leads to the generation of the 3-D conductive network which assists in improving the electron kinetic rates. Apart from this, the developed nanocomposites showcase enhanced stability and catalytic activity in combination with conductivity and dispersibility as well owing to the high surface-to-volume ratio of $Ni(OH)_2$ nanoparticles. The $Ni(OH)_2$ nanoparticles directly produced on highly conductive graphene sheets also exhibits superior performance toward electrochemical oxidation of insulin as compared to the combination of pure graphene, pure $Ni(OH)_2$ nanoaggregates, and $Ni(OH)_2$ nanoaggregates physically blended with graphene sheets. The developed electrochemical sensor showed a detection limit of 200nM and a highly stable state current response that ameliorates linearly with insulin concentration ranging from 800 nM to 6400 nM with a rapid response time < 2 s [96]. Lu et al. reported the development of an electrochemical sensor for insulin detection that was based on nickel nanoparticle (NiNP)/carbon nanotube (CNT)-modified carbon fiber microelectrode (NiNPs/CNTs/ CFME), which was designed via a facile two-step electroless plating/chemical vapor deposition technique as shown in Figure 11.2. The overall structure of the composite formed consisted of CNTs present in the middle layer and NiNPs nearby the tip of the CNTs in the outer layer on the surface of the CFs. The developed electrochemical sensor exhibited agreeable stability (approx. 8 weeks), was eminently successful toward detecting very minimal insulin concentrations with high sensitivity (1.11 nA μM^{-1}) and a low-level detection limit (270nM) [97].

A recent study conducted by Šišoláková and co-workers combined transition metal nanoparticles along with MWCNTs leading to the development of CoNPs/ chitosan-MWCNTs/SPCE and CuNPs/ chitosan-MWCNTs/SPCE, whose analytical properties along with sensitivity and selectivity toward insulin electrochemical determination were compared. Superior stability and analytical properties with very less or no notable effect of interferences toward electrochemical insulin determination were revealed at CoNPs/chitosan-MWCNTs/SPCE. Moreover, the developed electrodes exhibited good sensitivity (0.031 mA μM^{-1}), low-level detection limit (25 nM), along stable and wide linear range (0.05 μM to 5 μM) toward electrochemical determination of insulin. Šišoláková et al. furthermore designed an advanced electrochemical sensor toward insulin detection by utilizing screen-printed carbon electrodes (SPCEs) designed by combining chitosan, multi-walled carbon nanotubes (MWCNTs), and zinc nanoparticles (ZnNPs). The fabricated electrochemical sensor showcased

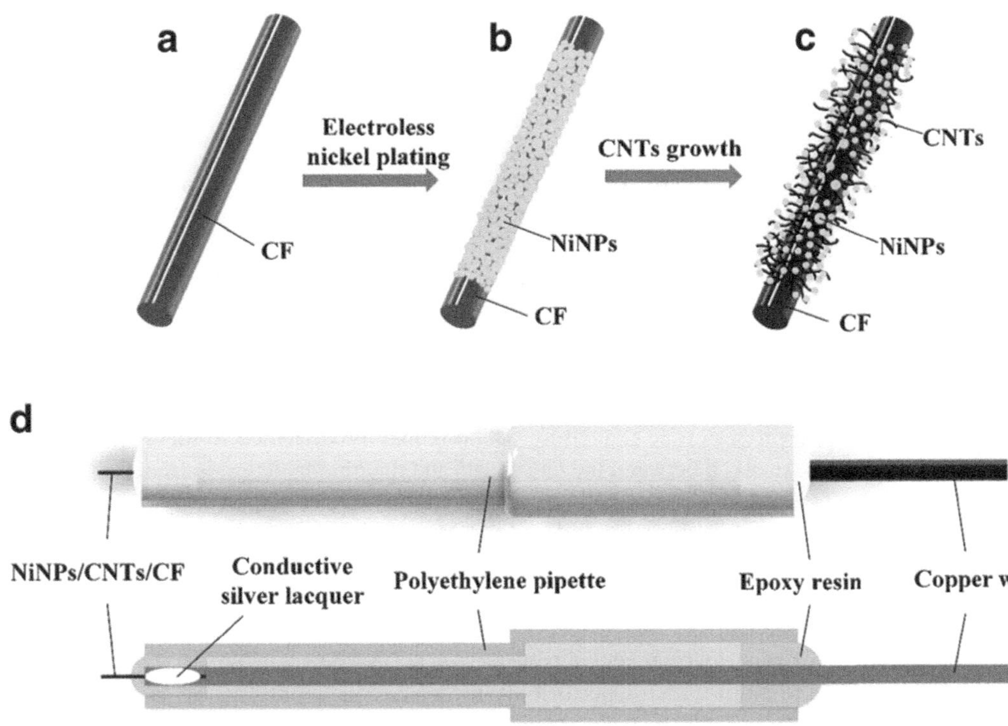

FIGURE 11.2 Preparation of a NiNPs/ CNTs/CFME. (a) A bare CF. (b) A NiNP-modified CF. (c) A NiNP/ CNT-modified CF. (d) Image of the structure and cross-sectional view of a NiNPs/CNTs/CFME. Reproduced with permission from Reference 97.

high efficiency and rapid detection and was economical in comparison with the already-reported sensors in the literature. Based on the findings, the developed sensor ensured a low-level detection limit (0.23 μM), superior sensitivity and selectivity (0.19 mA/μM), and a wide linear range (0.5 μM – 5 μM) [98].

(B) Electrochemical Sensors for Progesterone

Progesterone (4–pregnone–3,20 dione) is a principal biomolecule that is naturally found in organs, including ovaries and adrenal glands, containing an unsaturated α, β-ketone (pregn-4-ene-3,20-dione) structure [99, 100] It assists in maintaining a balance during pregnancy in mammals, plays a paramount role during the entire ovulatory cycle and even maintains various hormones. It also enacts a chief role in the synthetic route followed by most of the steroids [101] and also is part of hormone replacement therapies for women experiencing the post-menopausal period [102, 103]. Even though it is considered a female hormone, progesterone acts as a precursor for other hormones, including estrogen and testosterone, and helps in maintaining stable health [104]. Monitoring of variations in progesterone concentration is highly significant as an imbalance in the normal levels can lead to various issues in the reproductive system, and can cause serious mental illness such as mood swings, anxiety, and even depression [105]. Akshaya and co-workers proposed the designing of an effective and facile electrochemical sensor for Progesterone detection. The electrochemical sensor consisted of onion peel wastes acquired from carbon nanospheres coated on carbon fiber paper (CFP) electrodes The method was considered to be highly economical and environmentally friendly, and the utilization of natural resources derived CNS how cases ameliorated electrocatalytic activity and also assists in reducing the operating potential. The developed sensor was capable of electrochemical progesterone determination with an ultra low-level detection limit (0.012 nM) and heightened sensitivity (19590 μA μM^{-1}).

Aravind and Hemmati developed a facile electrochemical sensor with good stability and sensitivity by utilizing nanocomposite graphene quantum dots (GQDs) along with f–MWCNTs designed glassy carbon electrode (Fe_3O_4@GQD/f–MWCNTs/GCE). Moreover, the resultant 3D hybrid network, in addition to the synergic and electrocatalytic characteristics, enhanced active surface area, leading to an enhancement of electrochemical catalytic activity toward progesterone oxidation. The analytical properties of the designed electrochemical sensor under optimal conditions exhibited a low detection limit (2.18 nM), high sensitivity (16.84 μA μM^{-1}) with a wide linear range, varying in the range of 0.01–0.5 and 0.5–3.0 μM [106]. Gevaerd and co-workers, in a novel attempt, utilized imidazole-functionalized graphene oxide (GO-IMZ) as an artificial enzymatic active site toward progesterone quantification. The developed sensor showcased a notable improvement toward electrochemical progesterone reduction, which is mainly attributed to the synergic effect that seems to be evolved between the GO sheets and the imidazole groups tethered onto the backbone. The efficient sensor exhibited enhanced sensitivity (426 nA L μmol^{-1}), low-level detection limit, and determination (64 and 210 nmol L^{-1} respectively), thereby facilitating stable electrochemical determination of progesterone [107].

Tin nanorods were template-free electrodeposited on GCE and the resulting electrochemical sensor was analyzed for progesterone sensing by Das et al. as shown in Figure 11.3. DPV and chronoamperometry were employed for performing the electrochemical sensing of progesterone in the presence of CTAB surfactant in pure solution and interfering effect from real pharmaceutical samples including testosterone, 17β-estradiol, creatinine, and uric acid (UA), and ascorbic acid (AA) was also investigated. The fabricated electrode showcases excellent response toward progesterone sensing with high sensitivity, a low-level detection limit of 0.12 μM, and a wide linear range (40 to 600 μM) [108].

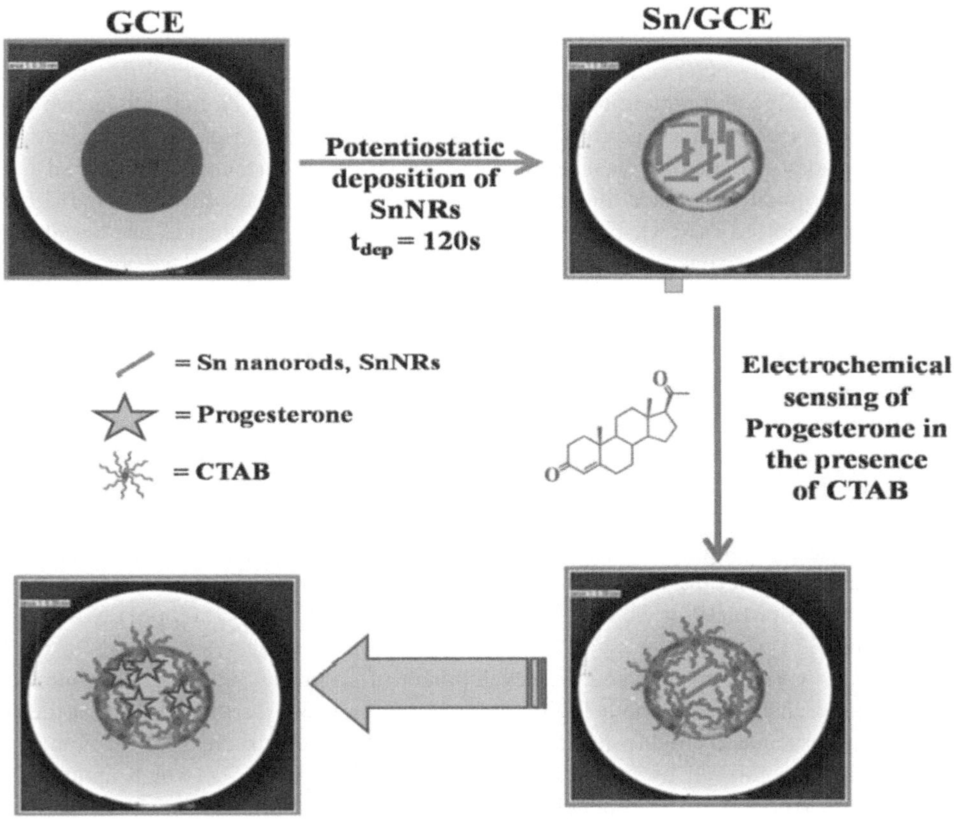

FIGURE 11.3 A sensitive electrochemical detection of progesterone using tin-nanorods modified glassy carbon electrodes: Voltammetric and Computational studies. Reproduced with permission from Reference 108.

Naderi and Jalali pioneered the fabrication of a biocompatible electrochemical sensor elicited by modifying GCE with MWCNTs, Au nanoparticles, and poly-L-serine. The fabricated electrode showcased an enhancement in the oxidation current peak of progesterone as compared to the non-modified electrode. Also, the electrochemical sensor exhibited high sensitivity, a wide linear range of 0.001–2.0 μM (0.31 to 636 ng ml^{-1}) and low-level detection of 0.2 nM (0.063 ng ml^{-1}) under optimum conditions. Later, the developed electrode was successfully utilized for progesterone determination in human blood serum samples of women, men, and pharmaceutical preparations [109].

(C) Electrochemical Sensors for Cortisol

Cortisol, a steroid hormone synthesized by the adrenal gland, acts as a vital biomarker for various diseases and assists in controlling blood pressure, glucose levels, and carbohydrate metabolism [110-112] It is produced by the adrenal gland and acts as a life-supporting hormone that is regularly released during stressful periods [113]. Cortisol, commonly known as the body's stress hormone also serves as a life-supporting hormone [114]. The release of cortisol in the bloodstream anticipates stress in humans. Irregular and very high levels of cortisol in the human body affect the cardiovascular and emotional systems and even leads to Cushing's disease and Addison's disease [115]. The presence of glucocorticoid receptors in most of the tissues in the human body system increases the possibilities of dysfunction in cardiovascular, endocrine, respiratory, and nervous systems [116]. Therefore, the determination of cortisol for understanding and studying the onset of various psychological disorders including stress, mood swings, changes in behavioral patterns is desirable for better functioning of the human biological system. Clinical laboratories and the healthcare system currently practice the determination of the total sum of cortisol that is defined as the total of free and protein-bound fractions. Nonetheless, the biologically active fraction that can also be defined as free cortisol is responsible for cortisol-related events in the human body [117]. Analytical techniques that are being currently administered for cortisol determination include fluorescent enzyme-linked immunosorbent assays (ELISA) or radioimmunoassay (RIA). However, the above-mentioned techniques are revealed to be complicated, less sensitive, time-consuming, require a large sample size and a lab technician, and are highly expensive [118-120].

A biosensor, capable of detecting both salivary and sweat cortisol, designed the utilization of label-free transition metal oxide, and was developed by Zhou et al. An electrostatic approach by means of isoelectric-point of cortisol antibodies (4.50) and β-MnO_2 (7.30), was employed for immobilizing anti-cortisol antibodies produced in rabbit (ACAs onto the GCE/β-MnO_2 CNs electrode as shown in Figure 11.4. Efficient specific interaction was seen between the ACAs/beta-MnO_2 cacti-shaped nanostructured (β-MnO_2 CNs) electrode and cortisol molecules that advance significantly fast and are highly responsive electrochemical cortisol sensors. The fabricated electrode showcased excellent binding with cortisol as compared to other glucocorticoids, a low limit of detection of 0.023 pM (S/N = 3), and a wide linear range (0.1 pM to 1500 pM) with a limit of detection (LOD) [121].

Madhu and co-workers investigated the development of a non-invasive and real-time electrochemical immunosensor having both high selectivity and sensitivity toward cortisol detection involving a yarn-associated, highly flexible and superwettable electrochemical immunosensing technique. This was achieved by utilization of ZnONRs combined CCY which anticipated the immobilization of Anti-Cmab for producing an immunosensing platform for responsive cortisol determination as shown in Figure 11.5. The developed immunosensor showcased high sensitivity (2.12 μA/(g mL^{-1}), stability, reproducibility and ensures rapid detection, low detection limits ranging between 0.45 and 0.098 fg/mL, and wide linear detection range from 1 fg/mL to 1 μg/mL [122]. An electrochemical immunosensor for cortisol determination elicited on Au nanoparticles and magnetic functionalized reduced graphene oxide (AuNPs/MrGO) was developed by Sun et al. The designing and construction of the base of the immunosensor involved the fabrication of MrGO by chemical cross-linking technique and was later utilized for modifying the nafion-pretreated GCE. The utilization of AuNPs/MrGO led to an improvement in the electrical conductivity, higher surface area, and enhanced biocompatibility, and also assisted in rapid response and determination of cortisol. The highly efficient immunosensor showcased significant analytical capabilities toward cortisol detection with a wide linear range (0.1 to 1000 ng/mL) and a low-level detection limit (0.05 ng/mL at 3σ) [123].

Fabrication of electrochemical immunosensor for ultrasensitive cortisol determination established using gold microelectrode arrays functionalized with dithiobis (succinimidyl propionate) self-assembled monolayer (SAM) was developed by Arya and co-workers. The design involved covalent immobilization of cortisol-specific monoclonal antibody (C-Mab) on the surface of the gold microelectrode array. Once the binding was completed, unreacted active groups of DTSP were blocked utilizing ethanol amine (EA). The developed biosensor showcased a wide linear range of 1 pM–100 nM, the LOD of 1 pM, and was even capable of detection of cortisol to 1 pM within the 40 min analysis time [124]. Another label-free paper-based biosensor chip was fabricated by combining anti-cortisol antibody (anti-CAB) on gold (Au) microelectrodes utilizing 3,3′-Dithiodipropionic acid di(N-hydroxysuccinimide ester (DTSP) as SAM agent. For enhancing the immune response of the developed electrode, poly(styrene)-blockpoly (acrylic acid) (PS67-b-PAA27) polymer and graphene nanoplatelets (GP) suspension was coated on filter paper. Finally, a handheld biosensor

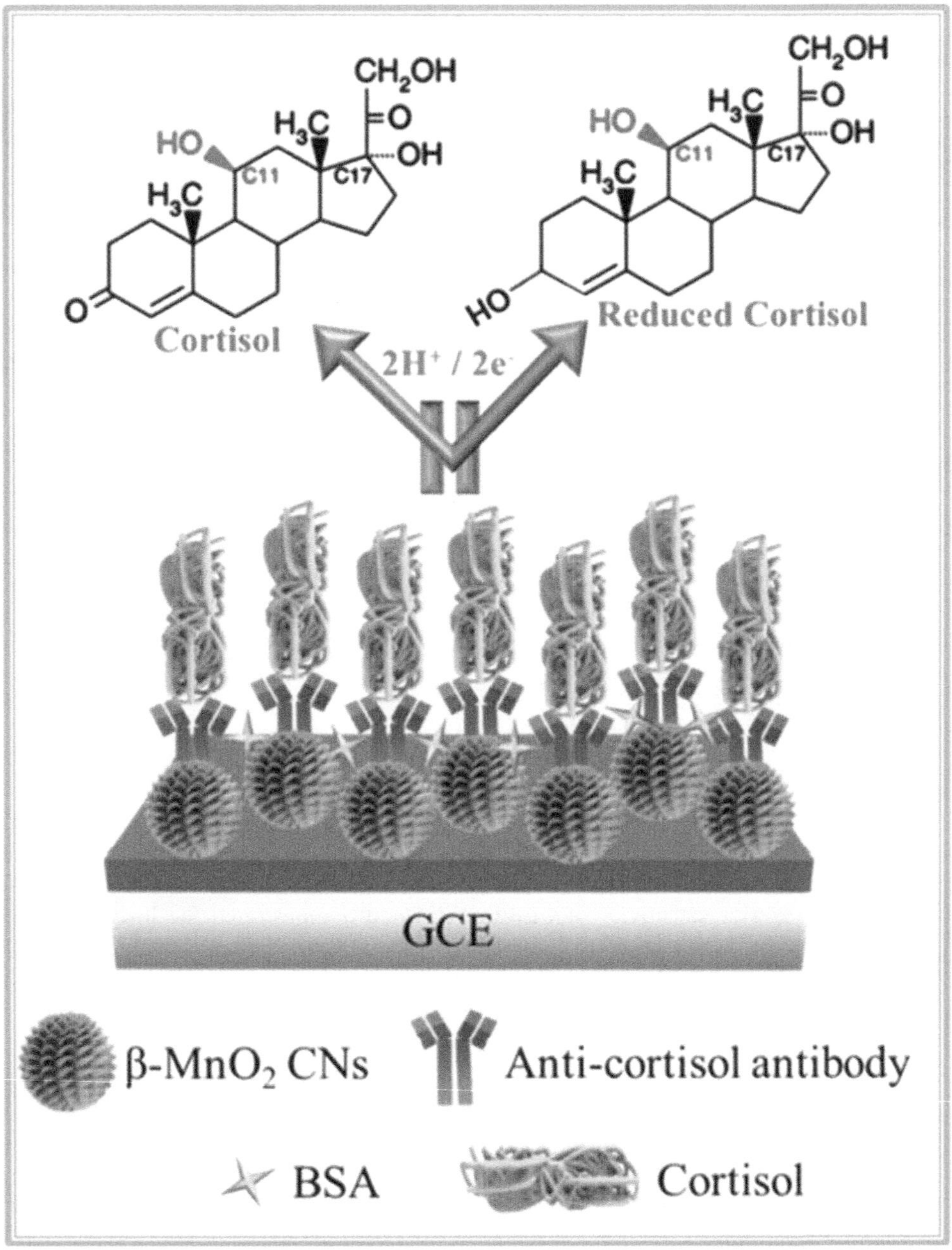

FIGURE 11.4 Pictorial display showing the stepwise fabrication of cortisol biosensor assay. Reproduced with permission from Reference 121.

was developed by combining a biosensor chip along with an economic lab-built miniaturized printed circuit board (PCB) for providing electrical connection to MATLAB. The developed biosensor chip showcased a significant sensitivity of 50 Ω $(pg\ mL^{-1})^{-1}$ and a wide linear range (3 pg/mL to 10 μg/mL) value of 0.9951 [125].

A study made use of sonochemical method for the preparation of 1D ZnO nanorods (ZnO-NRs) and two-dimensional 2D ZnO nanoflakes utilizing Au-coated substrates toward the designing of

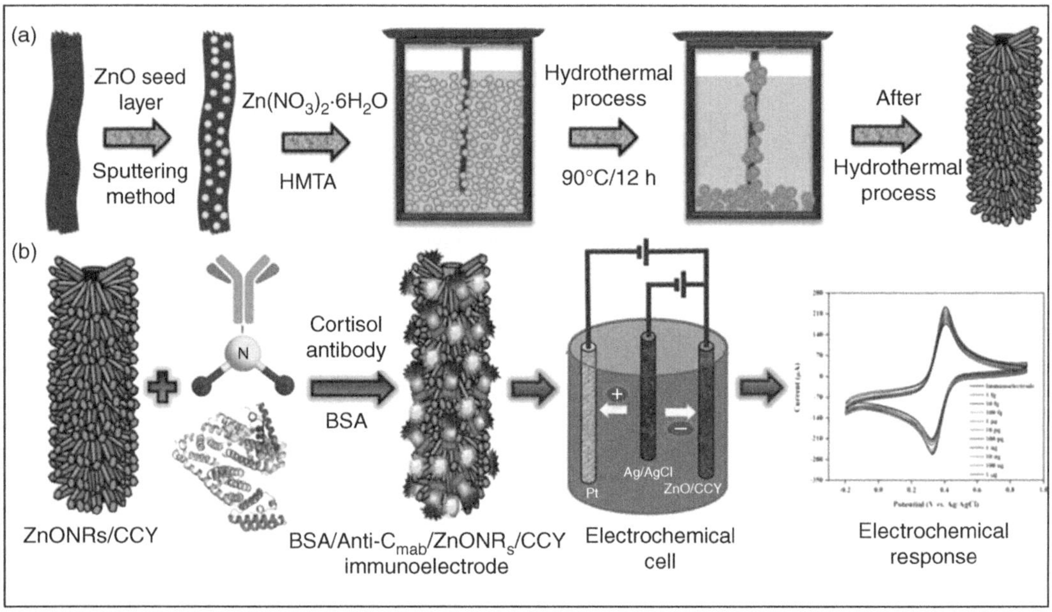

FIGURE 11.5 Schematic illustration of (a) integration of ZnONRs on CCY, and (b) stepwise fabrication of immunosensor using ZnONRs/CCY and its electrochemical responses toward cortisol.

particularly selective and responsive immunosensor for cortisol. The addition of 1D ZnO-NRs and 2D ZnO-NFs yields superior sensing abilities, including superior chemical stability, biocompatibility, and improved catalytic activity. 1D-NSs embodies a very high surface area to volume ratio and similarly, the superior surface charge density and a large area in polarized (0001) plane of 2D-NSs lead to enhanced Anti-Cab loading eventually ameliorated sensing abilities. The immunosensor could accomplish superior sensitivity of 11.86 mA/M and demonstrated the low-level detection limit (1 pM) toward cortisol detection that is revealed to be 100 times better as compared to standard ELISA [126].

(D) Electrochemical Sensors for Serotonin

Serotonin (5-hydroxytryptamine, 5-HT) is a monoamine neurotransmitter produced in the serotonergic neurons of the central nervous system and has been instrumental in regulating emotional behaviors which include aggression, mood swings, and sexual activity [127–129]. 5-HT is majorly distributed in the central nervous system, circulatory system, spinal cord, and gastrointestinal tract as well [130]. Even though the physiological effects of 5-HT in the biological system are not well understood, variations in the levels of peripheral 5-HT have been revealed to be related to events including depression, kidney-related diseases, and hypertension [131–132]. The normal 5-HT level in a healthy human whole blood is in the range of 500 nM to 1200 nM, about 295–687 nM in urine samples, and approximately <0.0591 nM in cerebrospinal fluid [133–134]. Reduced 5-HT levels are commonly associated with various diseases, most of which affect mental health such as anxiety and depression [135]. A few of the drawbacks associated with the utilization of unmodified electrodes for electrochemical analysis include interference from biochemical compounds, fouling of electrodes, lowered sensitivity followed by reduced stability and reproducibility [136]. This has resulted in the designing of modified electrodes for simultaneous sensing of neurotransmitters [137]. Cesarino and co-workers reported the advancement of a facile sensor that was based on electrodeposition of MWCNT, polypyrrole (PPy) in combination of colloidal Ag NPs onto the platinum (Pt) electrode surface. The low detection limit of 0.15 μmol L^{-1} (26.4 μg L^{-1-1}) was attributed to the synergistic effect due to the formation of the MWCNT/PPy/AgNPs nanohybrid. A'so, the developed sensor, exhibited high reproducibility (2.2%) and repeatability (1.7%) [138]. Wang et al. performed the fabrication of the modified electrode by intercalating carbon nanotubes (CNT) on a graphite surface, which has been revealed to be promising for simultaneous detection of determining dopamine (DA) and (5-HT) simultaneously in the presence of AA, as shown in Figure 11.6. It was then revealed to be the structural properties of the CNTs, including the presence of porous interfacial layers, which greatly enhanced both the sensitivity and selectivity. The linear calibration range for DA in the presence of 0.5 mM AA and 5 mM 5-HT was found to be 0.5/10 mM, with a correlation coefficient of 0.9996, and a detection limit of 0.1 mM. The current sensitivities were 1.89 and 8.05 mA mM1 for 5-HT and DA, respectively. The developed electrode was found to be highly economic and facile, showcasing superior sensitivity and selectivity toward the determination of neurotransmitters, and was even seen as a promising candidate for in vivo and in vitro measurement of 5-HT and DA in presence of excess levels of AA [135].

Mazloum-Ardakani and Khoshroo utilized functionalized CNTs with TiO_2, 9-(1,3-dithiolan-2-yl)-6,7-dihydroxy-3,3-dimethyl-3,4-dihydrodibenzo[b,d]furan-1(2H)-one (benzofuran derivative (DDF)) and 1-butyl-3-methylimidazolium tetrafluoroborate (IL) for developing electrochemical sensor toward the simultaneous detection of both 5-HT and isoproterenol (IP). The enhancement in the electrocatalytic activity toward IP was majorly attributed to direct electron transfer kinetics occurring between DDF and the GCE. The DPV results revealed the linear dependence of anodic peak currents on 5-HT (1.0–650.0 μM) and IP (0.1–1300.0). The comparable sensitivity of the developed electrode toward IP in the absence and presence of 5-HT confirms the independent oxidation processes of IP and 5-HT, which eventually substantiates the interference-free simultaneous detection of both IP and 5-HT. The developed sensor showcased a wide linear range

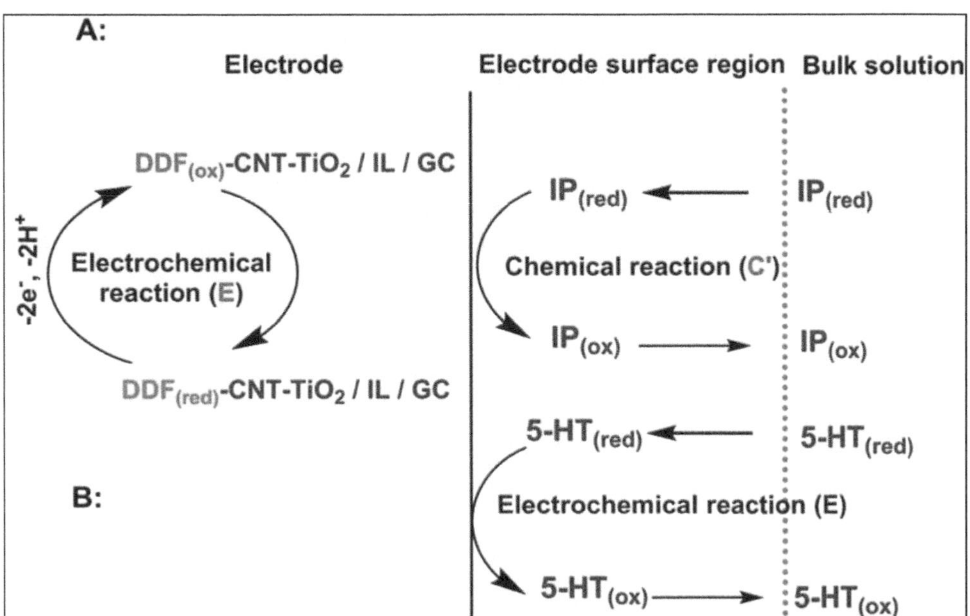

FIGURE 11.6 Electrocatalytic reactions mechanism for IP (A) and electrochemical reaction of 5-HT (B) at surface of DDF-CNT-TiO_2/IL/GC electrode. Reproduced with permission from Reference 139.

(1.0–650.0 μM) and a low-level detection limit (0.154 μM) toward 5-HT determination [139]. Sun et al. demonstrated the development of a state-of-the-art NiO/carbon nanotube (CNT)/poly(3,4-ethylenedioxythiophene) (PEDOT) composite with a coaxial tubular nanostructure deposited on GCE. The PEDOT matrix present on the outer surface has a heightened affinity toward DA, 5-HT, and tryptophan, while the CNT inner layer has superior electron-transfer properties which explain the utilization of NiO/CNT/PEDOT/GCE toward the simultaneous detection of biomolecules. The presence of synergistic effect generated due to the NiO, CNT, and PEDOT results in the formation of three distinct oxidation peaks in a diverse system consisting of DA, 5-HT, and Tryptophan; the 5-HT–DA and Tryptophan-5-HT, wherein the potential separations were 160 and 324 mV, respectively. The developed sensor showcased a wide linear range (DA, 5-HT, and Tryptophan in the 0.03–20, 0.3–35, and 1–41 μM) and low-level detection limits (DA, 5-HT, and Tryptophan - 0.026, 0.063, and 0.210 μM) and also ensured the simultaneous detection of three analytes [140]. Babaei and Taheri proposed the designing of an electrochemical sensor for simultaneous detection of DA and 5-HT in presence of AA. The electrochemical sensor was based on Nafion/$Ni(OH)_2$ nanoparticles-carbon nanotube composite designed GCE. The developed electrode assisted in the shuttling of electrons for DA and 5-HT in presence of AA. The results obtained from DPV signifies that the observed anodic peak currents were linearly proportional to the concentration in the range of 0.05–25 mol L−1 along with a low-level detection limit of 0.015 mol L^{-1} in the case of DA and the range of 0.008–10 mol L^{-1} and with a detection limit of 0.003 mol L^{-1} for 5-HT [141].

11.5 CONCLUSIONS AND FUTURE PERSPECTIVES

Because of its remarkable benefits of cost-viability, high sensitivity, simplicity of scaling down, dependability, and reproducibility, new advancements in nanomaterials electrochemical sensors toward the determination of vitamins and hormones have received a lot of attention. In this regard, our chapter examined up-to-date advances in the employment of nano materials in electrochemical sensors for the determination of hormones and vitamins. Electrochemical sensors based on recognizing water-soluble nutrients, including VB_1, VB_2, VB_6, VB_9 as well as insulin, progesterone, cortisol, and 5-HT, were extensively discussed.

Although not all topics are covered, an overview of the various approaches as well as a display of numerous electrochemical sensors for nutrients have been provided. According to the findings discussed here, well-designed and developed sensors could have high prospects in the time to come. Despite tremendous advancements in recent years, electrochemical sensors for nutrient detection still have opportunity for improvement. New materials that can improve sensitivity and selectivity, for example, are still being investigated and could be useful. In conclusion, the authors are confident that electrochemical sensors for vitamin and hormone detection will continue to improve and be fully utilized in the future.

REFERENCES

1. Dhanjai, X.B. Sinha, L.X. Lu, D.Q. Wu, Y. Tan, J.P. Li, R. Jain Chen, Voltammetric sensing of biomolecules at carbon based electrode interfaces: a review, *Trac. Trends Anal. Chem.* 98 (2018) 174–189.
2. E. Mehmeti, D.M. Stankovi ́c, S. Chaiyo, L. ˇ Svorc, ˇ K. Kalcher, Manganese dioxidemodified carbon paste electrode for voltammetric determination of riboflavin, *Microchimica Acta* 183 (5) (2016) 1619–1624.
3. P.K. Brahman, R.A. Dar, K.S. Pitre, DNA-functionalized electrochemical biosensor for detection of vitamin B1 using electrochemically treated multiwalled carbon nanotube paste electrode by voltammetric methods, Sensor. *Actuator. B Chem.* 177 (2013) 807–812.
4. T. Jamali, H. Karimi-Maleh, M.A. Khalilzadeh, A novel nanosensor based on Pt:Co nanoalloy ionic liquid carbon paste electrode for voltammetric determination of vitamin B-9 in food samples, *Lwt-Food Sci. Technol.* 57 (2) (2014) 679–685.

5. V. Arabali, M. Ebrahimi, M. Abbasghorbani, V.K. Gupta, M. Farsi, M.R. Ganjali, F. Karimi, Electrochemical determination of vitamin C in the presence of NADH using a CdO nanoparticle/ionic liquid modified carbon paste electrode as a sensor, *J. Mol. Liq.* 213 (2016) 312–316.
6. E.S. Sa, P.S. da Silva, C.L. Jost, A. Spinelli, Electrochemical sensor based on bismuth-film electrode for voltammetric studies on vitamin B-2 (riboflavin), Sensor. *Actuator. B Chem.* 209 (2015) 423–430.
7. Y.M. Issa, F.M. Abou Attia, O.E. Sherif, A.S.A. Dena, Potentiometric and surface topography studies of new carbon-paste sensors for determination of thiamine in Egyptian multivitamin ampoules, *Arab. J. Chem.* 10 (6) (2017) 751–760.
8. D.M. Stankovic, D. Kuzmanovic, E. Mehmeti, K. Kalcher, Sensitive and selective determination of riboflavin (vitamin B2) based on boron-doped diamond electrode, *Monatshefte Fur Chemie* 147 (6) (2016) 995–1000.
9. S.B. Revin, S.A. John, Simultaneous determination of vitamins B2, B9 and C using a heterocyclic conducting polymer modified electrode, *Electrochim. Acta* 75 (2012) 35–41.
10. L. Garai, Improving HPLC analysis of vitamin A and E: use of statistical experimental design, in: P. Koumoutsakos, M. Lees, V. Krzhizhanovskaya, J. Dongarra, P. Sloot (Eds.), International Conference on Computational Science (2017), pp. 1500–1511.
11. D.C. Woollard, A. Bensch, H. Indyk, A. McMahon, Determination of vitamin A and vitamin E esters in infant formulae and fortified milk powders by HPLC: use of internal standardisation, *Food Chem.* 197 (2016) 457–465.
12. B.J. Petteys, E.L. Frank, Rapid determination of vitamin B-2 (riboflavin) in plasma by HPLC, *Clin. Chim. Acta* 412 (1–2) (2011) 38–43.
13. S. Vidovic, B. Stojanovic, J. Veljkovic, L. Prazic-Arsic, G. Roglic, D. Manojlovic, Simultaneous determination of some water-soluble vitamins and preservatives in multivitamin syrup by validated stability-indicating high-performance liquid chromatography method, *J. Chromatogr. A.* 1202 (2) (2008) 155–162.
14. C.K. Markopoulou, K.A. Kagkadis, J.E. Koundourellis, An optimized method for the simultaneous determination of vitamins B-1, B-6, B-12, in multivitamin tablets by high performance liquid chromatography, J. *Pharmaceut. Biomed. Anal.* 30 (4) (2002) 1403–1410.
15. Li, Q. Yang, X. Wang, M. Arabi, H. Peng, J. Li, H. Xiong, L. Chen, Facile approach to the synthesis of molecularly imprinted ratiometric fluorescence nanosensor for the visual detection of folic acid, *Food Chem.* 319 (2020).
16. M. Wang, Y. Liu, G. Ren, W. Wang, S. Wu, J. Shen, Bioinspired carbon quantum dots for sensitive fluorescent detection of vitamin B12 in cell system, *Anal. Chim. Acta* 1032 (2018) 154–162.
17. L. Ding, H. Yang, S. Ge, J. Yu, Fluorescent carbon dots nanosensor for label-free determination of vitamin B-12 based on inner filter effect, Spectrochim. *Acta Mol. Biomol. Spectrosc.* 193 (2018) 305–309.
18. K. Yang, Y. Wang, C. Lu, X. Yang, Ovalbumin-directed synthesis of fluorescent copper nanoclusters for sensing both vitamin B-1 and doxycycline, *J. Lumin.* 196 (2018) 181–186.
19. X. Wang, K. Li, L. Yao, C. Wang, A. Van Schepdael, Recent advances in vitamins analysis by capillary electrophoresis, *J. Pharmaceut. Biomed. Anal.* 147 (2018) 278–287.
20. S. Dziomba, P. Kowalski, T. Baczek, Field-amplified sample stacking-sweeping of vitamins B determination in capillary electrophoresis, J. *Chromatogr. A* 1267 (2012) 224–230.
21. M. Shabangi, J.A. Sutton, Separation of thiamin and its phosphate esters by capillary zone electrophoresis and its application to the analysis of water-soluble vitamins, *J. Pharmaceut. Biomed. Anal.* 38 (1) (2005) 66–71.
22. J.-A. Lopez-Pastor, A. Martinez-Sanchez, J. Aznar-Poveda, A.-J. Garcia-Sanchez, J. Garcia-Haro, E. Aguayo, Quick and cost-effective estimation of vitamin C in multifruit juices using voltammetric methods, *Sensors* 20 (3) (2020).
23. I.A. Mir, K. Rawat, P.R. Solanki, H.B. Bohidar, ZnSe core and ZnSe@ZnS coreshell quantum dots as platform for folic acid sensing, *J. Nanoparticle Res.* 19 (7) (2017).
24. A. Kowalczyk, M. Sadowska, B. Krasnodebska-Ostrega, A.M. Nowicka, Selective and sensitive electrochemical device for direct VB2 determination in real products, *Talanta* 163 (2017) 72–77.
25. H. Filik, A.A. Avan, S. Aydar, Electrochemical determination of vitamin B-12 in food samples by poly(2,2'-(1,4-phenylenedivinylene) bis-8 hydroxyquinaldine)/ multi-walled carbon nanotube-modified glassy carbon electrode, Food Anal. *Methods* 9 (8) (2016) 2251–2260.

26. R. Gupta, P.K. Rastogi, U. Srivastava, V. Ganesan, P.K. Sonkar, D.K. Yadav, Methylene blue incorporated mesoporous silica microsphere based sensing scaffold for the selective voltammetric determination of riboflavin, *RSC Adv.* 6 (70) (2016) 65779–65788.
27. S.S. Khaloo, S. Mozaffari, P. Alimohammadi, H. Kargar, J. Ordookhanian, Sensitive and selective determination of riboflavin in food and pharmaceutical samples using manganese (III) tetraphenylporphyrin modified carbon paste electrode, *Int. J. Food Prop.* 19 (10) (2016) 2272–2283.
28. R.-W. Si, Y. Yang, Y.-Y. Yu, S. Han, C.-L. Zhang, D.-Z. Sun, D.-D. Zhai, X. Liu, Y.- C. Yong, Wiring bacterial electron flow for sensitive whole-cell amperometric detection of riboflavin, *Anal. Chem.* 88 (22) (2016) 11222–11228.
29. M. Arvand, M. Dehsaraei, A simple and efficient electrochemical sensor for folic acid determination in human blood plasma based on gold nanoparticles-modified carbon paste electrode, *Mater. Sci. Eng. C-Mater. Biol. Appl.* 33 (6) (2013) 3474–3480.
30. L. Carlucci, G. Favero, C. Tortolini, M. Di Fusco, E. Romagnoli, S. Minisola, F. Mazzei, Several approaches for vitamin D determination by surface plasmon resonance and electrochemical affinity biosensors, Biosens. *Bioelectron.* 40 (1) (2013) 350–355.
31. L. Lu, Recent advances in synthesis of three-dimensional porous graphene and its applications in construction of electrochemical (bio)sensors for small biomolecules detection, Biosens. *Bioelectron.* 110 (2018) 180–192.
32. S.Z. Mohammadi, H. Beitollahi, Z. Dehghan, R. Hosseinzadeh, Electrochemical determination of ascorbic acid, uric acid and folic acid using carbon paste electrode modified with novel synthesized ferrocene derivative and core-shell magnetic nanoparticles in aqueous media, *Appl. Organomet. Chem.* 32 (12) (2018).
33. L. Wang, C. Gong, Y. Shen, W. Ye, M. Xu, Y. Song, A novel ratiometric electrochemical biosensor for sensitive detection of ascorbic acid, Sensor. *Actuator. B Chem.* 242 (2017) 625–631.
34. N. Shadjou, M. Hasanzadeh, A. Omari, Electrochemical quantification of some water soluble vitamins in commercial multi-vitamin using poly-amino acid caped by graphene quantum dots nanocomposite as dual signal amplification elements, Anal. *Biochem.* 539 (2017) 70–80.
35. M. Behpour, A.M. Attaran, M.M. Sadiany, A. Khoobi, Adsorption effect of a cationic surfactant at carbon paste electrode as a sensitive sensor for study and detection of folic acid, *Measurement* 77 (2016) 257–264.
36. H. Zhang, Y. Gao, H. Xiong, Sensitive and selective determination of riboflavin in milk and soymilk powder by multi-walled carbon nanotubes and ionic liquid BMPi PF6 modified electrode, *Food Anal. Methods* 10 (2) (2017) 399–406.
37. S. Yan, X. Li, Y. Xiong, M. Wang, L. Yang, X. Liu, X. Li, L.A.M. Alshahrani, P. Liu, C. Zhang, Simultaneous determination of ascorbic acid, dopamine and uric acid using a glassy carbon electrode modified with the nickel(II)-bis(1,10- phenanthroline) complex and single-walled carbon nanotubes, *Microchimica Acta* 183 (4) (2016) 1401–1408.
38. A. Nezamzadeh-Ejhieh, P. Pouladsaz, Voltammetric determination of riboflavin based on electrocatalytic oxidation at zeolite-modified carbon paste electrodes, *J. Ind. Eng. Chem.* 20 (4) (2014) 2146–2152.
39. T. Nie, J.-K. Xu, L.-M. Lu, K.-X. Zhang, L. Bai, Y.-P. Wen, Electroactive speciesdoped poly(3,4-ethylenedioxythiophene) films: enhanced sensitivity for electrochemical simultaneous determination of vitamins B-2, B-6 and C, *Biosens. Bioelectron.* 50 (2013) 244–250.
40. M. Mazloum-Ardakani, M.A. Sheikh-Mohseni, M. Abdollahi-Alibeik, A. Benvidi, Electrochemical sensor for simultaneous determination of norepinephrine, paracetamol and folic acid by a nanostructured mesoporous material, Sensor. *Actuator. B Chem.* 171 (2012) 380–386.
41. A.A. Ensafi, E. Heydari-Bafrooei, M. Amini, DNA-functionalized biosensor for riboflavin based electrochemical interaction on pretreated pencil graphite electrode, *Biosens. Bioelectron.* 31 (1) (2012) 376–381.
42. O.J. D'Souza, R.J. Mascarenhas, A.K. Satpati, S. Detriche, Z. Mekhalif, J. Delhalle, A. Dhason, High electrocatalytic oxidation of folic acid at carbon paste electrode bulk modified with iron nanoparticle-decorated multiwalled carbon nanotubes and its application in food and pharmaceutical analysis, *Ionics* 23 (1) (2017) 201–212.
43. W. Wu, Z. Sun, W. Zhang, Simple and rapid determination of vitamin C in vegetables and fruits by a commercial electrochemical reader, *Food Anal. Methods* 9 (11) (2016) 3187–3192.

44. J. Guo, Smartphone-powered electrochemical biosensing dongle for emerging medical IoTs application, *IEEE Trans. Indus. Inf.* 14 (6) (2018) 2592–2597.
45. J. Guo, X. Ma, Simultaneous monitoring of glucose and uric acid on a single test strip with dual channels, *Biosens. Bioelectron.* 94 (2017) 415–419.
46. Brunetti, Recent advances in electroanalysis of vitamins, *Electroanalysis* 28 (9) (2016) 1930–1942.
47. S. Studzinska, B. Buszewski, Fast method for the resolution and determination of sex steroids in urine, *J. Chromatogr. B: Anal. Technol. Biomed. Life Sci.* 927 (2013) 158–163.
48. N. Tagawa, H. Tsuruta, A. Fujinami, Y. Kobayashi, Simultaneous determination of estriol and estriol 3-sulfate in serum by column-switching semi-micro high-performance liquid chromatography with ultraviolet and electrochemical detection, *J. Chromatogr. B.* 723 (1999) 39-45.
49. S. Wang, W. Huang, G. Fang, J. He, Y. Zhang, On-line coupling of solidphaseextraction to high-performance liquid chromatography for determination of estrogens in environment, *Anal. Chim. Acta* 606 (2008) 194–201.
50. Z.L. Li, S. Wang, N.A. Lee, R.D. Allan, I.R. Kennedy, Development of a solid-phase extraction – enzyme-linked immune sorbent assay method for the determination of estrone in water, *Anal. Chim. Acta* 503 (2004) 171–177.
51. Somayeh Tajik, Mohammad Ali Taher, Hadi Beitollahi, Application of a new ferrocenederivative modified-graphene paste electrode for simultaneous determination of isoproterenol, acetaminophen and theophylline, *Sens. Actuators B Chem.* 197 (2014) 228–236.
52. Zahra Taleat, Mohammad Mazloum Ardakani, Hossein Naeimi, Hadi Beitollahi, Maryam Nejati, Hamid Reza Zare, Electrochemical behavior of ascorbic acid at a 2,2¢-[3,6- Dioxa- 1,8-octanediylbis(nitriloethylidyne)]-bis-hydroquinone carbon paste electrode, *Anal. Sci.* 24 (2008) 1039–1044.
53. Mohammad Mehdi Foroughi, Hadi Beitollahi, Somayeh Tajik, Mozhdeh Hamzavi, Hekmat Parvan, Hydroxylamine Electrochemical sensor based on a modified carbon nanotube paste electrode: Application to determination of hydroxylamine in water samples, *Int. J. Electrochem. Sci.* 9 (2014) 2955–2965
54. P. Xun, H. Lin, R. Wang, Z. Huang, C. Zhou, W. Yu, Q. Huang, L. Tan, Y. Wang, J. Wang, Effects of dietary vitamin B-1 on growth performance, intestinal digestion and absorption, intestinal microflora and immune response of juvenile golden pompano (Trachinotus ovatus), *Aquaculture* 506 (2019) 75–83.
55. F. Gong, W. Zou, Q. Wang, R. Deng, Z. Cao, T. Gu, Polymer nanoparticles integrated with excited-state intramolecular proton transfer-fluorescent modules as sensors for the detection of vitamin B-1, *Microchem. J.* 148 (2019) 767–773.
56. Y. Chen, L. Wang, F. Shang, W. Liu, J. Lan, J. Chen, N.-C. Ha, C. Quan, K.H. Nam, Y. Xu, Structural insight of the 5-(Hydroxyethyl)-methylthiazole kinase ThiM involving vitamin B1 biosynthetic pathway from the Klebsiella pneumoniae, *Biochem. Biophys. Res. Commun.* 518 (3) (2019) 513–518.
57. Z. Zhou, C. Tan, Y. Zheng, Q. Wang, Electrochemical signal response for vitamin B1 using terbium luminescent nanoscale building blocks as optical sensors, Sensor. *Actuator. B Chem.* 188 (2013) 1176–1182.
58. N.M. Muppariqoh, W.T. Wahyuni, B.R. Putra, Iop, Detection of Vitamin b(1) (Thiamine) Using Modified Carbon Paste Electrodes with Polypyrrole, 3rd International Seminar on Sciences on Precision and Sustainable Agriculture, (2017).
59. B.B. Prasad, R. Singh, K. Singh, Development of highly electrocatalytic and electroconducting imprinted film using Ni nanomer for ultra-trace detection of thiamine, Sensor. *Actuator. B Chem.* 246 (2017) 38–45.
60. A. Smart, K.L. Westmacott, A. Crew, O. Doran, J.P. Hart, An electrocatalytic screen-printed amperometric sensor for the selective measurement of thiamine (vitamin B1) in food supplements, *Biosensors-Basel* 9 (3) (2019).
61. W.T. Wahyuni, B.R. Putra, F. Marken, Voltammetric detection of vitamin B1 (thiamine) in neutral solution at a glassy carbon electrode via in situ pH modulation, *Analyst* 145 (5) (2020) 1903–1909.
62. J. Lu, Y. Kou, X. Jiang, M. Wang, Y. Xue, B. Tian, L. Tan, One-step preparation of poly(glyoxal-bis(2-hydroxyanil))-amino-functionalized graphene quantum dotsMnO2 composite on electrode surface for simultaneous determination of vitamin B-2 and dopamine, Colloid. *Surface. Physicochem. Eng. Aspect.* 580 (2019).

63. M. Diniz, N. Dias, F. Andrade, B. Paulo, A. Ferreira, Isotope dilution method for determination of vitamin B2 in human plasma using liquid chromatographytandem mass spectrometry, *J. Chromatogr. B-Anal. Technol. Biomed. Life Sci.* 1113 (2019) 14–19.
64. L.C. Gribat, J.T. Babauta, H. Beyenal, N.A. Wall, New rotating disk hematite film electrode for riboflavin detection, *J. Electroanal. Chem.* 798 (2017) 42–50.
65. S. Selvarajan, A. Suganthi, M. Rajarajan, A facile synthesis of ZnO/Manganese hexacyanoferrate nanocomposite modified electrode for the electrocatalytic sensing of riboflavin, *J. Phys. Chem. Solid.* 121 (2018) 350–359.
66. R. Madhuvilakku, S. Alagar, R. Mariappan, S. Piraman, Green one-pot synthesis of flowers-like Fe3O4/rGO hybrid nanocomposites for effective electrochemical detection of riboflavin and low-cost supercapacitor applications, Sensor. *Actuator. B Chem.* 253 (2017) 879–892.
67. Z. Zhang, J. Xu, Y. Wen, T. Wang, A highly-sensitive VB2 electrochemical sensor based on one-step co-electrodeposited molecularly imprinted WS2-PEDOT film supported on graphene oxide-SWCNTs nanocomposite, *Mater. Sci. Eng. C-Mater. Biol. Appl.* 92 (2018) 77–87.
68. G. Muthusankar, C. Rajkumar, S.-M. Chen, R. Karkuzhali, G. Gopu, A. Sangili, N. Sengottuvelan, R. Sankar, Sonochemical driven simple preparation of nitrogen-doped carbon quantum dots/SnO2 nanocomposite: a novel electrocatalyst for sensitive voltammetric determination of riboflavin, *Sensor. Actuator. B Chem.* 281 (2019) 602–612.
69. A. Sangili, P. Veerakumar, S.-M. Chen, C. Rajkumar, K.-C. Lin, Voltammetric determination of vitamin B-2 by using a highly porous carbon electrode modified with palladium-copper nanoparticles, *Microchimica Acta* 186 (5) (2019).
70. Y.-Y. Yu, J.-X. Wang, R.-W. Si, Y. Yang, C.-L. Zhang, Y.-C. Yong, Sensitive amperometric detection of riboflavin with a whole-cell electrochemical sensor, *Anal. Chim. Acta* 985 (2017) 148–154.
71. G. Tigari, J.G. Manjunatha, C. Raril, N. Hareesha, Determination of riboflavin at carbon nanotube paste electrodes modified with an anionic surfactant, *Chemistryselect* 4 (7) (2019) 2168–2173.
72. S. Yao, G. Li, Y. Liu, S. Wei, H. Wang, X. Huang, Z. Luo, Electrochemical detection of pyridoxal using a sonoelectrochemical prepared highly-oxidized carbon nanosheets modified electrode, Sensor. *Actuator. B Chem.* 274 (2018) 324–330.
73. Manoj, S. Rajendran, J. Qin, E. Sundaravadivel, M.L. Yola, N. Atar, F. Gracia, R. Boukherroub, M.A. Gracia-Pinilla, V.K. Gupta, Heterostructures of mesoporous TiO2 and SnO2 nanocatalyst for improved electrochemical oxidation ability of vitamin B6 in pharmaceutical tablets, *J. Colloid Interface Sci.* 542 (2019) 45–53.
74. R. Rejithamol, S. Beena, Electrochemical quantification of pyridoxine (VB6) in human blood from other water-soluble vitamins, *Chem. Pap.* 74 (6) (2020) 2011–2020.
75. Garkani, H. Beitollahi, S. Tajik, S. Jahani, *La3+-doped Co3O4 Nanoflowers Modified Graphite Screen Printed Electrode for Electrochemical Sensing of Vitamin B6*, (2019) 69–79.
76. L.S. Porto, D.N. da Silva, M.C. Silva, A.C. Pereira, Electrochemical sensor based on multi-walled carbon nanotubes and cobalt phthalocyanine composite for pyridoxine determination, *Electroanalysis* 31 (5) (2019) 820–828.
77. V. Jimenez, C. Jambrina, E. Casana, V. Sacristan, S. Muñoz, S. Darriba, J. Rodó, C. Mallol, M. Garcia, X. León, S. Marcó, A. Ribera, I. Elias, A. Casellas, I. Grass, G. Elias, T. Ferré, S. Motas, S. Franckhauser, F. Mulero, M. Navarro, V. Haurigot, J. Ruberte, F. Bosch, *EMBO Mol. Med.* 10 (2018) 1–24.
78. Fu, Z., Gilbert, E.R., Pfeiffer, L., Zhang, Y., Fu, Y. and Liu, D., Genistein ameliorates hyperglycemia in a mouse model of nongenetic type 2 diabetes. *Applied Physiology, Nutrition, and Metabolism*, *37*(3) (2012) 480–488.
79. Yagati, A.K., Y. Choi, J. Park, J.-W. Choi, H.-S. Jun, and S. Cho. "Silver nanoflower – reduced graphene oxide composite based micro-disk electrode for insulin detection in serum." *Biosensors and Bioelectronics* 80 (2016) 307–314.
80. Lindsay, J.R., A.M. McKillop, M.H. Mooney, F.P.M. O'Harte, P.M. Bell, and P.R. Flatt. "Demonstration of increased concentrations of circulating glycated insulin in human Type 2 diabetes using a novel and specific radioimmunoassay." *Diabetologia* 46, no. 4 (2003) 475–478.
81. Borer-Weir, K. E., S. R. Bailey, N. J. Menzies-Gow, P. A. Harris, and J. Elliott. "Evaluation of a commercially available radioimmunoassay and species-specific ELISAs for measurement of high concentrations of insulin in equine serum." *American Journal of Veterinary Research* 73, no. 10 (2012) 1596–1602.

82. Wang, Y., D. Gao, P. Zhang, P. Gong, C. Chen, G. Gao, and L. Cai. "A near infrared fluorescence resonance energy transfer based aptamer biosensor for insulin detection in human plasma." *Chemical Communications* 50, no. 7 (2014) 811–813.
83. Even M.S., Sandusky, C.B., Barnard, N.D., Mistry, J., Sinha M.K. "Development of a novel ELISA for human insulin using monoclonal antibodies produced in serum-free cell culture medium". *Clin. Biochem* 40 (2007) 98–103.
84. Zhang, X, S. Zhu, C. Deng, and X. Zhang. "An aptamer based on-plate microarray for high-throughput insulin detection by MALDI-TOF MS." *Chemical Communications* 48, 21 (2012) 2689–2691.
85. Yu, Y., M. Guo, M. Yuan, W. Liu, and J. Hu. "Nickel nanoparticle-modified electrode for ultra-sensitive electrochemical detection of insulin." *Biosensors and Bioelectronics* 77 (2016) 215–219.
86. Gerasimov, J.Y., C.S. Schaefer, W. Yang, R.L. Grout, and R.Y. Lai. "Development of an electrochemical insulin sensor based on the insulin-linked polymorphicregion." *Biosensors and Bioelectronics* 42 (2013) 62–68.
87. Jaafariasl, M., E. Shams, and M.K. Amini. "Silica gel modified carbon paste electrode for electrochemical detection of insulin." *Electrochimica acta* 56, no. 11 (2011) 4390–4395.
88. Luo, Y., K. Huang, and H. Xu. "Application of electrospray ionization-mass spectrometry to screen extractants for determination of insulin in an emulsion system by HPLC-UV." *Analytica chimica acta* 553, no. 1–2 (2005) 64–72.
89. Murayama, H., N. Matsuura, T. Kawamura, T. Maruyama, N. Kikuchi, T. Kobayashi, F. Nishibe, and A. Nagata. "A sensitive radioimmunoassay of insulin autoantibody: reduction of non-specific binding of [125I] insulin." *Journal of Autoimmunity* 26, no. 2 (2006) 127–132.
90. Ortner, K., W. Buchberger, and M. Himmelsbach. "Capillary electrokinetic chromatography of insulin and related synthetic analogues." *Journal of Chromatography A* 1216, no. 14 (2009) 2953–2957.
91. Yılmaz, B., Y. Yaşar Kadıoğlu, and Y. Aksoy. "Simultaneous determination of gemcitabine and its metabolite in human plasma by high-performance liquid chromatography." *Journal of Chromatography B* 791, no. 1–2 (2003) 103–109.
92. Mizutani, F., E. Ohta, Y. Mie, O. Niwa, and T. Yasukawa. "Enzyme immunoassay of insulin at picomolar levels based on the coulometric determination of hydrogen peroxide." *Sensors and Actuators B: Chemical* 135, no. 1 (2008) 304–308.
93. Shen, G.-Y., H. Wang, T. Deng, G.-L. Shen, and R.-Q. Yu. "A novel piezoelectric immunosensor for detection of carcinoembryonic antigen." *Talanta* 67, no. 1 (2005) 217–220.
94. Wang, J. "Electrochemical biosensing based on noble metal nanoparticles." *Microchimica Acta* 177, no. 3 (2012) 245–270.
95. Noorbakhsh, A., and A.I.K. Alnajar. "Antifouling properties of reduced graphene oxide nanosheets for highly sensitive determination of insulin." *Microchemical Journal* 129 (2016) 310–317.
96. Lin, Y., Hu.L., L. Linbo, and K. Wang. "Facile synthesis of nickel hydroxide–graphene nanocomposites for insulin detection with enhanced electro-oxidation properties," *RSC Adv.* 86 (2014) 46208–46213.
97. Lu, L., L. Liang, Y. Xie, K. Tang, Z. Wan, and S. Chen. "A nickel nanoparticle/carbon nanotube-modified carbon fiber microelectrode for sensitive insulin detection." *Journal of Solid-State Electrochemistry,* 22, (2018) 825–833.
98. Šišoláková, I., J. Hovancová, F. Chovancová, R. Oriňaková, I. Maskaľová, A. Oriňak, and J. Radoňak. "Zn nanoparticles modified screen printed carbon electrode as a promising sensor for insulin determination." *Electroanalysis* 33, 3 (2021) 627–634.
99. M.S. Christian, R.L. Brent and P. Calda "Embryo–fetal toxicity signals for 17α-hydroxyprogesterone caproate in high-risk pregnancies: A review of the non-clinical literature for embryo–fetal toxicity with progestins." *The Journal of Maternal-Fetal & Neonatal Medicine*, 20, 2 (2007) 89–112.
100. Akshaya, K.B., Vinay, S. Bhat, A. Varghese, L. George, and G. Hegde. "Non-enzymatic electrochemical determination of progesterone using carbon nanospheres from onion peels coated on carbon fiber paper." *Journal of the Electrochemical Society* 166, no. 13 (2019) B1097.
101. Pucci, V., F. Bugamelli, R. Mandrioli, B. Luppi, and M.A. Raggi. "Determination of progesterone in commercial formulations and in non conventional micellar systems." *Journal of Pharmaceutical and Biomedical Analysis* 30, no. 5 (2003) 1549–1559.
102. Sitruk-Ware, R.. "Progestogens in hormonal replacement therapy: New molecules, risks, and benefits." *Menopause* 9, no. 1 (2002) 6–15.

103. Belchetz, P.E. "Hormonal treatment of postmenopausal women." *New England Journal of Medicine* 330, no. 15 (1994) 1062–1071.
104. Bahadır, E.B., and M.K. Sezgintürk. "Electrochemical biosensors for hormone analyses." *Biosensors and Bioelectronics* 68 (2015) 62–71.
105. Jiang, Y., M.G. Colazo, and Michael J. Serpe. "Poly (N-isopropylacrylamide) microgel-based sensor for progesterone in aqueous samples." *Colloid and Polymer Science* 294, no. 11 (2016) 1733–1741.
106. Arvand, M., and S. Hemmati. "Magnetic nanoparticles embedded with graphene quantum dots and multiwalled carbon nanotubes as a sensing platform for electrochemical detection of progesterone." *Sensors and Actuators B: Chemical* 238 (2017) 346–356.
107. Gevaerd, A., S.F. Blaskievicz, A.J.G. Zarbin, E.S. Orth, M.F. Bergamini, and L.H. Marcolino, Jr. "Nonenzymatic electrochemical sensor based on imidazole-functionalized graphene oxide for progesterone detection." *Biosensors and Bioelectronics* 112 (2018) 108–113.
108. Das, A., and M.V. Sangaranarayanan. "A sensitive electrochemical detection of progesterone using tin-nanorods modified glassy carbon electrodes: Voltammetric and computational studies." *Sensors and Actuators B: Chemical* 256 (2018) 775–789.
109. Naderi, P., and F. Jalali. "Poly-L-serine/AuNPs/MWCNTs as a platform for sensitive Voltammetric determination of progesterone." *Journal of The Electrochemical Society* 167, no. 2 (2020) 027524.
110. Kim, Y.-H., K. Lee, H. Jung, H. Kyung Kang, J. Jo, I.-K. Park, and H. H. Lee. "Direct immune-detection of cortisol by chemiresistor graphene oxide sensor." *Biosensors and Bioelectronics* 98 (2017) 473–477.
111. R. Fraser, M.C. Ingram, N.H. Anderson, C. Morrison, E. Davies, J.M.C. Connell, Hypertension. 33, (1999) 1364–1369.
112. Khan, M.S., K. Dighe, K.D. Zhen Wang, I. Srivastava, A.S. Schwartz-Duval, S.K. Misra, and D. Pan. "Electrochemical-digital immunosensor with enhanced sensitivity for detecting human salivary glucocorticoid hormone." *Analyst* 144, no. 4 (2019) 1448–1457.
113. Djuric, Z., C.E. Bird, A. Furumoto-Dawson, G.H. Rauscher, M.T. Ruffin IV, R.P. Stowe, K.L. Tucker, and C.M. Masi. "Biomarkers of psychological stress in health disparities research." *The Open Biomarkers Journal* 1 (2008) 7.
114. Steckl, A.J., and P. Ray. "Stress biomarkers in biological fluids and their point-of-use detection." *ACS sensors* 3, no. 10 (2018) 2025–2044.
115. Delahanty, D.L., A.J. Raimonde, and E. Spoonster. "Initial posttraumatic urinary cortisol levels predict subsequent PTSD symptoms in motor vehicle accident victims." *Biological Psychiatry* 48, no. 9 (2000) 940–947.
116. Yaribeygi, H., Y. Panahi, H. Sahraei, T.P. Johnston, and A. Sahebkar. "The impact of stress on body function: A review." *EXCLI journal* 16 (2017) 1057.
117. Le Roux, C.W., G.A. Chapman, W.M. Kong, W.S. Dhillo, J. Jones, and J. Alaghband-Zadeh. "Free cortisol index is better than serum total cortisol in determining hypothalamic-pituitary-adrenal status in patients undergoing surgery." *Journal of Clinical Endocrinology and Metabolism* 88, no. 5 (2003) 2045–2048.
118. Spano, G., S. Cavalera, F. Di Nardo, C. Giovannoli, L. Anfossi, and C. Baggiani. "Development of a biomimetic enzyme-linked immunosorbent assay based on a molecularly imprinted polymer for the detection of cortisol in human saliva." *Analytical Methods* 11, no. 17 (2019) 2320–2326.
119. Abdulsattar, J.O., and G.M. Greenway. "A sensitive chemiluminescence based immunoassay for the detection of cortisol and cortisone as stress biomarkers." *Journal of Analytical Science and Technology* 10, no. 1 (2019) 1–13.
120. Moore, T.J., and B. Sharma. "Direct Surface Enhanced Raman Spectroscopic Detection of Cortisol at Physiological Concentrations." *Analytical Chemistry* 92, no. 2 (2019) 2052–2057.
121. Zhou, Q., P. Kannan, B. Natarajan, T. Maiyalagan, P. Subramanian, Z. Jiang, and S. Mao. "MnO2 cacti-like nanostructured platform powers the enhanced electrochemical immunobiosensing of cortisol." *Sensors and Actuators B: Chemical* 317 (2020) 128134.
122. Madhu, S., A.J. Anthuuvan, S. Ramasamy, P. Manickam, S. Bhansali, P. Nagamony, and V. Chinnuswamy. "ZnO nanorod integrated flexible carbon fibers for sweat cortisol detection." *ACS Applied Electronic Materials* 2, no. 2 (2020) 499–509.

123. Sun, B., Y. Gou, Y. Ma, X. Zheng, R. Bai, A.A.A. Abdelmoaty, A. A. A. and F. Hu. "Investigate electrochemical immunosensor of cortisol based on gold nanoparticles/magnetic functionalized reduced graphene oxide." *Biosensors and Bioelectronics* 88 (2017) 55–62.
124. Arya, S.K., G. Chornokur, M. Venugopal, and S. Bhansali. "Dithiobis (succinimidyl propionate) modified gold microarray electrode based electrochemical immunosensor for ultrasensitive detection of cortisol." *Biosensors and Bioelectronics* 25, no. 10 (2010) 2296–2301.
125. Khan, M.S., S.K. Misra, Z. Wang, E. Daza, A.S. Schwartz-Duval, J.M. Kus, D. Pan, and D. Pan. "Based analytical biosensor chip designed from graphene-nanoplatelet-amphiphilic-diblock-co-polymer composite for cortisol detection in human saliva." *Analytical Chemistry* 89, no. 3 (2017) 2107–2115.
126. P.K. Vabbina, A. Kaushik, N. Pokhrel, S. Bhansali, N. Pala. "Electrochemical cortisol immunosensors based on sonochemically synthesized zinc oxide 1D nanorods and 2D nanoflakes." *Biosens. Bioelectron.* 63 (2015) 124–130.
127. Mühlbauer, H.D. "Human aggression and the role of central serotonin." *Pharmacopsychiatry* 18, no. 02 (1985) 218–221.
128. Peeters, M., F.J. Troost, B. Van Grinsven, F. Horemans, J. Alenus, M.S. Murib, D. Keszthelyi et al. "MIP-based biomimetic sensor for the electronic detection of serotonin in human blood plasma." *Sensors and Actuators B: Chemical* 171 (2012) 602–610.
129. Zen, J-M., I-L. Chen, and Y. Shih. "Voltammetric determination of serotonin in human blood using a chemically modified electrode." *Analytica chimica acta* 369, no. 1–2 (1998) 103–108.
130. Kema, I.P., E.G.E. de Vries, and F.A.J. Muskiet. "Clinical chemistry of serotonin and metabolites." *Journal of Chromatography B: Biomedical Sciences and Applications* 747, no. 1–2 (2000) 33–48.
131. Watts, S.W., S.F. Morrison, R.P. Davis, and S.M. Barman. "Serotonin and blood pressure regulation." *Pharmacological Reviews* 64, no. 2 (2012) 359–388.
132. Khoshnevisan, K., H. Maleki, E. Honarvarfard, H. Baharifar, M. Gholami, F. Faridbod, B. Larijani, R.F. Majidi, and M.R. Khorramizadeh. "Nanomaterial based electrochemical sensing of the biomarker serotonin: a comprehensive review." *Microchimica Acta* 186, no. 1 (2019) 49.
133. Brand, T., and G.M. Anderson. "The measurement of platelet-poor plasma serotonin: A systematic review of prior reports and recommendations for improved analysis." *Clinical Chemistry* 57, no. 10 (2011) 1376–1386.
134. I.J. Rognum, H. Tran, E.A. Haas, K. Hyland, D.S. Paterson, R.L. Haynes, K.G. Broadbelt, B.J. Harty, O. Mena, H.F. Krous, H.C. Kinney, J. Neuropathol. *Exp. Neurol.* 73 (2014) 115–122.
135. Wang, Z.-h., Q.-l. Liang, Y.-m. Wang, and G.-a. Luo. "Carbon nanotube-intercalated graphite electrodes for simultaneous determination of dopamine and serotonin in the presence of ascorbic acid." *Journal of Electroanalytical Chemistry* 540 (2003) 129–134.
136. Chen, J., and C.-s. Cha. "Detection of dopamine in the presence of a large excess of ascorbic acid by using the powder microelectrode technique." *Journal of Electroanalytical Chemistry* 463, no. 1 (1999) 93–99.
137. Yang, Z., G. Hu, X. Chen, J. Zhao, and G. Zhao. "The nano-Au self-assembled glassy carbon electrode for selective determination of epinephrine in the presence of ascorbic acid." *Colloids and Surfaces B: Biointerfaces* 54, no. 2 (2007) 230–235.
138. Cesarino, I., H.V. Galesco, and S.A.S. Machado. "Determination of serotonin on platinum electrode modified with carbon nanotubes/polypyrrole/silver nanoparticles nanohybrid." *Materials Science and Engineering: C* 40 (2014) 49–54.
139. M.-A., Mohammad, and A. Khoshroo. "Electrocatalytic properties of functionalized carbon nanotubes with titanium dioxide and benzofuran derivative/ionic liquid for simultaneous determination of isoproterenol and serotonin." *Electrochimica Acta* 130 (2014) 634–641.
140. Sun, D., H. Li, M. Li, C. Li, H. Dai, D. Sun, and B. Yang. "Electrodeposition synthesis of a NiO/CNT/PEDOT composite for simultaneous detection of dopamine, serotonin, and tryptophan." *Sensors and Actuators B: Chemical* 259 (2018) 433–442.
141. Babaei, A., and A.R. Taheri. "Nafion/Ni (OH) 2 nanoparticles-carbon nanotube composite modified glassy carbon electrode as a sensor for simultaneous determination of dopamine and serotonin in the presence of ascorbic acid." *Sensors and Actuators B: Chemical* 176 (2013) 543–551.

12 Economic Analysis, Environmental Impact, Future Prospects and Mechanistic Understandings of Nanosensors and Nanocatalysis

Libina Benny, Vinay Bhat, Ashlay George, Anitha Varghese, and Gurumurthy Hegde

CONTENTS

12.1 Introduction 271
12.2 Economic Analysis of Nanosensors and Nanocatalysts 272
12.3 Environmental Impact of Nanosensors and Nanocatalysts 275
12.4 Mechanistic Understanding 280
12.5 Future Outlook 283
12.6 Conclusion 284
12.7 Acknowledgment 285
References 285

12.1 INTRODUCTION

Nanotechnology is a large multidisciplinary field of study, development, and economic activity that has been rapidly expanding globally during the last decade. It is a multidisciplinary collection of physical, chemical, biological, engineering, and electronic processes, materials, applications, and concepts with size as the distinguishing feature [1]. They have recently developed and are now an important element of the majority of current analytical procedures for health, environment, food safety, and pharmaceutical analysis. Nano is a prefix that indicates one-billionth of a specific parameter. The prefix is typically given to the length scale to denote a very tiny magnitude of the length measurement. Individual atoms have a diameter of less than 1 nm, a DNA strand has a diameter of 2 nm, a virus has a diameter of around 100 nm, and a sheet of paper has a thickness of around 100,000 nm [2]. This is the accepted method of categorizing what belongs to the "nano" universe. The word "nanotechnology" might be confusing because it refers to more than one technology or scientific area. Nanotechnology is based on the notion that some objects, often smaller than 100 nm, display unique characteristics and behaviors that bulk matter of the same composition does not [3]. However, it is a transdisciplinary collection of physical, chemical, biological, engineering, and electronic processes, materials, applications, and concepts with size as the distinguishing feature.

DOI: 10.1201/9781003218708-14

Nanoparticles' intrinsic tiny size and peculiar optical, magnetic, catalytic, and mechanical properties, which are not present in bulk materials, enable the development of hitherto unimaginable technologies and applications. Nanoparticles, with their large surface area and exceptional physical/chemical/electronic characteristics, can improve interactions with biomolecules as compared to their bulk counterparts [4]. Nanoparticles and nanostructured materials have characteristics that differ dramatically from their visible size equivalents. Gold nanoparticles, for example, seem red rather than yellow, while aluminium, which is stable on a macro scale, becomes flammable on a nanoscale [5].

12.2 ECONOMIC ANALYSIS OF NANOSENSORS AND NANOCATALYSTS

Nanotechnology possesses greater potential and provides an informative framework to develop effective nanosensors and nanocatalysts. It also helps to analyze the implications and applications of green and sustainable chemistry in a safer and economical method. The use of non-toxic, renewable, energy efficient and cost-effective resources is the need of the hour. Technologies based on nanomaterials can be used to prevent and minimize the damage caused to the environment and human well-being [6]. The utilization of green methods entails the development of inexpensive, innocuous, and multifaceted efficient nanoproducts without the production of hazardous by-products [7]. Nanotechnology is now making significant progress in environmental technologies, and it is being investigated for its potential to give new solutions for managing and reducing air, water, and land pollution. Its goal is to improve the efficiency of traditional clean-up solutions [8].

Laboratory tests designed to detect pollutants in water are expensive as they require a skilled workforce and exorbitantly expensive tools to perform. In order to solve this difficulty, researchers are working on developing cost-effective sensors to detect pollutants using nanoparticles. Lu and others developed an optical sensor using carbon nanoparticles that were recycled from the waste of pomelo peels. Pomelo (Citrus maximus) is a popular fruit from Brazil to Southeast Asia, with a grapefruit-like flavor and thick peel. The quantum yield of carbon particles synthesized through the hydrothermal process has been reported to be around 6.9 percent. The mercury-induced fluorescence quenching of the carbon particles was used to evaluate these particles for application in mercuric ion detection. These particles showed excellent performance selectivity and sensitivity with a low limit of detection of 0.23 nM. These carbon nanoparticles also have successfully been used in detecting mercury in lake water (Figure 12.1) [9]. Abdelbasir and colleagues designed an electrochemical sensor for environmental protection applications using copper nanoparticles generated from electrical waste and laser-scribed graphene electrodes. The attachment of the particles to the graphene electrode improved electrochemical performance by lowering internal/charge transfer resistance. The sensor was designed to detect both dopamine and mercury. For dopamine, the sensor demonstrated a linear calibration curve between 300 nM and 5 uM, a limit of detection of 200 nM, a sensitivity of 30 nA μM^{-1} cm^{-2}, and a response time of 2.4 0.7 s, whereas for mercury, the sensor demonstrated a linear calibration curve between 0.02 and 2.5 ppm, a limit of detection of 25 ppb, a sensitivity of 10 nA ppm^{-1}, and a response time less than 3 minutes. The suggested method for the development of these sensors is economical as the wires are a common material [10]. Table 12.1 includes a list of various precursors from biomass and industrial wastes for the development of nanosensors and nanocatalysts.

Wang et al. utilized waste eggshell particles and synthesized biochar functionalized particles. The presence of eggshell particles on the surface and within the pore networks was discovered using scanning electron microscopy. Due to the presence of CaCO3 from the eggshells, biochar treated with eggshells was shown to be more effective for lead (Pb2+) adsorption than pristine biochar [11]. Peres and others synthesized silica nanoparticles from corn husk waste and utilized it for the effective adsorption of methyl blue. The microwave-synthesized silica nanoparticles had better surface area, pore volume and diameter, and also porosity in comparison with the pristine

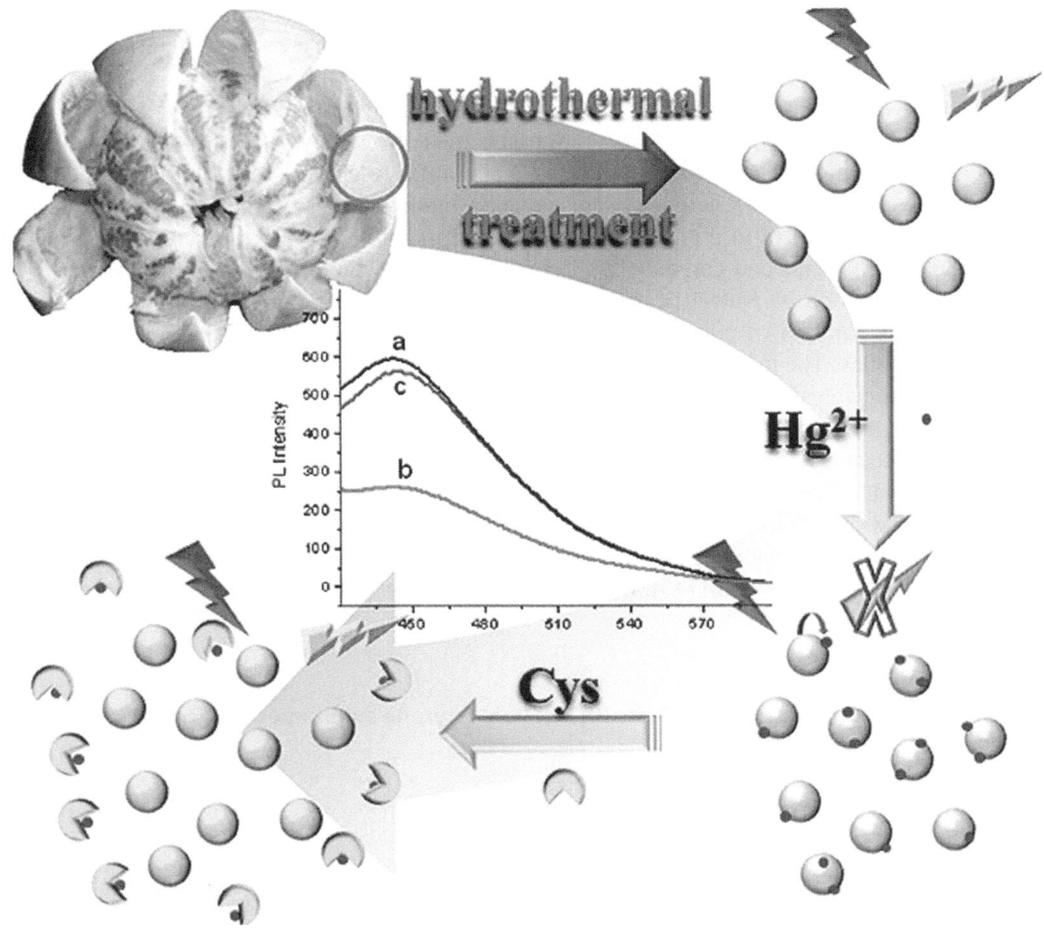

FIGURE 12.1 Graphical representation of simple preparative strategy towards fluorescent carbon nanoparticles using pomelo peel for effective detection of Hg^{2+} [9].

TABLE 12.1
List of Nanosensors and Nanocatalysts Developed from Various Biomass and Industrial Waste

Sl. No.	Electrode Material	Precursor Material	Analytes	References
Nanosensors				
1	CNPs	Pomelo peel	Mercury	9
2	Cu NPs-Graphene	Electrical waste	Dopamine Mercury	10
3	$CaCO_3$	Egg shell	Lead	11
4	Silica nanoparticles	Corn husk	Dye adsorption	12
5	Porous aerogels	Paper, cotton textiles, and plastic bottles	Oil adsorption	13
Nanocatalyst				
6	Silica-polystyrene sulfonic acid nanocomposite	Sulphonated polystyrene	Biomass valorization reactions	15
7	Cu_2O/TiO_2 nanocomposite	Printed circuit boards	Methylene blue degradation	16
8	Pt/silica NPs	Stainless steel slag	Oxidation of CO and VOC	17
9	Au and Ag nanoparticles	*Breynia rhamnoides* stem waste	Conversion of 4-nitrophenol to 4-aminophenol	18

silica nanoparticles. These silica nanoparticles possessed an adsorption capacity of 679.9 mg/g and showed 80 percent of removal. The study revealed that the nanosilica particles showed favorable spontaneous exothermic reaction [12].

Thai and his co-workers reviewed highly porous aerogels synthesized from recycled substances like paper, cotton textiles, and plastic bottles. Aerogels made from recycled paper were synthesized by sonicating a solution of recycled cellulose with plyamide-epichlorohydrin and freeze-drying it at 98° Celsius. Textile-derived aerogels were made by mixing pieces of cotton cloth in millipore water, then sonicating them in a plyamide-epichlorohydrin solution. The resultant dispersion was freeze-dried at 98°Celsius and cured at 120°Celsius. Polyethylene terephthalate (PET) fibers was immersed in a solution of NaOH heated to 80° Celsius, then washed with millipore water and combined with a solution of polyvinyl alcohol, glutaraldehyde, and HCl to make aerogels from recycled PET bottles. The mixture was then sonicated, heated to 80° Celsius, and freeze-dried [13].

Heterogenous catalysts have been used in chemical reactions to accelerate the rate of the reaction by lowering the activation energy. Nanomaterials dominate over the conventional catalyst due to its remarkable surface area-to-volume ratio and possess unique properties, which is not observed in the macroscopic counterparts. Nanomaterials obtained from bio-waste are highly targeted as they are cost-effective and show high catalytic activity [14]. Alonso and others studied the catalytic activity of waste-derived sulphonic acid nanocomposites for various applications that involved biomass valorization reactions (Figure 12.2a) [15]. Cu_2O/TiO_2 nanocatalyst was obtained via electroseparation

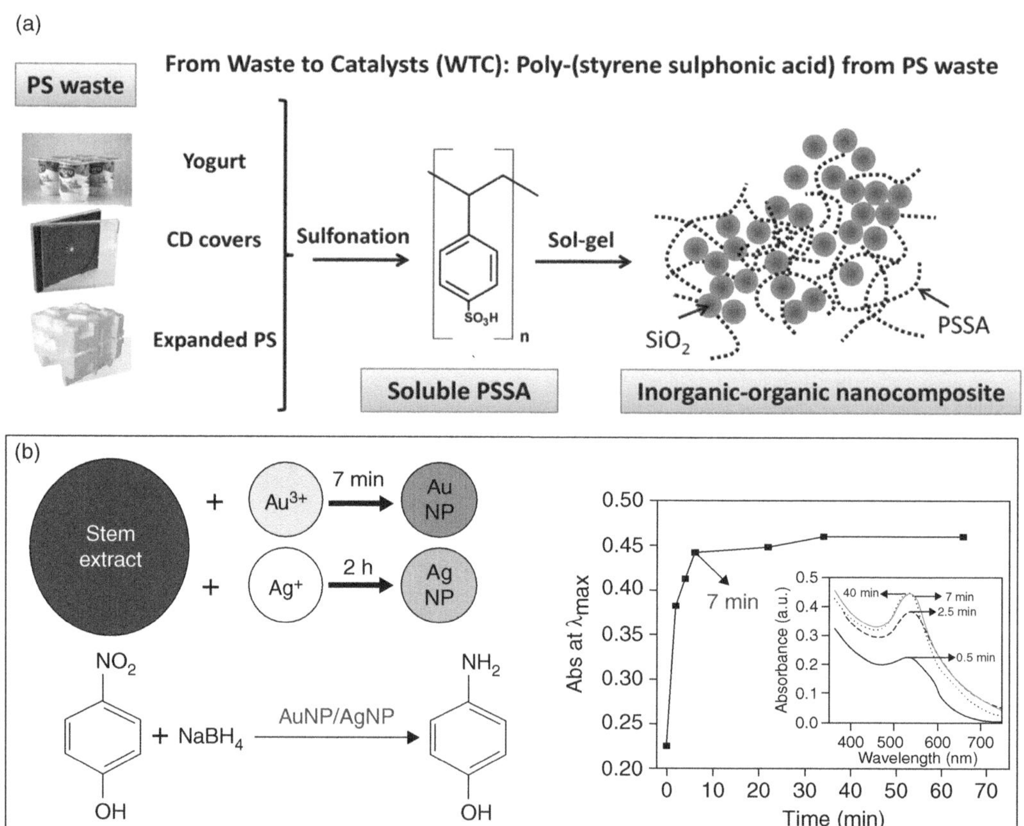

FIGURE 12.2 Schematic representation for **(a)** preparation of inorganic-organic nanocomposite using polystyrene waste [15] **(b)** synthesis of Gold (Au) and Silver (Ag) nanoparticles (NPs) from the stem extract of *Breynia rhamnoides* [18].

of copper from printed circuit boards by Xiu and Zhang. A surface area of 50 m^2/g of the developed catalyst enhanced the photocatalytic degradation of methylene blue. It was discovered that 4.5 wt% Cu doped TiO_2 could efficiently decompose 10 ppm methylene blue in less than an hour. Due to the presence of Cu_2O on the TiO_2 surface, the photocatalytic activity was considerably increased by speeding the rate of electron transfer to oxygen and creating a higher number of holes for dye degradation [16].

Moreover, Domingue et al. developed Pt impregnated silica NPs (9-20 nm) using stainless steel slag waste via hydrothermal method. The developed nanocatalyst was used for oxidation of carbon monoxide and volatile organic compounds like toluene, and so forth. Because of the obvious presence of transition metals in the sludge-derived support, the novel catalytic performance of CO oxidation was obtained in comparison to commercial Pt/SiO_2. Due to the decreased Pt dispersion over the waste support, however, Toluene conversion was poorer [17]. Plant biomass waste (*Breynia rhamnoides* stem) have been utilized for the synthesis of Au and Ag nanoparticles (Figure 12.2b). The phenolic glycosides and reducing sugars present in the stem extract were found to be the reason for metal ion reduction. They have utilized the developed nanoparticles in converting 4-nitrophenol to 4-aminophenol. The observed catalytic rate constant was $K= \sim 9.2 \times 10^{-3}\ s^{-1}$ and $K = \sim 4.06 \times 10^{-3}\ s^{-1}$ for Au and Ag nanoparticles respectively [18].

12.3 ENVIRONMENTAL IMPACT OF NANOSENSORS AND NANOCATALYSTS

Nanotechnology is a fast emerging and developing science that has piqued the interest of the media and the general public. To name a few, nanotechnology has made significant advances in agriculture, water purification, healthcare, transportation, energy, electronics, and environmental bioremediation. Nanosensors are the most recent detection technology in pollutant analysis, such as water quality assessments, food analysis, and quality control evaluation. Nanosensors also provide advantages such as mobility and the ability to be installed on-site [19]. Numerous nanomaterials, including metal and metal oxide semiconductor-based sensors have been established to fulfil the rising need for analytical sensors in a variety of sectors, including environment healthcare, food safety, and so on, during the last decade. Nano-enabled sensors, among many other technologies, have recently attracted a lot of attention [20]. These sensors are paving new paths in biological, chemical, and physical sensing, allowing for increased detection sensitivity, specificity, multiplexing capability, and mobility in a variety of health, safety, and environmental applications. The associated discoveries in nanotechnology have resulted in a variety of advancements, such as increased agricultural yields to satisfy the requirements of the growing population, more cost-effective water purification technologies, enhanced medical treatment, and a decrease in energy consumption of more than 10 percent, resulting in a cleaner environment [5]. The creation of enhanced chemical and biological sensors was one of the first uses of nanotechnology that has been accomplished [21]. Nanotechnology has the potential to drastically alter environmental sensing research requirements, using which nanosensors will let us precisely and rapidly identify and track the consequences of human actions on the environment [22].

Nanotechnology offers a chance to significantly impact environmental sensing research needs. for instance, nanotechnology makes it possible to develop parallel arrays of nanoscale sensor elements, which might result in increased sensitivity, accuracy, and spatial resolution within the simultaneous detection of an outsized number of compounds. Most sensors depend upon interactions occurring at the molecular level; hence, nanotechnology-enabled sensors can have an incredible effect on our capacity to watch and protect the environment [23]. At the nanoscale, material properties are subject to several changes. This has advantages that will bring good results, including compatibility with the environment. Considering the benefits and various applications of nanotechnology, it is often utilized in global plans in various environmental areas like controlling and reducing pollution, combating with water shortage problems, management of wastes, and other cases as modern science [22].

TABLE 12.2
Nanosensors for Different Environmental Pollutants

Nanosensor Material	Analytes	References
1. Carbon nanotube	Ammonia	[37]
2. Pd nanoparticles	Hydrogen sensor	[38]
3. Zr-based MOFs	Organophosphorus pesticides	[29]
4. Graphene	TNT	[27]
5. Single-walled carbon	Nitrogen oxide	[39]
nanotube	Trinitrotoluene	[40]
6. In-doped ZnO nanoparticles	NO_2 and NH_3 molecules	[41]
7. MoS_2	i). Monocrotophos	[31]
8. Au NPs	ii). Mercuric ion (Hg^{2+})	[28]
9. MOF-5	Organophosphate	[42]
10. CdTe QDs	Carbaryl	[32]

The primary goal of a nanosensor is to gather data/signals at the atomic level in order to achieve high detection sensitivity. A nanosensor's performance characteristics, such as detection sensitivity, detection selectivity, reaction time, detection limit and range, portability, cost, and possibly recyclability, may be assessed and evaluated. High sensitivity, dynamic range, selectivity, and stability, low detection limit, strong linearity, little hysteresis, rapid reaction time, and extended life cycle are all characteristics of an ideal sensor. When it comes to successful environmental application, nanosensors are thought to provide exceptional technical answers to many environmental challenges such as pollution, climate change, and clean drinking water. Proponents argue that they enable economic growth through improved product quality and new markets while drastically reducing our environmental footprint [24].

Nanosensors being utilized for improved sensing, treating, and remediating environmental contaminants are listed in Table 12.2. The one-of-a-kind attributes of nanomaterials utilized in nanoscale devices might be utilized to screen unexpected ecological issues. Consistent and profoundly explicit air contamination estimation is one of the essential vital developments for controlling climate contamination. [25]. Carbon-based nanosensors for label-free environmental pollution sensing account for the majority of this [26]. Nanosensors based on carbon nanotubes are being developed to detect even small amounts of poisonous and lethal substances [25]. Besides being compact and more sensitive than conventional sensors, these nanosensors are also more cost-effective due to their low power consumption and efficient operation.

Pesticides, herbicides, fertilizers, and other agrochemicals used indiscriminately are among the primary drivers of soil and water pollution, which eventually pollutes the whole biological system [25]. In the current scenario, toxic gas detection and monitoring are in high demand for environmental and national security, as a result of industrial and automobile exhausts, increased population, fossil fuel combustion, excessive use of chemicals in scientific, industrial, and agricultural fields, and explosives in the scientific, industrial, and agricultural fields. A. Dettlaff, et al. fabricated a novel sensor based on a boron-doped diamond/graphene nano-wall electrode prepared by a one-step chemical vapor deposition process to detect explosive TNT using differential pulse voltammetry (Figure 12.3) [27]. Jae-Seung Lee developed a sensor for the colorimetric detection of mercuric ion (Hg^{2+}) in aqueous media using DNA-functionalized gold nanoparticles [28].

Nanotechnology has the ability to reduce pesticide usage, improve animal and plant procreation, and create more nano–bio-industrial products. It provides higher productivity as well as lower input costs. Future precision agricultural techniques enabled by small sensors and monitoring systems will be greatly influenced by nanotechnology. Precision farming employs computers, GPS, and remote

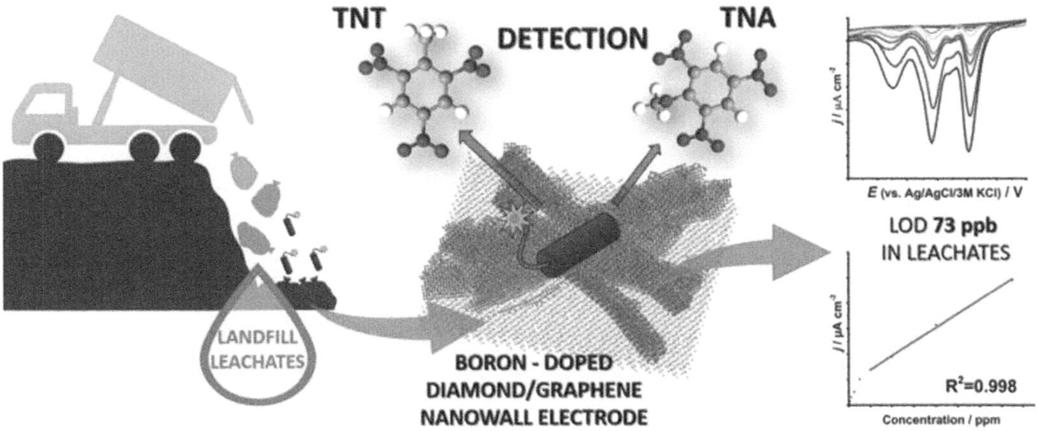

FIGURE 12.3 Schematic representation for novel boron-doped diamond/graphene nano-wall based sensor to detect the presence of TNT from landfill leachates [27].

sensing devices to assess highly localized environmental variables, allowing farmers to determine if crops are growing at their peak efficiency or accurately identifying the type and location of issues. Size- and morphology-controlled metal oxide nanoparticle syntheses have been a very active area in the development of material technology in recent decades, not only because of their unique size- and morphology-induced physical and chemical properties, which differ greatly from those of bulk materials, but also because of their many potential applications in industry and technique. Yong-Fei Zhang et.al developed magnetic Fe_3O_4@C/Cu and Fe_3O_4@CuO core–shell composite material from MOF-based substances, which were an ideal photocatalytic device for the photodegradation of organic contaminants in waste water in industry under visible light [29]. Pesticides containing organophosphates (OPs) are a broad group of chemicals that include insecticides, fungicides, and herbicides. Pawan Kumar et al. developed a sensitive method for the chemosensing of nitro groups containing organophosphate pesticides with MOF-5 [28].

Levna Chacko et al. fabricated MoS_2, MoS_2-ZnO, MoS_2-Ni and MoS_2-Pd nanostructures-based sensors for the detection of toxic gases [30]. Nanocomposites are conductive polymers containing noble metal nanoparticles, such as polyaniline. Based on the Raman intensity response, nanosensors containing gold/PANI nanocomposites were utilized to detect the presence of Hg^{2+} ions in different water samples. The device's high sensitive response (as low as 1011 M), repeatability, and selectivity were all impressive. These nanocomposites would be a new form of nanosensor, allowing for future developments in the identification of target molecules or ions for very ultrasensitive scientific tests in a variety of sensing contexts in the environment [19]. By immobilizing acetylcholinesterase on gold nanoparticles embedded in sol–gel film, Dan Du et al. developed a nano biosensor for the amperometric detection of organophosphorus pesticide [31]. Dan Du et al. have also worked on biosensors-based on CdTe quantum dots/gold nanoparticles modified chitosan microspheres interface [32]. The potential of nanobiosensors to detect a wide range of compounds at extremely low concentrations has sparked a lot of interest. As this area (and the wide range of technologies it includes) evolves, it is increasingly probable that today's diagnostic procedures will become obsolete, and a new class of low-cost, robust, dependable, easy-to-use, and ultrasensitive diagnostics will emerge [33].

Chemical analyses may be monitored using optical sensors. They are reliant on nanoparticles' optical characteristics. They can be used in a variety of fields, including the chemical industry, biotechnology, medicine, environmental sciences, and human safety [34].Optical nanosensors based on noble metal nanoparticles or semiconductor quantum dots have been developed using nanoparticles

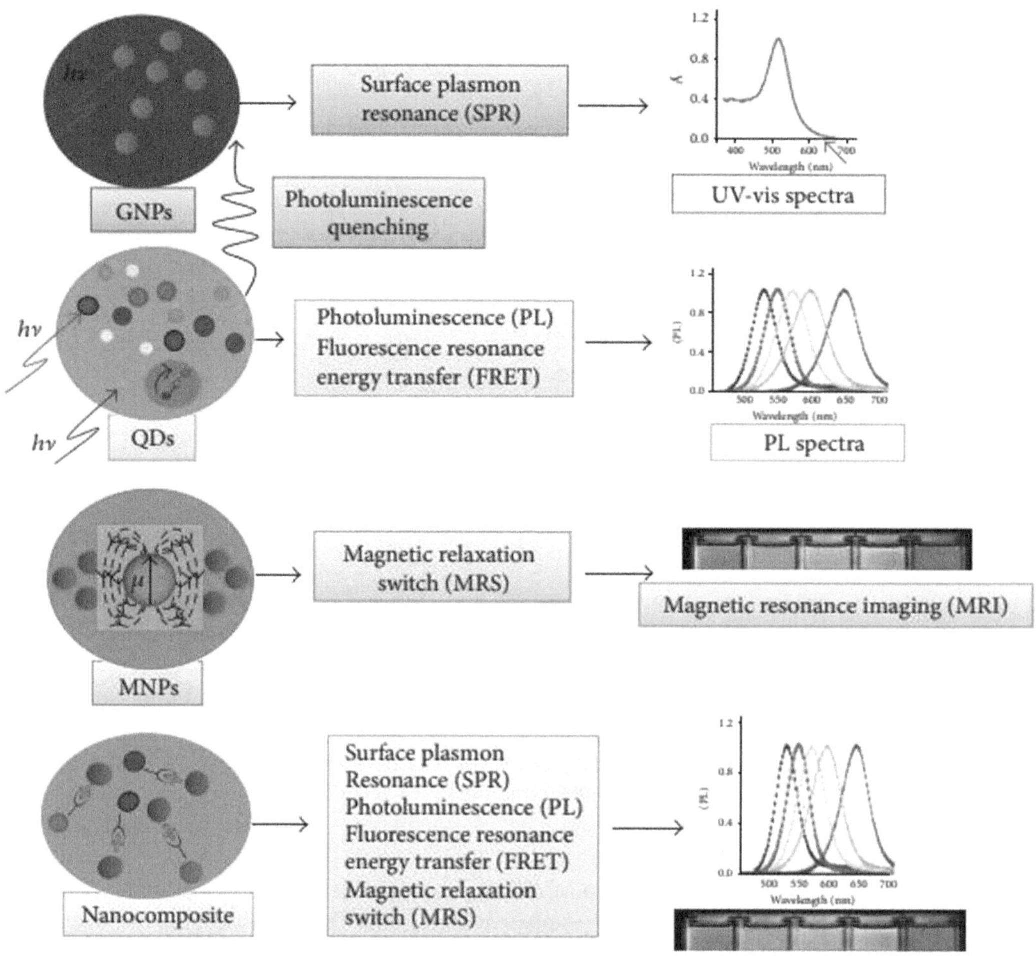

FIGURE 12.4 Schematic diagram illustrating different nanoparticles conferring optical (e.g., gold nanoparticles, GNPs), fluorescence (e.g., quantum dots, QDs), and magnetic (e.g., magnetic nanoparticles, MNPs) properties, and combinations between these particles as nanocomposites conferring multifunctionalities provide distinct advantages for environmental monitoring [35].

with excellent size-dependent optical characteristics [3]. The first optical nanosensor was used for pH monitoring and was based on fluorescein trapped within a polyacrylamide nanoparticle [35]. Optical nanosensors have been developed to make use of fluorescence's sensitivity to make quantitative measurements in the intracellular environment with devices tiny enough to be introduced into live cells with minimal physical disruption (Figure 12.4). The nanosensor matrix confers two key benefits over commonly used fluorescence dye-based methods: (1) protection of the sensing component from interfering species within the intracellular environment, and (2) protection of the intracellular environment from any toxic effects of the sensing component.[36].

The dangerous nature of today's water pollutants has definitely become a more essential issue in the design and practical use of nanocatalysts for the containment and removal of dangerous contaminants. During the last two decades, many metal nanoparticle-based catalysts have been synthesized utilizing simple and efficient methods [43]. These metal nanoparticle-based catalysts may conduct particular reactions due to their selective activity, stability, and porous architectures. Optimizing experimental settings to create diverse forms and structures can affect their development and chemical makeup. For in situ and ex situ applications, metal-based/metal oxide-based nanomaterials with high stability, controllable form, and dispersability have been produced [34]. The high specific surface area and surface energy of nanoscale catalysts result in strong catalytic activity. The use of a nanocatalyst increases reaction selectivity by allowing reactions to take place at lower temperatures, minimizing the incidence of side reactions, increasing recycling rates, and recovering energy usage. As a result, they are widely employed in green chemistry, environmental cleanup, biomass conversion efficiency, renewable energy development, and other fields [44].

By applying green and sustainable nanocatalyst technology, nanomaterials play a vital role in catalytic reactions, offering convenient reaction and low temperature requirements for transformation, as well as minimum creation of undesirable reagent wastes. For the production of nanocatalysts, a variety of metals, combination of metals, and their oxides, polymers, carbon-based materials, and silica have been used [34]. The significance of bimetallic catalysts in the breakdown and removal of various organophosphorus and organochlorine pesticides has been established by several studies [45, 46]. Catalysts made of metal oxides are used in the manufacture of petrochemicals, as well as in energy and environmental applications [47].

A photocatalyst is a substance that, when exposed to sunlight, produces a chemical reaction without being altered. Photocatalysis, air and water remediation, self-cleaning and antibacterial coatings, next-generation solar cells, hydrogen production, sensing, and cultural heritage protection might all benefit from semiconductor-assisted photocatalytic oxidation processes [48]. These materials do not participate directly in oxidation and reduction processes; instead, they supply the necessary circumstances for the reaction to occur [49]. Titanium dioxide possesses almost all of the properties of an excellent photocatalyst. Titanium dioxide nanoparticles have a greater surface-to-volume ratio than bigger particles, resulting in more favorable photocatalytic characteristics (Figure 12.5) [48]. To remediate leachate from waste material landfills, this material is utilized as

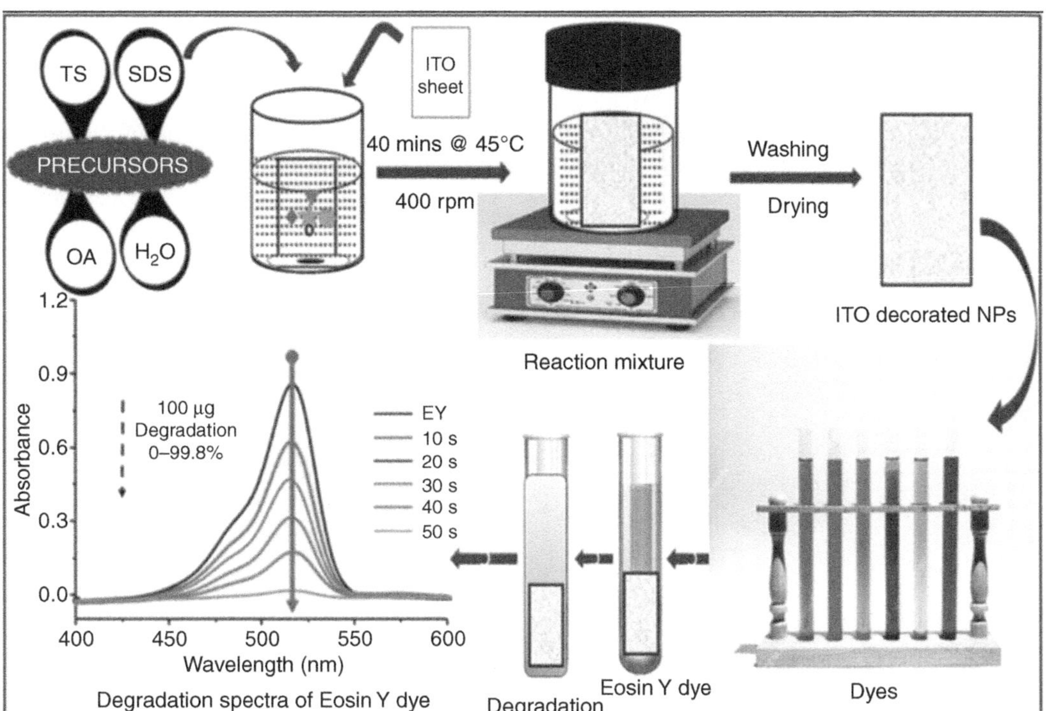

FIGURE 12.5 Schematic representation of TiO_2@ITO nanostructure synthesis.[48].

a cover for fixed membranes, nanocrystalline microspheres, and membranes coupled with silica. To convert trash into ethanol, porous nanocatalysts can be utilized. In the presence of porous nanocatalysts, syngas is transformed to ethanol. Ethanol is produced by carbon monoxide molecules in syngas. Suitable conditions for ethanol production are supplied by improving the absorption of these molecules by nanocatalysts [22]. Photo catalytically active hybrid nanocatalysts might open the path for large-scale photocatalysis for environmental clean-up, answering the growing need for materials that respond to visible light.

12.4 MECHANISTIC UNDERSTANDING

The new environmental applications based on nanomaterials demonstrate improved performances but necessitate different processes that are of unique mechanisms. The newer mechanisms could depend on numerous variables like size, shape, composition, surface feature, or hybridizations of the materials. Some interesting mechanisms are discussed below.

(A) Ag nanostructures as the surface plasmon resonance (SPR)-based sensors
The Ag nanoparticles (Ag NPs) are one of the widely researched nanomaterials and find applications in numerous fields [50]. Ag NPs based sensors work on analysis methods depending on light's interaction and nano materials; that is, the surface plasmon resonance (SPR) phenomenon. When wavelength of incident electromagnetic wave is higher than metallic nanoparticle's size, the electrons in the conduction band start to oscillate with a characteristic "resonant frequency" which is known as surface plasmon resonance (SPR) (Figure 12.6). A localized surface plasmon resonance (LSPR) is generally noticed in metallic nanoparticles where electrons in conduction band are affected by an electromagnetic radiation [50, 51]. As per Drude model, intensity of the produced LSPR is a function of dielectric property as depicted by the equation (12.1) below.

$$\varepsilon = \varepsilon_r + i\varepsilon_i \tag{12.1}$$

where, ε_r and ε_i correspond to real and imaginary parts of the dielectric function indicating that, every metal has its own characteristic interaction with electromagnetic radiation [52].

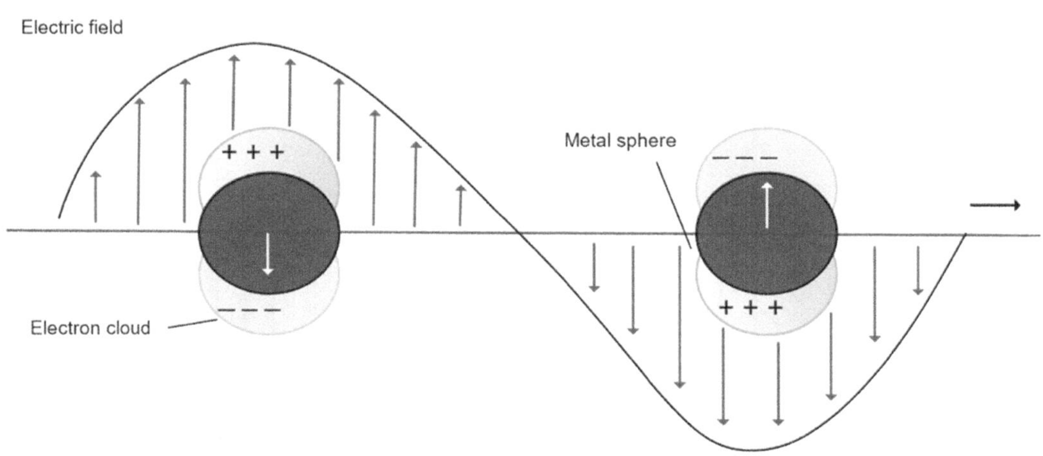

FIGURE 12.6 Illustration of LSPR during interaction between metal nanoparticles and electromagnetic wave. (Adapted with permission from reference [50], *Sens. Actuators B Chem*, Elsevier).

Hence, any modifications or alterations in Ag NPs could change its SPR, which is the basis of a chemical sensor. For example, any attachment of chemical moiety (or analyte) to the surface of Ag NPs leading to distinguishable signals in its SPR. Three distinct possibilities could arise when analytes are present [50]:

(a) Difference in morphological features of Ag NPs
(b) Oxidation of Ag NPs
(c). Accumulation/anti-accumulation of Ag NPs

On considering morphological changes, the presence of some analytes may alter the morphology or dimension of Ag nanoparticles. It is worth noting that when one-dimension Ag nanomaterials (nanowires/nanorods) are converted to "zero" dimension Ag nanoparticles, certain peaks in the optical spectrum vanish, and new peaks develop in their place (Figure 12.7a). Along with these, the color of the solution too changes [50]. Some other conversions like nanoprism to nanodisk or nanospherical to nanoprism could also be possible. Sung et al. [53], showed that a particular morphology of Ag NPs only had the ability to detect Co^{2+}. Only the Ag nanorod has the ability to detect Co^{2+} among the sphere, plate, and rod-shaped Ag NPs.

Some analytes, cause Ag NPs to become oxidized and, as a consequence, the number of metallic Ag NPs reduces, resulting in changes in the color as well as the absorbance intensity (Figure 12.7b). This was shown by the work of Devadiga et al. [54], by using biosynthesized Ag NPs as a colorimetric sensor for the detection of Hg^{2+} with a detection limit of 0.85 g mL-1. The creation of Ag-Hg amalgams on the surface of Ag NPs results from the reduction of Hg^{2+} to Hg° at the expense of Ag° oxidation to Ag^{+}, which has a significant impact on the SPR properties.

In case of analyte detection by aggregation of Ag NPs, adding of analytes to the colloidal solution of Ag NPs decreases the distance between the nanoparticles. This induces accumulation and consequently a difference in SPR of Ag NPs. Concentration of metallic ions could be gauged using spectrophotometric methods [50]. The color changes in the system, and the red–shift in the SPR peak is the typical characteristic of this mechanism. Kailasa et al. [55] demonstrated the part of ligands towards selectivity of the Ag NPs sensors with regard to heavy metal ions. Melamine functionalized Ag NPs showed selectivity to Hg^{2+} while citrate capped Ag NPs detected Cr^{3+} (Figure 12.7c). The sensing mechanism in both the cases, was the aggregation of Ag NPs. Using FTIR spectroscopy, it was showed that, hydroxyl and carboxylic groups of citrate-capped Ag NPs were accountable for the interaction with Cr^{3+} while primary amine group of melamine strongly interacted with Hg^{2+}. The advantage of this research is the selective and simultaneous detection of Cr^{3+} and Hg^{2+}.

(B) Carbon Catalysis

Advanced oxidation processes (AOPs) are emerging as a powerful technique in environmental remediation. They involve deprivation of toxic organic contaminants into harmless mineralized salts, carbon dioxide, and water by producing highly reactive oxygen species (ROS) from numerous superoxides [56].

Reduced graphene oxide (rGO) was used to activate peroxymonosulphate (PMS) for generating $SO_4^{?-}$ toward phenolic and dye degradation [57]. The edges and C=O moieties of rGO are chemically reactive. Carbon nanospheres and rGO were tailored with –C= O, –OH and –COOH groups for their catalytic efficacy and activity was observed to be stemming from carbonyl moieties. Density functional theory (DFT) calculations indicated PMS ($HO–OSO^{3-}$) could be cleaved into $SO_4^{?-}$on carbonyl groups of carbon clusters highlighting the important role played by quinoidic redox centers [58,59]. As can be seen from Figure 12.8, the quinone functional groups situated at terminals have a lone pair of electrons interacting with PMS, which weakens peroxide bond (O – O). Electrons are then supplied from hydrogen bonding to PMS through inner-sphere charge transfer producing

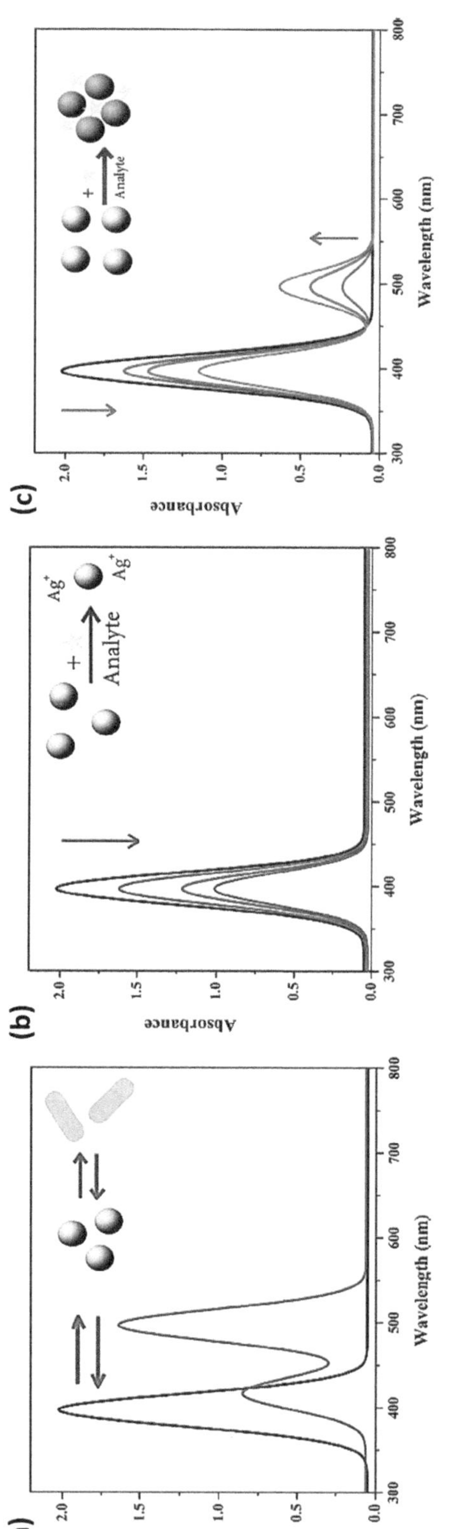

FIGURE 12.7 Schematic representation of the changes occurred in UV–vis spectra and colloidal system upon dimensional change (a), upon oxidation (b) and inter-particle distance of colloidal system upon the aggregation of particles (c). (Reproduced with permission from reference [50], *Sens. Actuators B Chem*, Elsevier).

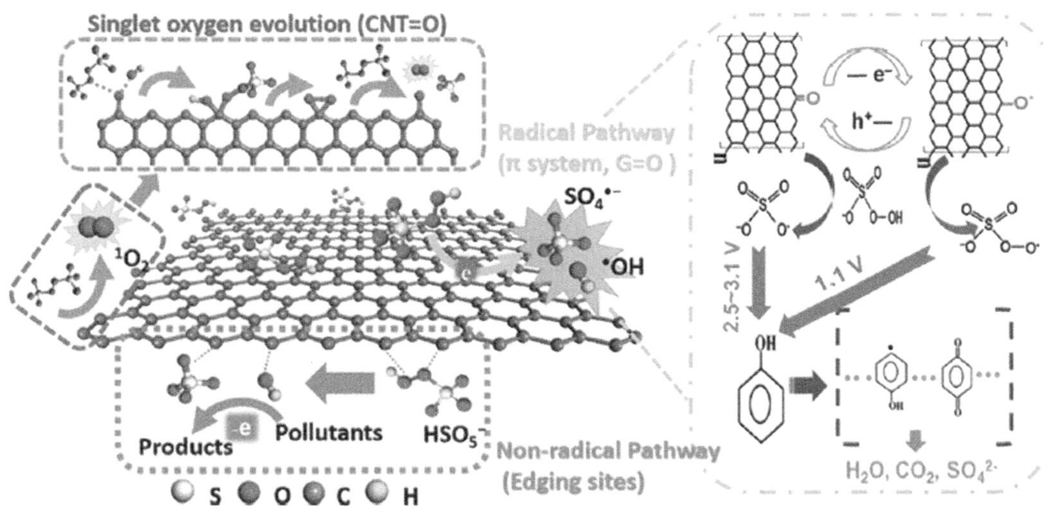

FIGURE 12.8 Schematic illustration of activation of PMS via carbocatalysis. (Reproduced with permission from reference [56], *Acc. Chem. Res*, 2018, American Chemical Society).

$SO_4^{?-}$[58,59]. The oxygen centre that now becomes metastable and positively charged regains charges from other PMS molecules to produce $SO_5^{?-}$thereby restoring the ketone site completing the redox process [59]. It was proposed from theoretical modelling that the edge sites and vacancies are more reactive than basal planes involving persulfate interaction. Carbon groups along the graphene boundaries maintaining sp^2 hybridization and confined partial п electrons with a localized state are believed to be highly catalytically active [59].

Introduction of foreign atoms (B, N, S, and P) into the carbon matrix is an easier method to create point defects and thus alter the physicochemical properties of graphene-based materials. The foreign atoms possess a distinct atomic radius and orbitals, electron density, and electronegativity, which break the chemical inertness of graphene by disrupting the electronic and spin culture of the sp^2-hybridized carbons [56]. The nonradical pathway could be utilized for better mineralization of organic pollutants in real waste water containers and natural systems with the presence of diverse radical scavengers. Figure 12.9 shows a promising platform towards environmental remediation where the redox potential of the carbon/persulfate system is maneuverable from a nonradical pathway to a radical pathway by altering the composite components or doping/defect level [56, 60, 61].

12.5 FUTURE OUTLOOK

An enormous volume of studies has been reported regarding nanomaterials for environmental sustainability, mature, cost-effective and reliable manufacturing of such materials for safer applications, yet remains unconquered [62]. Some future outlook and challenges which have to be addressed are below.

- Reactivity and selectivity: Enhanced efficacy in adsorption or catalysis is owed to delicate structure and tuned porosity of the materials [63, 64]. Selective detection/adsorption or selective degradation plays a critical role in successful application of nanomaterials. Harsh environmental conditions far deviated from ideal laboratory situations could hamper selectivity

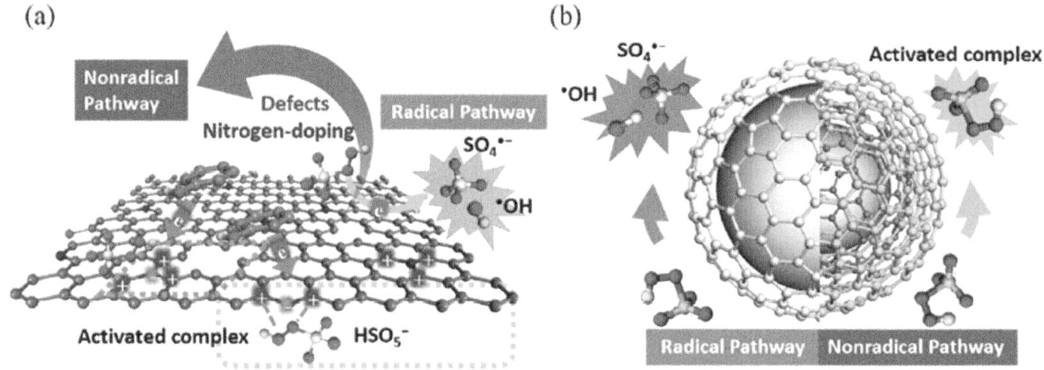

FIGURE 12.9 Schematic representations of radical and non-radical oxidation on (a) graphene and (b) graphitized nanodiamond. (Reproduced with permission from reference [56], *Acc. Chem. Res*, 2018, American Chemical Society).

and easily destroy the designed morphology and structure. Catalytic poisoning could easily reduce reactivity of the nanocatalysts and affect sensing capabilities. Rational designing and specific control over applications must consider reactivity, selectivity and also the stability of nanomaterials [65].

- Mechanistic understandings: Application of nanomaterials often produces enhanced catalytic efficiency and improved sensitivity and selectivity, which could arise due modified size, shape, structure, hybridization, bonding and material nature [56]. A rational and thorough reasoning is the need of the hour to understand the functioning and results generated. A combination of traditional experiments coupled with computational and chemical engineering are required for practical environmental applications.
- Nanotoxicity: Nanotechnology and nanomaterials have made it possible for novel applications in biological, energy, sensors and many other realms [62, 66-71], which could not have been possible with bulk materials. This widespread applicability of nanomaterials itself is posing a great risk to the environment. Implications and health hazards on exposure to such materials need to be tackled.

12.6 CONCLUSION

Environmental sustainability is becoming major concern in techniques and processes involved in production and manufacturing industries. Recent research on nanotechnology and nanomaterials have played a huge role in advancement towards environmental remediation, either through designing novel materials or by innovation of newer and improving efficiency of existing techniques. Public awareness and government involvement through policy and collaboration will play an important role in economic aspects involved with newer technologies for environmental remediation. Bulk materials when reduced to nanoscale, exhibit very intriguing physicochemical properties due to the higher surface area and quantum confinement. This creates opportunities and challenges alike in terms of material properties, such as shape, structure, optical, magnetic, electrical, electrochemical, thermal and so forth. The current chapter provides inputs regarding availability of some novel nanomaterials for sensing and catalytic applications along with environmental impact, cost factors and so forth. Some intriguing mechanisms for sensing and catalysis are also discussed. The efficacy of such nanomaterials can be improved by modifying them in terms of microstructure (pore and surface), composition (heterojunction and doping) and morphology (particle size and shape).

12.7 ACKNOWLEDGMENT

Gurumurthy Hegde acknowledges DST-Nanomission, Govt of India for providing research grant on "Biowaste based porous nano materials for efficient low-cost energy storage devices" with file no SR/NM/NT-1026/2017.

REFERENCES

1. D. Drobne, "Nanotoxicology for safe and sustainable nanotechnology," *Arh. Hig. Rada Toksikol.*, vol. 58, no. 4, pp. 471–478, 2007.
2. R.J. Aitken, K.S. Creely, and C.L. Tran, "Nanoparticles: An Occupational Hygiene Review," *Heal. Saf. Exec.*, p. 113, 2004.
3. J. Riu, A. Maroto, and F.X. Rius, "Nanosensors in environmental analysis," *Talanta*, vol. 69, no. 2 SPEC. ISS., pp. 288–301, 2006.
4. A. Joshi and K.H. Kim, "Recent advances in nanomaterial-based electrochemical detection of antibiotics: Challenges and future perspectives," *Biosens. Bioelectron.*, vol. 153, p. 112046, 2020.
5. D.E. Babatunde, I.H. Denwigwe, O.M. Babatunde, S.L. Gbadamosi, I.P. Babalola, and O. Agboola, "Environmental and Societal Impact of Nanotechnology," *IEEE Access*, vol. 8, pp. 4640–4667, 2020.
6. E. Hood, "Nanotechnology: looking as we leap". *Environ Health Perspect.* 112(13): A740, 2004.
7. D. Dornfeld, C. Yuan, N. Diaz, T. Zhang, A. Vijayaraghavan, "Introduction to green manufacturing," In: *Green Manufacturing*. Springer, Boston, pp 1–23, 2013.
8. P. Shapira, and J. Youtie, "The economic contributions of nanotechnology to green and sustainable growth," In: *Green Processes for Nanotechnology*. Springer, Cham, pp 409–434, 2015.
9. W. Lu, X. Qin, S. Liu, G. Chang, Y. Zhang, Y. Luo, et al. (2012). "Economical, green synthesis of fluorescent carbon nanoparticles and their use as probes for sensitive and selective detection of mercury(II) ions," *Anal. Chem.* 84, 5351–5357, 2012.
10. S.M. Abdelbasir, S.M. El-Sheikh, V.L. Morgan, H. Schmidt, L.M. Casso-Hartmann, D. C. Vanegas, et al. "Graphene-anchored cuprous oxide nanoparticles from waste electric cables for electrochemical sensing," *ACS Sustain. Chem. Eng.* 6, 12176–12186, 2018.
11. H. Wang, B. Gao, J. Fang, Y. S. Ok, Y. Xue, K. Yang, et al. "Engineered biochar derived from eggshell-treated biomass for removal of aqueous lead," *Ecol. Eng.* 121, 124–129, 2018.
12. E. C. Peres, J. C. Slaviero, A. M. Cunha, A. Hosseini-Bandegharaei, G. L. Dotto, "Microwave synthesis of silica nanoparticles and its application for methylene blue adsorption," *J. Environ. Chem. Eng.* 6, 649–659, 2018.
13. Q.B. Thai, D.K. Le, T.P. Luu, N. Hoang, D. Nguyen, H.M. Duong, "Aerogels from wastes and their applications," *JOJ Mater. Sci.* 5, 555663, 2019.
14. P. Samaddar, Y.S. Ok, K.H. Kim, E.E. Kwon, D.C. Tsang, "Synthesis of nanomaterials from various wastes and their new age applications," *J. Clean. Prod.*, 197,1190–1209, 2018.
15. N. Alonso-Fagúndez, V. Laserna, A.C. Alba-Rubio, M. Mengibar, A. Heras, R. Mariscal, M. López Granados, "Poly-(styrene sulphonic acid): An acid catalyst from polystyrene waste for reactions of interest in biomass valorization," *Catal.*, 234, 285–294, 2014.
16. F.-R. Xiu, F.-S. Zhang, "Preparation of nano-Cu2O/TiO2 photocatalyst from waste printed circuit boards by electrokinetic process," *J. Hazard. Mater.*, 172, 1458–1463, 2009.
17. M.I. Domínguez, I. Barrio, M. Sánchez, M.Á. Centeno, M. Montes, J.A. Odriozola, "CO and VOCs oxidation over Pt/SiO2 catalysts prepared using silicas obtained from stainless steel slags," *Catal.*, *133*, 467–474, 2008.
18. A. Gangula, R. Podila, L. Karanam, C. Janardhana, A.M. Rao, "Catalytic reduction of 4-nitrophenol using biogenic gold and silver nanoparticles derived from Breynia rhamnoides," *Langmuir*, *27*, 15268–15274, 2011.
19. M. Graboski, J. Martinazzo, S.C. Ballen, J. Steffens, and C. Steffens, *Nanosensors for Water Quality Control*. Elsevier, 2020.
20. R. Konwarh, G. Gollavelli, and S.B. Palanisamy, *Designing of Novel Nanosensors for Environmental Aspects*. Elsevier, 2020.
21. R. Yonzon, D.A. Stuart, X. Zhang, A.D. McFarland, C.L. Haynes, and R.P. Van Duyne, "Towards advanced chemical and biological nanosensors – An overview," *Talanta*, 67, 438–448, 2005.

22. M. Taran, M. Safaei, N. Karimi, and A. Almasi, "Benefits and application of nanotechnology in environmental science: an overview," *Biointerface Res. Appl. Chem.*, 11, 7860–7870, 2021.
23. Karn, "Nanotechnology and the Environment: Applications and Implications," 391, 2005.
24. A. El Moussaouy, "Environmental Nanotechnology and Education for Sustainability: Recent Progress and Perspective," *Handb. Environ. Mater. Manag.*, 1–27, 2018.
25. S. Das, B. Sen, and N. Debnath, "Recent trends in nanomaterials applications in environmental monitoring and remediation," *Environmental Science and Pollution Research*, 18333–18344, 2015.
26. F. Tan, N.M. Saucedo, P. Ramnani, and A. Mulchandani, "Label-Free Electrical Immunosensor for Highly Sensitive and Specific Detection of Microcystin-LR in Water Samples," *Environ. Sci. Technol.*, 49, 9256–9263, 2015.
27. A. Dettlaff et al., "Electrochemical determination of nitroaromatic explosives at boron-doped diamond/graphene nanowall electrodes: 2,4,6-trinitrotoluene and 2,4,6-trinitroanisole in liquid effluents," *J. Hazard. Mater.*, 387, 121672, 2020.
28. J.S. Lee, M.S. Han, and C.A. Mirkin, "Colorimetric detection of mercuric ion (Hg2+) in aqueous media using DNA-functionalized gold nanoparticles," *Angew. Chemie – Int.* Ed., 46, 4093–4096, 2007.
29. Y.F. Zhang, L.G. Qiu, Y.P. Yuan, Y.J. Zhu, X. Jiang, and J.D. Xiao, "Magnetic Fe3O4@C/Cu and Fe3O4@CuO core-shell composites constructed from MOF-based materials and their photocatalytic properties under visible light," *Appl. Catal. B Environ.*, 144, 863–869, 2014.
30. L. Chacko, E. Massera, and P.M. Aneesh, "Enhancement in the Selectivity and Sensitivity of Ni and Pd Functionalized MoS 2 Toxic Gas Sensors Enhancement in the Selectivity and Sensitivity of Ni and Pd Functionalized MoS 2 Toxic Gas Sensors," *J. Electrochem. Soc.*, 167, 106506, 2020.
31. Du, S. Chen, J. Cai, and A. Zhang, "Immobilization of acetylcholinesterase on gold nanoparticles embedded in sol-gel film for amperometric detection of organophosphorous insecticide," *Biosens. Bioelectron.*, 23, 130–134, 2007.
32. Du, S. Chen, D. Song, H. Li, and X. Chen, "Development of acetylcholinesterase biosensor based on CdTe quantum dots/gold nanoparticles modified chitosan microspheres interface," *Biosens. Bioelectron.*, 24, 475–479, 2008.
33. L.M. Bellan, D. Wu, and R.S. Langer, "Current trends in nanobiosensor technology," *Wiley Interdiscip. Rev. Nanomedicine Nanobiotechnology*, 3, 229–246, 2011.
34. R. Bhadouria et al., *Chapter 16 – Nanocatalyst types and their potential impacts in agroecosystems: An overview*. Elsevier. 2020.
35. P. Koedrith, T. Thasiphu, J.I. Weon, R. Boonprasert, K.Tuitemwong, and P., Tuitemwong, "Recent trends in rapid environmental monitoring of pathogens and toxicants: potential of nanoparticle-based biosensor and applications." *The Scientific World Journal*, 2015.
36. J.W. Aylott, "Optical nanosensors - An enabling technology for intracellular measurements," *Analyst*, 128, 309–312, 2003.
37. A. Modi, N. Koratkar, E. Lass, B. Wei, and P.M. Ajayan, "Miniaturized gas ionization sensors using carbon nanotubes," *Nature*, 424, 171–174, 2003.
38. C. Walter, F. Favier, and R.M. Penner, "Palladium mesowire arrays for fast hydrogen sensors and hydrogen-actuated switches," *Anal. Chem.*, 74, 1546–1553, 2002.
39. J.E. Huang, X.H. Li, J.C. Xu, and H.L. Li, "Well-dispersed single-walled carbon nanotube/polyaniline composite films," *Carbon N. Y.*, 41, 2731–2736, 2003.
40. Y. Ge, Z. Wei, Y. Li, J. Qu, B. Zu, and X. Dou, "Highly sensitive and rapid chemiresistive sensor towards trace nitro-explosive vapors based on oxygen vacancy-rich and defective crystallized In-doped ZnO," *Sensors Actuators, B Chem.*, 244, 983–991, 2017.
41. R. Cao, B. Zhou, C. Jia, X. Zhang, and Z. Jiang, "Theoretical study of the NO, NO2, CO, SO2, and NH3 adsorptions on multi-diameter single-wall MoS2 nanotube," *J. Phys. D. Appl. Phys.*, 49, 45106, 2015.
42. P. Kumar, A.K. Paul, and A. Deep, "Sensitive chemosensing of nitro group containing organophosphate pesticides with MOF-5," *Microporous Mesoporous Mater.*, 195, 60–66, 2014.
43. S. Sonkaria et al., "Ionic liquid-induced synthesis of a graphene intercalated ferrocene nanocatalyst and its environmental application," *Appl. Catal. B Environ.*, 182, 326–335, 2016.
44. N. Sharma, H. Ojha, A. Bharadwaj, D.P. Pathak, and R.K. Sharma, "Preparation and catalytic applications of nanomaterials: a review," *RSC Adv.*, 5, 53381–53403, 2015.

45. K.L. Marcelo, A.R. Means, and B. York, "The Ca2+/Calmodulin/CaMKK2 Axis: Nature's Metabolic CaMshaft," *Trends Endocrinol. Metab.*, 27, 706–718, 2016.
46. A. Morales-Marín, J.L. Ayastuy, U. Iriarte-Velasco, and M.A. Gutiérrez-Ortiz, "Nickel aluminate spinel-derived catalysts for the aqueous phase reforming of glycerol: Effect of reduction temperature," *Appl. Catal. B Environ.*, 244, 931–945, 2019.
47. D.G. Rickerby and M. Morrison, "Nanotechnology and the environment: A European perspective," *Sci. Technol. Adv. Mater.*, 8, 19–24, 2007.
48. A.H Pato, A. Balouch, F.N. Talpur, P. Panah, A.M. Mahar, M.S. Jagirani, S. Kumar, and S., Sanam, Fabrication of TiO 2@ ITO-grown nanocatalyst as efficient applicant for catalytic reduction of Eosin Y from aqueous media. *Environmental Science and Pollution Research, 28*, 947–959, 2021.
49. N.D. Cuong and D.T. Quang, "Progress through synergistic effects of heterojunction in nanocatalysts – Review," *Vietnam J. Chem.*, 58, 434–463, 2020.
50. A. Amirjani, D.F. Haghshenas, Ag nanostructures as the surface plasmon resonance (SPR)-based sensors: A mechanistic study with an emphasis on heavy metallic ions detection, *Sens. Actuators B Chem*, 273, 1768–1779, 2018.
51. M. Rycenga, C. Cobley, J. Zeng, W. Li, C. Moran, Q. Zhang, et al., Controlling the synthesis and assembly of silver nanostructures for plasmonic applications, *Chem. Rev.*, 111 (6), 3669–3712, 2011.
52. R. Ishida, S. Yamazoe, K. Koyasu, T. Tsukuda, Repeated appearance and disappearance of localized surface plasmon resonance in 1.2 nm gold clusters induced by adsorption and desorption of hydrogen atoms, *Nanoscale*, 8 (5), 2544–2547, 2016.
53. Sung, S. Oh, C. Park, Y. Kim, Colorimetric detection of Co^{2+} ion using AgNPs with spherical, plate, and rod shapes, *Langmuir*, 29 (28), 8978–8982, 2013.
54. A. Devadiga, K. Vidya Shetty, M. Saidutta, Highly stable AgNPs synthesized using Terminalia catappa leaves as antibacterial agent and colorimetric mercury sensor, *Mater. Lett.*, 207, 66–71, 2017.
55. S. Kailasa, M. Chandel, V. Mehta, T. Park, Influence of ligand chemistry on silver nanoparticles for colorimetric detection of Cr^{3+} and Hg^{2+} ions, *Spectrochim. Acta A. Mol. Biomol. Spectrosc.*, 195, 120–127, 2018.
56. Xiaoguang Duan, Hongqi Sun, and Shaobin Wang, Metal-Free Carbocatalysis in Advanced Oxidation Reactions, *Acc. Chem. Res.*, 51 (3), 678–687, 2018.
57. Sun, H.Q. Liu, S.Z. Zhou, G.L. Ang, H.M. Tade, M.O., Wang, S.B, Reduced Graphene Oxide for Catalytic Oxidation of Aqueous Organic Pollutants, *ACS Appl. Mater. Interfaces*, 4, 5466–5471, 2012.
58. Wang, Y.X. Ao, Z.M. Sun, H.Q. Duan, X.G. Wang, S.B., Activation of Peroxymonosulfate by Carbonaceous Oxygen Groups: Experimental and Density Functional Theory Calculations. *Appl. Catal., B*, 198, 295–302, 2016.
59. Duan, X.G., Sun, H.Q. Ao, Z.M. Zhou, L. Wang, G.X. Wang, S.B. Unveiling the Active Sites of Graphene-Catalyzed Peroxymonosulfate Activation. *Carbon*, 107, 371–378, 2016.
60. Li, D.; Duan, X.; Sun, H.; Kang, J.; Zhang, H.; Tade, M. O., Wang, S. Facile Synthesis of Nitrogen-Doped Graphene via Low-Temperature Pyrolysis: The Effects of Precursors and Annealing Ambience on Metal-Free Catalytic Oxidation. *Carbon*, 115, 649–658, 2017.
61. Duan, X.G. Ao, Z.M.; Zhang, H.Y.; Saunders, M.; Sun, H.Q., Shao, Z.P.; Wang, S.B. Nanodiamonds in sp2/sp3 Configuration for Radical to Nonradical Oxidation: Core-Shell Layer Dependence. *Appl. Catal. B*, 222, 176–181, 2018.
62. Vinay, S. Bhat et al, Review – Biomass Derived Carbon Materials for Electrochemical Sensors, *J. Electrochem. Soc.* 167, 037526, 2020.
63. Supriya, S., Bhat, V.S., Jayeoye, T.J. et al. An investigation on temperature-dependant surface properties of porous carbon nanoparticles derived from biomass. *J Nanostruct Chem*, https://doi.org/10.1007/s40097-021-00427-4, 2021.
64. Supriya, S, Ananthnag, G.S., Shetti, V.S., Nagaraja, B.M., Hegde, G. Cost-effective bio-derived mesoporous carbon nanoparticles-supported palladium catalyst for nitroarene reduction and Suzuki–Miyaura coupling by microwave approach. *Appl Organometal Chem.*, 34, e5384, 2020.
65. Sun Hongqi, Grand Challenges in Environmental Nanotechnology, *Front. Nanosci.*, 1, 2, 2019.
66. Bhat, V.S., Krishnan, S.G., Jayeoye, T.J. et al. Self-activated "green" carbon nanoparticles for symmetric solid-state supercapacitors. *J Mater Sci*, 56, 13271–13290, 2021.

67. Kanagavalli, P., Pandey, G.R., Bhat, V.S. et al. Nitrogenated-carbon nanoelectrocatalyst advertently processed from bio-waste of Allium sativum for oxygen reduction reaction. *J Nanostruct Chem*, 11, 343–352, 2021.
68. Akshaya K.B. et al. Non-Enzymatic Electrochemical Determination of Progesterone Using Carbon Nanospheres from Onion Peels Coated on Carbon Fiber Paper, *J. Electrochem. Soc*. 166, B1097, 2019.
69. Roy, P., Bhat, V.S., Saha, S. et al. Mesoporous carbon nanospheres derived from agro-waste as novel antimicrobial agents against gram-negative bacteria. *Environ Sci Pollut Res* 28, 13552–13561, 2021.
70. Bhat V.S., and Kanagavalli, P. et al. Low cost, catalyst free, high-performance supercapacitors based on porous nano carbon derived from agriculture waste, *J. Energy Storage*, 32, 101829, 2020.
71. Bhat, V.S., Hegde, G. and Nasrollahzadeh, M. A sustainable technique to solve growing energy demand: Porous carbon nanoparticles as electrode materials for high-performance supercapacitors. *J Appl Electrochem* 50, 1243–1255, 2020.

Index

A

advanced TiO_2 103, 114, 126
aluminophosphates 49–50, 52–3

B

biomedical 159, 181, 195
biosensors 38, 160, 165–6, 173, 181, 207

C

conducting polymers 166, 172
CQDs 163–4, 173
CVD 133, 147

E

economic 57, 59, 145, 252, 254, 259, 271–2, 284
electrodes 134, 171, 217
energy harvesting 37

F

FDCM 84–5
fluorophores 182, 187, 190, 207–213
FRET 194, 209, 211–12, 253

G

GQDs 163–5, 171, 173

H

heteroatoms 66, 68, 184
heterojunctions 103, 108–10
HMF 85
hydrogen evolution 131, 138, 140, 146
hydrothermal 8, 52, 132, 162, 183

L

limit of detection 250, 252, 258, 272
liquid phase oxidation 65, 67
luminescent 164, 170, 184, 203, 207

M

mechanistic 117, 271, 280, 284
mesoporous 3, 52–3
metal-organic framework 48, 108, 123
metal oxides 4, 15
microwave 75, 132, 185

N

nano porous 3, 170

O

organic transformations 4, 6, 11, 13, 15, 56

P

palladium 3, 65, 68
photocatalysis 15, 103–4, 119
pH-sensitive 187, 212
plasmonic 110–11
platinum 38, 65, 68
PVD 134
pyridoxine 253

R

ruthenium 65, 81

S

self-assembled 208, 210–13
sol-gel 52, 118, 132
sonochemical 132–3, 147, 225
sulfoxidation 27, 30
sulfur-redox 26
supra molecular 5, 208, 211
surface properties 49, 56, 172

U

ultrasonic 40, 164, 185

V

vitamins 87, 163, 192, 249–50, 262